Urban Transportation: Future Perspectives

Urban Transportation: Future Perspectives

Edited by Julian Custer

Clanrye International
www.clanryeinternational.com

Clanrye International,
750 Third Avenue, 9th Floor,
New York, NY 10017, USA

ISBN: 978-1-63240-932-4

Cataloging-in-Publication Data

Urban transportation : future perspectives / edited by Julian Custer.
p. cm.
Includes bibliographical references and index.
ISBN 978-1-63240-932-4
1. Urban transportation. 2. Urban transportation policy. 3. City planning. 4. Urban policy.
5. Urbanization. I. Custer, Julian.

HE305 .U73 2020

388.4--dc23

For information on all Clanrye International publications visit our website at www.clanryeinternational.com

CLANRYE INTERNATIONAL

Printed in China.

Contents

Preface

Urban transportation is concerned with providing transport services to the passengers in and around urban and sub-urban regions. City buses, trolleybuses, personal automobiles, taxis, passenger trains and ferries are a few of the most commonly used modes of urban transportation. The aim of the urban transportation is to mitigate the fuel consumption by encouraging urban population to use public transport. It can also result in a significant reduction in the carbon footprint of the user. Researches are being conducted to find environment-friendly fuel alternatives such as hydrogen-powered vehicles. This book discusses the fundamentals as well as modern approaches of urban transportation. It explores all the important aspects of this field in the present day scenario. The topics included in this book on urban transportation are of utmost significance and bound to provide incredible insights to readers.

The researches compiled throughout the book are authentic and of high quality, combining several disciplines and from very diverse regions from around the world. Drawing on the contributions of many researchers from diverse countries, the book's objective is to provide the readers with the latest achievements in the area of research. This book will surely be a source of knowledge to all interested and researching the field.

In the end, I would like to express my deep sense of gratitude to all the authors for meeting the set deadlines in completing and submitting their research chapters. I would also like to thank the publisher for the support offered to us throughout the course of the book. Finally, I extend my sincere thanks to my family for being a constant source of inspiration and encouragement.

Editor

Minimizing the Impact of Large Freight Vehicles in the City: A Multicriteria Vision for Route Planning and Type of Vehicles

Sara Ezquerro, Juan P. Romero, Jose L. Moura, Juan Benavente, and Ángel Ibeas

G.I.S.T. Group of Research in Transport Systems, University of Cantabria, Cantabria, Spain

Correspondence should be addressed to Sara Ezquerro; sara.ezquerro@unican.es

Academic Editor: Edward Chung

The impact of freight transport in cities is significant, and as such correct planning and management thereof help reduce their enormous negative impact. Above all, the special large vehicles have a greater impact than the remainder of freight vehicles, so a special attention should be paid to them. The vehicles which supply or pick up large amounts of goods at specific points throughout the city are an example of this type of vehicles. The aim of this paper is to minimize the cost of this freight transport type from a social, economic, and environmental viewpoint. To this effect, an optimization model has been proposed based on bilevel mathematical programming which minimizes the total system costs. City network model data are obtained on the lower level such as vehicle flow and travelling times, which are then used on the upper level to calculate total system costs. The model has been applied to a real case in Santander (Spain), whose final result shows the size and typology of the fleet of vehicles necessary to have the least impact on the city. The greater the vehicles size is (i.e., using fewer trucks), the less the cost of the freight transport is.

1. Introduction

Freight transport in the urban environment has a great impact on the city, that is, increase in congestion, pollution, costs, and so on. Many researchers have studied the impact of different policies in freight transport [1–3]. In some cases, access to the city centre is restricted to reduce social, environmental, and economic impacts [4].

Different criteria can be used to choose which typology of vehicles can or cannot enter the city centre, such as weight limits in Santander [5] or vehicle emissions in the case of Rome [6]. However, there are exceptional situations where this kind of vehicle is allowed to drive through the city centre due to specific circumstances (e.g., material delivery and pick-up on a construction site). This is why in those situations where the presence of freight transport increases at a specific point albeit due to increase in vehicle frequency and/or typology, consideration should be given to the reduction of its impact on the city. This paper studies the impact of the presence of these vehicles, less common in city centres.

This paper builds upon prior research by the authors on freight transport simulation and optimization models. The aim is go one step further and take a closer approach to the reality extant in cities; hence, unlike in prior studies [7], this one researched the impact of an increase in freight transport during a specific period of time, taking into account variable vehicle capacity and fleet.

This problem can be resolved via application of the supply chain problem when materials need delivering to a certain point following a specific schedule. Numerous researchers have also modelled and simulated the supply chain [8]. Reiner and Trcka [9] modelled and designed the supply chain structure for a food company. Guo et al. [10] carried out a study where they had to minimize the system costs (construction, operation, information processing, and transport costs and likewise coal tax) for a route-planner model for fresh food e-commerce companies. Others, however, have undertaken more innovative studies, such as Smart supply chains [11] or how Smart Cities affect the supply chain [12].

Furthermore, a bilevel methodology has been applied to minimize the impact of large freight vehicles. Bilevel optimization is a kind of optimization that allows one problem to be nested within another, so it can explicitly represent the mutual action between the upper level and lower level. In addition, it allows solving two decision levels. For example, in the upper level, the freight systems planner makes the decisions, and, in the lower level, the private transport users make decisions based on the decisions of the first level. In essence, the bilevel model allows considering how heavy truck traffic affects traffic throughout the city. Moreover, bilevel programming can be used to analyse two different objectives and can reflect some practical problems better [13]. Bilevel optimization methodology has been used by many transport researchers to study different subjects, such us public transport optimization (e.g., best bus stops locations [14], the relationship between transport and residence [15]), or it has been used in freight transport optimization for planning the delivery of supplies to large public infrastructure works [16]. In this study, as in Romero et al. [7], the bilevel optimization takes into account in the upper level environmental, economic, and social costs, while there are other studies where only the economic cost has been considered [17].

Freight vehicles are the land transport vehicles with the greatest impact on pollution in cities; nevertheless, the movement of all vehicles must be considered regarding environmental contamination in the urban area. Transport significantly contributes to the major pollutant emissions (NOX, NMVOC, CH4, PM, and CO2), which is why they have been considered when studying the environmental impact of transport, as Romero et al. [7] did. It is worth highlighting that pollutant emissions from all transport sectors have dropped considerably since 1990, despite the general increase in vehicle movement [18]. Adams et al. [19] described existing initiatives which endeavoured to improve the quality of data explaining air quality and pollutant emissions; air quality monitoring is an example [20]. However, knowing the vehicle emission models is important when planning transport routes if we wish to reduce pollutant emissions [21]. Wang et al. [22] researched gas dispersion characteristics and particles due to traffic in Hong Kong, considering CO and PM 2.5.

Freight transport optimization models can be studied considering one or multiple viewpoints. Behrends et al. [23] explain that sustainable urban freight transport (SUFT) must take into account three points of view: social, economic, and environmental. For that reason, these three points have been considered in this paper, as well as in Romero et al. [7]. But, in other studies, one or two of these points have been applied. Yan et al. [24] designed a model whose sole aim was to minimize operation costs, excluding social and environmental costs. Nevertheless, many other studies also did consider social and environmental costs. He et al. [25] built a multipurpose model on logistics network planning to minimize total logistics network costs and likewise total carbon emissions. Browne et al. [2] reviewed the initiatives implemented by local authorities to reduce the environmental and social impact of urban freight transport in cities in 4 countries (UK, Japan, the Netherlands, and France).

To summarise, this paper studies the impact of the increase of large freight vehicles movement during a certain period of time within a city when they have a specific goods' delivery/pick-up point. The study has been carried out from an economic (operation costs), social (congestion), and environmental (emission of pollutants) point of view. This problem has been studied by other researchers. Moura et al. [16] proposes an optimization–simulation model for planning the transport of supplies to large public infrastructure works located in congested urban areas. But, the difference lies in that the methodology developed in this paper considers different types of the fleet—in size and typology—to minimize the total cost of the system, and nevertheless Moura et al. do not consider it

The introduction and State of the Art have been presented in this section, and the rest of the paper is arranged as follows: Section 2 describes the methodology used, then Section 3 applies the proposed methodology in a city along with presentation of main results, and, finally, Section 4 sets out the main conclusions.

2. Methodology

This paper presents a model to optimize large freight vehicle management and planning in city centres (lorries) used to load and unload large quantities of goods. The model considers a number of potential routes, which are defined by the restrictions of space and turning radii required by vehicles of this kind.

The purpose of the model is to determine journey distribution via different routes by defining fleet capacity and size, to minimize economic, social, and environmental costs. The lorries interact with the other transport means travelling in the city, like cars, buses, and smaller freight vehicles (vans or light trucks) and should be considered in the network modelling and calibration to which the optimization model will be applied. The optimization model considers total cost of the system, that is, social costs comprising bus and car user costs, bus operation and freight vehicle costs, and likewise environmental costs of all vehicles.

The optimization model is based on application of a bilevel mathematical program (Figure 1) [7, 16, 29] to find the best alternatives from an economic, social, and environmental viewpoint. From the lower level, via the city network model, vehicle flow, access, waiting and travelling times, and so on are obtained and then used on the upper level. At the upper level, an exhaustive search optimization is used to evaluate all the possibilities, to afterwards obtain those solutions that minimize total system cost:

$$\text{Min (System costs)} \tag{1}$$

$$\text{System Cost} = \text{Cu} + \text{Cop} + \text{Cma} \tag{2}$$

$$\text{Cu} = \text{Cu}_{\text{C}} + \text{Cu}_{\text{B}} \tag{3}$$

$$\text{Cu}_{\text{C}} = \varphi_{\text{Travel,C}} \cdot T_{\text{Travel,C}} \tag{4}$$

$$\begin{aligned}\text{Cu}_{\text{B}} = {} & \varphi_{\text{Acc,B}} \cdot T_{\text{Acc,B}} + \varphi_{\text{Egr,B}} \cdot T_{\text{Egr,B}} + \varphi_{\text{Wai,B}} \cdot T_{\text{Wai,B}} \\ & + \varphi_{\text{Travel,B}} \cdot T_{\text{Travel,B}} + \varphi_{\text{Tra,B}} \cdot T_{\text{Tra,B}},\end{aligned} \tag{5}$$

where Cu is total users cost; Cu_C is car users cost; Cu_B is bus users cost; $T_{\text{Travel,C}}$ is car travel time; $\varphi_{\text{Travel,C}}$ is Car travel time worth; $T_{\text{Acc,B}}$ is bus access time; $\varphi_{\text{Acc,B}}$ is bus access time worth; $T_{\text{Egr,B}}$ is bus egress time; $\varphi_{\text{Egr,B}}$ is bus egress time worth;

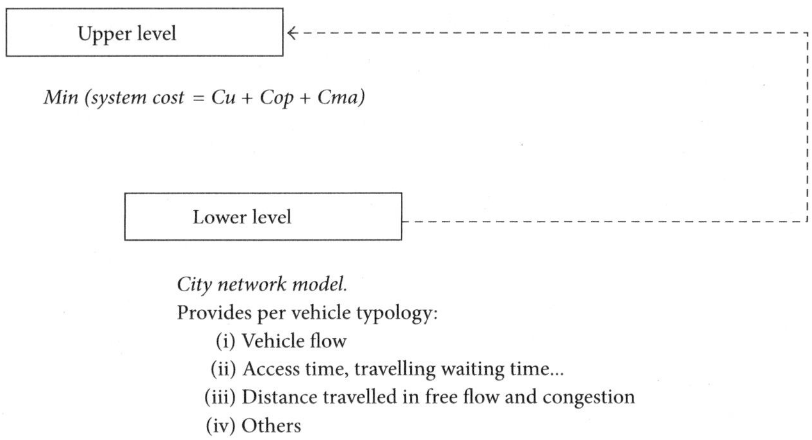

FIGURE 1: Bilevel methodology.

$T_{\mathrm{Wai,B}}$ is Bus waiting time; $\varphi_{\mathrm{Wai,B}}$ is bus waiting time worth; $T_{\mathrm{Travel,B}}$ is bus travel time; $\varphi_{\mathrm{Travel,B}}$ is bus travel time worth; $T_{\mathrm{Tra,B}}$ is bus transfer time; $\varphi_{\mathrm{Tra,B}}$ is bus transfer time worth.

Operating costs are calculated using the following formulation:

$$\begin{aligned}
\mathrm{Cop} &= \mathrm{Cop_B} + \mathrm{Cop_{Tr}},\\
\mathrm{Cop_B} &= \mathrm{CR} + \mathrm{CP} + \mathrm{CF},\\
\mathrm{CR} &= \varphi_{\mathrm{CR}} \cdot \text{Total Km.},\\
\mathrm{CP} &= \varphi_{\mathrm{CP}} \cdot \text{Person hours},\\
\mathrm{CF} &= \varphi_{\mathrm{CF}} \cdot \mathrm{N^{\circ}\ Buses},\\
\mathrm{Cop_{Tr}} &= \sum_i T_i \cdot f_i \cdot C_u,\\
T &= T_{\mathrm{outward}} + T_{\mathrm{return}} + T_{\mathrm{loading}} + T_{\mathrm{unloading}},
\end{aligned} \tag{6}$$

where Cop is total operating costs; $\mathrm{Cop_B}$ is bus operating cost; $\mathrm{Cop_{Tr}}$ is truck operating cost.

Bus operating costs ($\mathrm{Cop_B}$) are made up of three factors: cost proportional to travelled distance (CR), personnel costs (CP), and fixed costs (CF).

Total cost due to the distance travelled by the buses is equal to

$$\mathrm{CR} = \varphi_{\mathrm{CR}} \cdot \text{Total Km.}, \tag{7}$$

where φ_{CR} is unit cost per kilometer covered by bus.

$$\text{Total Km.} = \sum_i L_i \cdot f_i, \tag{8}$$

where L_i is length of route i; f_i is frequency of route i.

Employee costs are calculated considering only the personnel who are really working on the buses:

$$\mathrm{CP} = \varphi_{\mathrm{CP}} \cdot \text{Person hours}, \tag{9}$$

where φ_{CP} is the hourly employee cost.

$$\text{Person hours} = \sum_i \frac{tc_i}{h_i}, \tag{10}$$

where tc_i is time of a round trip; h_i is headway on route i.

Fixed costs are calculated with the following formula that only considers the buses that are really circulating:

$$\mathrm{CF} = \varphi_{\mathrm{CF}} \cdot \mathrm{N^{\circ}\ buses}, \tag{11}$$

where φ_{CF} is fixed cost per hour of bus.

$$\mathrm{N^{\circ}\ buses} = \sum_i \frac{tc_i}{h_i}, \tag{12}$$

where tc_i is time of a round trip; h_i is headway on route i.

Truck operating cost ($\mathrm{Cop_{Tr}}$) is estimated as

$$\begin{aligned}
\mathrm{Cop_{Tr}} &= \sum_i T_i \cdot f_i \cdot C_u,\\
T &= T_{\mathrm{outward}} + T_{\mathrm{return}} + T_{\mathrm{loading}} + T_{\mathrm{unloading}},
\end{aligned} \tag{13}$$

where T_{outward} is truck outward time; T_{return} is truck return time; T_{loading} is truck loading time; $T_{\mathrm{unloading}}$ is truck unloading time; C_u is cost per hour of truck use; f_i is truck flow.

The environmental costs (Cma) were calculated for different alternatives considering 5 types of pollutants (p): NOx, NMVOC, CH4, PM2.5, and CO2, as well as the difference vehicle typologies (v): petrol cars, diesel cars, buses, heavy lorries (HGV), medium lorries, and lightweight lorries.

$$\begin{aligned}
\mathrm{Cma} &= \sum_p Q_p \cdot C_p\\
Q_p &= \sum_v (\text{km cong}_v \cdot \text{Consum.cong}_v + \text{km uncong}_v\\
&\quad \cdot \text{Consum.uncong}_v) \cdot \delta_v \cdot \text{Emmisions}_{p,v},
\end{aligned} \tag{14}$$

where Cma is total environmental cost. Q_p is amount of pollutant p. C_p is environmental cost of pollutant p. km cong_v is km routed with congested network per vehicle type v. km uncong_v is km routed with uncongested network per vehicle type v. Consum.cong_v is consumption per vehicle type v with congested network. Consum.uncong_v is consumption per vehicle type v with uncongested network. δ_v is conversion factor kg/litres of fuel per vehicle type v. $\text{Emmisions}_{p,v}$ is emissions of type p pollutants per vehicle typology v.

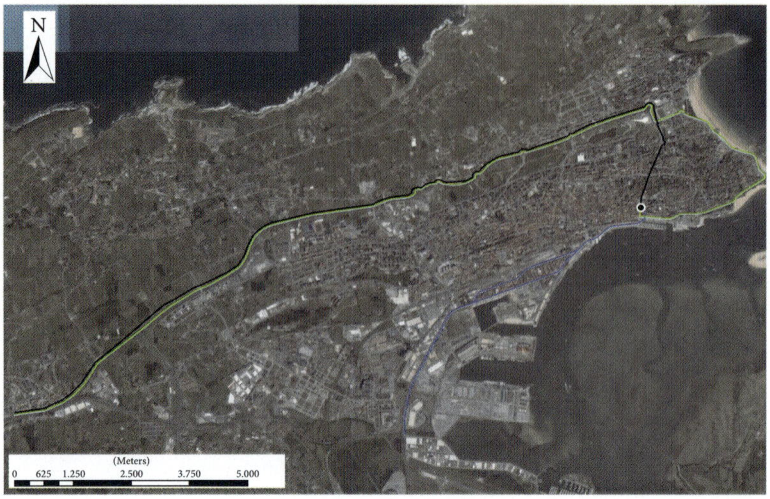

• Constructions
—— Route 1
—— Route 2
—— Route 3

FIGURE 2: Considered routes.

TABLE 1: Vehicle's consumption rates (litres/Km) and vehicle's emission rates (g of pollutant/Kg of fuel) [26, 27].

	Emissions (g of pollutant/Kg of fuel)					Consumption (litres/Km)		δ (Kg/litre)
	NOx	NMVOC	CH4	PM	CO2	Congested	Uncongested	
Gasoline cars	10.89	13.44	1.19	0.03	3180.00	0.08	0.06	0.680
Diesel cars	11.12	0.61	0.07	0.80	3140.00	0.07	0.05	0.850
Buses	32.67	0.99	0.24	0.81	3140.00	0.34	0.26	0.850
Heavy trucks						0.34	0.26	
Medium trucks	34.84	3.06	0.38	1.34	3140.00	0.31	0.22	0.850
Light trucks						0.29	0.21	

TABLE 2: Environmental cost of each kind of pollutant (€/ton) [28].

	NOx	NMVOC	CH4	PM	CO2
€/ton of pollutant	3.60	800.00	775.00	114.00	29.00

Firstly, fuel consumption per alternative is obtained. Once the emission of pollutants is known, it is calculated per type (Table 1). The emission of pollutants depends on how congested the network is, distance travelled, and vehicle typology [26]. Finally, this is converted into monetary terms (Table 2).

3. Case Study

The methodology proposed has been applied to a real case in Santander, a city in northern Spain. Santander is a medium-sized coastal city with approx. 180,000 inhabitants, and approx. 75% of the city is bordered by water, so it has few access points.

A large construction project in the southeast of the city required delivery of 180 cubic metres of building material during the rush hour. This is the reason for studying the economic, social, and environmental impact caused by an increase in demand for large freight vehicles. The building material would be transported via a fleet of homogenous vehicles, selected from several typologies characterized by their maximum capacity and speed (Table 3).

TABLE 3: Fleet typologies.

Truck size	Cargo capacity (tons)	Maximum speed (km/h)
Light	9	10
Medium	12	15
Heavy	18	20

Beginning with Santander network model data [29], three different routes will be considered: the first one (R1) goes through roads with two lanes per direction, with the exception of an 800-metre tunnel at the end of it; the second route (R2) differs only from R1 in that instead of going through the tunnel, it borders the city using coastal 1-lane per direction roads; and the third route (R3) travels through some streets that, though they have two or three lanes per direction, are also the most congested ones in the city. See Figure 2.

Each route has two legs, which contribute to its total travel time and length: an urban one (inside Santander) and an external one (from the quarry to the outskirts of the city).

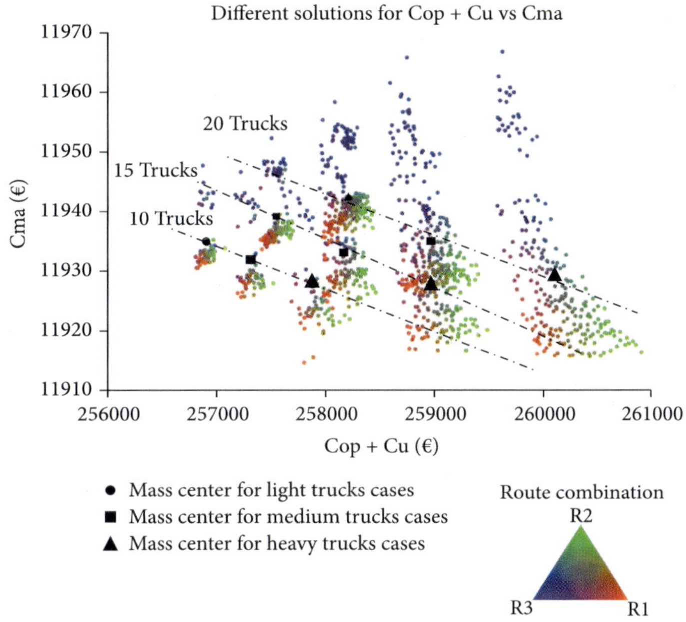

FIGURE 3: Different solutions for Cop + Cu versus Cma.

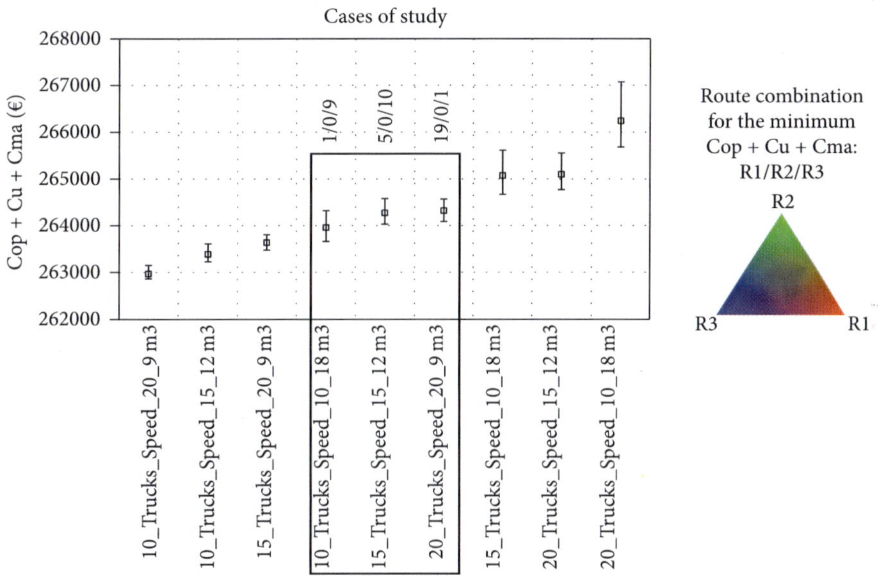

FIGURE 4: Total cost value for the cases of study.

First of all, model behaviour was verified studying all possible means of moving the vehicle fleet from the quarry to building site via 3 predefined routes. The vehicle fleet may be of different typologies as already mentioned (Table 3); furthermore, 3 different fleet sizes were considered, that is, 10, 15, and 20 vehicles. Application of the model resulted in the obtainment of operation, user, and environmental costs per scenario in monetary terms (Figure 3). The difference between each scenario is the number of vehicles travelling along each route.

With this figure we can see that the model behaves as expected; that is, for the same size fleet, the combined user and operator costs increase the heavier and slower the lorry is, whereas environmental costs increase when the fleet consists of smaller lorries. Nevertheless, as might be expected costs increased with the size of the fleet for the same lorry typology. Moreover, as can be seen per scenario, if the majority of the fleet takes route 3 (R3) (blue dots at the top of the graph), environmental costs are higher than if routes 1 and/or 2 are taken; whereas should the majority of the fleet use route 2 (R2), the combined operation and user costs are higher. In terms of both combined user and operator costs and environmental costs, the best solutions correspond to those scenarios where most of the fleet uses route 1 (red dots).

Figure 4 shows the total costs for the different fleet size and typology combinations, indicating the total average

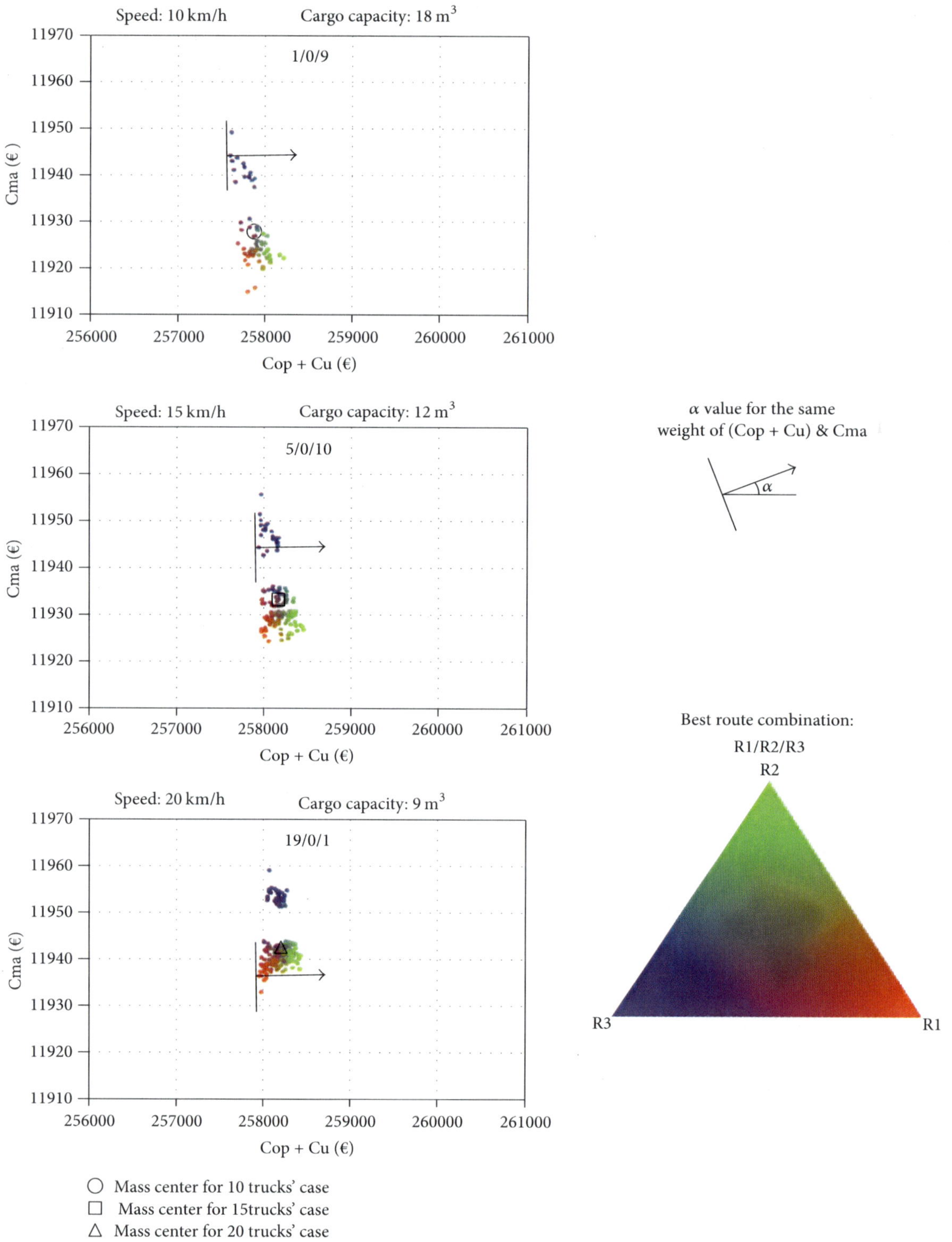

FIGURE 5: Possible ways to transport the load.

cost per combination and likewise the total maximum and minimum cost.

Getting back to our study case where the building site required delivery of 180 cubic metres during the rush hour, per vehicle typology, the solutions with lower costs are shown in Figure 4. Figure 5 shows these 3 possibilities in greater detail together with the number of vehicles per route for the average cost.

If the planner needs the solution that minimizes environmental costs, he should move in a line parallel to the x-axis ($\alpha = 0^{\circ}$), starting at Cma = 0, and choose the first solution touched by it, which corresponds to sending 10 heavy trucks

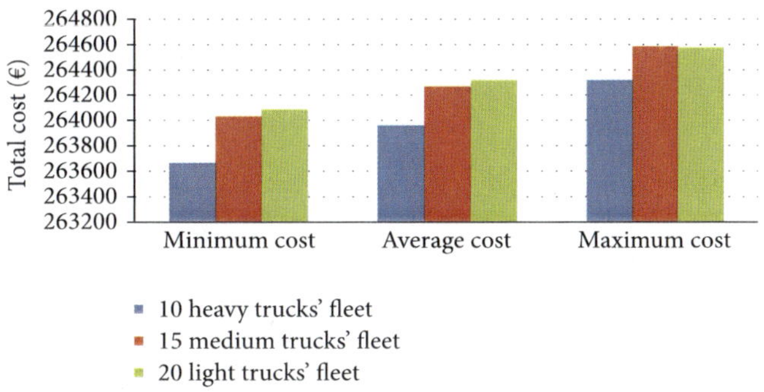

FIGURE 6: Minimum, average, and maximum cost for the three vehicle typologies.

through R1. If our objective were to minimize Cop + Cu, the process would be similar, but using a line parallel to the y-axis instead ($\alpha = 90^\circ$). In this case, we should also use 10 heavy trucks, with ten of them going through R3. If only medium or light trucks were available, to minimize Cop + Cu most of the fleet should travel through R3 or R1, respectively. Other intermediate Pareto solutions, with different emphases in the environmental or in the user and operator costs, can be found using values of α between 0° and 90°.

Figure 5 shows which alpha value to use if Cma and (Cop + Cu) had the same importance. Below, the following is shown and represented (Figure 6): the 3 vehicle typologies (heavy, medium, and light), the minimum, average, and maximum costs necessary to move the building material from the quarry to the site, and likewise the number of lorries which would use each of the 3 routes available.

(i) 10 heavy trucks' fleet:

(a) Minimum cost: 263.661,08 € (R1/R2/R3: 1/0/9)
(b) Average cost: 263.956,73 €
(c) Maximum cost: 264.325,87 € (R1/R2/R3: 0/10/0)
(d) Difference between max. and min. costs: 664,79 €

(ii) 15 medium trucks' fleet:

(a) Minimum cost: 264.027,50 € (R1/R2/R3: 5/0/10)
(b) Average cost: 264.267,95 €
(c) Maximum cost: 264.582,49 € (R1/R2/R3: 0/15/0)
(d) Difference between max. and min. costs: 554,91 €

(iii) 20 light trucks' fleet:

(a) Minimum cost: 264.081,94 € (R1/R2/R3: 19/0/1)
(b) Average cost: 264.315,18 €
(c) Maximum cost: 264.570,07 € (R1/R2/R3: 0/19/1)

(iv) Difference between max. and min. costs: 488,12 €

Regarding total costs, it can be seen that using a fleet of 10 heavy trucks results in significantly lower total costs (minimum as well as average or maximum). There is little difference between using a fleet of 15 medium trucks or/and 20 light trucks, the former being cheaper than the latter in terms of minimum and average costs, while it is slightly more expensive when considering maximum costs. In conclusion, if a fleet of 10 heavy trucks is available, it is the option less costly from the environmental, economic, and social points of view.

4. Conclusions

This paper presents a model to optimize management and planning of special freight vehicles coming from outside the city to a specific point in the centre. The proposed model minimizes freight vehicle impact for those occasions when there is movement of large quantities of goods. The strength of the model is that it is based on bilevel mathematical programming, making a more realistic reflection of the practical problem possible by allowing variables that represent aspects of the city network to be nested within the problem of cost minimization. Furthermore, the model minimizes freight vehicle impact for those occasions when there is movement of large quantities of goods.

The model was applied to Santander (Spain) for the case where a large quantity of building materials had to be delivered to a specific point in the city characterized by a high mobility rate of both private and public transport, due to a large building site. In addition to selecting their typology, vehicles may also be assigned to different routes to minimize total system costs.

The model provides solutions for different vehicle typologies. The most economical solution considering the total system costs, that is, user, operation, and environmental costs, would be to use 10 HGVs, 1 vehicle via route R1 and the other 9 via route R3. However, if the purpose is to find the best solution from an environmental viewpoint, then these 10 HGVs should use route R1. Nevertheless, should only lightweight vehicles be available, then the best environmental solution would be to assign 15 vehicles to route R1 and 5 to route R3; and regarding total costs the best solution would be 10 vehicles via R1, 9 via R2, and 1 via R1.

In conclusion, we have learnt from the model proposed that large vehicle management and planning reduce impact on cities, from a social, economic, and/or environmental viewpoint. Selection of both vehicle typology and routes to be used and followed is important in the reduction of the costs which the movement of said vehicles causes.

Conflicts of Interest

The authors declare that there are no conflicts of interest regarding the publication of this paper.

References

[1] J. Holguín-Veras, K. Ozbay, A. Kornhauser et al., "Overall impacts of off-hour delivery programs in New York city metropolitan area," *Journal of The Transportation Research Record*, vol. 2238, pp. 68–76, 2011.

[2] M. Browne, J. Allen, T. Nemoto, D. Patier, and J. Visser, "Reducing social and environmental impacts of urban freight transport: a review of some major cities," *Procedia - Social and Behavioral Sciences*, vol. 39, pp. 19–33, 2012.

[3] J. Visser, T. Nemoto, and M. Browne, "Home delivery and the impacts on urban freight transport: a review," *Procedia - Social and Behavioral Sciences*, vol. 125, pp. 15–27, 2014.

[4] H. J. Quak and M. De Koster, "Urban distribution: the impacts of different governmental time-window schemes," ERIM Report Series Reference ERS-2006-053-LIS, 2006.

[5] E. a. d. Santander, "Ordenanza de circulación del excmo. ayuntamiento de Santander," *Boletin Oficial de Cantabria*, vol. 103, 2016.

[6] R. Capitale, Ordinanza Sindicale n.88 del 27 ottobre, 2016.

[7] J. P. Romero, J. Benavente, J. L. Moura, A. Ibeas, and B. Alonso, "An environmental approach to optimize urban freight transport systems," *Advances in Intelligent Systems and Computing*, vol. 262, pp. 127–136, 2014.

[8] T. Schoenherr, "Logistics and supply chain management applications within a global context: an overview," *Journal of Business Logistics*, vol. 30, no. 2, pp. 1–25, 2009.

[9] G. Reiner and M. Trcka, "Customized supply chain design: problems and alternatives for a production company in the food industry. a simulation based analysis," *International Journal of Production Economics*, vol. 89, no. 2, pp. 217–229, 2004.

[10] J. Guo, X. Wang, S. Fan, and M. Gen, "Forward and reverse logistics network and route planning under the environment of low-carbon emissions: A case study of Shanghai fresh food E-commerce enterprises," *Computers & Industrial Engineering*, vol. 106, pp. 351–360, 2017.

[11] L. Wu, X. Yue, A. Jin, and D. C. Yen, "Smart supply chain management: a review and implications for future research," *International Journal of Logistics Management*, vol. 27, no. 2, pp. 395–417, 2016.

[12] C. Öberg and G. Graham, "How smart cities will change supply chain management: a technical viewpoint," *Production Planning and Control*, vol. 27, no. 6, pp. 529–538, 2016.

[13] J. F. Bard, "Optimality conditions for the bilevel programming problem," *Naval Research Logistics Quarterly*, vol. 31, no. 1, pp. 13–26, 1984.

[14] Á. Ibeas, L. Dell'Olio, B. Alonso, and O. Sainz, "Optimizing bus stop spacing in urban areas," *Transportation Research Part E: Logistics and Transportation Review*, vol. 46, no. 3, pp. 446–458, 2010.

[15] J. S. Chang and R. L. Mackett, "A bi-level model of the relationship between transport and residential location," *Transportation Research Part B: Methodological*, vol. 40, no. 2, pp. 123–146, 2006.

[16] J. L. Moura, A. Ibeas, and L. dell'Olio, "Optimization-simulation model for planning supply transport to large infrastructure public works located in congested urban areas," *Networks and Spatial Economics*, vol. 10, no. 4, pp. 487–507, 2010.

[17] H. Sun, Z. Gao, and J. Wu, "A bi-level programming model and solution algorithm for the location of logistics distribution centers," *Applied Mathematical Modelling*, vol. 32, no. 4, pp. 610–616, 2008.

[18] EEA, *Emissions of Air Pollutants from Transport*, European Environment agency, 2013.

[19] M. Adams, S. Cryan, A. Mourelatou, and K. Nytorv, *Air emission inventory data in europe: new perspectives*, 2009.

[20] P. Kumar, L. Morawska, C. Martani et al., "The rise of low-cost sensing for managing air pollution in cities," *Environment International*, vol. 75, pp. 199–205, 2015.

[21] E. Demir, T. Bektaş, and G. Laporte, "A review of recent research on green road freight transportation," *European Journal of Operational Research*, vol. 237, no. 3, pp. 775–793, 2014.

[22] J. S. Wang, T. L. Chan, Z. Ning, C. W. Leung, C. S. Cheung, and W. T. Hung, "Roadside measurement and prediction of CO and PM2.5 dispersion from on-road vehicles in Hong Kong," *Transportation Research Part D: Transport and Environment*, vol. 11, no. 4, pp. 242–249, 2006.

[23] S. Behrends, M. Lindholm, and J. Woxenius, "The impact of urban freight transport: a definition of sustainability from an actor's perspective," *Transportation Planning and Technology*, vol. 31, no. 6, pp. 693–713, 2008.

[24] S. Yan, W. Lai, and M. Chen, "Production scheduling and truck dispatching of ready mixed concrete," *Transportation Research Part E: Logistics and Transportation Review*, vol. 44, no. 1, pp. 164–179, 2008.

[25] Q. He, L. Hu, and H. Qian, "Low-carbon sales logistics network planning method based on multi-objective planning," *Systems Engineering*, vol. 7, pp. 37–43, 2013.

[26] EMEP/EEA, "EMEP/EEA air pollutant emission inventory guidebook 2009," Technical guidance to prepare national emission inventories, 2009.

[27] IDAE, "Guía de vehículos turismo de venta en España, con indicación de consumos y emisiones de CO_2, 2013".

[28] H. van Essen et al., *External Costs of Transport in Europe. Update Study for 2008*, 2011.

[29] J. P. Romero, A. Ibeas, J. L. Moura, J. Benavente, and B. Alonso, "A Simulation-optimization approach to design efficient systems of bike-sharing," *Procedia - Social and Behavioral Sciences*, vol. 54, pp. 646–655, 2012.

2

Urban Motion Planning Framework based on N-Bézier Curves Considering Comfort and Safety

Ray Lattarulo, [1] **Leonardo González**, [2] **Enrique Martí**, [1] **José Matute**, [1] **Mauricio Marcano**, [1] **and Joshue Pérez** [1]

[1] *Department of Automotive, Industry and Transportation Division, Tecnalia Research and Innovation, 48160 Derio, Vizcaya, Spain*
[2] *Department of Cyber-Security, ICT Division, Tecnalia Research and Innovation, 48160 Derio, Vizcaya, Spain*

Correspondence should be addressed to Ray Lattarulo; rayalejandro.lattarulo@tecnalia.com

Academic Editor: Irena I. Otković

In last decades, great technology advances have been done related to the automotive sector, especially in Advanced Driver Assistance Systems (ADAS) developed to improve mobility in terms of comfort and safety during driving process; hence, automated driving is presented as an evolution of those systems in the present and upcoming years. The aim of this work is to present a complete framework of motion planning for automated vehicles, considering different constraints with parametric curves for lateral and longitudinal planners. Parametric Bézier curves are used as the core approach for trajectory design in intersections, roundabouts, and lane change maneuvers. Additionally, a speed planner algorithm is presented using the same parametric curve approach, considering comfort and safety. A simulation environment is used for testing the planning method in urban conditions. Finally, tests with the real platform in automated mode have been performed showing goods results.

1. Introduction

Every year 1.2 million people die in road related accidents. The NHTSA conducted a study concluding that 94% of accidents are due to human errors, where a great amount of these crashes (33%) is related to wrong decisions during driving process [1]. In 2009, 2.4% of fatalities in the United States were related to a common problem for drivers, i.e., drowsiness [2]. Considering these points, automated driving technology promises a great reduction in crash incidents, adding efficiency in fuel consumption, reduction of parking spaces (ride-sharing), and inclusion of elderly and people with disabilities [3].

In last decades, a considerable amount of institutions, research centers, and companies is seeking to improve this technology faster [4], to make automated driving a reality in public roads as soon as possible [5]. Lateral and longitudinal controllers [6–8], as well as perception (sensors) [9, 10] and communications [11], are some of the topics mostly studied.

There are some other topics; those have received less attention than the ones named before, i.e., vehicle decision algorithm. This area is one of the most challenging topics in automated driving because it must deal with a great number of task, as generation of smooth trajectories [12], speed profiles based on fuel consumption and comfort [13], obstacles avoidance [14], and even doing coordination among other participants to execute a maneuver [15].

Some methods explored in automated vehicle decision are in [16]; the authors have studied a novel method for autonomous vehicles decision task. They considered dynamic maneuvers satisfying traffic norms and road limitations based on a data-driven vehicle dynamic model and an optimization process using curve arcs. Other authors have explored methods based on generation of multiple curves, i.e., clothoids, setting a possible group of tentacles (possible trajectories) to be tracked by the vehicle, and the optimal solution is obtained using a Markov Decision Process (MDP) [17]. Other approaches are related to Partially Observable Markov Decision Process (POMDP) for generation of high level decision policies (following the lane, lane change, parking, etc.) [18]. Rapidly exploring Random Tree (RRT) is another method currently used for trajectory planning; some authors as [19] are testing the integration of RRT and visual driver behavior in driving to improve the decision process. The problem is that those methods have been developed to support specific scenarios considering a bunch of specific conditions; sometimes they did not even consider online deployment [20].

Other authors, as [21], are tackling the problem considering the importance of dealing with the general case of driving (especially in urban scenarios). The idea is handling situations that goes from, for example, cruise control to being able of respecting traffic lights or signs, among other things. The problem with this approach was that the authors only considered the longitudinal part of the vehicle's decision. On the other hand, [22, 23] have presented works about general methods for trajectory generation considering intersections and roundabouts, but the speed profile generation was considered later on [24].

In [25] a method combining parametric curves of Bézier is presented; the goal is to deal with the general case of trajectory planning (in urban environments), combining this with a speed profile generation based on Model Predictive Control. The problem with the approach is, basically, considerably more difficult for using two different techniques to resolve the planning problem (they are not unified).

In such a way, this work presents a unified motion planning framework (trajectory planning and speed profile) based on 4th and 5th degree parametric Bézier curves. It specially focuses on urban scenarios and their geometrical design. It leans on the construction of two-dimensional curves for both (lateral and longitudinal) planning methods, simplifying the possible designing parameters to the most considerable ones without degrading the curve capacities. The work specially focuses on giving a tool-set for trajectory designing under typical scenarios as intersections and lane changes and to a more specific and complex scenario as roundabouts. This, last scenario is especially interesting for its set of possible parameters and conditions; it is relevant to consider that the roundabout complexity has not been sufficiently studied for other authors in the past [26].

Finally, the contribution will be organize as follows: Section 2 has the detailed explanation of three urban scenarios considered: intersections, roundabout, and lane changes (the rest can be modeled as straight line or arcs). A definition of relevant Bézier curve properties is carried on, then a global planning approach is developed to satisfy some requirements of the local planning approach (reducing the total amount of map points); these concepts will be followed by the trajectory and speed planning approach. Section 3 presents the general components used in the software architecture (all modules considered for simulation and real vehicle tests). In the following, it presented an urban test case based on simulation (Section 4) and a real case of study (Section 5); both specially focus on roundabout considerations. Finally, Section 6 contains the conclusions of this article summarizing the contributions and future works.

2. Motion Planning Framework

This section explains all the information related to the trajectory planning approach based on real-time parametric 4th and 5th degree Bézier curves. The approach's core will be focused on using these curves in path planning generation for urban scenarios (intersections, roundabouts, and lane changing) and speed planning for comfortable and safe behaviors. This approach is not limited to urban scenarios, although it is the main target of this work.

The planning framework explanation will be divided into three parts: (i) global planning using a global map (based on a point to define the structure of intersections, lane change or roundabouts) describing the route with reduced amount of points, (ii) local planning approach based on Bézier and specially focus on intersections, lane changes, and roundabouts (the straight and arc segment can be resolved as sequences of points), and (iii) the speed planning approach based on the same type of curves.

2.1. N-Bézier Curve Basis. Bézier is a type of parametric curve, which has been commonly used for computer graphics, animations, and path generation in robotics [27]. In general, they are good for real-time implementation, and the computational cost of designing them is lower than clothoids and splines curves [28]. Bézier curves are described by

$$\mathbf{B}(t \mid n, P_0, \dots, P_n) = \sum_{i=0}^{n} b_i \mathbf{P}_i, \quad b_i = \binom{n}{i} t^i (1-t)^{n-i} \qquad (1)$$

where $\{b_i \in \mathbb{R}\}$ is the Bernstein polynomial, $\{\mathbf{P}_i \in \mathbb{R}^2\}$ are the control points used to generate the curve, $\{n \in \mathbb{N}^+\}$ is the Bézier order, and $\{t \in \mathbb{R}, t = [0,1]\}$ is the parameter for curve construction.

These curves have several properties relevant for the purpose of the current work:

(i) The starting point of the Bézier curve corresponds with control point $\mathbf{P}_0$, and the ending point corresponds with $\mathbf{P}_n$.

(ii) The first point tangent vector (at $t = 0$) will be given by $\overrightarrow{P_0P_1}$ and the last point tangent vector (at t = 1) will be given by $\overrightarrow{P_{n-1}P_n}$.

(iii) The curve will lie into the convex hull formed by the control points.

(iv) Bézier curves are continuous geometrically and in curvature $\{C^n \; \& \; G^n, \forall n \in \mathbb{N}^+\}$, and this continuity can be preserved in joints of two different curves.

(v) Bézier curves are symmetric, and the generated curve for $t : 0 \longrightarrow 1$ is equal to the one for $t : 1 \longrightarrow 0$.

This work uses 4th and 5th grade Bézier curves (some considerations with 3rd degree will be done). Higher order curves are not considered because they do not have additional benefits within current approach but increase the complexity [29].

Rewriting the general Bézier equation (see (1)) a more compact representation is obtained:

$$\mathbf{B}(t) = \mathbf{K}_0 + \mathbf{K}_1 t + \mathbf{K}_2 t^2 + \cdots + \mathbf{K}_n t^n \qquad (2)$$

where each $\mathbf{K}_i$ will be a function of the control points (fixed position) and curve order. The values of $\mathbf{K}_i$ coefficients are

TABLE 1: Compact Bézier polynomial Coefficients "C".

C	Bézier curve order 3rd Order	4th Order	5th Order
$\mathbf{K}_0$	$\mathbf{P}_0$	$\mathbf{P}_0$	$\mathbf{P}_0$
$\mathbf{K}_1$	$-3\mathbf{P}_0 + 3\mathbf{P}_1$	$-4\mathbf{P}_0 + 4\mathbf{P}_1$	$-5\mathbf{P}_0 + 5\mathbf{P}_1$
$\mathbf{K}_2$	$3\mathbf{P}_0 - 6\mathbf{P}_1 + 3\mathbf{P}_2$	$6\mathbf{P}_0 - 12\mathbf{P}_1 + 6\mathbf{P}_2$	$10\mathbf{P}_0 - 20\mathbf{P}_1 + 10\mathbf{P}_2$
$\mathbf{K}_3$	$-\mathbf{P}_0 + 3\mathbf{P}_1 - 3\mathbf{P}_2 + \mathbf{P}_3$	$-4\mathbf{P}_0 + 12\mathbf{P}_1 - 12\mathbf{P}_2 + 4\mathbf{P}_3$	$-10\mathbf{P}_0 + 30\mathbf{P}_1 - 30\mathbf{P}_2 + 10\mathbf{P}_3$
$\mathbf{K}_4$	–	$\mathbf{P}_0 - 4\mathbf{P}_1 + 6\mathbf{P}_2 - 4\mathbf{P}_3 + \mathbf{P}_4$	$5\mathbf{P}_0 - 20\mathbf{P}_1 + 30\mathbf{P}_2 - 20\mathbf{P}_3 + 5\mathbf{P}_4$
$\mathbf{K}_5$	–	–	$-\mathbf{P}_0 + 5\mathbf{P}_1 - 10\mathbf{P}_2 + 10\mathbf{P}_3 - 5\mathbf{P}_4 + \mathbf{P}_5$

given in Table 1. $\mathbf{K}_1$ and $\mathbf{K}_2$ determine curvature at starting and ending points of the curve, this property is going to be used repeatedly on explanations. Using the information from 3rd to 5th degree Bézier, $\mathbf{K}_1$ and $\mathbf{K}_2$ are defined as follows:

$$\mathbf{K}_1 = n\vec{v_1}$$
$$\mathbf{K}_2 = \frac{n(n-1)}{2}\left(\vec{v_2} - \vec{v_1}\right) \tag{3}$$

where $\vec{v_1} = \mathbf{P}_1 - \mathbf{P}_0$ and $\vec{v_2} = \mathbf{P}_2 - \mathbf{P}_1$ are introduced and n is the degree of the curve.

In general terms the curvature in $\mathbb{R}^2$ is defined as [30]

$$k(t) = \frac{\dot{B}_x(t)\ddot{B}_y(t) - \ddot{B}_x(t)\dot{B}_y(t)}{\sqrt{\left(\dot{B}_x(t)^2 + \dot{B}_y(t)^2\right)^3}} = \frac{\dot{\mathbf{B}}(t) \times \ddot{\mathbf{B}}(t)}{\left\|\dot{\mathbf{B}}(t)\right\|^3} \tag{4}$$

and in this case, function B is the Bézier equation. Evaluating at $t = 0$, it is obtained:

$$k(t=0) = 2\frac{\mathbf{K}_1 \times \mathbf{K}_2}{\left\|\mathbf{K}_1\right\|^3} = \frac{(n-1)}{n}\frac{\vec{v_1} \times \vec{v_2}}{\left\|\vec{v_1}\right\|^3} \tag{5}$$

The previous equation explicitly shows that if the three starting points in a curve are colinear, generated Bézier curve will have zero curvature at its starting point ($k(t = 0) = 0$); this can be extended to the three ending points due to symmetry. This property is useful for designing the intersections and lane changes, and for the entrance and exit part of the roundabout.

2.2. Global Planning Environment Used for the Approach. In this work, global planning is used as base route to apply the local planning approach using Bézier. A simple map containing a point to define intersections, roundabouts, or lane changes (point with "x" in Figures 1, 2 and 3) is the input of the global method, decreasing the amount of points and keeping geometric qualities of road.

2.2.1. Intersections. Figure 1 depicts the description of an intersection using a single point (junction of 2 or more

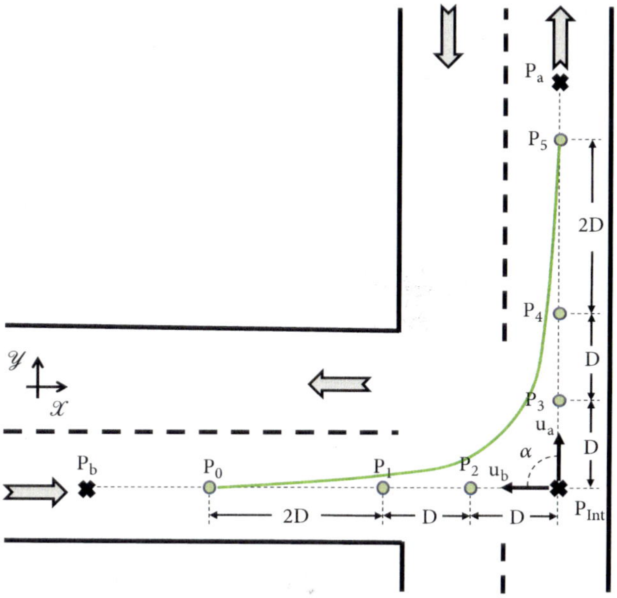

FIGURE 1: Trajectory planning on intersection.

straight paths). The unitary vectors defining the road (in intersection case) are given by

$$\vec{u_b} = \frac{\mathbf{P}_b - \mathbf{P}_i}{\left\|\mathbf{P}_b - \mathbf{P}_i\right\|}$$
$$\vec{u_a} = \frac{\mathbf{P}_a - \mathbf{P}_i}{\left\|\mathbf{P}_a - \mathbf{P}_i\right\|} \tag{6}$$

where $\vec{u_a}$ is an unitary vector that goes from intersection $\mathbf{P}_i$ to past global point $\mathbf{P}_b$ and $\vec{u_b}$ is other unitary vector formed with the next global points $\mathbf{P}_a$ and $\mathbf{P}_i$. It is relevant to see that the angle α (Figure 1) can be different from 90°, modeling a great variety of intersections. The parameter D will be used later in local planning and for the definition of $\mathbf{P}_i$.

2.2.2. Lane Changes. In this case, the approach considers straight segments during the maneuver (Figure 2). The vectors for path description defined in (6) can be used for lane change and they have the following relation:

$$\vec{u_b} = -\vec{u_a} \tag{7}$$

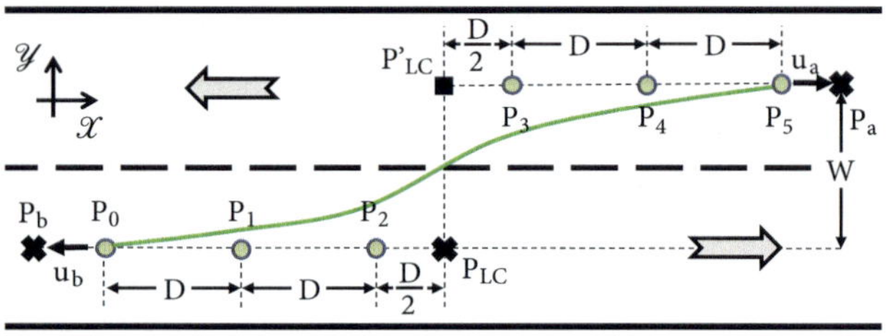

FIGURE 2: Trajectory planning on lane change.

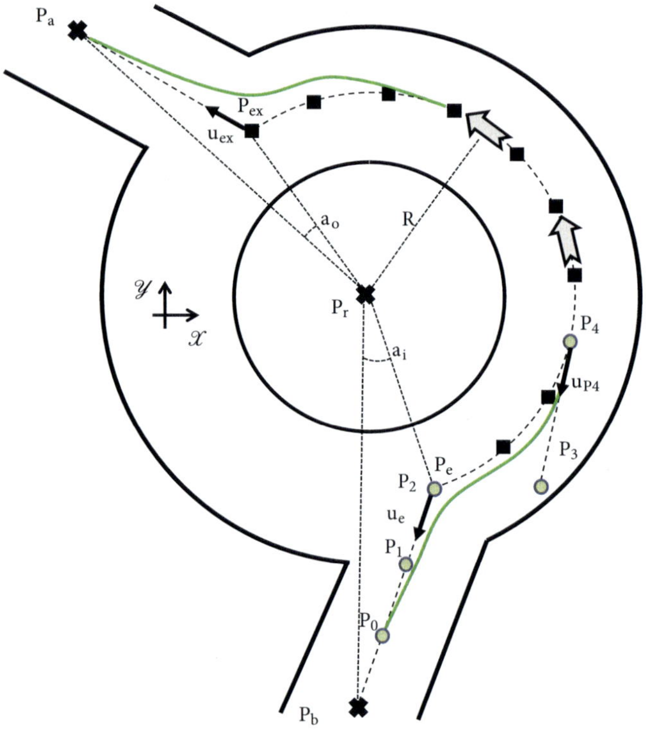

FIGURE 3: Trajectory planning on roundabouts.

Additionally, in this type of curve a new point $\mathbf{P}'_{LC}$ is introduced to handle the maneuver in local planning. This point is defined by

$$\mathbf{P}'_{LC} = \mathbf{P}_{LC} + w \begin{bmatrix} \cos\left(\tan^{-1}\left(\dfrac{u_{a_y}}{u_{a_x}}\right) \pm \dfrac{\pi}{2}\right) \\ \sin\left(\tan^{-1}\left(\dfrac{u_{a_y}}{u_{a_x}}\right) \pm \dfrac{\pi}{2}\right) \end{bmatrix} \quad (8)$$

where w is the road width and $\overrightarrow{u_a}$ is the vector presented in (7). The symbol ± is negative when the lane change is done to the right side and positive to the left side.

2.2.3. Roundabouts. Figure 3 shows the definition of a roundabout using a simple point. It uses roundabout center point $\mathbf{P}_r$, its radius R, entrance angle a_i, and exit angle a_o. Using these parameters, the roundabouts can be modeled with an approximated circular shape.

Roundabout entrance and exit points are defined by

$$\mathbf{P}_e = \mathbf{P}_r + R \begin{bmatrix} \cos\left(\tan^{-1}\left(\dfrac{u_{b_y}}{u_{b_x}}\right) \pm a_i\right) \\ \sin\left(\tan^{-1}\left(\dfrac{u_{b_y}}{u_{b_x}}\right) \pm a_i\right) \end{bmatrix}$$
$$\mathbf{P}_{ex} = \mathbf{P}_r + R \begin{bmatrix} \cos\left(\tan^{-1}\left(\dfrac{u_{a_y}}{u_{a_x}}\right) \mp a_o\right) \\ \sin\left(\tan^{-1}\left(\dfrac{u_{a_y}}{u_{a_x}}\right) \mp a_o\right) \end{bmatrix} \quad (9)$$

where $\mathbf{P}_e$ is roundabout entrance coordinates, $\mathbf{P}_{ex}$ is roundabout exit coordinates, values of $\overrightarrow{u_a}$ and $\overrightarrow{u_b}$ are defined by (6), but the intersection point $\mathbf{P}_i$ is substituted by roundabout points $\mathbf{P}_{ex}$ and $\mathbf{P}_e$, respectively. In these equations, the upper sign is used when traffic is defined counter clockwise and lower one for clockwise traffic [31].

In addition to these two points, n_r points will be included (points with square shape in Figure 3) defining the route inside the roundabout. Following this procedure, the total amount of points substituting basic center point $\mathbf{P}_r$ will be given by $n_r + 2$.

2.3. Local Planning Method Based on Curvature. The main contribution of the current paper is in the generation of smooth and continued curvature trajectories by definition in automatic vehicles. The principal goal is urban scenarios, but it is not limited to them; hence the section will be divided into three scenarios: intersections, roundabout, and lane changes.

2.3.1. Intersections. Figure 1 shows an intersection defined using a single point. This configuration reduces the amount of point in map but demands a trajectory method to connect the two straight segments smoothly. Bézier curves of 5th order will be used to solved this problem.

As it is shown in (5), three Bézier control points (from $\mathbf{P}_0$ to $\mathbf{P}_2$) set over the same line will ensure curvature zero in the starting point of the curve $\{K(t = 0) = 0 : P_{y_{0,1,2}} = mP_{x_{0,1,2}} + b\}$; this means that the segment starts as a straight path. The same equation can be used to define curvature 0 in the ending part of the curve ($t = 1$) with control points $\mathbf{P}_n$ to $\mathbf{P}_{n-2}$ (symmetry criteria).

Using this definition is inferred that the minimum Bézier order that can be used in an intersection is 4th (5 control points), but in this case control point $\mathbf{P}_2$ will be set in the position of the intersection point $\mathbf{P}_i$. Hence, the 5th order is picked as a solution without the addition of many control points and less restrictive than 4th order. The position of the control points is shown in Table 2 where the vectors $\overrightarrow{u_b}$ and $\overrightarrow{u_a}$ are given by (6).

The distance of control points has been set to ensure that maximum curvature of Bézier curve is reached in middle point of the trajectory, in order to know if peak curvature is under the vehicle's mechanical limits. Figure 4 shows the normalized curvature $K/K_{\max}$ (curvature of the generated segment divided by the maximum curvature value) with respect to normalized trajectory distance $d/d_{\max}$ (total

TABLE 2: Bézier control points for lateral and longitudinal approach for intersections "Int", roundabout entrance "RE", roundabout exit "Rex", lane change maneuver "LC", and speed planning.

Planning		Control points n-order Bézier					
		P_0	P_1	P_2	P_3	P_4	P_5
Lateral	Int	$4D\overrightarrow{u_b} + \mathbf{P}_i$	$2D\overrightarrow{u_b} + \mathbf{P}_i$	$D\overrightarrow{u_b} + \mathbf{P}_i$	$D\overrightarrow{u_a} + \mathbf{P}_i$	$2D\overrightarrow{u_a} + \mathbf{P}_i$	$4D\overrightarrow{u_a} + \mathbf{P}_i$
	RE	$\frac{3D}{2}\overrightarrow{u_e} + \mathbf{P}_e$	$\frac{D}{2}\overrightarrow{u_e} + \mathbf{P}_e$	$\mathbf{P}_e$	$D_{P_3}\overrightarrow{u_{P_4}} + \mathbf{P}_4$	$\mathbf{P}_r + R\overrightarrow{u_{\theta_e}}$ *	-
	REx	$\mathbf{P}_r + R\overrightarrow{u_{\theta_{ex}}}$ **	$D_{P_1}\overrightarrow{u_{P_0}} + \mathbf{P}_0$	$\mathbf{P}_{ex}$	$\frac{D}{2}\overrightarrow{u_{ex}} + \mathbf{P}_{ex}$	$\frac{3D}{2}\overrightarrow{u_{ex}} + \mathbf{P}_{ex}$	-
	LC	$\frac{5\overrightarrow{u_b}D}{2} + \mathbf{P}_{LC}$	$\frac{3\overrightarrow{u_b}D}{2} + \mathbf{P}_{LC}$	$\frac{\overrightarrow{u_b}D}{2} + \mathbf{P}_{LC}$	$\frac{\overrightarrow{u_a}D}{2} + \mathbf{P}'_{LC}$	$\frac{3\overrightarrow{u_a}D}{2} + \mathbf{P}'_{LC}$	$\frac{5\overrightarrow{u_a}D}{2} + \mathbf{P}'_{LC}$
Longitudinal	Speed	$\begin{bmatrix} s_0 \\ v_0 \end{bmatrix}$	$\begin{bmatrix} s_0 + D \\ v_0 \end{bmatrix}$	$\begin{bmatrix} s_0 + 2D \\ v_0 \end{bmatrix}$	$\begin{bmatrix} s_0 + 3D \\ v_0 + W \end{bmatrix}$	$\begin{bmatrix} s_0 + 4D \\ v_0 + W \end{bmatrix}$	$\begin{bmatrix} s_0 + 5D \\ v_0 + W \end{bmatrix}$

* $\theta_e = \tan^{-1}((P_{e_y} - P_{r_y})/(P_{e_x} - P_{r_x}))$, $\overrightarrow{u_{\theta_e}} = \left[\cos(\theta_e + D/R) \quad \sin(\theta_e + D/R)\right]$

** $\theta_{ex} = \tan^{-1}((P_{ex_y} - P_{r_y})/(P_{ex_x} - P_{r_x}))$, $\overrightarrow{u_{\theta_{ex}}} = \left[\cos(\theta_{ex} - D/R) \quad \sin(\theta_{ex} - D/R)\right]$

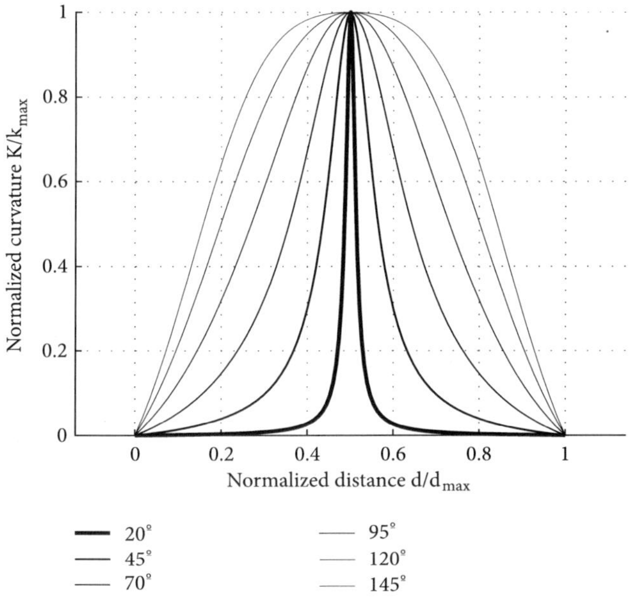

FIGURE 4: Curvature analysis for different intersection angles.

distance of the generated segment divided by the maximum one), considering angles between straight segments α from 20 degrees to 145 degrees.

With the purpose of finding the position of maximum curvature, it is applied derivative to (4):

$$\frac{d}{dt}K(t) = \frac{\left(\dot{\mathbf{B}} \times \dddot{\mathbf{B}}\right)\left(\dot{\mathbf{B}} \cdot \dot{\mathbf{B}}\right) - 3\left(\dot{\mathbf{B}} \times \ddot{\mathbf{B}}\right)\left(\dot{\mathbf{B}} \cdot \ddot{\mathbf{B}}\right)}{\left\|\dot{\mathbf{B}}\right\|^5} \tag{10}$$

evaluating the derivative of Bézier curve in $t = 0.5$:

$$\begin{aligned} \dot{\mathbf{B}}(t = 0.5) &= \frac{15}{4}D\left(\overrightarrow{u_2} - \overrightarrow{u_1}\right) \\ \ddot{\mathbf{B}}(t = 0.5) &= 10D\left(\overrightarrow{u_2} + \overrightarrow{u_1}\right) \\ \dddot{\mathbf{B}}(t = 0.5) &= 0 \end{aligned} \tag{11}$$

where D is the design parameter for trajectory and $\overrightarrow{u_{1,2}}$ are unitary vectors defined by (3). From this set of equations it is obtained that $\dot{\mathbf{B}} \times \dddot{\mathbf{B}} = \dot{\mathbf{B}} \cdot \ddot{\mathbf{B}} = 0$ resulting in curvature equal to 0 in $t = 0.5$ (maximum value).

2.3.2. Roundabouts. Figure 3 shows the roundabout using simple point description. In such a way, roundabouts are modeled using two Bézier curves, each one is used to connect the straight road with the inner part, and the roundabout middle part is modeled as a circle with constant radius. The criteria used for design roundabouts trajectory have three considerations:

(i) The curvature at entrance in $t = 0$ or at exit in $t = 1$ points must be 0 (straight paths).

(ii) Generated segments at the entrance and exit of roundabout must fit its inner part with a curvature equal to the inverse of the radius.

(iii) Bézier joining point angle must be the same of the circle arc angle (continuous trajectory).

In order to follow the aforementioned criteria, 3 colinear control points are designed to ensure the first consideration; another 2 points are selected to form a tangent segment to the circle, which assures curvature and direction continuity (second and third criteria) along the roundabout (three colineal points will generate curvature 0 and not R^{-1}). Hence, a Bézier curve 4th order is used to ensure curvature zero in straight path side and different of 0 in the inner roundabout side for entrance and exit segments (Figure 3).

Table 2 shows the control points used for trajectory generation at entrance (RE) and exit (REx) of roundabouts. Control points from $\mathbf{P}_0$ to $\mathbf{P}_2$ in the case of entrance ($\mathbf{P}_2$ to $\mathbf{P}_4$ for exit) are designed using similar criteria as in intersections, with D as a designing parameter. The points $\mathbf{P}_4$ at entrance or $\mathbf{P}_0$ at exit are selected using the distance D in the arc of the circle from entrance or exit depending on the case.

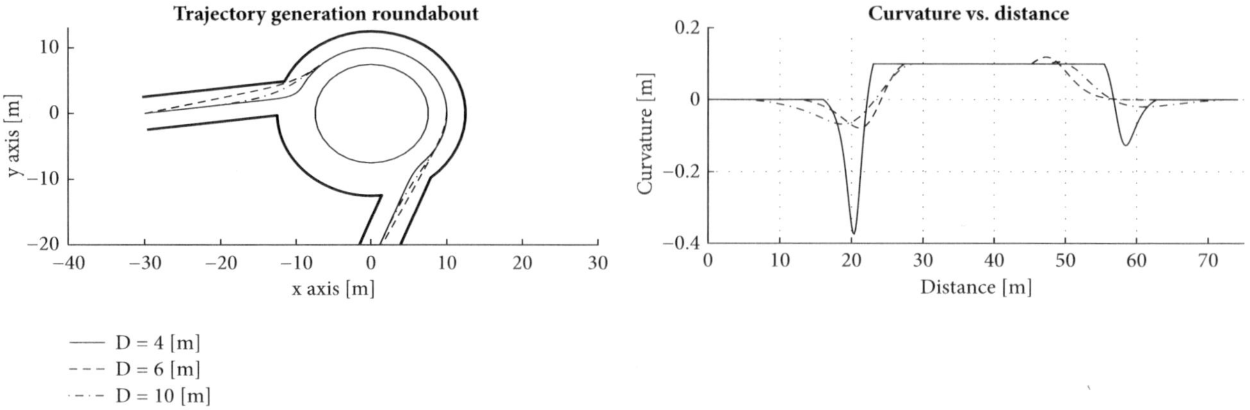

(a) Generated trajectories

(b) Curvature of the different trajectories

FIGURE 5: Examples of roundabout trajectories for different D values.

The point $\mathbf{P}_3$, in entrance trajectory, is separated of point $\mathbf{P}_4$, a distance given by

$$D_{P_3} = \sqrt{\frac{3}{4}\frac{\left\|\overrightarrow{u_{P_4}} \times (\mathbf{P}_2 - \mathbf{P}_4)\right\|}{K_r}} \quad (12)$$

where $\overrightarrow{u_{P_4}}$ is the tangent vector (with the circle) in the point $\mathbf{P}_4$ and K_r is the roundabout curvature. In the case of exit, the point $\mathbf{P}_4$ is substitute by $\mathbf{P}_0$ and $\mathbf{P}_3$ by $\mathbf{P}_1$.

Figure 5 shows the behavior of trajectory generation and its curvature using the specifications given in the planning method. Figure 5(a) depicts 3 different trajectories generated for different distances D and Figure 5(b) shows the trajectories curvature fitting perfectly the curvature radius $0.1[m^{-1}]$.

2.3.3. Lane Change. The approach is done using similar criteria as intersections, but control points are aligned with lane axis and equidistant by a distance D (Figure 2). Where the unitary vectors u_b and u_a are explained in (7), w is the width of the road and D is the separation in u_a axis between each pair of control points $\{D \in \mathbb{R}, D = \|P_{n-1} - P_n\|_{u_a}\}$.

The location of control points is given in Table 2 (LC). The minimum value of D is set to w in our design; these criteria yield to a maximum curvature in $t \approx 0.20$ and $t \approx 0.80$ in the curve definition. If a vehicle is capable of dealing with this curvature considering dynamic limitations in steering wheel angle and lateral acceleration, it will be also capable of handling any trajectory with $D > W$.

The overtaking is considered as a special case composed by two lane changes; a first lane change is using the proposed method and when the first lane change is finished, this lane will be kept until the overtaking process is done. The returning will be done applying symmetry criteria and same propositions used in first lane change.

2.4. Speed Planning Based on Comfort and Vehicle Constraints. The speed control is directly affected by sudden changes in reference speed, and if these changes are smooth and continuous the reference speed tracking will be done with a better performance. In such a way a speed planner is proposed based on Bézier curves that permits:

(i) Anticipate future conditions in the speed of the road.

(ii) Applying physical constraints in vehicle acceleration and deceleration.

(iii) Keeping safely the vehicle velocity under speed limits.

(iv) Incorporating comfort of passengers.

The speed planner approach uses 5th order Bézier curve to keep advantages of a higher degree and symmetry. Figure 7 depicts the location of control points for a speed profile curve that is function of distance in the path. The formula related to each control point is presented in Table 2.

Define the separation between consecutive control points P_{i-1} and P_i as $\{D \in \mathbb{R}, D = distance_x(P_{i-1}, P_i)\ \forall i = 1,2,3,4,5\}$. These conditions generate Bézier curve x-coordinates proportional to the parameter t given by the equation

$$B_x(t) = 5Dt = s \quad (13)$$

with this equation is seeing that the distance in x-axis and the parameter $t \in [0,1]$ have a proportional relationship. It is introduced $\{s \in [s_0, s_0 + 5D]\}$ to make the calculation easier (total distance of the speed curve).

The physical constraints of the vehicle (acceleration and deceleration process) will have a direct relation with the generated curve. In this case, it is considered the general equation that relates speed to acceleration and doing a variable change $v_l = ds/dt$, the resulting equation will be

$$a_l(t) = \frac{dv_l(t)}{dt} \Longrightarrow a_l(s) = v_l(s)\frac{dv_l(s)}{ds} \quad (14)$$

where the variable a_l is longitudinal acceleration, v_l is longitudinal speed, t is time, and s position. Doing a nomenclature change from $v_l = B_y$ and $ds = dB_x = 5Ddt$ will result in the equation

$$a_l(s) = B_y\frac{dB_y}{dB_x} = B_y\frac{dB_y}{ds}, \quad s \in [s_0, s_0 + 5D] \quad (15)$$

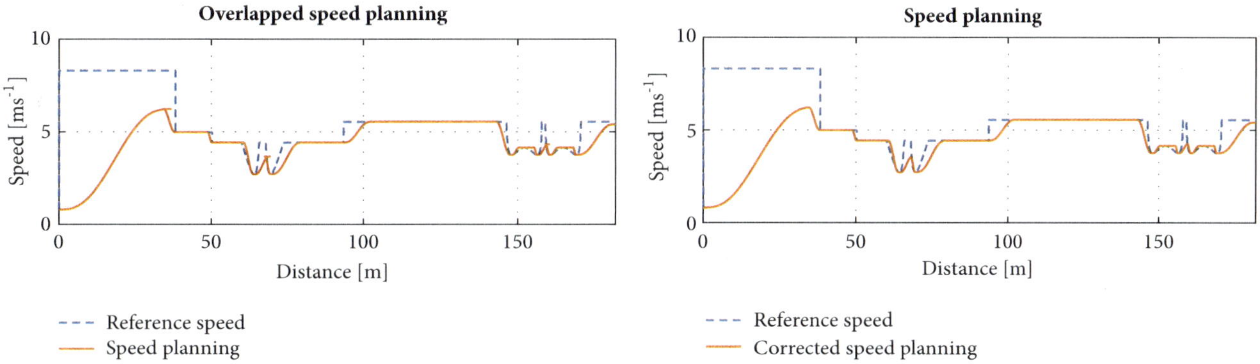

(a) Speed planning with overlapped intersections

(b) Speed planning without overlapped parts

FIGURE 6: Stages for speed planning approach.

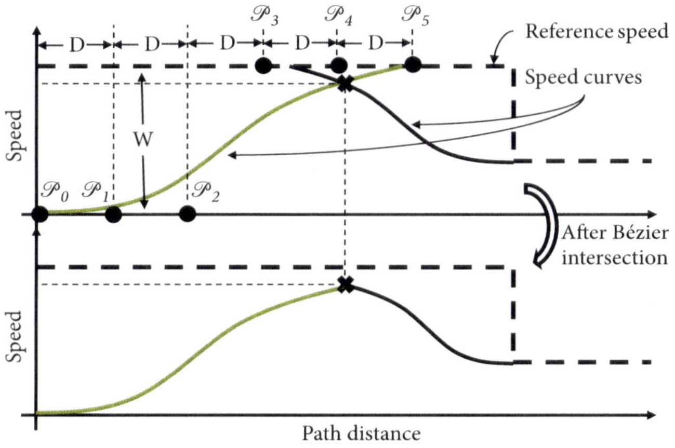

FIGURE 7: Speed planning designing method.

Applying derivative to (15) maximum points of acceleration can be found. The resulting equation is

$$\frac{da_l}{ds} = \left(\frac{dB_y}{ds}\right)^2 + B_y \frac{d^2 B_y}{ds^2} \tag{16}$$

From (16), introducing the 5th order Bézier equation and doing some approximations, the following roots can be obtained:

$$s_{a_{max}} = 5D \begin{cases} \dfrac{0.49\,(v_o/W) + 0.88}{v_o/W + 1.41}, & \text{if } \dfrac{v_0}{W} > 0 \\ \dfrac{0.52\,(v_o/W) - 0.03}{v_o/W - 0.46}, & \text{if } \dfrac{v_0}{W} < -1 \end{cases} \tag{17}$$

providing a direct solution for the maximum acceleration in a speed curve.

The first step in the speed planning method is applying the comfort speed limitation (reference speed) presented in [32] with the equation

$$a_\omega = \sqrt{(1.4a_x)^2 + (1.4{v_l}^2K)^2 + {a_z}^2} \tag{18}$$

where a_ω is the total acceleration felt by passengers (comfort criteria given by ISO2631-1); longitudinal acceleration a_x and vertical acceleration a_z contributions are approximated to zero (small changes ≈ 0), v_l is longitudinal speed, and K is curvature in each point of the path.

Figure 6(a) depicts speed planning approach with some segments overlapped one with the other. In this stage Bézier curves will be generated using the maximum acceleration criteria for each upward and downward (ensured to be under the speed limit) reference speed step. The intersections between curves will be calculated. In the example, these intersections appeared three times. Calculating the Bézier curve for each speed step (current segment is independent of the segment before) will ensure that numerical instabilities (accumulation of numerical errors by concatenated curves) will not be present a long time.

Finally, Figure 6(b) depicts the last stage in the speed planning calculation that will regenerate all the Bézier curves considering the intersections in speed profile, generating a smoother profile. This behavior can be easily understood looking at Figure 7. The profile will be constrained to the acceleration limits of the vehicle, considering comfort during driving process and keeping speed safely under road limits.

3. Integration on Automated Vehicle Architecture

The architecture presented in [22] is a versatile and modular framework composed by six abstraction blocks which defines the major areas of influence in automated driving applications: acquisition, perception, communication, decision, control, and actuation. This section gives an introduction to the aforementioned architecture combined with details of simulated and experimental platforms used in this approach. For further general details refer to [33].

Figure 8 shows a brief summary of each block that is contained within the architecture. Acquisition is in charge of gathering raw input data coming from real or simulated sensors in the vehicle and obtaining information such as position, velocity, acceleration, obstacles around, driver status, actuators position, and others. Perception has the goal

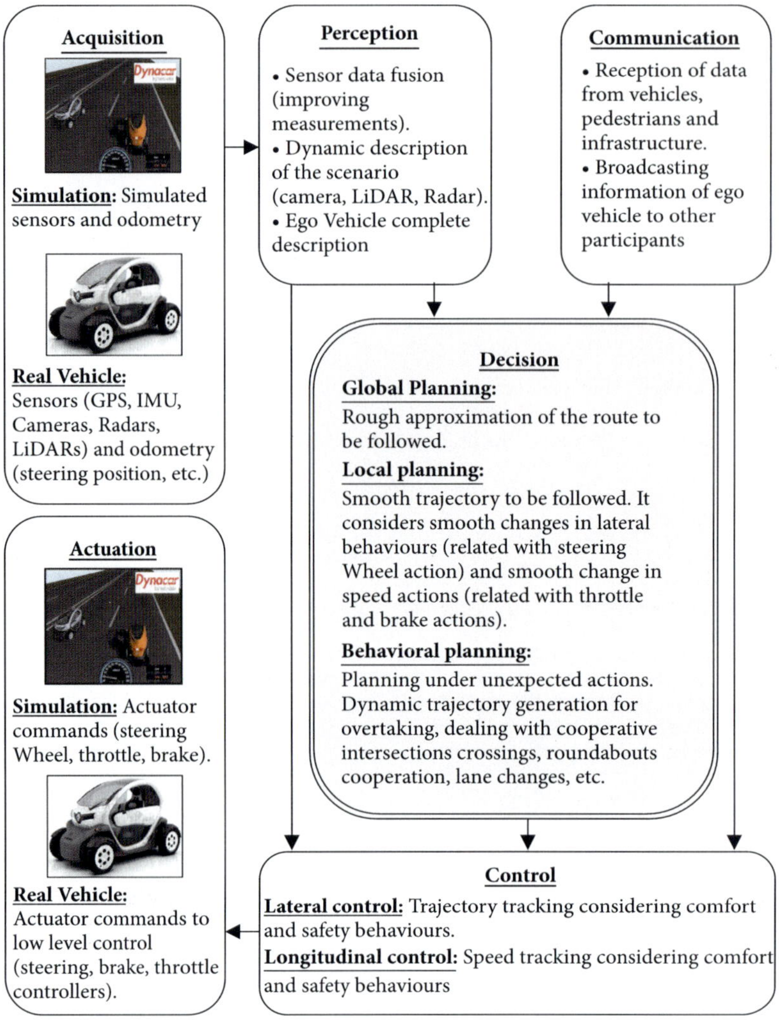

FIGURE 8: Control architecture of automated vehicles.

of postprocessing the information coming from acquisition generating descriptions on physical unit of the environment and the ego-vehicle. Communication refers to the capacities of exchanging information with other participants during driving (infrastructure, pedestrians, or other vehicles) to cooperate or knowing current states of other participants. Control is constituted for the lateral (steering) and longitudinal (throttle and brake) tracking controllers of the vehicle, and actuation contains the low level controllers of each actuator.

The decision module is the main target of the current work (grey coloured module in Figure 8) and it is divided into 3 main subtasks: those are as follows: (i) the global planning is in charge of doing a first approximation of the route; in the current work one general simple point for each roundabout, intersection, or lane change will be used; (ii) the local planning generates smooth and continues trajectories to be tracked by the vehicle; the major contributions of this work are done in this block with the Bézier trajectory designing conditions and the speed profile based on the same type of curves; finally, (iii) the behavioral planning deals with all the unexpected conditions occurring while driving, e.g., obstacle avoidance, overtaking, etc.

4. Simulation Test Case

The current section explains the approach with a numerical example using the simulation tool Dynacar [34]. The test location corresponds with an urban scenario of the Basque Country in Spain. This is depicted in Figure 9(a) and it is constituted by 2 roundabouts, 3 right turns, and 6 left turns.

The total amount of points used in the map definition will be divided among: 1 starting point, 1 ending point, 9 intersections, and 2 roundabouts, resulting in 13 points defining all the route. Table 3 shows the x-y coordinates in meters (simulator absolute coordinate system), the reference speed in $[m/s]$, and the type of point (1 for intersections and 2 for roundabouts). For roundabout global planning, additional information for its description is required: the radius (meters) and the angles of entrance and exit (radians).

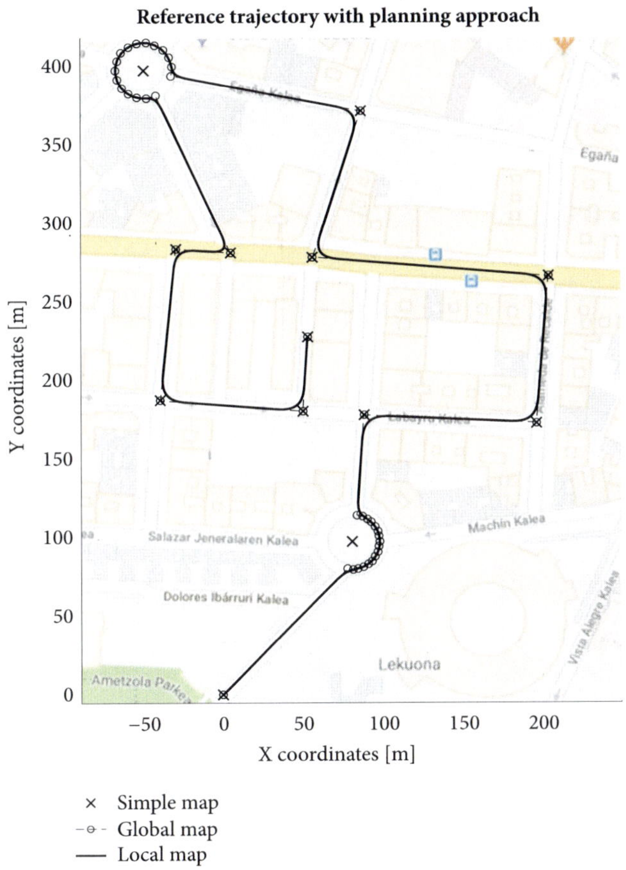

(a) Trajectory generated with the approach

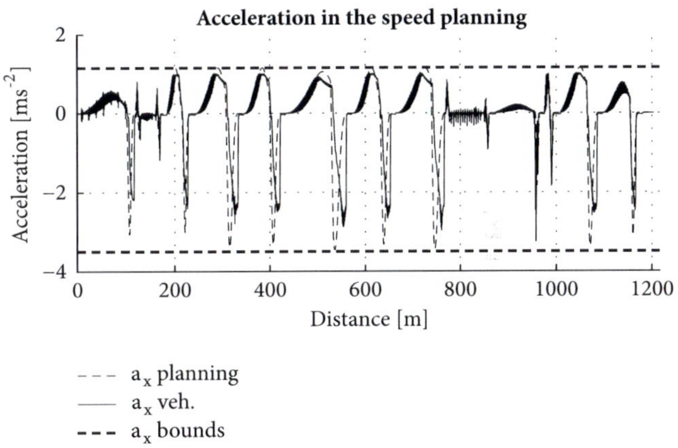

(b) Comparison of the longitudinal acceleration

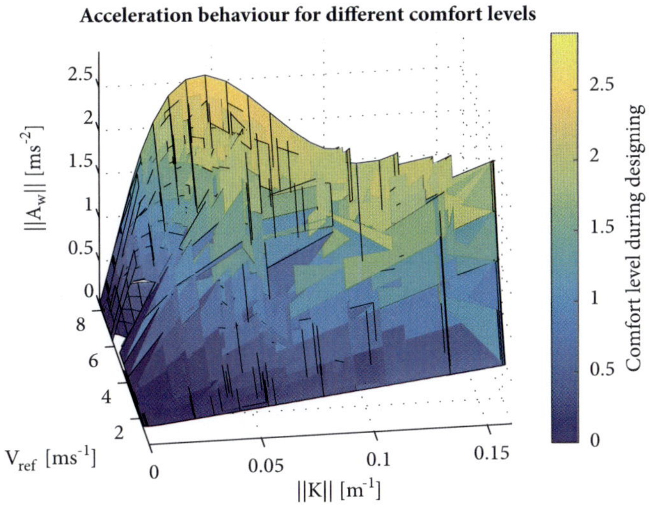

(c) Comparison of the path curvature and simulated vehicle one

FIGURE 9: Simulated results based on real urban scenario.

The previously described points are shown in Figure 9(a) with a cross marker.

From this map a global plan is generated, which gives a better description of roundabouts in the bend segment (completion of the circle arc). In case of intersections, points in global planner coincide with those used to define the route in the scenario. All the global planner points are depicted as circles in Figure 9(a).

The soft and continuous black line represents the trajectory generated using the Bézier approach. The values of D (designing parameter of the approach) are set by hand but they can be selected using optimization algorithm as in [31, 35].

TABLE 3: Point description of the map used for global planning.

X [m]	Y [m]	V [m/s]	Type []	R [1/m]	a_i [rad]	a_o [rad]
0.00	0.00	11.11	1	–	–	–
80.48	97.09	11.11	2	17.29	0.52	0.09
88.04	177.90	11.11	1	–	–	–
196.21	172.89	11.11	1	–	–	–
203.72	266.16	11.11	1	–	–	–
56.55	278.46	11.11	1	–	–	–
86.24	371.34	11.11	1	–	–	–
−49.30	397.61	11.11	2	17.76	0.00	0.00
5.08	281.16	11.11	1	–	–	–
−29.37	283.73	11.11	1	–	–	–
−39.17	187.34	11.11	1	–	–	–
50.16	180.36	11.11	1	–	–	–
53.19	227.94	11.11	1	–	–	–

An interesting example of the entrance/exit angle in the roundabouts is the one located in the absolute coordinates [80.48; 97.09]. This point is an entrance of a roundabout and it uses the definition of the entrance angle for its description (see (9)):

$$P_e = \begin{bmatrix} 80.48 \\ 97.09 \end{bmatrix} + 17.29 \begin{bmatrix} \cos(-2.26) \\ \sin(-2.26) \end{bmatrix} = \begin{bmatrix} 69.4 \\ 83.8 \end{bmatrix} \tag{19}$$

The entrance point to present a numerical example is shown, but the exit point can be calculated using the same data and (9).

The speed planning defined for that trajectory is shown in Figure 9(b). The continuous line is the reference acceleration given by the speed profile of the urban scenario and the application of a comfort parameter $a_\omega = 0.5[m/s^2]$; based on [32] this amount of acceleration a_w corresponds with a value between "not uncomfortable" and "a little uncomfortable" for vehicle passengers. Results show that vehicle succeed on tracking the generated speed profile (by tracking the acceleration profile shown in dashed line). One major cause was the consideration of the maximum acceleration and deceleration (red bounds of Figure 9(b)).

Figure 9(c) shows a surface plot representing the different values of acceleration a_w with respect to the speed profile generated and the curvature of the path. Five different values of a_ω were used: 0.5, 1.0, 1.5, 2.0, and 2.5$[m/s^2]$ to generate the surface. Its objective is verifying the smooth and continued relationship with the Bézier speed profile and the level of comfort of the passenger.

5. Real Circuit Tests

The current section shows the trajectory planning approach used with the real vehicle. In this sense, two possible definitions of the route were implemented (Figure 10(a)). For both the center of the lane is used (taken from a manual driving record using GPS) as guide for the implementation of trajectories.

The first case is shown in the left side of Figure 10(a), where the trajectory described by the manual driving is modeled using intersection points in the 180[°] turn (roundabout). In this case, the error in the turning points (top and bottom of the circuit) is greater than using roundabout for the definition of the curve (right side Figure 10(a)).

The trajectory in the right side was tested with 5 different speed profiles and the behavior of the vehicle was analyzed with the curvature. Figure 10(b) depicts the results for these tests. Curvature of the vehicle in the trajectory is represented with a grey line and calculated according to the formula:

$$K(t) = \frac{\omega(t)\cos(\Phi(t))}{V_l(t)} \tag{20}$$

where K is the curvature, ω is the rate of change of yaw, Φ is the roll angle of the vehicle, and V_l is the longitudinal speed. Curvature of the path is depicted in black and the vehicle is able to replicate the behavior of the planned curvature (the overshoots in curvature are caused by controllers).

Figure 10(c) shows vehicle lateral acceleration (as function of the vehicle speed and the curvature of the circuit) using different speed profiles, generated with a_w equal to 0.5, 1.0, 1.5, 2.0, and 2.5$[m/s^2]$. Values of total acceleration slightly overpass the limit of 2.5 at some points because the actuators delay is not considered for longitudinal control and the consideration of $a_x \approx 0$ in speed planning. Additionally, this figure shows the lateral acceleration in relationship with the curvature and the speed but without considering the total acceleration because in the speed planner those are assumed as zero. The lack of data taken, considerations over longitudinal acceleration, and controllers delay during tracking generate some surface holes without data. Table 4 shows the maximum acceleration obtained with the approach after filtering all the values $a_x > 0.005[m/s^2]$ (the approach considers $a_x \approx 0$). It is relevant to understand that those values must be filtered because actions like control behavior, delay of the controllers, or even the longitudinal acceleration are not considered and the magnitudes can differ. In general,

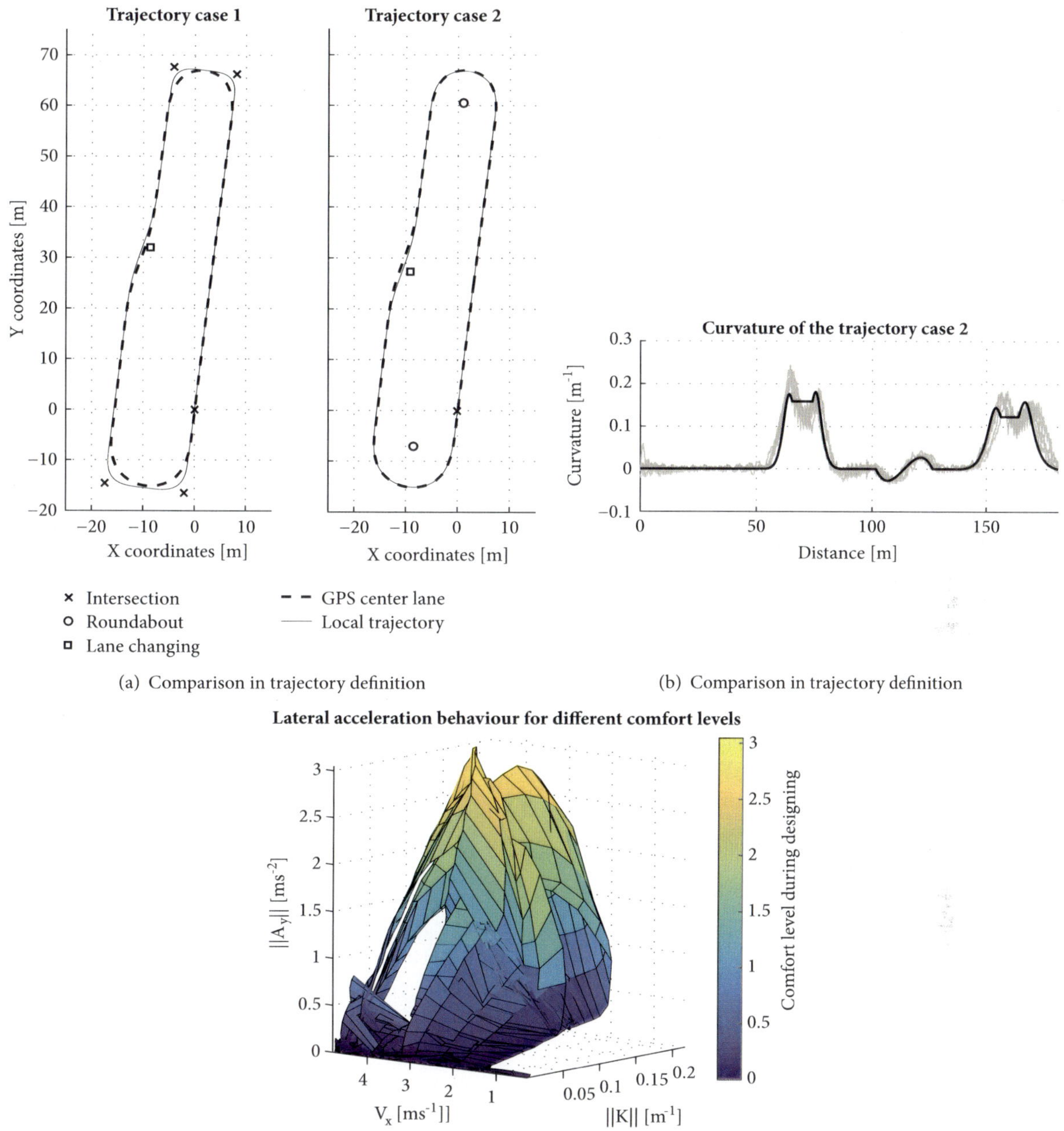

(a) Comparison in trajectory definition

(b) Comparison in trajectory definition

(c) Comparison in curvature for the trajectory definition case 2

FIGURE 10: Real life validation of the Bézier planning approach.

the results obtained are close to proposed values with parameter a_ω (designing).

6. Conclusion and Future Works

This work presents a trajectory and speed planning approach based on Bézier curves used in the decision module of an automated vehicle control architecture. The approach considers urban scenarios with intersections, roundabouts, and lane changes but it is not limited to those; the approach could also be used in scenarios such as highways. In the case of the speed planner, a parametric curve approach to adapt future changes in the speed limit considering physical constraints of the vehicle and traffic rules is presented (i.e., considering the maximum speed limit of the road).

Simplified mathematical models for trajectory planning using Bézier curves on intersections, roundabouts, and lane changes are presented reducing the time of calculation and allowing the use of these definitions for other applications, i.e., trajectory optimization process with computational efficiency. Additionally, the approach is evaluated using a simulated urban scenario based on a location of Basque Country showing that a great variety of urban scenario can be managed using this approach.

Table 4: Maximum lateral acceleration obtained in the real experiment using speed profile.

Uncomfortable levels	Not	A little	Fairly	Very	Extreme
Uncomfortable magnitudes $a_\omega [m/s^2]$	0.5	1.0	1.5	2.0	2.5
Filtered max. values $[m/s^2]$	0.45	1.03	1.54	2.00	3.02

The trajectory planner was tested for automated driving, using a comparison of the trajectory with the Bézier curves and the one generated using a manual GPS recording in manual driving. Additionally the generated curvature in the vehicle for different conditions (variations in the speed profile) was compared with the curvature of the trajectory planned showing that it is feasible to be tracked by the vehicle.

The speed planning approach shows the capability of using parametric curves to generate a profile to be tracked correctly by the vehicle, giving a prediction of future behaviors. Consideration of longitudinal acceleration in curve segments is one of the future improvements related to the speed planner of this work.

This approach is considered better than others due to the fact that it includes comfort and safety criteria during trajectory and speed planning. For comfort the ISO2631-1 standard to generate speed profiles based on geometric information given by the trajectory planning was considered. Comfortable speed profiles are related to trajectories and this can be associated with safety during driving (vehicle lateral forces are related to possible dangerous situations), and additionally a continuous curvature of the trajectory planned can reduce possible unexpected situations given by vehicle' automatic control (increasing safety).

Some of the future works are related to using the presented method to generate possible optimal solutions in terms of the parameter D, road geometry, and vehicle dynamics and in such way, substituting the definition by hand of parameter D during trajectory planning.

Conflicts of Interest

The authors declare that they have no conflicts of interest.

Acknowledgments

This work was partly supported by ECSEL Project ENABLE-S3 with Grant Agreement no. 692455-2 and the AutoDrive ECSEL Project with Grant Agreement no. 737469.

References

[1] J.-w. Choi, R. E. Curry, and G. H. Elkaim, "Path planning based on Bézier curve for autonomous ground vehicles," in *Proceedings of the IEEE Advances in Electrical and Electronics Engineering—IAENG Special Edition of the World Congress on Engineering and Computer Science 2008, WCECS 08*, pp. 158–166, October 2008.

[2] L. Han, H. Yashiro, H. T. N. Nejad, Q. H. Do, and S. Mita, "Bézier curve based path planning for autonomous vehicle in urban environment," in *Proceedings of the 2010 IEEE Intelligent Vehicles Symposium, IV 2010*, pp. 1036–1042, USA, June 2010.

[3] L. Labakhua, U. Nunes, R. Rodrigues, and F. S. Leite, "Smooth trajectory planning for fully automated passengers vehicles: Spline and clothoid based methods and its simulation," *Informatics in control automation and robotics*, pp. 169–182, 2008.

[4] R. Lattarulo, J. Pérez, and M. Dendaluce, "A complete framework for developing and testing automated driving controllers," *IFAC-PapersOnLine*, vol. 50, no. 1, pp. 258–263, 2017.

[5] R. Lattarulo, M. Marcano, and J. Pérez, "Overtaking maneuver for automated driving using virtual environments," in *Proceedings of the Springer International Conference on Computer Aided Systems Theory*, pp. 446–453, 2018.

[6] J. P. Rastelli, R. Lattarulo, and F. Nashashibi, "Dynamic trajectory generation using continuous-curvature algorithms for door to door assistance vehicles," in *Proceedings of the 25th IEEE Intelligent Vehicles Symposium, IV 2014*, pp. 510–515, IEEE, June 2014.

[7] D. González, J. Pérez, R. Lattarulo, V. Milanés, and F. Nashashibi, "Continuous curvature planning with obstacle avoidance capabilities in urban scenarios," in *Proceedings of the 2014 17th IEEE International Conference on Intelligent Transportation Systems, ITSC 2014*, pp. 1430–1435, China, October 2014.

[8] R. Lattarulo, E. Marti, M. Marcano, J. Matute, and J. Perez, "A Speed Planner Approach Based On Bézier Curves Using Vehicle Dynamic Constrains and Passengers Comfort," in *Proceedings of the 2018 IEEE International Symposium on Circuits and Systems (ISCAS)*, pp. 1–5, Florence, Italy, May 2018.

[9] D. González, J. Pérez, V. Milanés, and F. Nashashibi, "A Review of Motion Planning Techniques for Automated Vehicles," *IEEE Transactions on Intelligent Transportation Systems*, vol. 17, no. 4, pp. 1135–1145, 2016.

[10] US Department of Transportation, *Critical Reasons for Crashes Investigated in the National Motor Vehicle Crash Causation Survey*, NHTSA Traffic Safety Facts Crash Stats, February 2015.

[11] J. Dokic, B. Müller, and G. Meyer, "European roadmap smart systems for automated driving," *European Technology Platform on Smart Systems Integration*, January 2015.

[12] G. Meyer and S. Deix, "Research and innovation for automated driving in germany and europe," pp. 71–81, Springer Road Vehicle Automation, 2014.

[13] E. Adell, A. Várhelyi, and M. D. Fontana, "The effects of a driver assistance system for safe speed and safe distance - A real-life field study," *Transportation Research Part C: Emerging Technologies*, vol. 19, no. 1, pp. 145–155, 2011.

[14] J. Ni and J. Hu, "Dynamics control of autonomous vehicle at driving limits and experiment on an autonomous formula racing car," *Mechanical Systems and Signal Processing*, vol. 90, pp. 154–174, 2017.

[15] M. Zhu, H. Chen, and G. Xiong, "A model predictive speed tracking control approach for autonomous ground vehicles," *Mechanical Systems and Signal Processing*, vol. 87, pp. 138–152, 2017.

[16] R. Domínguez, E. Onieva, J. Alonso, J. Villagra, and C. González, "LIDAR based perception solution for autonomous vehicles," in *Proceedings of the 2011 11th International Conference on Intelligent Systems Design and Applications, ISDA'11*, pp. 790–795, Spain, November 2011.

[17] J.-P. Jodoin, G.-A. Bilodeau, and N. Saunier, "Urban Tracker: Multiple object tracking in urban mixed traffic," in *Proceedings of the 2014 IEEE Winter Conference on Applications of Computer Vision, WACV 2014*, pp. 885–892, USA, March 2014.

[18] L. Li, D. Wen, and D. Yao, "A survey of traffic control with vehicular communications," *IEEE Transactions on Intelligent Transportation Systems*, vol. 15, no. 1, pp. 425–432, 2014.

[19] D. J. Fagnant and K. Kockelman, "Preparing a nation for autonomous vehicles: Opportunities, barriers and policy recommendations," *Transportation Research Part A: Policy and Practice*, vol. 77, pp. 167–181, 2015.

[20] N. Wan, A. Vahidi, and A. Luckow, "Optimal speed advisory for connected vehicles in arterial roads and the impact on mixed traffic," *Transportation Research Part C: Emerging Technologies*, vol. 69, pp. 548–563, 2016.

[21] W. Cao, M. Mukai, T. Kawabe, H. Nishira, and N. Fujiki, "Cooperative vehicle path generation during merging using model predictive control with real-time optimization," *Control Engineering Practice*, vol. 34, pp. 98–105, 2015.

[22] T. Shim, G. Adireddy, and H. Yuan, "Autonomous vehicle collision avoidance system using path planning and model-predictive-control-based active front steering and wheel torque control," *Proceedings of the Institution of Mechanical Engineers, Part D: Journal of Automobile Engineering*, vol. 226, no. 6, pp. 767–778, 2012.

[23] T. Gu, J. Snider, J. M. Dolan, and J.-W. Lee, "Focused Trajectory Planning for autonomous on-road driving," in *Proceedings of the 2013 IEEE Intelligent Vehicles Symposium, IEEE IV 2013*, pp. 547–552, Australia, June 2013.

[24] A. Best, S. Narang, D. Barber, and D. Manocha, "AutonoVi: Autonomous vehicle planning with dynamic maneuvers and traffic constraints," in *Proceedings of the 2017 IEEE/RSJ International Conference on Intelligent Robots and Systems (IROS)*, pp. 2629–2636, Vancouver, BC, September 2017.

[25] H. Mouhagir, R. Talj, V. Cherfaoui, F. Guillemard, and F. Aioun, "A Markov Decision Process-based approach for trajectory planning with clothoid tentacles," in *Proceedings of the 2016 IEEE Intelligent Vehicles Symposium, IV 2016*, pp. 1254–1259, Sweden, June 2016.

[26] M. Du, T. Mei, H. Liang, J. Chen, R. Huang, and P. Zhao, "Drivers' Visual Behavior-Guided RRT Motion Planner for Autonomous On-Road Driving," *Sensors*, vol. 16, no. 1, p. 102, 2016.

[27] US Department of Transportation, *A Brief Statistical Summary: Drowsy Driving*, NHTSA: Traffic Safety Facts Crash Stats, March 2011.

[28] A. G. Cunningham, E. Galceran, R. M. Eustice, and E. Olson, "MPDM: multipolicy decision-making in dynamic, uncertain environments for autonomous driving," in *Proceedings of the IEEE International Conference on Robotics and Automation (ICRA '15)*, pp. 1670–1677, Seattle, Wash, USA, May 2015.

[29] K. Zhou, L. Yu, Z. Long, and S. Mo, "Local path planning of driverless car navigation based on jump point search method under urban environment," *Future Internet*, vol. 9, no. 3, 2017.

[30] C. Hubmann, M. Aeberhard, and C. Stiller, "A generic driving strategy for urban environments," in *Proceedings of the 19th IEEE International Conference on Intelligent Transportation Systems, ITSC 2016*, pp. 1010–1016, Brazil, November 2016.

[31] X. Qian, I. Navarro, A. De La Fortelle, and F. Moutarde, "Motion planning for urban autonomous driving using bezier curves and MPC," in *Proceedings of the 19th IEEE International Conference on Intelligent Transportation Systems, ITSC 2016*, pp. 826–833, Brazil, November 2016.

[32] V. Desaraju, H. C. Ro, M. Yang, E. Tay, S. Roth, and D. Del Vecchio, "Partial order techniques for vehicle collision avoidance: Application to an autonomous roundabout test-bed," in *Proceedings of the 2009 IEEE International Conference on Robotics and Automation, ICRA '09*, pp. 82–87, Japan, May 2009.

[33] S. Glaser, B. Vanholme, S. Mammar, D. Gruyer, and L. Nouvelière, "Maneuver-based trajectory planning for highly autonomous vehicles on real road with traffic and driver interaction," *IEEE Transactions on Intelligent Transportation Systems*, vol. 11, no. 3, pp. 589–606, 2010.

[34] H. Delingette, M. Herbert, and K. Ikeluchi, "Trajectory generation with curvature constraint based on energy minimization," in *Proceedings of the IEEE/RSJ International Workshop on Intelligent Robots and Systems (IROS '91)*, vol. 1, pp. 206–211, Osaka, Japan, November 1991.

[35] J. Cuadrado, D. Vilela, I. Iglesias, A. Martín, and A. Peña, "A multibody model to assess the effect of automotive motor in-wheel configuration on vehicle stability and comfort," in *Proceedings of the ECCOMAS Thematic Conference on Multibody Dynamics*, pp. 1083–1092, 2013.

3

Optimal Design of Transportation Networks with Automated Vehicle Links and Congestion Pricing

Yipeng Ye[1] **and Hua Wang**[1,2]

[1]*School of Economics and Management, Tongji University, Shanghai 200092, China*
[2]*Department of Civil & Environmental Engineering, National University of Singapore, Singapore 117576*

Correspondence should be addressed to Hua Wang; hwang191901@gmail.com

Academic Editor: David Z. W. Wang

We propose a bi-level network design model comprising automated vehicle (AV) links and congestion pricing to improve traffic congestion. As upper-level road planners strive to minimize total travel-time costs by optimizing both the network design and the congestion pricing, lower-level travelers make choices about their routes to minimize their individual travel costs. Our proposed model integrates a network design and congestion pricing to improve traffic congestion and we use a relaxation-based method to solve the model. We conducted a series of numerical tests to analyze the proposed model and solution method. Our results indicate that network design is more effective than congestion pricing when the AV market penetration is high and the opposite is true when AV penetration is low. More importantly, we find that a network design of automated vehicle links with congestion pricing is superior to a single network design or congestion pricing, especially when both AVs and conventional vehicles have a relatively large market penetration.

1. Introduction

Urban transportation plays an important role in economic activity throughout the world. However, road congestion has become a serious problem in many metropolitan cities and has led to various problems, including economic inefficiency, traffic accidents, pollution, and excessive energy consumption. Network design and congestion pricing are classic and effective instruments for solving traffic congestion and have been studied for decades. Some studies have proposed network design strategies [1, 2]. Also, some comprehensive reviews have summarized the results of studies of the transportation network design problem (NDP) [3, 4]. Other studies worldwide have proposed and tested practical congestion pricing [5–7].

Automated vehicles (AV) have attracted increasing attention in recent years and they are expected to improve the future safety levels and traffic congestion of existing urban transportation systems through improvements in traffic safety, road capacity, travel time, energy consumption, and pollution levels. Two of the anticipated advantages (relative to human-driver vehicle control) are increased road network capacity and the freeing up of driver-occupant time to engage in leisure activities or economically productive (nondriving) tasks [8]. Jiang [9] studied optimal signal design for mixed equilibrium networks with autonomous and regular vehicles. With the rapid increase of car ownership, the problem of traffic emission and lack of land for parking becomes a serious issue [10], and AV can substantially reduce traffic emission and save land for parking. AVs can also help traffic safety. Zhu and Ukkusuri [11] indicated that connected vehicle technology would potentially reduce 81% of all-vehicle target crashes, 83% of all light-vehicle target crashes, and 72% of all heavy-truck target crashes annually. Tientrakool et al. [12] and Chen et al. [13] stated that the capacity of AV links may approximately triple due to the benefits resulting from vehicle-to-vehicle communication. It follows that congestion pricing may become unnecessary when conventional vehicles (CV) are completely phased out. Despite these exciting developments regarding AVs, CVs cannot be phased out immediately and will continue to be widely used for many years. As such, it is necessary to consider the NDP and congestion pricing during this transitional period characterized by heterogeneous AV and CV travel modes.

Among the transportation NDP and congestion pricing studies addressing AV and CV modes, few focus on the simultaneous optimization of the transportation system NDP and congestion pricing. An NDP with congestion pricing can be viewed as a Stackelberg game between the road planner and travelers, in which the road planner acts as the leader and travelers are the followers. In this paper, we investigate transportation networks involving both AV and CV modes and present an optimal and simultaneous network design and congestion pricing for these networks. We assume that some links in the network belong to the candidate link set and all links in this set can be converted to AV links or remain as regular links. AV links would allow only AVs to operate on them and their capacities could triple, as reported in Tientrakool et al. [12] and Chen et al. [13]. Regular links in the candidate link set allow both AVs and CVs to operate on them. To induce traffic flow and promote AV market penetration, the road planner can implement congestion pricing for CVs on the regular links in the candidate link set. Once the AV links are established and congestion pricing is deployed, we assume that AVs and CVs will follow the Wardrop equilibrium principle with respect to route choice to minimize their individual travel costs and yield a multiclass user equilibrium (UE). The results of this study show that the proposed NDP with congestion pricing can be an effective method for minimizing total travel time cost in a network comprising both AV and CV modes. More importantly, we find that NDP with congestion pricing outperforms a single NDP or congestion pricing, especially when both AVs and CVs have fairly large market penetrations. This means that the NDP with congestion pricing is an effective approach for alleviating traffic congestion in the transitional period characterized by heterogeneous AV and CV travel modes.

1.1. Literature Review. Although AVs are expected to greatly improve the safety and efficiency of transportation networks, proper management instruments must be applied in the transitional period during which there are heterogeneous AV and CV travel modes. Both the NDP and the congestion pricing are considered to be effective instruments for alleviating traffic congestion and they are typically employed to maximize network performance (e.g., minimize total travel-time cost, maximize social welfare) while travelers' route choices are taken into account. The literature includes several comprehensive reviews of NDP, including those by Yang and H. Bell [4] and Farahani et al. [3]. Congestion pricing has also been extensively studied by both road managers and researchers and has long been considered to be an effective method for managing traffic demand and increasing tax revenue. Zhang and Yang [14] developed a method of simultaneously determining optimal locations and toll levels of cordon-based congestion pricing. Zhang et al. [15] studied the cooperation and competition congestion pricing policy among multiple regions. We refer interested readers to Yang and Huang [16] for details regarding developments in road pricing research. In this section, we briefly summarize recent studies related to our work, including AV research, joint road transportation management, and traffic optimization methods.

(1) Research on AVs. Many studies have highlighted the potential social benefits of AVs. Le Vine et al. [8] identified two benefits of road vehicle automation: (I) increased road capacity and (II) the freeing up the drivers' in-car time for a wide range of economically valuable activities. Chen et al. [13] indicated that there are potential benefits of AVs to increase traffic safety, driver productivity, road capacity, and travel speed. Ji et al. [17] pointed out that connected AVs can be easily operated due to the application of information technology in AVs.

Although existing AV developments and emerging innovations in AV technology indicate the huge potential for improving traffic safety and congestion, applications of AV technologies are as yet limited. In general, AV technologies are being developed to sense and make assessments about the environment in which the AV is traveling (e.g., other travelers, road signs, and traffic flow density) and to initiate the appropriate actions in response. However, these assessments fully depend on the proper functioning of the sensory devices, which remain in the research-and-development phase with numerous ongoing experiments [18]. Some studies have sought to improve AV performance via technical innovation. Häne et al. [19] developed an obstacle detection system for self-driving cars, using monocular cameras and wheel odometry. Aparicio et al. [20] aimed to improve the effectiveness of safety systems for active vulnerable road users (VRUs) that are currently on the market by expanding the scope of scenarios addressed by these systems and improving the overall system performance.

Although currently immature AV technologies are being rapidly developed, it will be many years before AVs are widely adopted. As such, a long-term period of mixed AV and CV traffic flow is inevitable. Some researchers have suggested the expansion of dedicated AV areas to improve overall system traffic performance. For example, several studies have suggested the conversion of some traffic lanes into dedicated AV lanes to alleviate traffic congestion and improve traveler safety, since existing managed lanes are already equipped with communication systems [21]. Chen et al. [13] proposed the application of a time-dependent AV-lane design model on a general network for both AVs and CVs. Godsmark and Kakkar [22] proposed that AV areas can be designed to maximize the benefits realized from AVs, as well as to promote AV adoption. In addition, Chen et al. [23] proposed a mathematical framework for designing an optimal AV zone to achieve these goals.

(2) Joint Managements for Road Transportation. There has been an abundance of literatures published on the NDP and congestion pricing over the past half century. Readers can refer to Yang and H. Bell [4], Farahani et al. [3], and Yang and Huang [16] for comprehensive reviews and detailed discussions regarding the NDP and congestion pricing. Researchers also studied other kinds of joint implementation, such as

joint management of road tolls and road credits. Jiang et al. [24] investigated simultaneous optimization of road tolls and tradable road credits in public-private mixed traffic networks. Wang and Zhang [7] examined the joint implementation of tradable credit and road pricing in public-private partnership networks considering UE-CN mixed equilibrium behaviors. In this section, we review recent studies related to the joint management of road transportation.

Considering the development of network modelling to simultaneously determine network design strategy and road pricing, Koh et al. [25] studied joint optimal pricing and road capacity investment problems ranging from policy to technology perspectives. Zhang and van Wee [26] proposed a simultaneous toll-location and toll-level optimization model to maximize the network reserve capacity, which differs substantially from previous studies that aimed to minimize total travel time or maximize total social welfare. Wang et al. [27] proposed a bi-level programming model comprising a joint optimal-link-based tradable-credit-charging scheme and road capacity improvement. Their study results indicated that the synergistic effect of link-based credit charging and road capacity improvement enhanced urban transportation network performance. Xu et al. [28] proposed a discrete network design model with a road pricing scheme for minimizing total travel time under budgetary constraints. This optimal road capacity improvement and toll-level scheme considered the available budget levels and toll revenues.

(3) Traffic Optimization Methods. In recent years, a number of studies have focused on the development of algorithms for NDP and road pricing, including sensitivity-analysis-based [9, 29], cutting constraint [30], linearization [31, 32], and relaxation [27, 33] algorithms. Some metaheuristic algorithms have also been applied to solve these kinds of problems. For example, Meng et al. [5] proposed the hybrid genetic algorithm- (GA-) cost averaging (CA) method to solve the optimal-distance-based toll design problem for cordon-based congestion pricing schemes. Sun et al. [34] used a particle swarm optimization algorithm to design reliability-based traffic networks with advanced traveler information systems.

1.2. Positioning and Objective. Some researchers have considered the use of infrastructure adaption planning for AVs to reduce traffic congestion and improve traveler safety [21, 22, 35]. However, these works offered no integrated methods for solving traffic congestion. With this paper, we fill this gap by proposing a simultaneous optimization model of the NDP and congestion pricing for transportation networks with mixed AV and CV flow.

Although network design and congestion pricing have long been subjects of study, to the best of our knowledge, no studies have offered a specialized solution method for a discrete NDP with continuous pricing. In this paper, we propose a relaxation-based method for solving this kind of problem.

Specifically, our objectives in this paper are twofold. First, we propose a bi-level network design model of AV links and congestion pricing for improving traffic congestion, while considering traveler behavior with respect to route choice. Secondly, we propose a relaxation-based method for solving the above model, which can also be applied to solve a discrete NDP with continuous pricing.

1.3. Challenges, Our Solution Method, and Contributions. To achieve the above objectives, we have formulated an NDP with congestion pricing using bi-level programming, while also considering traveler behavior regarding route choice. Not surprisingly, this model is computationally challenging. The first difficulty is that the bi-level model is a NP-hard problem, even if it involves bi-level linear programming [36]. The second difficulty is that the variables in the upper level of the bi-level programming contain integer (network design variables) and continuous (pricing variables) aspects, which some previously developed bi-level programming algorithms for the NDP and pricing (e.g., Wang et al. [33], Wang et al. [27]) are unable to solve.

To tackle these computational challenges, we decomposed the bi-level programming process into a number of subproblems. The subproblems for determining the upper bounds can be solved by the relaxation algorithm proposed by Wang et al. [27]. The subproblems for determining the lower bounds can be solved by the outer-approximation algorithm proposed by Wang et al. [33]. As the number of iterations increases, the gap between the lower and upper bounds decreases sharply and finally converges to zero.

Contributions. This paper makes the following contributions:

(1) We propose a bi-level programming to formulate the NDP with congestion pricing for transportation networks with mixed AV and CV traffic flow. To the best of our knowledge, this is the first model that incorporates both network design and congestion pricing for this kind of network.

(2) We propose a relaxation-based method for solving the model, which can also be applied to solve a discrete NDP with continuous pricing.

Managerial Insights. This paper also contributes to the following managerial insights:

(1) If we compare congestion pricing with network design, we find that congestion pricing is effective when AV market penetration is low and network design is effective when AV penetration is high. As such, road planners must apply different strategies to network with different AV market penetrations.

(2) Network design becomes more effective as the origin–destination (OD) demand is increasing, so it follows that AV promotion and AV link design may be the solution for alleviating traffic congestion when traffic demand continually increases.

(3) In the transitional period characterized by heterogeneous AV and CV travel modes, the NDP with congestion pricing strongly outperforms either network design or congestion pricing when both AVs and CVs have fairly large market penetrations. This

finding indicates that the NDP with congestion pricing represents an effective management method for the transitional period when both AVs and CVs are operating.

This paper is organized as follows. In Section 2, we list the symbols used and propose our bi-level model for transportation systems in which both AV and CV modes are being used. In Section 3, we propose the use of the relaxation-based method for solving the bi-level model. In Section 4, we describe the numerical tests we conducted to analyze the proposed model and solution method. In Section 5, we conclude this paper with a brief summary and suggestions for future work.

2. Model Formulation

The network design problem (NDP) can be represented as a Stackelberg game between the road planner and travelers. The road planner can induce but not control traveler route choices. This game can be formulated as a bi-level optimization model, in which the planner in the upper level determines the network design and congestion pricing to minimize the total travel cost in the transportation network and the travelers determine their routes to minimize their individual travel costs. We list the set, parameter, and decision variable symbols as follows:

Sets

N: Set of nodes in the transportation network

A: Set of links in the transportation network

$\widetilde{A}$: Set of candidate AV links in the transportation network, $\widetilde{A} \subseteq A$

W: Set of origin–destination (OD) pairs in the transportation network

K_w^{AV}: Path set for AV mode between OD pair $w \in W$

K_w^{CV}: Path set for CV mode between OD pair $w \in W$

Parameters

$\delta_{a,k}$: A binary coefficient which equals 1 if path k uses link $a \in A$; otherwise $\delta_{a,k} = 0$

c_a: Link capacity of link $a \in A \setminus \widetilde{A}$

c_a^1: Link capacity of link $a \in \widetilde{A}$ if link a is set to be AV link

c_a^2: Link capacity of link $a \in \widetilde{A}$ if link a remains to be regular link

D_w^{AV}: Demands of AV mode for OD pair $w \in W$

D_w^{CV}: Demands of CV mode for OD pair $w \in W$

$t_a(v_a, C_a)$: Travel-time cost function on link $a \in A$ when the link flow and capacity are v_a and C_a, respectively

L: A positive large number

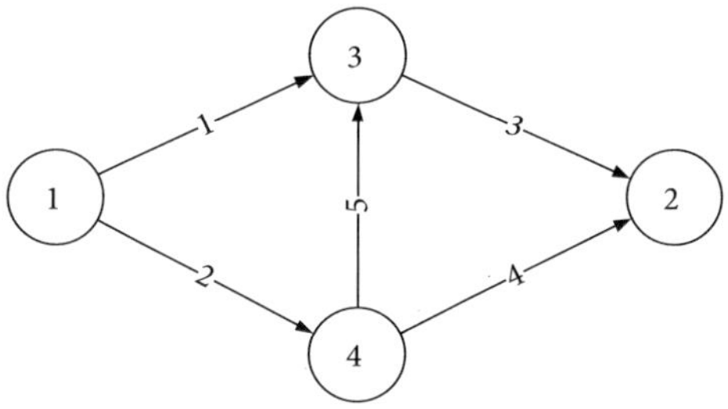

FIGURE 1: Schematic of a network with four nodes and five links.

Decision Variables

x_a: A binary decision variable. $x_a = 1$ if link $a \in \widetilde{A}$ is set to be AV link. Otherwise $x_a = 0$

τ_a: Link-based pricing on $a \in \widetilde{A}$ for CV

$\mathbf{x}$: A vector defined as $\mathbf{x} = (x_a, a \in \widetilde{A})$

$\boldsymbol{\tau}$: A vector defined as $\boldsymbol{\tau} = (\tau_a, a \in \widetilde{A})$

v_a^{AV}: Link flow of AV mode on link a

v_a^{CV}: Link flow of CV mode on link a

v_a: Link flow on link a, which equals to $v_a^{AV} + v_a^{CV}$

$\mathbf{v}$: A vector defined as $\mathbf{v} = (v_a, a \in A)$

f_k^{AV}: Path flow of AV mode on path k

f_k^{CV}: Path flow of CV mode on path k.

In this study, we assume that the road planner will aim at minimizing the total travel-time cost in the network by the network design and congestion pricing. After being presented with a designed network and congestion pricing for CVs, all travelers will strive to minimize their individual travel costs and their route choices can be characterized by the UE principle.

In addition, we assume that all links in the network are regular. However, links in candidate link set $\widetilde{A}$ can be converted to AV links and only AVs can operate on AV links. We note that the capacity of a link can be tripled by converting it from a regular to an AV link [12, 13]. It follows that pure AV flow will involve much less travel time when it passes through AV links. If some links in the candidate link set $\widetilde{A}$ are not converted to AV links, then the road planner can implement congestion pricing for CVs on these links. For example, we use the schematic in Figure 1 to illustrate network design and congestion pricing for a network comprising four nodes and five links.

In this network, we consider link 1 and link 4 to be candidate links that can be converted to AV links. That is, $A = \{1, 2, 3, 4, 5\}$ and $\widetilde{A} = \{1, 4\}$. However, we only choose link 1 to be an AV link and $x_1 = 1$ and $x_4 = 0$. It follows that link 4 is still a regular link and can be a tolled link for CVs that run on it.

Based on the above assumptions, we can formulate the NDP bi-level programming (BLP) as shown below:

[BLP]

$$\min_{\mathbf{x},\boldsymbol{\tau}} \sum_{a\in A\setminus\widetilde{A}} t_a\left(v_a(\mathbf{x},\boldsymbol{\tau}), c_a\right) v_a(\mathbf{x},\boldsymbol{\tau}) + \sum_{a\in\widetilde{A}} x_a t_a\left(v_a(\mathbf{x},\boldsymbol{\tau}), c_a^1\right) v_a(\mathbf{x},\boldsymbol{\tau}) + \sum_{a\in\widetilde{A}} (1-x_a) t_a\left(v_a(\mathbf{x},\boldsymbol{\tau}), c_a^2\right) v_a(\mathbf{x},\boldsymbol{\tau}) \quad (1)$$

s.t.

$$x_a \in \{0,1\}, \quad \forall a \in \widetilde{A}, \quad (2)$$

$$0 \le \tau_a \le L(1-x_a), \quad \forall a \in \widetilde{A}, \quad (3)$$

where flow $\mathbf{v}$ solves the following UE problem using link tolls and flow constraints for CVs:

$$\min_{\mathbf{v}} \sum_{a\in A\setminus\widetilde{A}} \int_0^{v_a} t_a(\theta, c_a)\,d\theta + \sum_{a\in\widetilde{A}} x_a \int_0^{v_a} t_a\left(\theta, c_a^1\right) d\theta + \sum_{a\in\widetilde{A}} (1-x_a) \int_0^{v_a} t_a\left(\theta, c_a^2\right) d\theta + \sum_{a\in\widetilde{A}} \tau_a v_a^{\mathrm{CV}} \quad (4)$$

s.t.

$$\sum_{k\in K_w^{\mathrm{AV}}} f_k^{\mathrm{AV}} = D_w^{\mathrm{AV}}, \quad \forall w \in W, \quad (5)$$

$$\sum_{k\in K_w^{\mathrm{CV}}} f_k^{\mathrm{CV}} = D_w^{\mathrm{CV}}, \quad \forall w \in W, \quad (6)$$

$$v_a^{\mathrm{AV}} = \sum_{w\in W} \sum_{k\in K_w^{\mathrm{AV}}} \delta_{a,k} f_k^{\mathrm{AV}}, \quad \forall a \in A, \quad (7)$$

$$v_a^{\mathrm{CV}} = \sum_{w\in W} \sum_{k\in K_w^{\mathrm{CV}}} \delta_{a,k} f_k^{\mathrm{CV}}, \quad \forall a \in A, \quad (8)$$

$$v_a = v_a^{\mathrm{AV}} + v_a^{\mathrm{CV}}, \quad \forall a \in A, \quad (9)$$

$$f_k^{\mathrm{AV}} \ge 0, \quad \forall k \in K_w^{\mathrm{AV}}, \forall w \in W, \quad (10)$$

$$f_k^{\mathrm{CV}} \ge 0, \quad \forall k \in K_w^{\mathrm{CV}}, \forall w \in W. \quad (11)$$

$$x_a v_a^{\mathrm{CV}} = 0, \quad \forall a \in \widetilde{A}. \quad (12)$$

In the above BLP, (1) and (4) are objective functions of upper-level and lower-level programs, respectively. The road planner in the upper level determines the network design variable $\mathbf{x}$ and congestion pricing variable $\boldsymbol{\tau}$, based on the responses of travelers being stable at UE. Constraint (2) ensures that any candidate AV link can be either an AV or regular link. Constraint (3) ensures that only regular links can be tolled links for CVs. Constraints (5)-(11) ensure flow balance in the network. Constraint (12) ensures that only regular links allow CVs to run on them.

Suppose that the link travel-time cost function is the following Bureau of Public Roads (BPR) function $t_a(v_a, C_a) = t_a^0(1 + 0.15(v_a/C_a)^4)$, where t_a^0 is the free-flow travel-time cost of link a. For a given network design $\mathbf{x}$ and congestion pricing $\boldsymbol{\tau}$, the lower-level programming is strictly convex and has a unique optimal solution for link flows. It follows that the lower-level programming can be replaced with its Karush–Kuhn–Tucker (KKT) conditions. The KKT conditions for the lower-level programming are as follows:

$$\sum_{k\in K_w^{\mathrm{AV}}} f_k^{\mathrm{AV}} = D_w^{\mathrm{AV}}, \quad \forall w \in W, \quad (13)$$

$$\sum_{k\in K_w^{\mathrm{CV}}} f_k^{\mathrm{CV}} = D_w^{\mathrm{CV}}, \quad \forall w \in W, \quad (14)$$

$$v_a^{\mathrm{AV}} = \sum_{w\in W} \sum_{k\in K_w^{\mathrm{AV}}} \delta_{a,k} f_k^{\mathrm{AV}}, \quad \forall a \in A, \quad (15)$$

$$v_a^{\mathrm{CV}} = \sum_{w\in W} \sum_{k\in K_w^{\mathrm{CV}}} \delta_{a,k} f_k^{\mathrm{CV}}, \quad \forall a \in A, \quad (16)$$

$$v_a = v_a^{\mathrm{AV}} + v_a^{\mathrm{CV}}, \quad \forall a \in A, \quad (17)$$

$$x_a v_a^{\mathrm{CV}} = 0, \quad \forall a \in \widetilde{A}, \quad (18)$$

$$\sum_{a\in\widetilde{A}} x_a t_a\left(v_a, c_a^1\right)\delta_{a,k} + \sum_{a\in\widetilde{A}} (1-x_a) t_a\left(v_a, c_a^2\right)\delta_{a,k} + \sum_{a\in A\setminus\widetilde{A}} t_a(v_a, c_a)\delta_{a,k} - c_w^{\mathrm{AV}} \ge 0, \quad f_k^{\mathrm{AV}} \ge 0, \forall k \in K_w^{\mathrm{AV}}, \forall w \in W, \quad (19)$$

$$\left(\sum_{a\in\widetilde{A}} x_a t_a\left(v_a, c_a^1\right)\delta_{a,k} + \sum_{a\in\widetilde{A}} (1-x_a) t_a\left(v_a, c_a^2\right)\delta_{a,k} + \sum_{a\in A\setminus\widetilde{A}} t_a(v_a, c_a)\delta_{a,k} - c_w^{\mathrm{AV}}\right) f_k^{\mathrm{AV}} = 0, \quad \forall k \in K_w^{\mathrm{AV}}, \forall w \in W, \quad (20)$$

$$\sum_{a\in\widetilde{A}} x_a t_a\left(v_a, c_a^1\right)\delta_{a,k} + \sum_{a\in\widetilde{A}} (1-x_a) t_a\left(v_a, c_a^2\right)\delta_{a,k} + \sum_{a\in\widetilde{A}} (\tau_a + \lambda_a x_a)\delta_{a,k} + \sum_{a\in A\setminus\widetilde{A}} t_a(v_a, c_a)\delta_{a,k} - c_w^{\mathrm{CV}} \ge 0, \quad f_k^{\mathrm{CV}} \ge 0, \forall k \in K_w^{\mathrm{CV}}, \forall w \in W, \quad (21)$$

$$\left(\sum_{a\in\widetilde{A}} x_a t_a\left(v_a, c_a^1\right)\delta_{a,k} + \sum_{a\in\widetilde{A}} (1-x_a) t_a\left(v_a, c_a^2\right)\delta_{a,k} + \sum_{a\in\widetilde{A}} (\tau_a + \lambda_a x_a)\delta_{a,k} + \sum_{a\in A\setminus\widetilde{A}} t_a(v_a, c_a)\delta_{a,k} - c_w^{\mathrm{CV}}\right) f_k^{\mathrm{CV}} = 0, \quad \forall k \in K_w^{\mathrm{CV}}, \forall w \in W. \quad (22)$$

3. Relaxation-Based Method for Network Design with Congestion Pricing

Solving the NDP with congestion pricing is a challenge since the decision variables of the road planner contain both discrete (AV link design) and continuous (pricing) variables. Wang et al. [33] proposed bi-level programming and a relaxation method for the discrete network design problem (DNDP). Our work extends the DNDP approach since we also consider continuous pricing variables. Here, we combine efficient and effective relaxation algorithms for the NDPs addressed in Wang et al. [27] and Wang et al. [33] and propose a relaxation-based method to solve the network design with congestion pricing.

Let

$$\Omega_{\mathbf{x},\boldsymbol{\tau}} = \{\mathbf{x},\boldsymbol{\tau} : \mathbf{x} \text{ and } \boldsymbol{\tau} \text{ satisfy constraints } (2)-(3)\},$$
$$\Omega_{\mathbf{v}}^{\mathrm{RP}} = \{\mathbf{v} : \mathbf{v} \text{ satisfy constraints } (5)-(12)\}, \quad (23)$$
$$\Omega_{\mathbf{v}}^{\mathrm{UE}} = \{\mathbf{v} : \mathbf{v} \text{ satisfy constraints } (13)-(22)\}.$$

We can formulate the relaxed problem (RP) for the network design with congestion pricing when the traffic assignment follows the system optimal (SO) principle as follows. This enables us to obtain the optimal solution $\mathbf{x}^*$ with SO traffic assignment.

[RP]

$$\min_{(\mathbf{x},\mathbf{0})\in\Omega_{\mathbf{x},\boldsymbol{\tau}},\mathbf{v}\in\Omega_{\mathbf{v}}^{\mathrm{RP}}} \sum_{a\in A\setminus\widetilde{A}} t_a(v_a,c_a)v_a + \sum_{a\in\widetilde{A}} x_a t_a(v_a,c_a^1)v_a + \sum_{a\in\widetilde{A}}(1-x_a)t_a(v_a,c_a^2)v_a \quad (24)$$

In the relaxation-based method, we solve [RP] by iteratively excluding the solution $\mathbf{x}^*$. We do so by adding the following constraint to [RP] iteratively:

$$\sum_{\text{all } a\in\widetilde{A} \text{ that satisfying } x_a^*=0} x_a + \sum_{\text{all } a\in\widetilde{A} \text{ that satisfying } x_a^*=1} (1-x_a) \geq 1 \quad (25)$$

Here, we define the pricing problem (PP) when the network design variable $\mathbf{x}$ is given as $\mathbf{x}^*$.

[PP]

$$\min_{(\mathbf{x}^*,\boldsymbol{\tau})\in\Omega_{\mathbf{x},\boldsymbol{\tau}},\mathbf{v}\in\Omega_{\mathbf{v}}^{\mathrm{UE}}} \sum_{a\in A\setminus\widetilde{A}} t_a(v_a,c_a)v_a + \sum_{a\in\widetilde{A}} x_a t_a(v_a,c_a^1)v_a + \sum_{a\in\widetilde{A}}(1-x_a)t_a(v_a,c_a^2)v_a \quad (26)$$

The relaxation-based method is as follows:

Step 0. Define a set $\overline{\Omega}_{\mathbf{x}} = \varnothing$ that contains all generated solutions for generating the cuts given by constraint (25). Define the upper bound $UB = +\infty$, incumbent optimal solution $(\mathbf{x}^{\mathrm{opt}},\boldsymbol{\tau}^{\mathrm{opt}})$.

Step 1. Reformulate [RP] as [RP-1], as follows:

[RP-1]

$$\min_{(\mathbf{x},\mathbf{0})\in\Omega_{\mathbf{x},\boldsymbol{\tau}},\mathbf{v}\in\Omega_{\mathbf{v}}^{\mathrm{RP}}} \sum_{a\in A\setminus\widetilde{A}} t_a(v_a,c_a)v_a + \sum_{a\in\widetilde{A}} t_a(v_a^1,c_a^1)v_a^1 + \sum_{a\in\widetilde{A}} t_a(v_a^2,c_a^2)v_a^2 \quad (27)$$

s.t.

$$\begin{aligned}
&v_a = v_a^1 + v_a^2, \quad \forall a\in\widetilde{A},\\
&0 \leq v_a^1 \leq Lx_a, \quad \forall a\in\widetilde{A},\\
&0 \leq v_a^2 \leq L(1-x_a), \quad \forall a\in\widetilde{A},\\
&0 \leq \tau_a \leq Lx_a, \quad \forall a\in\widetilde{A},\\
&x_a \in \{0,1\}, \quad \forall a\in\widetilde{A},
\end{aligned} \quad (28)$$

where v_a^1 and v_a^2 are auxiliary variables. Solve [RP-1] with the following constraints using the outer-approximation algorithm presented in Wang et al. [33]:

$$\sum_{a\in\widetilde{A}}(1-\overline{x}_a)x_a + \sum_{a\in\widetilde{A}}\overline{x}_a(1-x_a) \geq 1, \quad \forall\overline{\mathbf{x}}\in\overline{\Omega}_{\mathbf{x}}. \quad (29)$$

If the problem is infeasible, we have enumerated all feasible solutions and hence $(\mathbf{x}^{\mathrm{opt}},\boldsymbol{\tau}^{\mathrm{opt}})$ is the optimal solution and we stop. Otherwise, obtain the provisional optimal design denoted by $\mathbf{x}^*$ and the provisional optimal value denoted by Obj_{RP} of [RP] under constraint (29). If $UB \leq Obj_{\mathrm{RP}}$, $(\mathbf{x}^{\mathrm{opt}},\boldsymbol{\tau}^{\mathrm{opt}})$ is the optimal solution, stop. Otherwise, go to step 2.

Step 2. Solve [PP] by fixing $\mathbf{x}$ at $\mathbf{x}^*$ and apply the relaxation algorithm presented in Wang et al. [27]. Then, we can obtain the optimal value Obj_{PP}, optimal pricing $\boldsymbol{\tau}^*$, and vehicle flow $\mathbf{v}^*$. If $UB \leq Obj_{\mathrm{PP}}$, then $(\mathbf{x}^*,\boldsymbol{\tau}^*)$ is not better than $(\mathbf{x}^{\mathrm{opt}},\boldsymbol{\tau}^{\mathrm{opt}})$ and, hence, we set $\overline{\Omega}_{\mathbf{x}} := \overline{\Omega}_{\mathbf{x}} \cup \{\mathbf{x}^*\}$ and go to step 1. Else, set $UB = Obj_{\mathrm{PP}}$ and $(\mathbf{x}^{\mathrm{opt}},\boldsymbol{\tau}^{\mathrm{opt}}) = (\mathbf{x}^*,\boldsymbol{\tau}^*)$. If $UB \leq Obj_{\mathrm{RP}}$ and $(\mathbf{x}^{\mathrm{opt}},\boldsymbol{\tau}^{\mathrm{opt}})$ is the optimal solution, stop. Otherwise, set $\overline{\Omega}_{\mathbf{x}} := \overline{\Omega}_{\mathbf{x}} \cup \{\mathbf{x}^*\}$ and go to step 1.

4. Numerical Tests

4.1. Network Description. In the following sections, to evaluate the effectiveness of the proposed model and solution method, we present some numerical examples for a nine-node network with 18 links [37]. Suppose that the travel-time cost functions of all links are defined as having the following BPR form:

$$t_a(v_a,C_a) = t_a^0\left(1 + 0.15\left(\frac{v_a}{C_a}\right)^4\right), \quad (30)$$

where t_a^0 is the free-flow travel-time cost on link a and C_a is the capacity of link a. Figure 2 shows the initial network without converting regular links to AV links. The tuple above

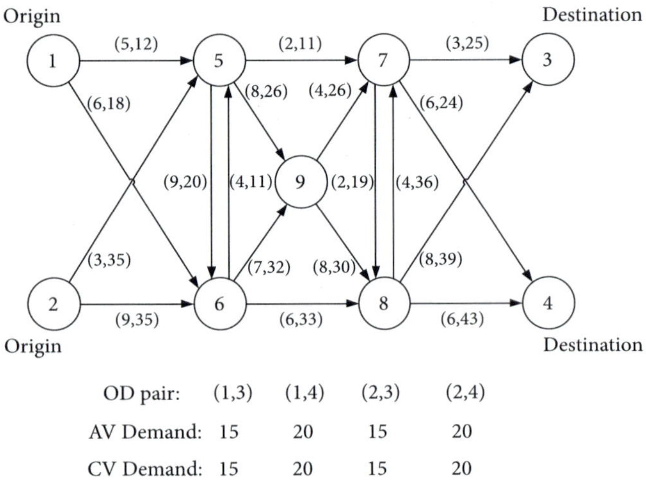

Figure 2: Nine-node network.

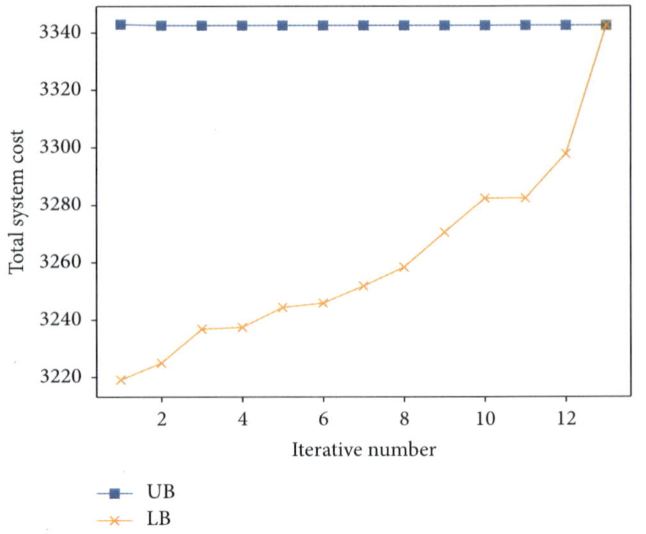

Figure 3: Lower and upper bounds at each iteration of the relaxation-based method.

each link a denotes the free-flow travel-time cost and initial capacity when all links are regular. There are four OD pairs in the network, and the OD demands of all the travel modes are shown at the bottom of Figure 2.

In the initial network, all links are regular links and all vehicles are allowed to operate on these links without congestion pricing. We can obtain vehicle flows in the initial network by solving UE problems (4)-(12); when setting candidate AV link set $\widetilde{A} = \varnothing$ without congestion pricing. After solving this UE problem for vehicle flows, we find that no AVs run on links (5,6), (6,5), (6,9), (7,8), (8,7), and (8,3) in the initial network. So we exclude these links from the candidate link set and form the following candidate link set:

$$\widetilde{A} = \{(1,5),(1,6),(2,5),(2,6),(5,7),(5,9),(6,8),(9,7),(9,8),(7,3),(7,4),(8,4)\}. \quad (31)$$

If any link $a \in \widetilde{A}$ is converted to an AV link, then the capacity of link a is tripled and only AVs are allowed to run on it. For any link $a \in \widetilde{A}$ that is not converted to an AV link, the road planner can implement congestion pricing for CVs.

In the following tests, we used a personal computer with an Intel Core (TM) i7 4700MQ CPU, 16GB RAM, and Windows 7 Professional operating system. We coded the proposed solution method using Python and called Gurobi 7.5.2 and Ipopt 3.12.7 to solve the NDP with congestion pricing. We ran the codes for all cases in the following tests and the longest computation time was less than five minutes.

4.2. Convergence of Relaxation-Based Method. We tested the convergence of the relaxation-based method, and Figure 3 shows the lower and upper bounds at each iteration of this method. We can see that the gap between lower and upper bounds decreases sharply and finally converges to zero. Table 1 shows the optimal network design with congestion pricing for this case. Table 2 shows the link flows in the network when the optimal network design with congestion pricing is implemented.

Table 1: Optimal network design with congestion pricing.

Candidate link	x_a	τ_a
(1,5)	1	0.000
(1,6)	0	4.756
(2,5)	0	3.272
(2,6)	0	5.821
(5,7)	1	0.000
(5,9)	0	3.350
(6,8)	0	8.732
(9,7)	0	3.132
(9,8)	0	1.532
(7,3)	1	0.000
(7,4)	1	0.000
(8,4)	0	4.593

4.3. Model Comparison. In this section, we use four models to demonstrate the efficiency of the NDP with congestion pricing. The first model is to do nothing, in that we set no links for AVs and no tolls for CVs. The second model is a pricing scheme for CVs without any network design. The third model is an NDP without congestion pricing. The fourth model is the NDP with congestion pricing, as proposed. The total travel-time costs associated with these four models are 5886.276, 5440.112, 3507.934, and 3342.627, respectively. We can clearly see that using both network design and congestion pricing can improve the traffic conditions with heterogeneous AV and CV travel modes. In addition, the road planner can implement these methods simultaneously, which realizes better results than implementing the network design or congestion pricing separately.

4.4. Sensitivity Analyses. Since many parameters can impact the total travel-time cost in a network, here, we perform sensitivity analyses. All tests in this section are based on the network proposed in Section 4.1.

TABLE 2: Link flows in the network upon implementation of the optimal network design with congestion pricing.

Link	AV link flow	CV link flow
(1,5)	35.000	0.000
(1,6)	0.000	35.000
(2,5)	35.000	0.000
(2,6)	0.000	35.000
(5,6)	0.000	0.000
(6,5)	0.000	0.000
(5,7)	55.774	0.000
(5,9)	14.226	0.000
(6,8)	0.000	42.768
(6,9)	0.000	27.232
(9,7)	4.450	18.119
(9,8)	9.775	9.113
(7,8)	0.000	18.119
(8,7)	0.000	0.000
(7,3)	30.000	0.000
(7,4)	30.225	0.000
(8,3)	0.000	30.000
(8,4)	9.775	40.000

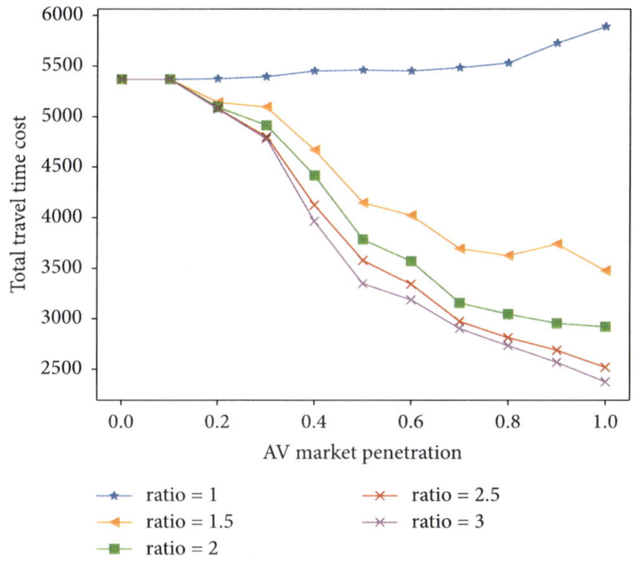

FIGURE 4: Variances of total travel-time cost with AV market penetrations.

Figure 4 shows variances in the total travel-time cost against AV market penetrations. "ratio = 3" indicates that the AV link capacity is tripled when it is converted from a regular to an AV link. We can see that a greater adoption of AVs always leads to less total travel-time cost for ratio = 2,2.5,3. However, this is not true for ratio = 1,1.5. This means that AVs should be promoted only when the AV technique is mature and AV link capacity can be increased significantly; otherwise, AV promotion may cause the opposite effect.

When the AV technique matures, more people will choose the AV mode to benefit from its higher capacity and

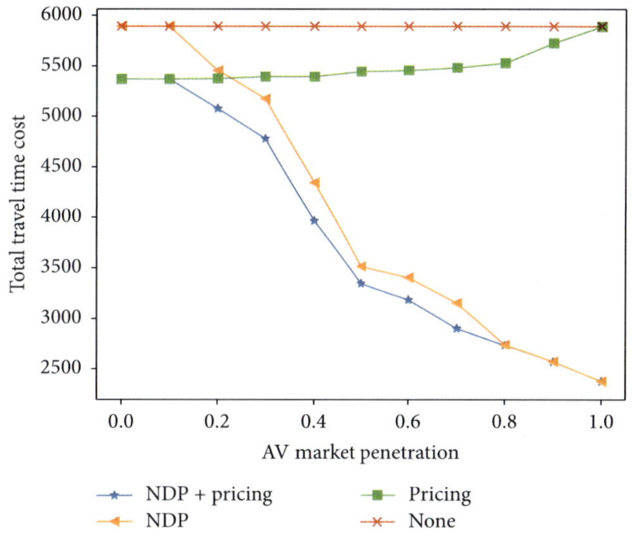

FIGURE 5: Four models at different AV market penetrations.

lower travel time. By fixing the total demand of the network, we tested the models described in Section 4.3 at different AV market penetrations. Figure 5 shows a comparison of the four models, in which we can see that the NDP with congestion pricing is always the most effective. Also, we can see that congestion pricing is effective when the AV market penetration is low and network design is effective when AV penetration is high. The reason for this is clear: when AV market penetration is low, congestion pricing to control CVs is effective. When AV penetration is high, a network design for AVs is effective. This means that road planners should concentrate on different strategies for different AV market penetrations. In addition, we find that the NDP with congestion pricing strongly outperforms either NDP or pricing when both AVs and CVs have large market penetrations. This shows that the proposed NDP with congestion pricing can be an effective method for alleviating traffic congestion in the transitional period characterized by heterogeneous AV and CV travel modes.

Assuming that the initial demand is that shown in Figure 2, Figures 6(a)–6(c) show variances in the total travel-time costs for the four models with increasing percentages of OD demand and AV market penetration. Given a fixed AV market penetration, we can see that the effect of pure congestion pricing becomes worse with increasing demand. However, the effect of a pure network design improves with increasing demand. The gap between the total travel-time cost by the NDP with congestion pricing and that by an NDP alone is almost the same even if the OD demand increases. Also, we can see that the NDP becomes more effective as OD demand increases, but congestion pricing does not. As such, we can infer that AV promotion and AV link design may be the best solution for alleviating traffic congestion as traffic demand continues to grow.

Since high congestion pricing for CVs would generate public resentment, we set an upper bound for pricing τ_a. Figure 7 shows the variance of total travel-time cost with

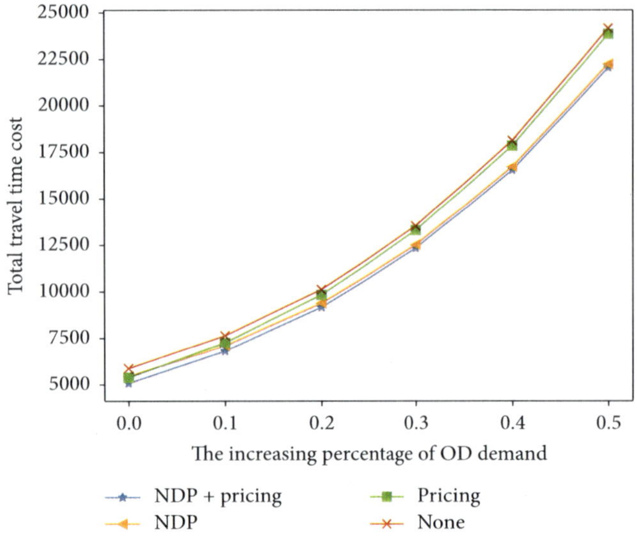

(a) Variances of total travel-time cost with increasing percentage of OD demand when AV market penetration = 20%

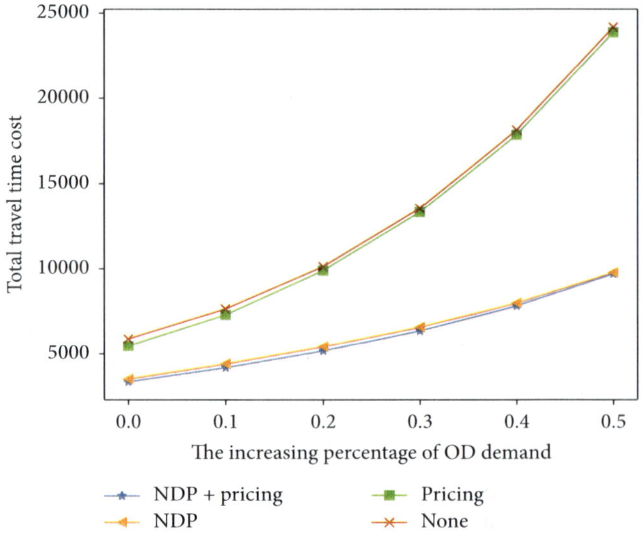

(b) Variances of total travel-time cost with increasing percentage of OD demand when AV market penetration = 50%

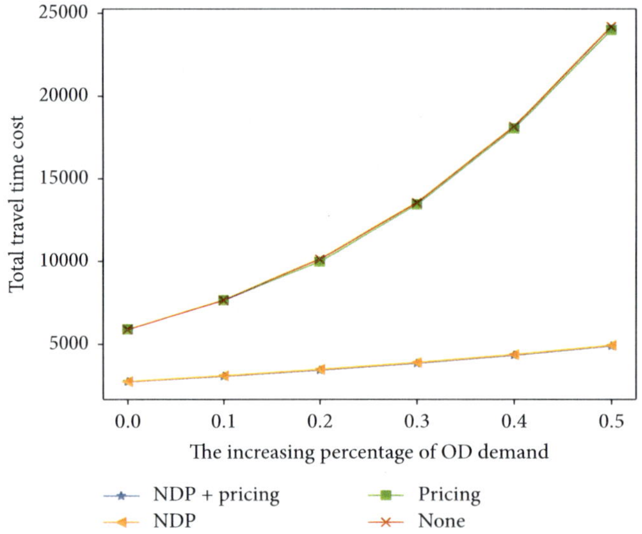

(c) Variances of total travel-time cost with increasing percentage of OD demand when AV market penetration = 80%

FIGURE 6

respect to the upper bound for congestion pricing. Clearly, setting a reasonable upper bound for congestion pricing cannot only reduce public discontent, but also can effectively reduce total travel-time cost.

5. Conclusions and Future Research

In this paper, we proposed a bi-level network design model comprising AV links and congestion pricing to alleviate traffic congestion when AVs and CVs are both operating in a transportation network. In this model, the upper-level road planner chooses the optimal network design for AV links and congestion pricing while lower-level travelers choose their routes based on their individual travel costs and thereby achieve user equilibrium. We conducted numerical tests on a nine-node network and the results reveal the feasibility of the proposed model and solution method. Our main findings are shown as follows:

(1) Both network design and congestion pricing can alleviate traffic congestion. The integrated optimization of a network design and congestion pricing can achieve better traffic conditions than either a single network design or a congestion pricing.

(2) If we compare the performance outcome of pure congestion pricing with that of a network design, we can see that congestion pricing is effective when AV market penetration is low and a network design is effective when AV penetration is high. As such, the

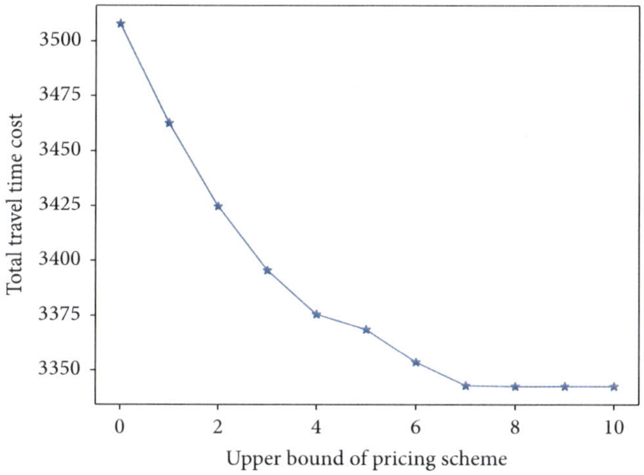

FIGURE 7: Variance of total travel-time cost with respect to the upper bound of congestion pricing.

road planner should employ different strategies for different AV market penetrations.

(3) Since network design becomes more effective as OD demand increases, it follows that AV promotion and AV link design may be the best solution for alleviating traffic congestion when traffic demand becomes increasingly large.

(4) In the transitional period characterized by heterogeneous AV and CV travel modes, the NDP with congestion pricing strongly outperforms either network design or congestion pricing alone when both AVs and CVs have fairly large market penetrations. In addition, if we add an appropriate upper bound for congestion pricing, we can reduce public discontent without losing the effectiveness of the NDP with congestion pricing.

Based on our proposed model and results, we suggest the consideration of future studies as follows:

(1) We suggest NDP with a tradable credit scheme to alleviate traffic congestion. Although the proposed NDP with congestion pricing can alleviate traffic congestion and an appropriate upper bound for congestion pricing can reduce public discontent without losing its effectiveness, congestion pricing can be viewed as unfair or as a flat tax [38]. If we consider the NDP with a tradable credit scheme, then the problem of unfairness is fully resolved.

(2) Multiclasses of travelers can be considered. We can infer the relationship between the classes of travelers and the perceived value of their travel time, based on the population income distribution. This could be helpful in tackling real-world situations and yield interesting findings.

(3) We suggest optimal design with congestion pricing considering variational AV market penetration. In this paper, we considered AV market penetration to be exogenous. However, AV market penetration can be viewed as an endogenous variable and the evolution of this penetration can be forecast using prediction models, e.g., the diffusion model [13, 39]. Based on this assumption, a time-dependent NDP with congestion pricing could be formulated to alleviate traffic congestion over a long period.

(4) The optimal structure of intermodal transportation network is very important for both passenger traffic and logistics [40]; network design of intermodal transportation network with AVs might be an important research topic in the future.

Conflicts of Interest

The authors declare that there are no conflicts of interest regarding the publication of this paper.

Acknowledgments

The authors would like to thank Professor Xiaoning Zhang of Tongji University for his time in proofreading the paper. This study has been substantially supported by the research grants from the National Natural Science Foundation Council of China (nos. 71531011, 71601142, and 71471134), Shanghai Shuguang Program (13SG23), and Shanghai Pujiang Program (16PJC090).

References

[1] H. Wang, W. H. K. Lam, X. Zhang, and H. Shao, "Sustainable transportation network design with stochastic demands and chance constraints," *International Journal of Sustainable Transportation*, vol. 9, no. 2, pp. 126–144, 2015.

[2] H. Wang and X. Zhang, "Game theoretical transportation network design among multiple regions," *Annals of Operations Research*, vol. 249, no. 1-2, pp. 97–117, 2017.

[3] R. Z. Farahani, E. Miandoabchi, W. Y. Szeto, and H. Rashidi, "A review of urban transportation network design problems," *European Journal of Operational Research*, vol. 229, no. 2, pp. 281–302, 2013.

[4] H. Yang and M. G. H. Bell, "Models and algorithms for road network design: a review and some new developments," *Transport Reviews*, vol. 18, no. 3, pp. 257–278, 1998.

[5] Q. Meng, Z. Liu, and S. Wang, "Optimal distance tolls under congestion pricing and continuously distributed value of time," *Transportation Research Part E: Logistics and Transportation Review*, vol. 48, no. 5, pp. 937–957, 2012.

[6] Z. Liu, Y. Zhang, S. Wang, and Z. Li, "A trial-and-error method with autonomous vehicle-to-infrastructure traffic counts for cordon-based congestion pricing," *Journal of Advanced Transportation*, vol. 2017, Article ID 9243039, pp. 1–8, 2017.

[7] H. Wang and X. Zhang, "Joint implementation of tradable credit and road pricing in public-private partnership networks considering mixed equilibrium behaviors," *Transportation Research Part E: Logistics and Transportation Review*, vol. 94, pp. 158–170, 2016.

[8] S. Le Vine, A. Zolfaghari, and J. Polak, "Autonomous cars: The tension between occupant experience and intersection capacity," *Transportation Research Part C: Emerging Technologies*, vol. 52, pp. 1–14, 2015.

[9] N. Jiang, "Optimal signal design for mixed equilibrium networks with autonomous and regular vehicles," *Journal of Advanced Transportation*, vol. 2017, pp. 1–13, 2017.

[10] J. Wang, X. Zhang, and H. M. Zhang, "Parking permits management and optimal parking supply considering traffic emission cost," *Transportation Research Part D: Transport and Environment*, 2016.

[11] F. Zhu and S. V. Ukkusuri, "A linear programming formulation for autonomous intersection control within a dynamic traffic assignment and connected vehicle environment," *Transportation Research Part C: Emerging Technologies*, vol. 55, pp. 363–378, 2015.

[12] P. Tientrakool, Y.-C. Ho, and N. F. Maxemchuk, "Highway capacity benefits from using vehicle-to-vehicle communication and sensors for collision avoidance," in *Proceedings of the IEEE 74th Vehicular Technology Conference, VTC Fall*, The Hilton San Francisco Union Square, San Francisco, USA, September 2011.

[13] Z. Chen, F. He, L. Zhang, and Y. Yin, "Optimal deployment of autonomous vehicle lanes with endogenous market penetration," *Transportation Research Part C: Emerging Technologies*, vol. 72, pp. 143–156, 2016.

[14] X. Zhang and H. Yang, "The optimal cordon-based network congestion pricing problem," *Transportation Research Part B: Methodological*, vol. 38, no. 6, pp. 517–537, 2004.

[15] X. Zhang, H. M. Zhang, H.-J. Huang, L. J. Sun, and T.-Q. Tang, "Competitive, cooperative and stackelberg congestion pricing for multiple regions in transportation networks," *Transportmetrica*, vol. 7, no. 4, pp. 297–320, 2011.

[16] H. Yang and H. J. Huang, *Mathematical and Economic Theory of Road Pricing*, Elsevier, Oxford, UK, 2005.

[17] Y. Ji, M. Xu, H. Wang, and C. Tan, "Commute equilibrium for mixed networks with autonomous vehicles and traditional vehicles," *Journal of Advanced Transportation*, vol. 2017, Article ID 6218363, pp. 1–10, 2017.

[18] S. Deb, L. Strawderman, D. W. Carruth, J. DuBien, B. Smith, and T. M. Garrison, "Development and validation of a questionnaire to assess pedestrian receptivity toward fully autonomous vehicles," *Transportation Research Part C: Emerging Technologies*, vol. 84, pp. 178–195, 2017.

[19] C. Häne, T. Sattler, and M. Pollefeys, "Obstacle detection for self-driving cars using only monocular cameras and wheel odometry," in *Proceedings of the IEEE/RSJ International Conference on Intelligent Robots and Systems, IROS 2015*, pp. 5101–5108, Germany, October 2015.

[20] A. Aparicio, L. Sanz, G. Burnett et al., *Advancing Active Safety Towards the Protection of Vulnerable Road Users: The Prospect Project*, 2017.

[21] M. W. Levin and S. D. Boyles, "A multiclass cell transmission model for shared human and autonomous vehicle roads," *Transportation Research Part C: Emerging Technologies*, vol. 62, pp. 103–116, 2016.

[22] P. Godsmark and G. Kakkar, "Why automated vehicle zones make sense," *The Canadian Automated Vehicles Centre of Excellence*, 2014.

[23] Z. Chen, F. He, Y. Yin, and Y. Du, "Optimal design of autonomous vehicle zones in transportation networks," *Transportation Research Part B: Methodological*, vol. 99, pp. 44–61, 2017.

[24] N. Jiang, X. Zhang, and H. Wang, "Simultaneous optimization of road tolls and tradable credits in public-private mixed networks," *PROMET - Traffic and Transportation*, vol. 29, no. 6, p. 603, 2017.

[25] A. Koh, S. Shepherd, and A. Sumalee, "Second best toll and capacity optimisation in networks: Solution algorithm and policy implications," *Transportation*, vol. 36, no. 2, pp. 147–165, 2009.

[26] X. Zhang and B. van Wee, "Enhancing transportation network capacity by congestion pricing with simultaneous toll location and toll level optimization," *Engineering Optimization*, vol. 44, no. 4, pp. 477–488, 2012.

[27] G. Wang, Z. Gao, M. Xu, and H. Sun, "Joint link-based credit charging and road capacity improvement in continuous network design problem," *Transportation Research Part A: Policy and Practice*, vol. 67, pp. 1–14, 2014.

[28] M. Xu, G. Wang, S. Grant-Muller, and Z. Gao, "Joint road toll pricing and capacity development in discrete transport network design problem," *Transportation*, vol. 44, no. 4, pp. 731–752, 2017.

[29] Y. Lim, B. G. Heydecker, and S. Lee, "A continuous network design model in stochastic user equilibrium based on sensitivity analysis," *Journal of Advanced Transportation*, vol. 39, no. 1, pp. 63–79, 2005.

[30] G. M. Wang, Z. Y. Gao, and M. Xu, "An MPEC formulation and its cutting constraint algorithm for continuous network design problem with multi-user classes," *Applied Mathematical Modelling*, vol. 38, no. 5-6, pp. 1846–1858, 2014.

[31] D. Z. W. Wang and H. K. Lo, "Global optimum of the linearized network design problem with equilibrium flows," *Transportation Research Part B: Methodological*, vol. 44, no. 4, pp. 482–492, 2010.

[32] H. Liu and D. Z. W. Wang, "Global optimization method for network design problem with stochastic user equilibrium," *Transportation Research Part B: Methodological*, vol. 72, pp. 20–39, 2015.

[33] S. Wang, Q. Meng, and H. Yang, "Global optimization methods for the discrete network design problem," *Transportation Research Part B: Methodological*, vol. 50, pp. 42–60, 2013.

[34] H. Sun, J. Wu, W. Wang, and Z. Gao, "Reliability-based traffic network design with advanced traveler information systems," *Information Sciences*, vol. 287, pp. 121–130, 2014.

[35] Y. Kong, S. Le Vine, and X. Liu, "Capacity impacts and optimal geometry of automated cars' surface parking facilities," *Journal of Advanced Transportation*, vol. 2018, pp. 1–13, 2018.

[36] J. F. Bard, "Some properties of the bilevel linear programming," *Journal of Optimization Theory and Applications*, vol. 2, pp. 146–164, 1991.

[37] D. W. Hearn and M. V. Ramana, "Solving congestion toll pricing models," in *Equilibrium and Advanced Transportation Modelling*, pp. 109–124, Springer, 1998.

[38] H. Yang and X. Wang, "Managing network mobility with tradable credits," *Transportation Research Part B: Methodological*, vol. 45, no. 3, pp. 580–594, 2011.

[39] H. Yang and Q. Meng, "Modeling user adoption of advanced traveler information systems: Dynamic evolution and stationary equilibrium," *Transportation Research Part A: Policy and Practice*, vol. 35, no. 10, pp. 895–912, 2001.

[40] H. Wang, X. Wang, and X. Zhang, "Dynamic resource allocation for intermodal freight transportation with network effects: Approximations and algorithms," *Transportation Research Part B: Methodological*, vol. 99, pp. 83–112, 2017.

4

Clustering Algorithm for Urban Taxi Carpooling Vehicle based on Data Field Energy

Xiao Qiang[1,2] **and Yao Shuang-Shuang**[1]

[1] *School of Economics and Management, Lanzhou Jiao Tong University, Lanzhou 730070, China*
[2] *School of Traffic and Transportation, Lanzhou Jiao Tong University, Lanzhou 730070, China*

Correspondence should be addressed to Xiao Qiang; lzjt_xq@126.com

Academic Editor: Giuseppe Musolino

A clustering algorithm for urban taxi carpooling based on data field energy and point spacing is proposed to solve the clustering problem of taxi carpooling on urban roads. The data field energy function is used to calculate the field energy of each data point in the passenger taxi offpoint dataset. To realize the clustering of taxis, the central point, outlier, and data points of each cluster subset are discriminated according to the threshold value determined by the product of each data point field values and point spacing. The classical algorithm and proposed algorithm are compared and analyzed by using the compactness, separation, and Dunn validity index. The clustering results of the proposed algorithm are better than those of the classical clustering algorithm. In the case of cluster numbers 25, 249, 409, and 599, the algorithm has good clustering results for the taxi trajectory dataset with certain regularity in space distribution and irregular distribution in time distribution. This algorithm is suitable for the clustering of vehicles in urban traffic roads, which can provide new ideas and methods for the cluster study of urban traffic vehicles.

1. Introduction

The acceleration of urbanization and the rapid increase in the number of travel vehicles have caused traffic congestion on urban roads to become a major problem that must be solved in urban development. Accordingly, several scholars have suggested the idea of "carpool" which has rapidly stimulated academic resonance. Empirical data show that this concept demonstrates policy effectiveness in alleviating urban traffic congestion and solving the plight of vehicle operation. Researchers from the Massachusetts Institute of Technology (MIT) analyzed a week-long run of New York City taxis in Manhattan in March 2013. Out of the 13,600 taxis in New York, approximately 10,000 were used during rush hours. Manhattan requires only 3,000 shared taxis to satisfy its 98% ride requirements [1]. This study showed that an effective ride-sharing system cannot only alleviate the traffic congestion in cities but also increase the passenger-carrying rate of operating vehicles and the operating income of drivers. In addition, energy consumption and environmental pollution must be reduced [2]. Therefore, implementing carpool is an effective means of improving the quality of urban traffic [3, 4].

The study of the Vehicle Routing Problem (VRP) has given rise to major developments in the fields of exact algorithms and heuristics. In particular, highly sophisticated exact mathematical programming decomposition algorithms and powerful metaheuristics for the VRP have been put forward in recent years [5]. In 2014, Lin [6] provided a classification of Green Vehicle Routing Problem (GVRP) that categorizes GVRP into Green-VRP, Pollution Routing Problem, and VRP in Reverse Logistics and suggests research gaps between its state and richer models describing the complexity in real-world cases. In 2018, Musolino [7] presented a procedure for the solution of the Vehicle Routing Problem (VRP) based on reliable link travel times, the equation of reliable link travel times is composed of a congestion term, expressing the traditional congested link travel times (or generalized costs), and a reliability term, which depends on the fundamental diagram of the link and the Network Fundamental Diagram (NFD) of the homogeneous cluster of adjacent links; NFDs data are used in the proposed link travel time function to calculate reliable travel times. The reliable link travel times are used for the solution of VRP to obtain optimal routes of freight vehicles.

Taxi carpooling as a way to alleviate taxi traffic congestion has been adopted and implemented in some cities. Therefore, as a hot topic of urban transportation research, carpooling has attracted the attention of many scholars. In the research of carpool service problem (CSP), in 2011, Agatz [8] studied the matching problem of drivers and passengers in a dynamic environment, and optimization method was proposed to minimise vehicle running mileage and individual travel cost. Shinde [9] proposed a genetic algorithm for multiobjective optimization based on the carpooling path matching. The algorithm effectively reduces the computational complexity and processing time and improves the carpooling effect. In 2015, Pelzer [10] proposed a dynamic decision algorithm based on network partition; the algorithm divides and numbers the road network and uses the spatial routing search algorithm to realize the matching of passengers and vehicles. In 2015, Jiau [11] implemented the carpooling path matching in a short time through genetic algorithm and realized the carpooling path matching scheme of low complexity and low memory. In 2015, Huang [12] propose a fuzzy-controlled genetic-based carpool algorithm by using the combined approach of the genetic algorithm and the fuzzy control system, with which to optimize the route and match assignments of the providers and the requesters in the intelligent carpool system. In 2016, Chou [13] developed a particle swarm carpool algorithm based on stochastic set-based particle swarm optimization (PSO); the set-based PSO (S-PSO) can be realized by local exploration. Method yielded the best result in a statistical test and successfully obtained numerical results for meeting the optimization objectives of the CSP.

In the research of taxi carpool problem (TCP), In 2013, Cheng [14] took the benefit of travellers and drivers as the optimization objective and established a multidynamic taxi carpooling model. The genetic algorithm was used to solve the problem of the carpooling algorithm. In 2013, Ma [15] proposed a large-scale taxi ride-sharing service; it efficiently serves real-time requests sent by taxi users and generates ride-sharing schedules that reduce the total travel distance significantly. In 2014, Xiao et al. [16] created membership function by using three factors of driving route, driving time, and the number of passengers to realize carpooling fuzzy clustering and recognition of passengers and taxis. In 2015, Ma [17] devised a mobile-cloud architecture based taxi-sharing system. Taxi riders and taxi drivers use the taxi-sharing service provided by the system via a smart phone App. In 2017, Zhang [18] presented the first systematic work to design a unified recommendation system for both the regular and the carpooling services, called CallCab, based on a data-driven approach; this recommendation system has been done to assist passengers to find a successful taxicab ride with carpooling.

It can be seen from the research of domestic and foreign carpool service problem and taxi carpool problem that the solution of the carpool problem is mainly realized by using multiobjective programming algorithm or intelligent algorithms. In practical application, because of the large amount of carpool data, the calculation time is very long when the algorithm is used to determine the carpool scheme. Therefore, a two-stage algorithm is proposed to solve the carpool problem in big data environment to reduce the computational time of the algorithm. Häme [19] proposed an adaptive insertion algorithm in 2011, using the idea of clustering to sort the order of passengers on and off, which greatly simplifies the complexity of the problem and the difficulty of the solution and is helpful in addressing the problem of taxi carpooling. In 2012, Manzini [20] proposed phased clustering model. Cluster factors such as route, distance, user information, and carpool were clustered separately. Then, the decision support system was used to judge the factors and provide decision support for the carpooling. In 2013, Shao [21–23] proposed a two-stage clustering heuristic matching strategy to solve the problem of multivehicle carpooling. In 2016, Yang [24] proposed a kind of carpooling model in distributed parallel environment and used the two-stage distributed estimation algorithm to solve the carpooling scheme.

From the above research, we can see that, in the two-stage algorithm, the first-stage carpool factors (such as route, distance, and time) are clustered. In the second stage, according to the results of the clustering, the algorithm of matching, multiobjective programming, and intelligent algorithm are used to solve the combinatorial problem. The main content of this paper is to study the first-stage clustering problem.

In the present application of the clustering algorithm, clustering centre and range are keys in the accurate application of the clustering algorithm, but the classic clustering algorithm on clustering effect is poor because of the city road network distribution. To realize taxi clustering in a city road, a clustering algorithm of urban taxi carpooling based on data field energy and point spacing is proposed for the clustering problem of taxi carpooling on urban roads. The data field energy function is used to calculate the field energy of each data point in the passenger taxi offpoint dataset. The central point, outlier, and data points of each cluster subset are discriminated according to the threshold value determined by the product of each data point field values and point spacing, and the taxi clustering can be realized to provide a basis for theoretical research on city taxis.

The rest of the paper is organized as follows. Section 2 introduces the data field theory. Section 3 proposes carpool taxi clustering algorithm based on data field energy and point spacing. Section 4 presents a case of Nanjing taxi data that illustrates the application of the proposed approach. Section 5 provides the conclusions.

2. Data Field Energy

2.1. Definition of Data Field Energy. From field theory, the field is used to describe the interaction between substances, such as gravitational field, electric field, and magnetic field. Each field is known to decay or increase with distance, and the distribution of field energy can be described by either a scalar or a vector function [25]. The method treats each object in space as a particle with a certain mass. There is a spherical symmetric virtual gravitational field around it, and any object in the field will be subjected to the joint action of other objects, so a data field is determined in the whole space. Similar to the vector strength function and scalar potential function description of physical field, this method introduces the definition of potential function and field strength function of data field and realizes the self-organizing hierarchical aggregation of data object by simulating the interaction and motion of object in data field. The proposed method is not dependent on the careful selection of user parameters and can identify nonspherical clustering of arbitrary size and density. It is insensitive to noise data and has an approximate linear time complexity. According to this principle, the data point in the large data set is considered a kind of data particle in space, where a virtual field is observed around the data particle. The data particles in the field will be subjected to field forces. Therefore, the data points interact to form a data field. In the data field, the distance between data points is near, and the interaction forces between the data are strong. Then, the data field can be largeand vice versa.

$D = \{x_1, x_2, \cdots x_n\}$, which is a set of data, where x_i refers to the data point from 1 to n. $I_s = \{1, 2, \cdots N\}$, which is a new index, where N refers to the number of data points. The data point of potential energy is defined as follows [26]:

$$E(x_i) = M \cdot \sum_{j \in I_s} e^{(-\|x_j - x_i\|/d_c)^R} \tag{1}$$

where $\|x_j - x_i\|$ is the distance between data points x_j and x_i, d_c is the interaction factor of data points, M is the quality of data objects (in this study, the value of M is 1), and R is the distance index. The data of spatial distribution mainly depend on the object interaction process or the radius of influence. Thus, spatial distribution has nothing to do with the specific form of the potential function or the selection of distance index, so distance index has minimal influence on the description of structural characteristics [25, 26]. Let $R = 2$, $E(x_i)$ be the Gauss nuclear field, and the data point of potential energy $E(x_i)$ is defined as follows:

$$E(x_i) = \sum_{j \in I_s} e^{(-\|x_j - x_i\|/d_c)^2} \tag{2}$$

Gauss field potential function is shown in Figure 1.

As shown in Figure 1, if $\|x_j - x_i\|$ gradually grows larger (which represents that the distance between data points x_j and x_i is becoming increasingly farther), then $E(x_i)$ is becoming smaller. By contrast, if $\|x_j - x_i\|$ gradually grows smaller (which represents that the distance between data points x_j and x_i is becoming increasingly closer), then $E(x_i)$ is becoming larger. From the data interaction factors, we can find the change; if d_c is small, then the distance of data points x_j and x_i interaction is smaller; if d_c is large, then the distance of data points x_j and x_i interaction is larger. Thus, parameter d_c can control the range and size of data clustering.

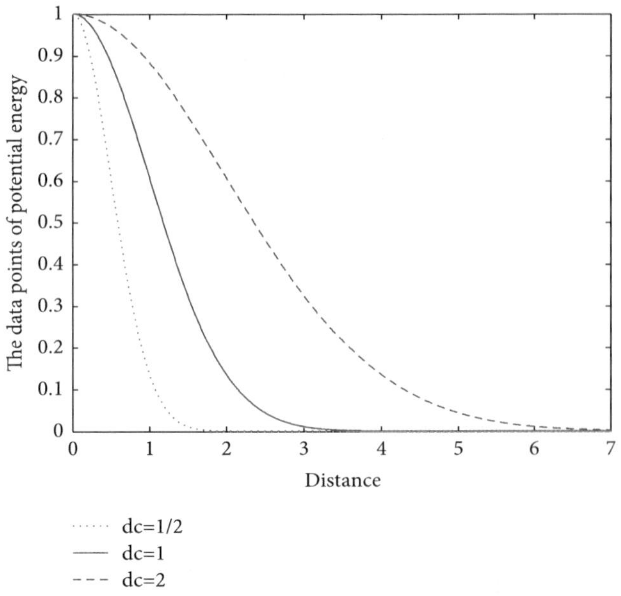

Figure 1: Gauss field potential function.

2.2. Distance between Data Points. The distance between data points reflects the degree of density between the data points and also the size of the data field between the data points. If the distance between data points is smaller, then the density of the data set is higher and the data field can be greater. By contrast, if the distance between data points is larger, then the density of the data set is low and the data field can be smaller. Thus, the data field energy of data points can be determined by the distance between data points. The distance d_{ij} between data points x_j and x_i is defined as follows:

$$d_{ij} = \|x_j - x_i\| = \sqrt{(x_j - x_i)^2} \tag{3}$$

Data point spacing matrix H is denoted as follows:

$$H = \begin{bmatrix} d_{11} & d_{12} & d_{13} & \cdots & d_{1n} \\ d_{21} & d_{22} & d_{23} & \cdots & d_{2n} \\ & & \vdots & & \\ d_{m1} & d_{m2} & d_{m3} & \cdots & d_{mn} \end{bmatrix} \tag{4}$$

In (4), $d_{mn} = d_{mn}, m \neq n$.

3. Carpool Taxi Clustering Algorithm

This section combines the aforementioned data field energy and the distance between data points to identify the clustering centres and cluster subsets of passenger taxi data by using GPS (Global Positioning System) trajectory data. In addition, the computing procedure is presented in detail.

We established passenger taxi GPS data set $D1$ at T_i time. $D1 = \{t_1, t_2, \cdots t_n\}$, where t_i is the i passenger taxi that can provide carpool. $t_i = \{long_{t_i}, lat_{t_i}, time_{t_i}\}$, where $long_{t_i}$ is the taxi longitude, lat_{t_i} is the taxi latitude, and $time_{t_i}$ is the taxi location time. $E = \{E(t_1), E(t_2), \cdots E(t_n)\}$, where set E represents the data points of potential energy and $E(t_1)$ represents the data points t_i of potential energy. These data points are sorted according to the size of potential energy [27–29]. $I_p = \{I_{p1}, I_{p2}, \cdots I_{pn}\}$, where set I_p is the ranking index of potential energy. $E_p = \{E_{I_{p1}}, E_{I_{p2}}, \cdots E_{I_{pn}}\}$, where set E_p is the ranking of potential energy. The distance u_i between different data points is defined as

$$u_i = \begin{cases} \min_{k \in I_p} \{d_{ik}\}, & E_{I_{pi}} < E_{I_{pk}} \quad I_{pk} \in I_p \\ \max_{k \in I_p} \{d_{ik}\}, & E_{I_{pi}} = \max \left\{ E_{I_{pk}} \; I_{pk} \in I_p \right\} \end{cases} \quad (5)$$

If $E_{I_{pi}} = \max\{E_{I_{pk}}\}_{I_{pk} \in I_p}$, then the potential energy of this data point t_{Ipi} will reach the maximum in the local data field, which means that u_i is the maximum distance between the cluster centre and the t_{Ipi} in local data set $D1$. If $E_{I_{pi}} < E_{I_{pk}}$, $I_{pk} \in I_p$ then in the local data field potential, t_{Ipi} is less than the maximum value of local data field, which means that u_i is the minimum distance between data points and t_{Ipi} in local data set $D1$.

Thus, we can obtain data pairs $(E(t_i), u_i), i \in I_t$. $w_i = E(t_i)u_i, i \in I_s$, where w_i is the discriminant value of the number of clusters in the dataset. According to formula (5), if t_i is the central point in a local data field, then $E(t_i)$ and u_i are the maximum value and the value of w_i is relatively large. If t_i is not the central point in a local data field, then the values of $E(t_i)$ and u_i are relatively small and the value of w_i is relatively small. According to the w_i value changes, the cluster centre number can be determined. $\{q_i\}_{i=1}^N$ represents a descending subscript order of $\{w_i\}_{i=1}^N$, and the number of clustering centre is defined as

$$g = \begin{cases} g+1 & w_{q_i} > w_{q_j}, \; q_j < q_i \\ 0 & w_{q_i} \approx w_{q_j}, \; q_j < q_i \end{cases} \quad (6)$$

where g is the number of clustering centres. If $w_{q_i} > w_{q_j}$, then a distinct jump is observed between w_{q_i} and w_{q_j}, which means that q_i is a data point clustering centre. If $w_{qi} \approx w_{qj}$, then no distinct jump is observed between w_{q_i} and w_{q_j}, which means that q_i is not a data point clustering centre. The threshold points between jump points and nonjumping points can be determined, and we can find the data potential values $E(t_g)$ and distance u_g corresponding to the threshold points.

The centre of clustering is determined by the number of cluster centres, and the subsets of each point are determined according to the distance from each point to the centre. A subset of C is defined as

$$C = \begin{cases} t_{ck} & u_i \le u_g, \; i \in I_t \\ t_{cli} & u_i > u_g, \; i \in I_t \end{cases} \quad (7)$$

where t_{ck} is a subset of the data t_{ci}, which indicates that the distance from the data point to the maximum point of the region data potential energy is less than the threshold u_g. t_{cli} is an outlier, which indicates that the distance from the data point to the maximum point of the region data potential energy is greater than the threshold u_g.

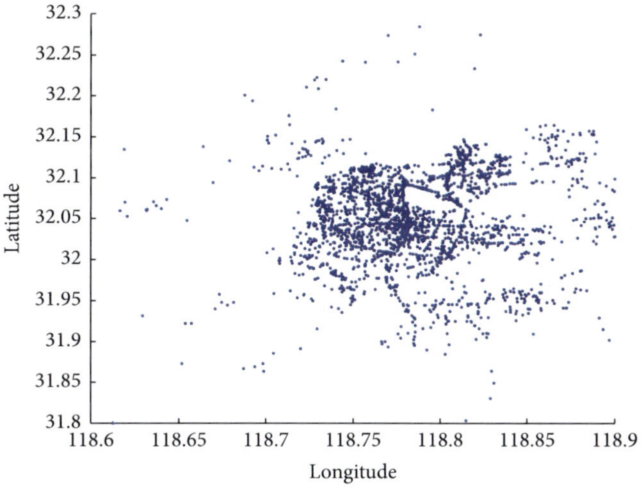

FIGURE 2: Passenger taxi to get-off point scatter diagram in Nanjing City between 9:00 and 9:10.

4. Case Study

4.1. System Description. This study uses a GPS history of Nanjing urban taxi data on September 1, 2014. We took the data on 3,391 passenger taxi drop-off points as experimental data from 9:00 to 9:10 and the taxi GPS trajectory data preprocessing to obtain the scatter distribution, as shown in Figure 2.

As shown in Figure 2, the main points of the passenger taxis in Nanjing are concentrated in the central area of the city at 9:00. If we move to the urban area, then fewer passengers get off the bus, thereby implying a lower aggregation rate. Moreover, as shown in Figure 2, although the get-off point distribution of the passenger taxi is out of order, the get-off point is the distribution of road networks in the city. Therefore, the dataset of the taxi get-off point has a certain spatial distribution, but it is disordered in the time distribution.

4.2. Experimental Process and Results. The horizontal axis is the data field energy and the vertical axis is the point spacing in Figure 3. As shown in the upper part of Figure 3, the longitudinal axis of the right data points in the data field to $E(t_i)$ and interval of u_i are large which shows the clustering centre data point of the local data subset. The potential energy $E(t_i)$ of the data point near the vertical axis is smaller, but the point spacing u_i is larger, which indicates that the point is the outlier in the data. As shown in the lower part of Figure 3, energy $E(t_i)$ and the point spacing u_i are less close to the horizontal axis, which shows the subset of data in the local clustering centres.

Figure 4 presents a descending w_i value chart of the Nanjing passenger taxi get-off point at 9:00–9:10. In the cluster centre above the threshold, w_i has a descending order from large to small. In the cluster centre below the threshold, the w_i value is almost equal. The threshold can be determined by the number of cluster centres and distance u_g. By calculation, u_g equals 1.116272.

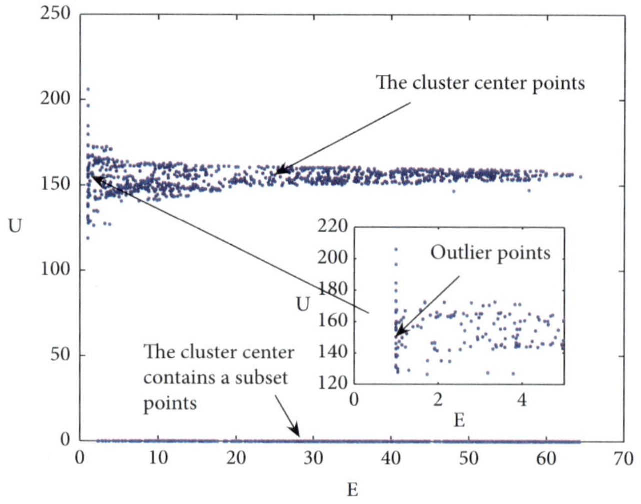

FIGURE 3: Data field energy and point spacing distribution diagram of passenger taxi to get-off point.

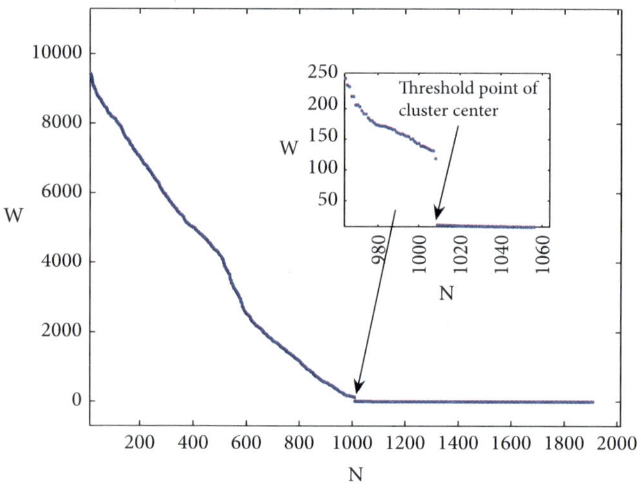

FIGURE 4: Schematic diagram of the w_i descending order.

As shown in Figure 5 and partial enlargement, the clustering algorithm proposed in this study can realize taxi clustering. The advantages of the algorithm are expressed in the following aspects:

(1) Based on the effectiveness of the algorithm, it can identify that the clustering centre is very good and, to determine the number of clusters, the clustering of arbitrary shape does not need to experience the selection of the number of cluster centres, which can be better applied to the trajectory data clustering.

(2) From the point of view of special treatment, the points are divided into a new cluster according to the number of edge points for concentrated or outliers, based on data field and distance discrimination.

(3) From the clustering result, the data clustering is basically distributed along the road network of the city to meet the requirements of cluster city taxi vehicles.

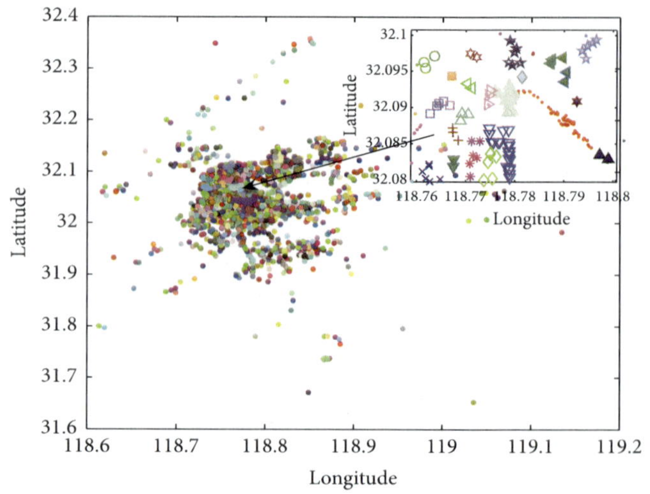

FIGURE 5: Clustering graph based on data field energy algorithm.

4.3. Algorithm Evaluation. We use classical clustering algorithm, which includes K-means algorithm, fuzzy C-means (FCM) clustering algorithm, and hierarchical clustering algorithm, and the proposed algorithm in this is used for comparative analysis. These algorithms are evaluated by compactness (CP), separation (SP), and Dunn validity index (DVI).

Figure 6 shows the clustering results of K-means algorithm, FCM algorithm, and hierarchical clustering algorithm. As shown in Figure 6(a), in the K-means algorithm, the number of initial centre points determines the number of clusters. For a class of clusters with a complex shape, it divides some edge points into near clusters. From the subgraph (a) of the partial enlargement, the K-means algorithm cannot accurately be clustered on the edge point. For subgraph (b), the clustering effect of the FCM algorithm is better than that of the K-means algorithm, but it is limited by the number of initial centres and the influence of isolated points. The FCM algorithm is similar to the K-means algorithm, but it cannot exactly cluster the edge points. From the subgraph, the hierarchical clustering algorithm is limited by the algorithm itself, and the clustering of the outlier data points is not accurate.

As shown in Table 1 and Figure 7 and on the basis of the K-means algorithm, FCM algorithm, hierarchical clustering algorithm, and data field energy clustering algorithm, the CP index will be reduced with the increase in the number of clusters. However, the CP index of the data field clustering algorithm is less than the CP index of other clustering algorithms. The proposed algorithm in this study is better than that of other clustering algorithms in the CP index. The SP index decreases with the increase of cluster numbers, which shows that the class spacing decreases as the number of clusters increases. However, the SP index is higher than that of other clustering algorithms, and the proposed algorithm in this study is better than other clustering algorithms in the SP index. Based on the overall clustering evaluation of the DIV index, the clustering algorithm in this study is better than other clustering algorithms in the clustering of GPS data in urban taxis.

Table 1: Comparison table of clustering effect evaluation index of each clustering algorithm.

Clustering number	Clustering algorithms	CP	SP	DVI
25	K-means	0.0392	0.3644	2.543×10^{-4}
	FCM	0.0201	0.1233	1.001×10^{-4}
	Hierarchical	0.0123	0.6069	2.717×10^{-4}
	Data field energy	0.0088	0.6666	1.32×10^{-2}
249	K-means	0.0047	0.1414	7.008×10^{-5}
	FCM	0.005	0.0597	2.749×10^{-5}
	Hierarchical	0.0042	0.181	2.724×10^{-5}
	Data field energy	0.0016	0.2319	5.152×10^{-4}
409	K-means	0.0034	0.1213	3.845×10^{-5}
	FCM	0.0033	0.0558	2.041×10^{-5}
	Hierarchical	0.0028	0.1398	2.1742×10^{-5}
	Data field energy	0.0013	0.1847	3.846×10^{-4}
599	K-means	0.0021	0.119	3.6076×10^{-5}
	FCM	0.0023	0.0569	2.3121×10^{-5}
	Hierarchical	0.0019	0.1242	1.086×10^{-5}
	Data field energy	0.001	0.1685	1.823×10^{-4}

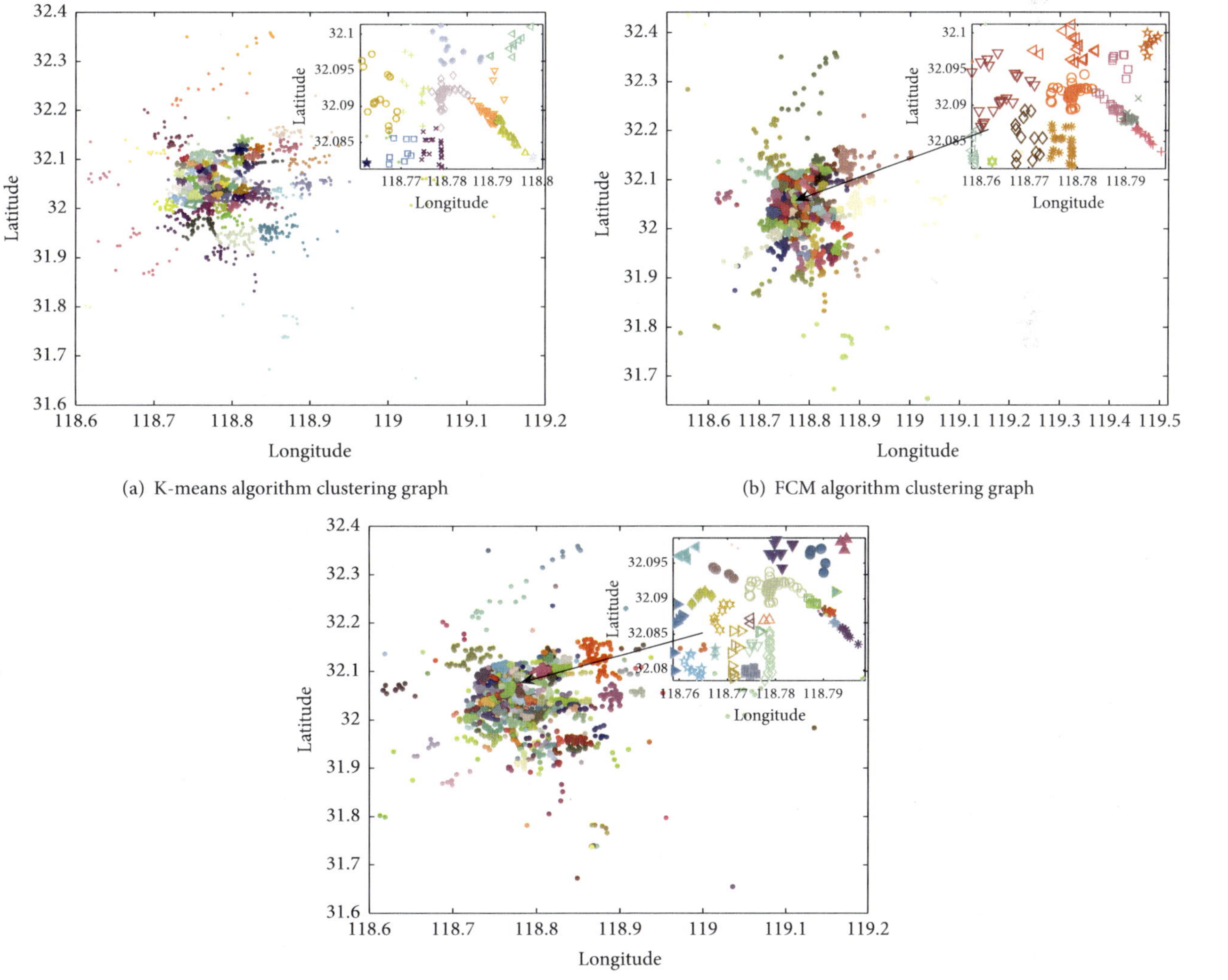

(a) K-means algorithm clustering graph

(b) FCM algorithm clustering graph

(c) Hierarchical clustering graph

Figure 6: Classical algorithm clustering graph.

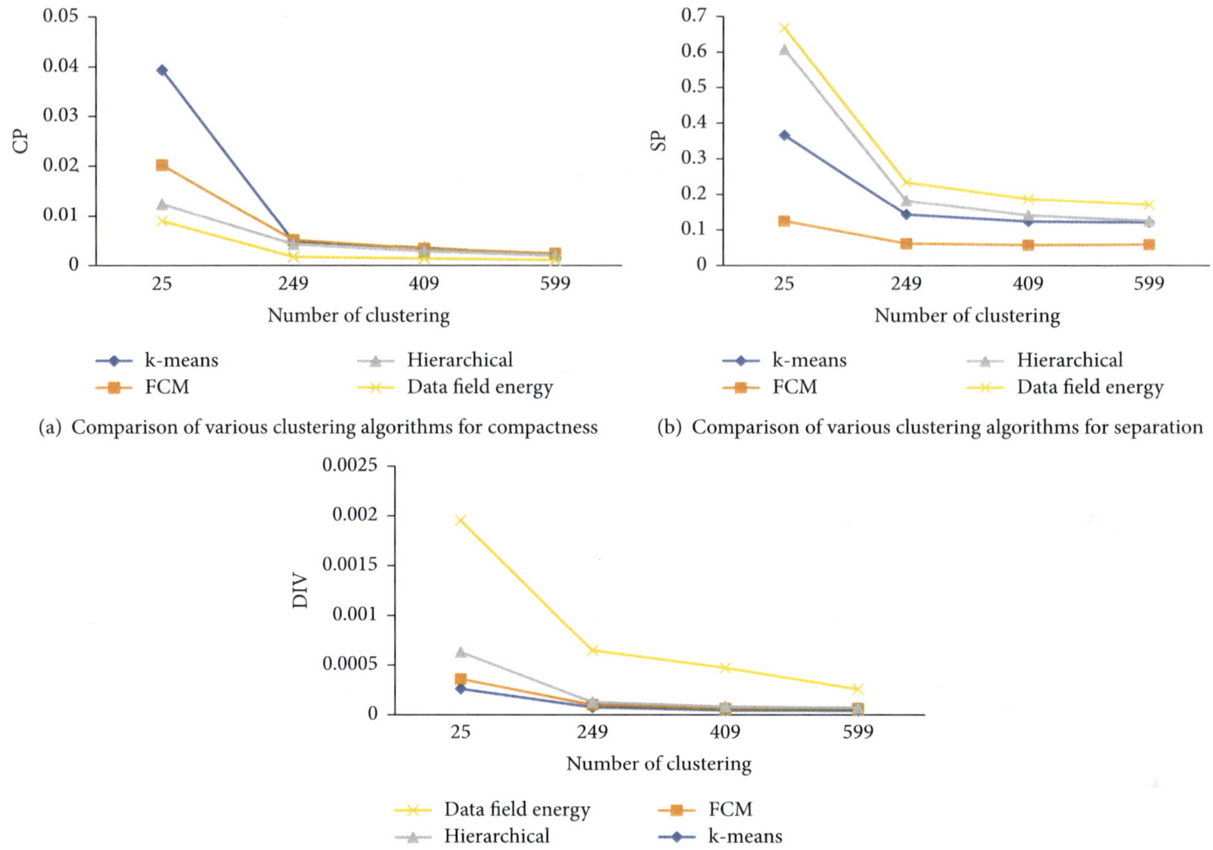

(a) Comparison of various clustering algorithms for compactness

(b) Comparison of various clustering algorithms for separation

(c) Comparison of various clustering algorithms for DIV

FIGURE 7: Comparison of various clustering algorithm evaluation index.

5. Conclusion

The main conclusions of this study are presented as follows:

(1) The use of field theory and city taxi clustering algorithm is established based on data field and space. The algorithm uses the dataset of taxi get-off points in Nanjing to achieve the clustering of taxis carpooling on urban roads.

(2) By comparing the classical clustering analysis, the proposed algorithm is found more suitable for the taxi data clustering, which has a certain spatial distribution shape. However, the time distribution of uncertainty provides new clustering methods and ideas for urban traffic data clustering.

(3) The clustering algorithm of urban taxi carpooling vehicles based on data field energy realized the taxi carpooling clustering. However, the taxi clustering data in this study are conducted in a static environment. In the future, we will discuss the clustering model and algorithm of taxi carpooling in a dynamic environment in combination with the dynamic change of the urban taxi get-off point.

Conflicts of Interest

The authors declare that there are no conflicts of interest regarding the publication of this paper.

Acknowledgments

This work has been supported by the National Natural Science Foundation of China (nos. 61364026 and 51408288)

References

[1] A. Afian, A. Odoni, and D. Rus, "Inferring Unmet Demand from Taxi Probe Data," in *Proceedings of the 18th IEEE International Conference on Intelligent Transportation Systems, ITSC 2015*, pp. 861–868, September 2015.

[2] W. He, K. Hwang, and D. Li, "Intelligent carpool routing for urban ridesharing by mining GPS trajectories," *IEEE Transactions on Intelligent Transportation Systems*, vol. 15, no. 5, pp. 2286–2296, 2014.

[3] S.-C. Huang, M.-K. Jiau, and C.-H. Lin, "A genetic-algorithm-based approach to solve carpool service problems in cloud computing," *IEEE Transactions on Intelligent Transportation Systems*, vol. 16, no. 1, pp. 352–364, 2015.

[4] S. Huang, M. Jiau, and Y. Liu, "An ant path-oriented carpooling allocation approach to optimize the carpool service problem with time windows," *IEEE Systems Journal*, vol. 99, pp. 1–12, 2018.

[5] G. Laporte, "Fifty years of vehicle routing," *Transportation Science*, vol. 43, no. 4, pp. 408–416, 2009.

[6] C. Lin, K. L. Choy, G. T. S. Ho, S. H. Chung, and H. Y. Lam, "Survey of green vehicle routing problem: past and future trends," *Expert Systems with Applications*, vol. 41, no. 4, pp. 1118–1138, 2014.

[7] G. Musolino, A. Polimeni, and A. Vitetta, "Freight vehicle routing with reliable link travel times: a method based on network fundamental diagram," *Transportation Letters-The International Journal of Transportation Research*, vol. 10, no. 3, pp. 159–171, 2018.

[8] N. A. H. Agatz, A. L. Erera, M. W. P. Savelsbergh, and X. Wang, "Dynamic ride-sharing: A simulation study in metro Atlanta," *Transportation Research Part B: Methodological*, vol. 45, no. 9, pp. 1450–1464, 2011.

[9] T. Shinde and B. Thombre, "An effective approach for solving carpool service problems using genetic algorithm approach in cloud computing," *International Journal of Advance Research in Computer Science and Management Studies*, vol. 3, no. 12, pp. 29–33, 2015.

[10] D. Pelzer, J. Xiao, D. Zehe, M. H. Lees, A. C. Knoll, and H. Aydt, "A partition-based match making algorithm for dynamic ridesharing," *IEEE Transactions on Intelligent Transportation Systems*, vol. 16, no. 5, pp. 1–12, 2015.

[11] M.-K. Jiau and S.-C. Huang, "Services-oriented computing using the compact genetic algorithm for solving the carpool services problem," *IEEE Transactions on Intelligent Transportation Systems*, vol. 16, no. 5, pp. 1–12, 2015.

[12] S.-C. Huang, M.-K. Jiau, and C.-H. Lin, "Optimization of the carpool service problem via a fuzzy-controlled genetic algorithm," *IEEE Transactions on Fuzzy Systems*, vol. 23, no. 5, pp. 1698–1712, 2015.

[13] S.-K. Chou, M.-K. Jiau, and S.-C. Huang, "Stochastic set-based particle swarm optimization based on local exploration for solving the carpool service problem," *IEEE Transactions on Cybernetics*, vol. 46, no. 8, pp. 1771–1783, 2016.

[14] J. Cheng, Z. Tang, J. Liu, and L. Zhong, "Research on dynamic taxipooling model based on genetic algorithm," *Journal of Wuhan University of Technology (Transportation Science and Engineering)*, vol. 37, no. 1, pp. 187–191, 2013.

[15] S. Ma, Y. Zheng, and O. Wolfson, "T-share: A large-scale dynamic taxi ridesharing service," in *Proceedings of the 29th International Conference on Data Engineering, ICDE 2013*, pp. 410–421, April 2013.

[16] Q. Xiao, R.-C. He, W. Zhang, and C.-X. Ma, "Algorithm research of taxi carpooling based on fuzzy clustering and fuzzy recognition," *Journal of Transportation Systems Engineering and Information Technology*, vol. 14, no. 5, pp. 119–125, 2014.

[17] S. Ma, Y. Zheng, and O. Wolfson, "Real-time city-scale taxi ridesharing," *IEEE Transactions on Knowledge and Data Engineering*, vol. 27, no. 7, pp. 1782–1795, 2015.

[18] D. Zhang, T. He, Y. Liu, S. Lin, and J. A. Stankovic, "A carpooling recommendation system for taxicab services," *IEEE Transactions on Emerging Topics in Computing*, vol. 2, no. 3, pp. 254–266, 2017.

[19] L. Häme, "An adaptive insertion algorithm for the single-vehicle dial-a-ride problem with narrow time windows," *European Journal of Operational Research*, vol. 209, no. 1, pp. 11–22, 2011.

[20] R. Manzini and A. Pareschi, "A decision-support system for the car pooling problem," *Journal of Transportation Technologies*, vol. 2, no. 2, pp. 85–101, 2012.

[21] Z. Z. Shao, H. G. Wang, H. Liu, and et al, "Research on service requirement distribution algorithms in carpooling problems," *Journal of Tsinghua University*, vol. 53, no. 2, pp. 252–258, 2013.

[22] Z. Z. Shao, H. G. Wang, H. Liu et al., "Single carpooling problem based on matching degree clustering algorithm," *Journal of Software*, vol. 23, supplemet 2, pp. 204–212, 2012.

[23] Z. Z. Shao, H. G. Wang, H. Liu, C. Song, C. Meng, and H. Yu, "Heuristic optimization algorithms of multi-carpooling problem based on two-stage clustering," *Journal of Computer Research and Development*, vol. 50, no. 11, pp. 2325–2335, 2013.

[24] Z. J. Yang, Z. Wang, Y. Wang et al., "Two-stage estimation of distribution algorithm to solve multi-vehicle carpooling problem," *Journal of Transportation Systems Engineering and Information Technology*, vol. 16, no. 2, pp. 164–169, 2016.

[25] W. Y. Gan, D. Y. Li, and J. M. Wang, "An hierarchical clustering method based on data fields," *Acta Electronica Sinica*, vol. 34, no. 2, pp. 258–262, 2006.

[26] N. Zhang, T. T. Chen, and X. X. He, "A single partition clustering algorithm based on data field," *Journal of Zhe Jiang University of Technology*, vol. 44, no. 1, pp. 52–57, 2016.

[27] A. Rodriguez and A. Laio, "Clustering by fast search and find of density peaks," *Science*, vol. 344, no. 6191, pp. 1492–1496, 2014.

[28] S. Wang, D. Wang, C. Li, Y. Li, and G. Ding, "Clustering by fast search and find of density peaks with data field," *Chinese Journal of Electronics*, vol. 25, no. 3, pp. 397–402, 2016.

[29] R. Mehmood, G. Zhang, R. Bie, H. Dawood, and H. Ahmad, "Clustering by fast search and find of density peaks via heat diffusion," *Neurocomputing*, vol. 208, no. 6191, pp. 210–217, 2016.

5

Estimation of Urban Link Travel Time Distribution using Markov Chains and Bayesian Approaches

Wenwen Qin and Meiping Yun

Key Laboratory of Road and Traffic Engineering of the Ministry of Education, Tongji University, 4800 Cao'an Highway, Shanghai 201804, China

Correspondence should be addressed to Meiping Yun; yunmp@tongji.edu.cn

Academic Editor: Sara Moridpour

Despite the wide application of Floating Car Data (FCD) in urban link travel time and congestion estimation, the sparsity of observations from a low penetration rate of GPS-equipped floating cars make it difficult to estimate travel time distribution (TTD), especially when the travel times may have multimodal distributions that are associated with the underlying traffic states. In this case, the study develops a Bayesian approach based on particle filter framework for link TTD estimation using real-time and historical travel time observations from FCD. First, link travel times are classified by different traffic states according to the levels of vehicle delays. Then, a state-transition function is represented as a Transition Probability Matrix of the Markov chain between upstream and current links with historical observations. Using the state-transition function, an importance distribution is constructed as the summation of historical link TTDs conditional on states weighted by the current link state probabilities. Further, a sampling strategy is developed to address the sparsity problem of observations by selecting the particles with larger weights in terms of the importance distribution and a Gaussian likelihood function. Finally, the current link TTD can be reconstructed by a generic Markov Chain Monte Carlo algorithm incorporating high weighted particles. The proposed approach is evaluated using real-world FCD. The results indicate that the proposed approach provides good accurate estimations, which are very close to the empirical distributions. In addition, the approach with different percentage of floating cars is tested. The results are encouraging, even when multimodal distributions and very few or no observations exist.

1. Introduction

Information of travel time distribution (TTD) is essential for route guidance and advanced traveler information systems. It reveals underlying traffic conditions for measuring travel time reliability, evaluating the level of service of intersection, and detecting the congestion and incident, which can be introduced to improve service reliability performance [1].

Up to now, the increasingly available vehicles equipped with GPS, named as floating cars, provide a significant amount of vehicle trajectory data in urban road networks. Many works have proposed analytical or data-driven models to infer mean link travel time from this data, but yet research on the estimation of link TTD is still evolving, especially when the travel times appear to be multistate features. In reality, the travel times of vehicles traversing an urban link are heavily affected by traffic lights, interaction among vehicles, and conflicting traffic from cross streets [1–5]. The characteristics of interrupted traffic flow would likely result in different clusters of travel times, which reflect how vehicles experienced traffic states on a link. Therefore, rather than to provide the mean value that is generally used in planning applications, a probability distribution, that fully characterizes different components of travel times for retaining maximum traffic state information, is preferred.

In order to estimate the link TTD, a relatively large number of floating cars is needed to generate useful travel times among various errors such as GPS errors, map-matching errors, and estimation errors. Unfortunately, the spatial distribution of floating cars is not uniform. Taxicabs, which have been suggested as the main types of floating cars [6], are used widely to collect travel time information for urban link travel time and congestion estimation [4, 5, 7]. However, taxicabs cover mainly densely populated and commercially well-developed areas, since these areas may have higher passenger flow and taxicabs are more likely to pick up passengers. For those areas with limited coverage by floating cars, it would raise inquiries about how to utilize the limited data sample from a small number of floating cars to produce an accurate link TTD, even for the multistate feature of travel times.

In this situation, it may be possible to combine real-time traffic information incorporating the current traffic states with prior knowledge from historical link TTDs that provide similar data sequences to the actual distributions to reconstruct the TTD. The objective of this study focuses on developing a methodology to estimate such probability distributions under situation when floating car density is quite low and link travel times may follow multimodal distributions.

The main contributions of this paper are as follows. (1) This research can be regarded as an initial attempt to estimate multistate link TTDs in the context of the sparse and incomplete observations from real-time FCD. The corresponding solutions are given by a Bayesian approach based on particle filter framework, in which the state-transition trends are modeled by historical trends to construct an initial solution providing past travel time patterns, as well as the sampling strategy that focuses on a fundamental issue of the irregular, unsmooth TTDs directly constructed by the only real-time FCD and allows much less real-time information. (2) The results of case study show that multistate link TTD models can be robustly realized by using our proposed approach for different percentage of floating cars, even for the condition with no data available. Moreover, the applications help provide in-depth understanding on the heterogeneity of link travel times that show multistate distributions and imply significantly different state combinations changing with time of day, in which the state converging most of travel times may play a dominant role on the whole link traffic conditions.

2. Literature Review

In review of literature, models on the estimation of travel time distribution (TTD) can be generally divided into two categories, namely, unimodal and multimodal/multistate distributions. Unimodal distributions, such as Weibull, exponential, Lognormal, or normal distributions, are often used for modeling travel time. For example, Hunter et al. [8] proposed a probabilistic model for inferring vehicle path and travel time based on low frequency GPS data in the arterial network. An expectation maximization (EM) algorithm for the proposed model is employed to simultaneously learn the likely paths and the parameters of lognormal link travel time distributions using the path travel time observations.

A recent work of the same group by Herring et al. [9] proposed a statistical approach to capture the evolution of traffic states with a coupled hidden Markov model, since the actual state of current link is unobserved (hidden) and depends on adjacent links (coupled). The authors assumed that travel times between successive links are independent, and only one traffic state on each link can exist in the same period. The authors went on to assume that the link travel time distribution is conditional on its hidden state. Likewise, an EM algorithm is used to iteratively update the parameters of the distributions and the state-transition matrix. The evaluation is done using sparse probe data from taxicabs in San Francisco, CA.

A similar approach was proposed by Westgate et al. [10]. In this work, a Markov Chain Monte Carlo (MCMC) algorithm was used to simultaneously estimate the likely paths, travel times, and parameters of link TTDs from GPS-equipped ambulances. The findings from these studies in general concluded that the lognormal distribution provides a superior fit over the alternative unimodal distributions. The main limitation of unimodal distributions is the assumption that travel time is the result of a single traffic state on a link. Hence unimodal distributions usually provide some poor fit results under complex traffic conditions, such as interrupted flow.

The multimodal travel time distribution comprising two or more distributions outperforms the single distribution by reflecting the coexistence of heterogeneous traffic states simultaneously [11, 12]. One of the most attractive features of the multimodal distribution provides a straightforward relationship for the probability of encountering state (e.g., congestion) and the component distribution accordingly under such state. In [11], a multistate model was proposed by using a mixture of normal distributions to fit travel times on freeways. The authors revealed that different traffic states can exist in the same analysis period such as free-flow and congested states. Besides each component distribution is associated with an underlying state. In the research by Park et al. [12], the authors attempted to quantify the impact of traffic incidents on travel time reliability using multistate (3 states) model. The state of a section on freeways describes the level of congestion. They concluded that the congested state corresponding to the third component distribution becomes more dominant since incidents increase the travel time variability.

While very few multimodal link TTD models using GPS data were proposed up to now, for the estimation of the parameters of multimodal distribution, Hofleitner et al. [13] developed a dynamic Bayesian network for the transition between link states using GPS probe data. A simulation-based EM algorithm is proposed to estimate the transition probabilities and the travel time distribution parameters. Feng et al. [7] integrated signal timing and arterial geometry information with real-time GPS data to dynamically update the parameters of the bimodal urban link TTDs based on MCMC simulation. Ji et al. [1] proposed a hierarchical Bayesian bimodal TTD to capture the interrupted nature of urban traffic flow as well as analyze traffic operations from probe vehicles. A further development using a hierarchical MCMC algorithm for producing the posterior distribution of Gaussian mixture models and the lognormal prior was presented in Li et al. [14]. The authors assumed that real-time GPS samples are distributed in each component distribution. These studies all adapted iterative procedures to estimate the parameters of link TTDs using maximum likelihood estimation and Bayesian approach. However, one limitation of their work is that travel times of probe vehicles are not readily available for each unobserved state since probe vehicles penetration rate is fairly low. Another limitation of their work is that they assumed the link travel times are independent. In reality, adjacent link travel times may have strong correlations. As an alternative, the multistate nature of travel times and their correlations between adjacent links can be modeled by a Markov chain methodology. Ramezani and Geroliminis [15] proposed a Markov model to estimate arterial route TTD. The model used a Transition Probability Matrix (TPM) constructed from simulated probe data to capture the spatial correlations. However, the approach conducted by Ramezani and Geroliminis [15] needs ample real-time observations conditional on link states to estimate the TPMs, which limits the application under urban areas with missing data (no or few data available).

In this paper, TPMs of the Markov chain between upstream and current links are calibrated from historical travel time observations to alleviate the dependence on the limited data sample. On this basis, the study develops a Bayesian approach based on particle filter framework for estimation of multistate link TTD by combining past information of link TTD represented as an importance distribution with real-time travel time information represented by a likelihood function. The importance distribution, used as an initial solution for approximating the actual distribution, is constructed by the summation of past TTDs conditional on link states weighted by the estimated state probabilities. The likelihood function is chosen as a normal distribution that describes how likely a particle (i.e., sample) tends to follow the real-time observations. As a result, the particles with high similarity are retained and replicated to address the sparsity problem of observations and then can be integrated into a Bayesian algorithm for producing a multistate link TTD.

3. Data Collection and Characteristics

3.1. Data Introduction. Nanjing is the capital of Jiangsu province, lying in the east of China, with over 8,236,000 inhabitants. The city has been rather well equipped with ITS infrastructures for nearly 10 years. Radio Frequency Identification Data (RFID) technology is applied in some main arterials. Currently, more than 0.7 million vehicles, including taxicabs and private vehicles, have been equipped with RFID tags. There are over 7700 GPS-equipped taxicabs to collect vehicle trajectory data as Floating Car Data (FCD). In this paper, FCD are used to evaluate the performance of the proposed approach, as well as RFID being chosen to produce the ground truth distributions compared to the corresponding estimated TTDs from FCD.

FCD vehicles generate a data record every 30 seconds including the vehicle's position, timestamp, speed, and status (with or without passengers). The process of matching the reported locations to road network links from raw data records has been accomplished by a widely used map-matching approach [16]. Since this is not the focus of this paper, we will not cover the detail here. Besides, the impact of different data polling frequency on the required sample size determination is not considered in this paper. In order to eliminate the effects from vacant taxi driving intentionally slowly, and searching for potential passengers, only FCD from hired taxis are used. As to RFID, several arterials are installed with both RFID sensors and video cameras at upstream and downstream locations of signalized intersections. RFID sensors scan the vehicle plate with high frequency of 20 milliseconds and mainly record plate number, timestamp, and station number when tagged vehicles pass by the RFID stations. More than 80% of vehicles on arterials are detected by RFID sensors [3–5]. Hence RFID can almost obtain the information of all the passing vehicles which constitutes a full-sample data.

As shown in Figure 1, the case road link divided by two consecutive RFID stations (i.e., the stations between 6004 and 6006) is around 1054-meter long. The road link has eight-lanes double way from Caochangmen Street and Qingliangmen Street all along and intersects with many other streets in between. The design speed of the selected studied link is 60 km/h.

For the source of FCD, the 1-month (November 2011) data for weekdays (Monday to Friday) between 7 a.m. and 8 p.m. were taken as the historical database. For both sources (FCD and RFID), data from 7 a.m. to 8 p.m. on Dec. 1, 2011 (Thursday), were used for the comparison and validation. Considering the word limit, we choose the south bound traffic flow to analyze the travel time.

3.2. Characterization of Travel Times. In order to estimate link travel time for individual floating car traversing a link, a travel time allocation method proposed by Sanaullah et al. [17] is applied here because of the simple calculation and high estimation accuracy. Then, mean link travel time can be calculated with an average of travel times for individual floating cars in an analysis interval. Note that an observation from FCD (RFID) is treated as the travel time in each link experienced by a vehicle. Besides, the time interval for estimation is set as 15 min to avoid unavailable observations.

To better understand the differences between the mean value and the probability distribution for explaining the trends or patterns of traffic states under interrupted flow, Figure 2 shows the temporal variation of travel times and the number of vehicles counted from both FCD and RFID in 15-min intervals. Apparently, the development of the mean travel times from FCD and RFID is quite consistent. The mean absolute percentage error (MAPE) is chosen for evaluating the accuracy of the estimated travel time. It turns out that the value of average MAPE is 8.46%, which appears to be quite acceptable.

However, Figure 2(a) shows an intuitive understanding on travel times, which can be clustered into two or more components from the scattered observations in a certain interval. This finding implies that the mean is not a good tracking index for capturing the multistate features of link travel times, once compared to TTD. On the other hand, Figure 2(b) illustrates that the proportions of floating cars are relatively low during time of day and that observations from FCD largely distribute at the high-speed regime (nearly between 25 and 50 km/h). These may likely result in inadequate sample to monitor the remaining travel time states. To

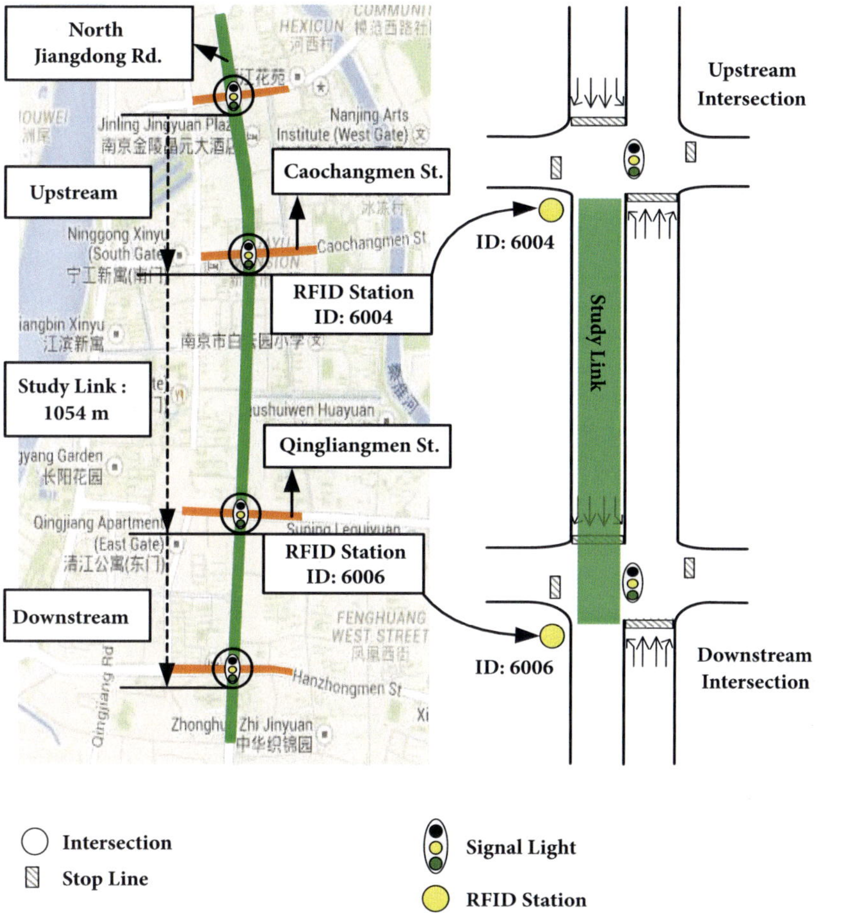

FIGURE 1: Layout of the study link selected in Nanjing, Jiangsu province, China.

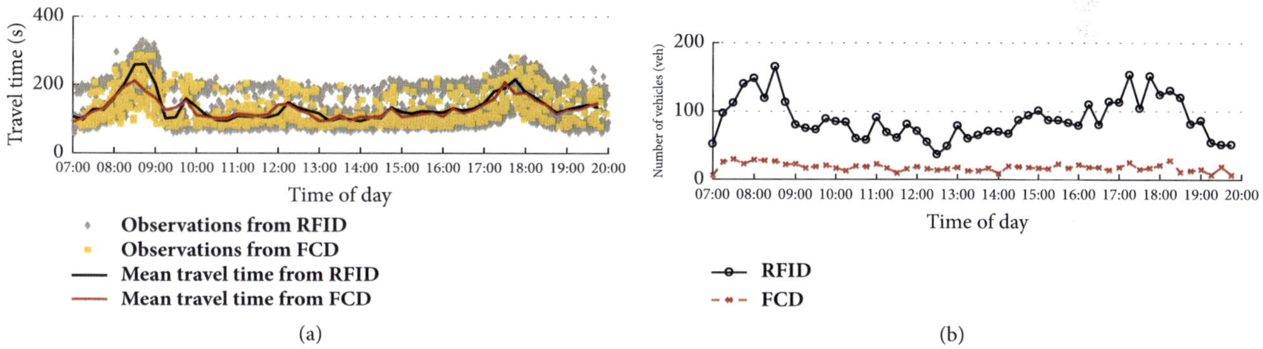

FIGURE 2: Variability by a 15-min interval: (a) mean travel times and (b) number of vehicles.

better model multistate TTD, a Bayesian approach based on particle filter framework is then introduced with limited FCD sample.

4. Methodology

The proposed solution framework, consisting of five major parts, is outlined in Figure 3. The state identification is presented first in this section, followed by a description of how the states transfer between two adjacent links. After that, the importance distribution is introduced to produce historical travel time patterns as prior knowledge. The following subsection presents the sampling strategy to generate a certain number of high weighted particles. Finally, the Gibbs sampling algorithm is applied to approximate the posterior distribution of Gaussian mixture model (GMM).

4.1. State Identification. The multistate nature of travel times on urban links is principally attributed to the road geometric features (e.g., roadway alignment, link length, lane width, and number of lanes), interaction among vehicles (e.g., weaving, diverging, and merging), and signal controls. With the results

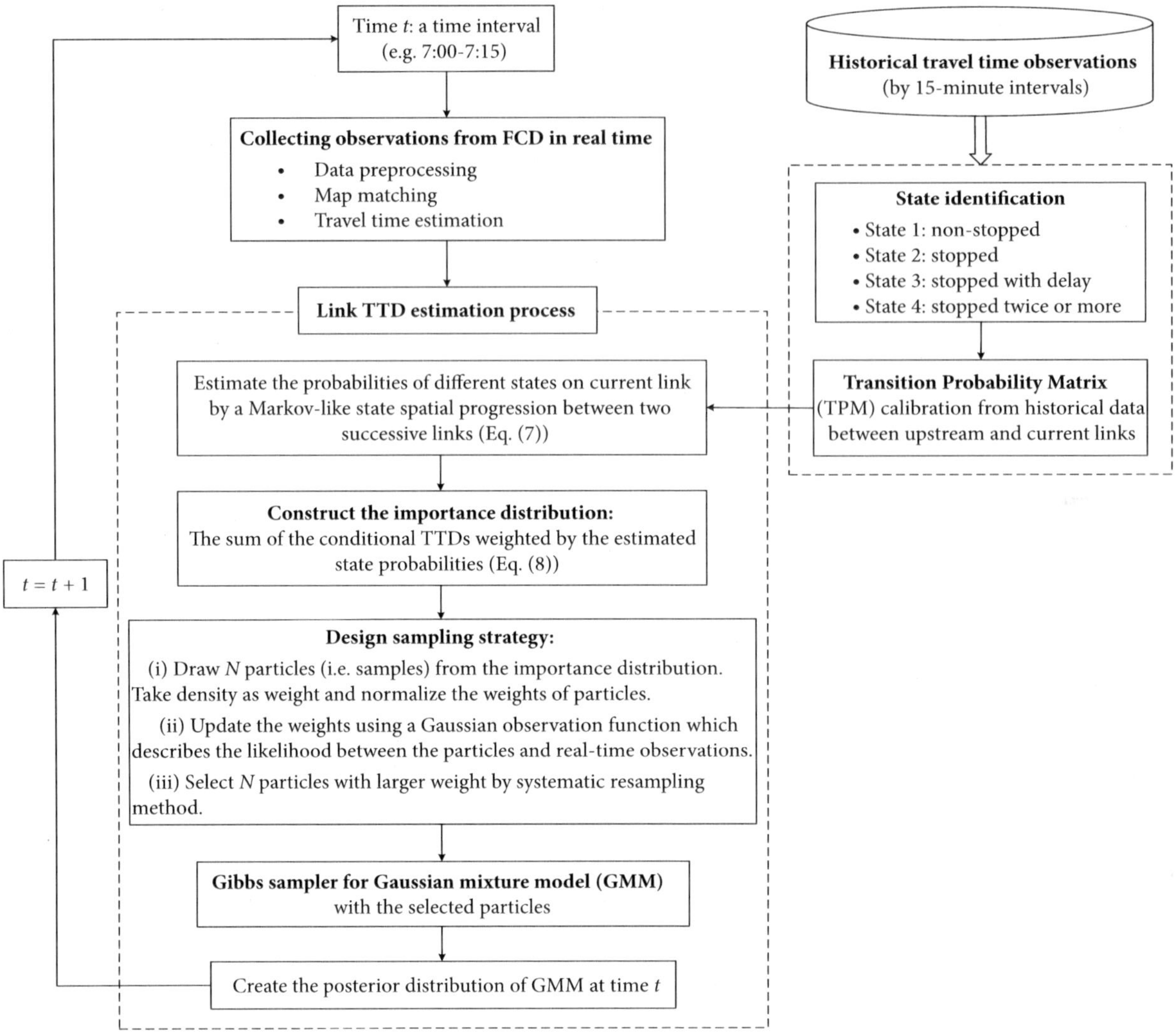

FIGURE 3: Link travel time distribution estimation framework. This figure is reproduced from a shorter version of this work presented at the 97th Annual Meeting of the Transportation Research Board [5].

of the interaction among these factors, four typical traffic states that exhibit travel times on a link experienced by vehicles are shown in Figure 4, and can be defined as follows:

(i) State 1 (nonstopped): vehicles traverse the whole link without stopping.

(ii) State 2 (stopped): vehicles stop at the end of the link for a red light.

(iii) State 3 (stopped with delay): vehicles traverse the whole link with delays caused by the interference among vehicles and signal control.

(iv) State 4 (stopped twice or more): vehicles have to stop twice or more due to the overflow queue at the signalized intersection.

In order to distinguish the heterogeneous traffic states, intuitive traffic engineering rules are applied to quantify state boundaries so each state represents the corresponding cluster of travel times. Here the boundaries between two neighbor states can be approximated as follows, respectively.

(i) The first boundary between state 1 and state 2 is to judge whether vehicles pass the signalized link without stopping.

Considering a free-flow traffic condition, the travel times of vehicles traversing a signalized link directly depend on the signal controls and their distribution appears to be bimodal [1–5, 7], which features two clusters of travel times corresponding to the states of nonstopped and stopped vehicles, respectively. To achieve the state partitions, a two-component mixed normal distribution, rather than the skewed mixture models (e.g., the components are both lognormal distributions), is adopted to obtain the travel times in state 1 and state 2 individually as a tradeoff between computation complexity and explanatory power [18]. Besides, the travel times between 02:00 a.m. and 04:00 a.m. are extracted from FCD for the free-flow condition [19]. Hence, the first travel time boundary (denoted as TT_f), determined by the point of intersection of two normal distributions [20], is stated mathematically as

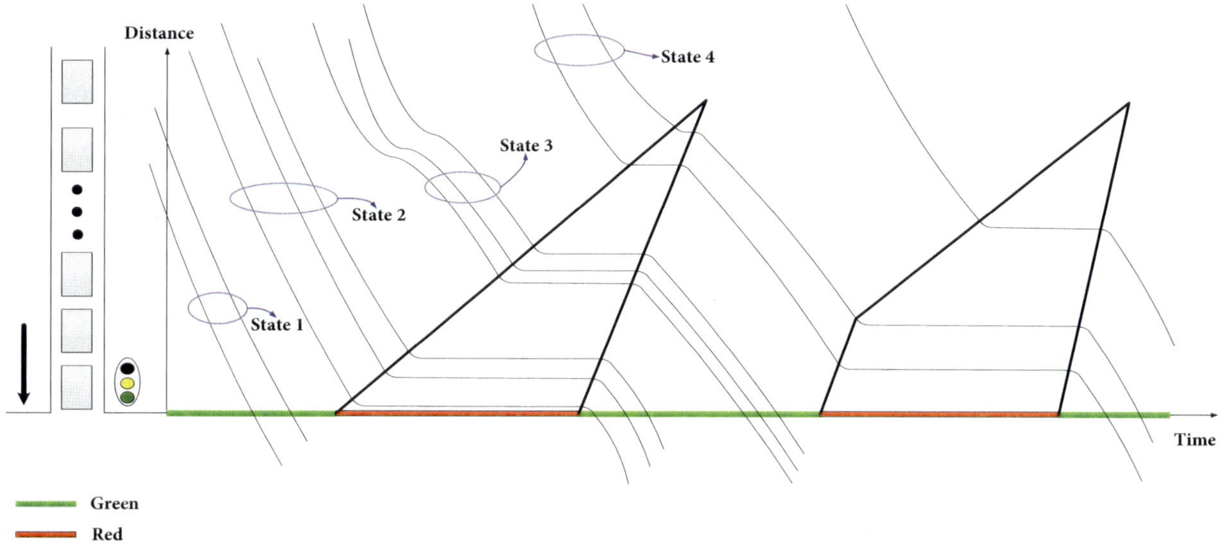

FIGURE 4: Four typical travel time states on urban signalized links.

$$TT_f = \mathrm{N}\left(TT \mid \mu_{f,1}, \sigma_{f,1}\right) = \mathrm{N}\left(TT \mid \mu_{f,2}, \sigma_{f,2}\right), \quad \mu_{f,1} < TT_f < \mu_{f,2} \tag{1}$$

where N(∗) is the normal distribution, TT is travel time variable, $\mu_{f,1}$ and $\mu_{f,2}$ are the mean of the first and second normal distribution, and $\sigma_{f,1}$ and $\sigma_{f,2}$ are the standard deviation of the first and second normal distribution.

(ii) The second boundary between state 2 and state 3 is to distinguish whether vehicles only experience intersection signal delay.

Signal delay is the primary factor that influences travel time patterns on an urban link [3]. Unfortunately, in most cases, the signal control plans at intersections are generally unknown. As an alternative, we use the red time for through movement (denoted as TT_{red}) as additional travel times to judge whether vehicles stop at the end of the link. This part of travel times is assumed to be only generated by the signal control, since it makes a major contribution to travel delay rather than traffic flow. Note that the means of $\mu_{f,1}$ and $\mu_{f,2}$ for state 1 and state 2 are closely related to the signal settings, and their difference between nonstopped and stopped vehicles can be used to approximately estimate the red time. Consequently, the second boundary by adding the red time to free-flow travel time (denoted as TT_{stop}) is represented as

$$TT_{stop} = TT_f + TT_{red} = TT_f + \mu_{f,2} - \mu_{f,1} \tag{2}$$

(iii) The last boundary is to judge whether vehicles experience oversaturated conditions.

The oversaturated conditions on urban principal arterials, characterized as extremely low speeds and extensive queuing, is defended as level of service F [21]. In this case, a residual queue is the most likely occurring at the signalized intersection in one cycle time. Hence, the last boundary for oversaturated conditions (denoted as TT_{OC}) where vehicles encounter two or more queues can be initially estimated by link length and 30% of speed limit:

$$TT_{oc} = \frac{L}{0.3 \times V_{limit}} \tag{3}$$

where L is the link length and V_{limit} is the speed limit of the arterial link.

4.2. State-Transition Function Modeling. This section adopts work of Ramezani and Geroliminis [15] to represent the spatial state transition between two adjacent links based on Markov chain framework. The spatial state transition exhibits how vehicles experience multiple traffic states on route's successive links whereas it differs from the traffic congestion propagation. For example, a vehicle that passed the upstream link without stopping would have a certain probability of stopping on the downstream link, and it would have another probability of not stopping on the downstream link. This simple case illustrates that the travel times on the current link are dependent on the upstream link state.

In this paper, there are four traffic states constituting a four by four TPM. The TPM $P_{n-1,n}(t)$ between any pair of successive links n-1 and n at time t is shown as follows:

$$P_{n-1,n}(t) = \begin{bmatrix} p_{11}(t) & \cdots & p_{14}(t) \\ \vdots & \ddots & \vdots \\ p_{41}(t) & \cdots & p_{44}(t) \end{bmatrix} \tag{4}$$

$$\sum_{j=1}^{4} p_{ij}(t) = 1 \quad \text{for } i = 1, \dots, 4, \; 0 \le p_{ij}(t) \le 1,$$

where $p_{ij}(t)$ denotes the transition probability from state i on link n-1 to state j on link n at time t and can be expressed mathematically as the corresponding observed frequencies:

$$p_{ij}(t) = Prob\{S_n(t) = j \mid S_{n-1}(t) = i\} = \frac{X_{ij}(t)}{\sum_{k=1}^{4} X_{ik}(t)} \quad (5)$$

where $S_n(t) = j$ indicates that the state on link n at time t is j and $X_{ij}(t)$ is the number of observations produced by the same vehicles experiencing state i on link n-1 and state j on link n at time t. Note that we adapt a relatively static TPM to relax the constraint of limited real-time observations and use the historical data during a 1-month period on workdays to calibrate the TPM at each time t from 7 a.m. to 8 p.m.

Given a set of real-time observations on link n-1 at time t, the state probability vector $\pi_{n-1}(t)$, consisting of the probabilities of four states, can be estimated by

$$\begin{aligned}\pi_{n-1}(t) &= \left[\pi_{n-1,t}^1 \ \pi_{n-1,t}^2 \ \cdots \ \pi_{n-1,t}^4\right] \\ &= \left[\frac{X_1(t)}{\sum_{k=1}^4 X_k(t)} \ \frac{X_2(t)}{\sum_{k=1}^4 X_k(t)} \ \cdots \ \frac{X_4(t)}{\sum_{k=1}^4 X_k(t)}\right]\end{aligned} \quad (6)$$

where $\pi_{n-1,t}^1$ is the probability of state 1 on link n-1 at time t and $X_k(t)$ is the number of observations of state k on link n-1 at time t. Finally, combined with (4)~(6), the current state probabilities $\pi_n(t)$ on link n at time t are calculated as

$$\pi_n(t) = \left[\pi_{n,t}^1 \ \pi_{n,t}^2 \ \cdots \ \pi_{n,t}^4\right] = \pi_{n-1}(t) P_{n-1,n}(t) \quad (7)$$

4.3. Importance Distribution Construction. A straightforward approach for estimation of link TTD is to take advantage of only real-time observations from FCD. However, under many circumstances, it is very difficult to represent the potentially multiple states with an incomplete collection of observations due to the low penetration rate of floating cars. Consequently, the chief challenge in our methodology is to construct an importance distribution (also often referred to as the proposal distribution) properly providing a high similarity to the posterior distribution, from which a series of samples with weights are randomly generated for the heterogeneous traffic states. More importantly, the importance distribution can be modified with the real-time observations to further approximate the posterior distribution.

Here we consider an importance distribution that consists of four historical normal distributions. And each historical normal distribution for the corresponding traffic state is fitted by historical observations located in the same state boundaries at time t. Combining with the estimated state probability vector $\pi_n(t)$, the importance distribution $p(x_{n,t} \mid x_{n-1,t})$ on link n incorporating the observation $x_{n-1,t}$ on link n-1 at time t can be adjusted as

$$p(x_{n,t} \mid x_{n-1,t}) = \sum_{i=1}^{4} \pi_{n,t}^i \mathrm{N}\left(x_{n,t} \mid \mu_{n,t}^i, \sigma_{n,t}^i\right) \quad (8)$$

where $\mu_{n,t}^i$ is the mean of historical normal distribution for state i on link n at time t and $\sigma_{n,t}^i$ is the standard deviation of historical normal distribution for state i on link n at time t.

4.4. Sampling Strategy Design. In this section, a sampling strategy is developed from the concept of the particle filter approach to generate a set of high weighted particles for representing the heterogeneous traffic states. Each particle refers to a travel time sample with a corresponding weight. Here a general particle filter approach mainly consists of the state update process showing a system evolving with time and the measurement update process showing the relationship between the system states and the system outputs at a certain time. Finally, a posterior distribution can be represented by a collection of weighted particles using the particle filter approach. More details about this approach are given in the work of Doucet et al. [22].

Similar to the particle filter approach, the proposed sampling strategy comprises three components, that is, state update, measurement update, and resampling scheme. First, the importance distribution based on the spatial TPMs is applied to produce the particles instead of using a temporal state-transition model in the state update process. Then, with the arrival of new observations, probability densities as weights are updated by the likelihood function with input of the particles and real-time observations. Note that, in Figure 2(a), each observation from FCD has an error distributed around the actual mean from RFID. More importantly, the means from FCD and RFID are quite close. If we take the means from FCD as ground truth, then it is reasonable to assume that errors between particles and the mean of observations in 15-minute interval can be described as a normal distribution. In this way, the smaller error between a particle and the mean achieves the larger weight assigned to the corresponding particle. Hence, the likelihood function is represented by a zero-mean normal distribution with a standard deviation of 50 seconds based on empirical evaluation (i.e., N(0, 50)). Finally, a resampling procedure is used to eliminate particles with low weights and concentrate on particles with large weights. Residual, stratified, and systematic resampling are three common resampling schemes. In this paper, systematic resampling is preferred to be easily implemented since it aims to minimize the Monte Carlo variation and save computational time.

To summarize, the sampling strategy is specified as follows.

Step 1 (state update). Set the number of particles $N = 100$ and generate N particles $y_{n,t}^i$ with the weights $w_{n,t}^i$ (the weight refers to the probability density of the random sample) according the importance distribution $p(x_{n,t} \mid x_{n-1,t})$ on link n at time t:

$$y_{n,t}^i \sim p(x_{n,t} \mid x_{n-1,t}), \quad i \in [1, N] \quad (9)$$

Step 2 (measurement update).

(2.1) Calculate the mean of M observations $x_{n,t}^j$ from FCD on link n at time t:

$$z_{mean}^t = \frac{\sum_j^M x_{n,t}^j}{M}, \quad j \in [1, M] \quad (10)$$

(2.2) Update the weights by likelihood function p_z:

$$w_{n,t}^{i} = w_{n,t}^{i} p_z \left(y_{n,t}^{i} - z_{mean}^{t} \right) \quad (11)$$

Step 3 (resampling). Normalize the weights to ensure that the sum of weights is equal to one. Thereafter, the systematic resampling method, similar to the roulette wheel strategy in a basic genetic algorithm, is applied to retain and replicate the particles with larger weights. For more details of the method the reader can refer to [23].

4.5. Travel Time Distribution Estimation. As mentioned in the literature review, travel times on urban links may have multimodal property which are associated with the heterogeneous traffic states. Hence, Gaussian mixture model (GMM), from which the flexible fitting structure and explanatory power can be provided, is preferred to fit multistate link TTD.

Specifically, a GMM for travel time with finite K components can be formulated as follows:

$$p(x \mid \Theta) = \sum_{k=1}^{K} \pi_k \mathrm{N}\left(x \mid \mu_k, \sigma_k^2\right) \quad (12)$$

where π_k is the mixture coefficient for the kth normal distribution such that $\pi_k \geq 0$ and $\sum_{k=1}^{K} \pi_k = 1$. μ_k and σ_k^2 are the mean and variance for the kth normal distribution, respectively. $\Theta = \{(\pi_1, \pi_2, \ldots, \pi_K), (\mu_1, \mu_2, \ldots, \mu_K), (\sigma_1^2, \sigma_2^2, \ldots, \sigma_K^2)\}$ is the set of GMM parameters. $x = (x_1, x_2, \ldots, x_N)$ is a vector of observations. In this paper, the number of components in GMM ranges from $K = 2$ to $K = 4$ in terms of four states in total.

To estimate the GMM parameters, EM algorithm and MCMC simulation are two commonly used methods. However, the EM algorithm based on the asymptotic theory needs samples, and it may result in a local maximum problem [24]. Unlike the EM algorithm, the MCMC simulation as a golden method for Bayesian inference can provide a posterior distribution of the parameters in the proposed model, which incorporates prior knowledge with smaller number of sample size in an iterative sampling procedure [25]. Such a sampling process would repeat many times until the estimated parameters tend to converge and produces the posterior distribution.

In Bayesian inference, the posterior distribution $p(\Theta \mid x)$ for GMM can be expressed in terms of the prior distribution $f(\Theta)$ and likelihood function $L_x(x \mid \Theta)$.

$$p(\Theta \mid x) = \frac{L_x(x \mid \Theta) f(\Theta)}{f(x)} \propto L_x(x \mid \Theta) f(\Theta) \quad (13)$$

$$L_x(x \mid \Theta) = \prod_{i=1}^{N} \sum_{k=1}^{K} \pi_k \mathrm{N}\left(x_i \mid \mu_k, \sigma_k^2\right) \quad (14)$$

$$f(\Theta) = \prod_{k=1}^{K} p\left(\mu_k \mid \sigma_k^2, \pi_k\right) p\left(\sigma_k^2 \mid \pi_k\right) p(\pi_k) \quad (15)$$

where according to the conjugacy property that describing the posterior distribution has the same parametric form as the prior distribution, the priors for distributions $\{\mu, \sigma, \pi\}$ are assumed to be as follows [24]:

$$\mu_k \mid \sigma_k^2, \pi_k \sim \text{Normal}\left(b, \frac{\sigma_k^2}{\tau}\right) \quad (16)$$

$$\frac{1}{\sigma_k^2} \mid \pi_k \sim \text{Gamma}(\alpha, \beta) \quad (17)$$

$$\pi_k \sim \text{Dirichlet}\left(\frac{1}{K}, \frac{1}{K}, \ldots, \frac{1}{K}\right) \quad (18)$$

where parameters $\{b, \tau, \alpha, \beta\}$ are taken as known 'hyperparameters'. In the initialization of the MCMC simulation, the hyperparameters can be set to $\{b = 150, \tau = 0.1, \alpha = 1, \beta = 2\}$. Finally, the pseudocode of the proposed MCMC method using Gibbs sampling algorithm is presented in Table 1.

Another important point is that the number of components in GMM is unknown. An information criterion, known as Bayesian Inference Criterion (BIC) for the goodness of fit assessment, is chosen to find the optimal number of components [26]. In use of the BIC, a smaller value of BIC indicates a better fitting model. Specifically, the BIC taking into account the log likelihood, the number of estimated parameters d, and sample size N is defined as

$$BIC(K) = -2 \ln L_x(x \mid \Theta) + d \ln N \quad (19)$$

5. Case Study and Verification

5.1. Multistate Travel Time Histograms. Before utilizing the proposed method for estimation of multistate link TTD, the simple histogram can exhibit intuitive information on the TTD. Figure 5 shows two or more clusters in travel time histograms for two different traffic conditions.

During off-peak period, Figure 5(a) gives a distinctly bimodal pattern for the observations from RFID, and most of the observations fall into state 1 and state 3. Conversely, during peak period, as shown in Figure 5(b), state 4 gathering over 80% of the observations plays a dominant role in the congested situation. In terms of the FCD, one (Figure 5(a)) is very small number of observations in state 1 and state 2 whereas there are no observations in state 3, and the other (Figure 5(b)) is the observations largely distributed beyond the two peaks in state 3 and state 4, which represent the major clusters of travel times.

Overall, this simple example implies that the issues of sparse and incomplete observations from FCD may result in irregular, unsmooth TTDs. To improve the estimation of link TTDs, it is suggested that historical information on prior travel time patterns is strongly needed.

5.2. Estimation of Link TTD. For each number of components, the sampling procedure was run in a total of 20,000 iterations and the first 10,000 iterations were discarded as burn-in, while the remaining samples were used for inferencing the posterior distribution. Convergence of the

TABLE 1: Gibbs sampling algorithm for a K component GMM combining with BIC.

Input: N particles (i.e. $x_1, x_2, \ldots, x_N$) with large weights selected by sampling strategy, K
Output: parameters $\Theta = \{(\pi_1, \pi_2, \ldots, \pi_K), (\mu_1, \mu_2, \ldots, \mu_K), (\sigma_1^2, \sigma_2^2, \ldots, \sigma_K^2)\}$, $BIC(K)$

Step 1. Initialization

Determine hyperparameters: b, τ, α, β
Set iteration $r = 0$
Set $e_k^{(r)} = 1/K$, draw $\pi_k^{(r)} \sim \text{Dirichlet}(e_1^{(r)}, \cdots, e_K^{(r)})$
Draw $\mu_k^{(r)} \sim \text{Normal}(b_k^{(r)}, \sigma_k^{2(r)}/\tau_k^{(r)})$
Draw $1/\sigma_k^{2(r)} \sim \text{Gamma}(\alpha_k^{(r)}, \beta_k^{(r)})$

Step 2. Gibbs sampling

For $r = 1$ to R
Update the mixing coefficients $\pi_k^{(r)}$:
Draw $\pi_k^{(r)} \sim \text{Dirichlet}(e_1^{(r)} + E_1^{(r)}, \cdots, e_K^{(r)} + E_K^{(r)})$, where, $E_K^{(r)}$ is the effective number of particles assigned to component K, and can be calculated by

$$E_k^{(r)} = \sum_{i=1}^{N} \frac{\pi_k^{(r-1)} N\left(x_i \mid \mu_k^{(r-1)}, \sigma_k^{2(r-1)}\right)}{\sum_{k=1}^{K} \pi_k^{(r-1)} N\left(x_i \mid \mu_k^{(r-1)}, \sigma_k^{2(r-1)}\right)}$$

For $k = 1$ to K
Update the variance $\sigma_k^{2(r)}$:
Draw $1/\sigma_k^{2(r)} \sim \text{Gamma}(\alpha_k^{(r)}, \beta_k^{(r)})$, where

$$\alpha_k^{(r)} = \alpha_k^{(r-1)} + \frac{E_k^{(r)}}{2}$$

$$\beta_k^{(r)} = \beta_k^{(r-1)} + \frac{1}{2}\left(\sum_{i=1}^{N} (x_i - \overline{x}_k)^2 + \frac{\left(b_k^{(r-1)} - \overline{x}_k\right)^2 \cdot E_k^{(r)} \cdot \tau_k^{(r-1)}}{E_k^{(r)} + \tau_k^{(r-1)}}\right)$$

Update the mean $\mu_k^{(r)}$:
Draw $\mu_k^{(r)} \sim \text{Normal}(b_k^{(r)}, \sigma_k^{(r)}/\tau_k^{(r)})$, where

$$b_k^{(r)} = \frac{\tau_k^{(r-1)} \cdot b_k^{(r-1)} + E_k^{(r)} \overline{x}_k}{\tau_k^{(r-1)} + N_k^{(r)}}$$

$$\tau_k^{(r)} = \tau_k^{(r-1)} + E_k^{(r)}$$

End for
End for

Step 3. BIC calculation

Calculate the BIC for a K component GMM using Eq. (19)

Note: we first execute the sampling with increasing number of components from $K = 2$ to $K = 4$.

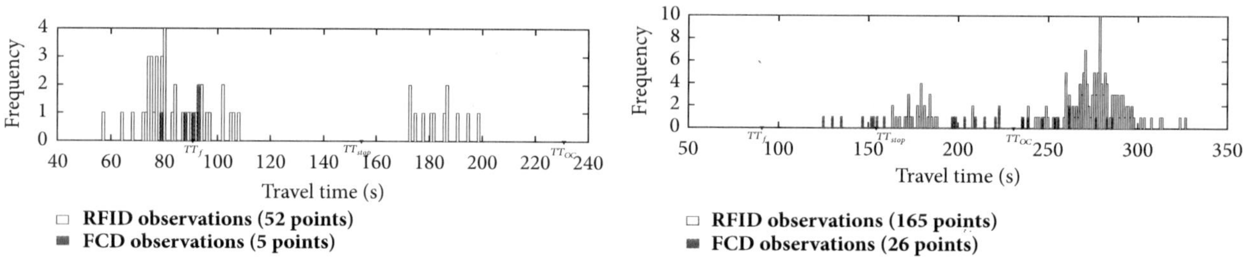

(a) a.m. off-peak period (07:00-07:15)
(b) a.m. peak period (08:30-08:45)

FIGURE 5: Travel time histograms of study link on Dec. 1, 2011 (Thursday).

estimated parameters was monitored by the trace plots of the generated sample values, the model log likelihood, and the autocorrelations [26].

To evaluate the performance of the proposed method, the Hellinger distance (HD) and Kolmogorov-Smirnov (KS) test are employed to assess the similarity between the estimated distribution from FCD and the empirical (actual) one from RFID. For discrete distributions, the HD between the estimated distribution TTD_{est} and the empirical one TTD_{emp} is defined as follows [26]:

$$H\left(\mathrm{TTD}_{est}, \mathrm{TTD}_{emp}\right) = \frac{1}{\sqrt{2}}\sqrt{\sum_{i=1}\left(\sqrt{f_{est}\left(x_i\right)} - \sqrt{f_{emp}\left(x_i\right)}\right)^2} \quad (20)$$

where $f_{est}(i)$ and $f_{emp}(i)$ are the estimated and observed probabilities for an observation x_i from RFID. Besides, the HD satisfies two requirements in (21).

$$H\left(\mathrm{TTD}_{est}, \mathrm{TTD}_{emp}\right) = H\left(\mathrm{TTD}_{emp}, \mathrm{TTD}_{est}\right), \quad 0 \le H\left(\mathrm{TTD}_{est}, \mathrm{TTD}_{emp}\right) \le 1 \quad (21)$$

It is indicated that HD has the nature of symmetry and nonnegativity. Furthermore, if $H(\mathrm{TTD}_{est}, \mathrm{TTD}_{emp}) = 0$, the two distributions will be close to each other, whereas $H(\mathrm{TTD}_{est}, \mathrm{TTD}_{emp}) = 1$ means that there exists the most difference between distributions.

The KS test is a common nonparametric test used to test if the estimated distribution passes the null hypothesis that the observations are distributed according to the estimated distribution. A larger p value provided from the test indicates a better goodness of fit. The estimated distribution is rejected when p value is smaller than 0.05 at a 95% confidence level.

Figure 6 shows a comparison of the estimated TTD and the empirical one. It indicates that if the importance distribution is not accurate enough as shown in Figure 6(a), the observations from FCD imply that there is no data in between the two peaks, and then Gaussian likelihood function tends to pull the posterior distribution in the right direction. Besides, the results of link TTD estimation show the advantage of the proposed approach with few data in terms of the calibration of TPMs from historical travel times.

If the importance distribution is quite accurate, as shown in Figure 6(b), however, the posterior distribution is close to the empirical one, the Bayesian posterior distribution says that some data with high probabilities are missing in the first peak (around 185 s). If we observe the travel time histogram at a.m. peak periods in Figure 5(b), eight observations from FCD located in between two major components from RFID, there should not be such an obvious gap between state 3 and state 4 under congested situation. Besides, the observations nearby a mean of 210 seconds in the Gaussian likelihood function will be assigned larger weights. Altogether, with the Bayesian update, the posterior distribution that incorporates information not only from the prior travel time pattern, but also from the real-time data, may provide a more convincing result.

Table 2 summarizes the model parameters and performance results. It can be seen that the GMM with three components provides a superior fit for empirical data during peak hours. This could be explained by the underlying situations that present two kinds of state combinations, that is, state 1 to state 3, and state 2 to state 4, while the larger weight implies the corresponding state in a state combination may have a leading impact on the whole link traffic conditions. These findings assist with insight into the heterogeneity in travel times. Moreover, the performance results show that of all 52 periods used for the KS test, only nine of them fail to pass the KS test, which help confirm the statistical superiority of the proposed method.

5.3. Effect of Floating Car Percentages on the Proposed Approach. To examine the robust of the proposed method, six different percentages of floating cars, i.e., floating cars% = 0%, 5%, 10%, 20%, 50%, and 100%, are conducted as an input of estimating the link TTDs. For each percentage of floating cars, the sampling procedure was run 100 times to randomly select the corresponding samples from observations and calculate the average values of HD and percentages of passing KS test in each 15-min interval. In particular, if the number of samples is zero, we use an alternative method with the steady-state probabilities and the means of historical travel times instead of the estimated state probabilities in (7) and the means of real-time observations in likelihood function.

According to the irreducible and ergodic properties of the Markov chain, the steady-state TPM $\Pi_n(t)$ on link n at time t can be defined as follows [27]:

$$\Pi_n(t) = \lim_{r \longrightarrow \infty} P^r_{n,n-1}(t) \quad (22)$$

where r is the number of steps. When r is big enough, each row of $\Pi_n(t)$ will all reach steady-state probabilities regardless of whatever the state probabilities on link n-1 are. Figure 7 shows the results of average HD and KS test for different percentages of floating cars during time of day.

It can be observed that, with the percentage of floating cars decreasing, the main trend of the average HD increases; on the contrary, the percentage of passing KS test in general decreases. Especially when the number of samples is zero, the performance of the alternative method (i.e., the average HD largely varying between 0.1 and 0.3 while the percentage of passing KS test mainly ranging from 70% to 90%) is superior to the proposed method with sample size less than five, which mainly generates the worst result. This is likely due to the random sampling resulting in higher travel time variability for estimation input, whereas the alternative method only depending on the historical travel time information is not sensitive to the input data.

Overall, the results confirm that the proposed method still performs well with small sample size. For the worst condition, i.e., floating cars% = 0%, the alternative method is preferred in terms of the estimation performance.

6. Conclusions

This paper presents a Bayesian approach based on particle filter framework to estimate link TTDs using the limited observations and historical data set from FCD. The main objective of the approach is to produce the multistate TTDs with the low penetrate rate of floating cars. Unlike previous studies that require a complete data in modeling the multistate TTDs, the proposed approach uses less observations to update the importance distribution. Concisely, the importance distribution is regarded as prior knowledge and provides past travel time patterns. On the other hand, limited observations are integrated into the particle filter framework and make the estimated TTDs further close to the empirical

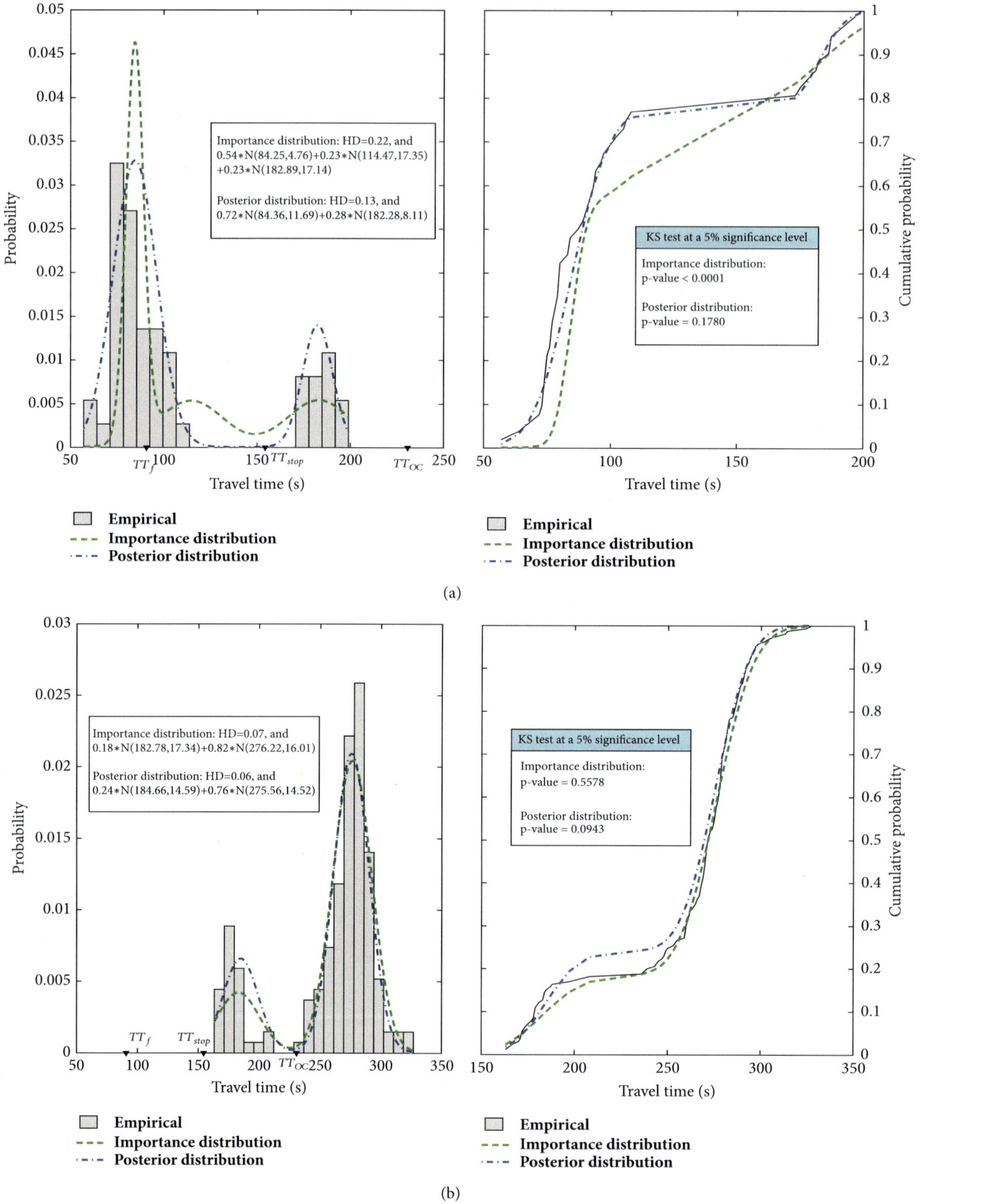

FIGURE 6: Comparison of the estimated TTD from FCD and the empirical TTD from RFID for the periods of (a) 07:00-07:15 a.m. and (b) 08:30-08:45 a.m.

Table 2: Link TTD parameters, HD, and KS tests. This table is reproduced from a shorter version of this work presented at the 97th Annual Meeting of the Transportation Research Board [5].

Time	Sample size (N)	Estimated parameters in posterior distribution: Weight (π)	Mean (μ)	Standard deviation (σ)	HD	KS test: h[a]	p value
a.m. Peak 7:00 - 9:00							
7:00-7:15	5	(0.72, 0.28)	(84.36, 182.28)	(11.69, 8.11)	0.13	0	0.1780
7:15-7:30	25	(0.68, 0.22, 0.1)	(84.29, 114.66, 181.87)	(10.01, 10.03, 10.055)	0.09	0	0.0865
7:30-7:45	29	(0.48, 0.2, 0.32)	(83.87, 115.81, 197.72)	(14.11, 19.11, 14.02)	0.08	1	0.0305
7:45-8:00	22	(0.46, 0.42, 0.12)	(96.48, 136.53, 203.14)	(10.23, 9.53, 9.5)	0.17	0	0.0797
8:00-8:15	28	(0.71, 0.17, 0.12)	(132.51, 183.23, 248.51)	(20.13, 20.01, 20.36)	0.08	0	0.2145
8:15-8:30	27	(0.53, 0.05, 0.42)	(143.34, 183.5, 269.59)	(20.21, 6.67, 10.17)	0.24	0	0.7334
8:30-8:45	26	(0.24, 0.76)	(184.66, 275.56)	(14.59, 14.52)	0.06	0	0.0943
8:45-9:00	21	(0.21, 0.1, 0.69)	(182.52, 222.52, 275.02)	(8.08, 12.01, 13.15)	0.11	1	0.0325
p.m. Peak 17:00 - 19:00							
17:00-17:15	17	(0.59, 0.41)	(104.72, 205.44)	(19.82, 13.07)	0.16	0	0.4162
17:15-17:30	24	(0.29, 0.56, 0.16)	(105.83, 207.03, 246.86)	(17.38, 9.23, 25)	0.09	0	0.4507
17:30-17:45	14	(0.24, 0.58, 0.18)	(123.7, 211.82, 251.41)	(9.55, 12.1, 32.82)	0.11	1	0.0022
17:45-18:00	16	(0.14, 0.48, 0.38)	(147.92, 218.88, 237.36)	(20.12, 6.93, 10.01)	0.14	0	0.1100
18:00-18:15	20	(0.35, 0.65)	(117.46, 216.85)	(20, 20.18)	0.06	0	0.0928
18:15-18:30	27	(0.43, 0.29, 0.28)	(108, 191.79, 206.68)	(21.3, 20, 20.39)	0.12	0	0.1824
18:30-18:45	10	(0.45, 0.13, 0.42)	(90.85, 129.37, 196.79)	(10.13, 20.01, 10.01)	0.28	0	0.7900
18:45-19:00	12	(0.7, 0.08, 0.22)	(95.04, 130.72, 193.38)	(12.57, 8.99, 17.53)	0.09	0	0.6123

Performance summary during time of day

Time of day	Average HD	Min HD	Max HD	Percentages of passing KS test (%)
7:00-20:00	0.121	0.024	0.355	83

[a]h = 0 indicates that the estimated distribution passes the KS test at a 5% significance level.

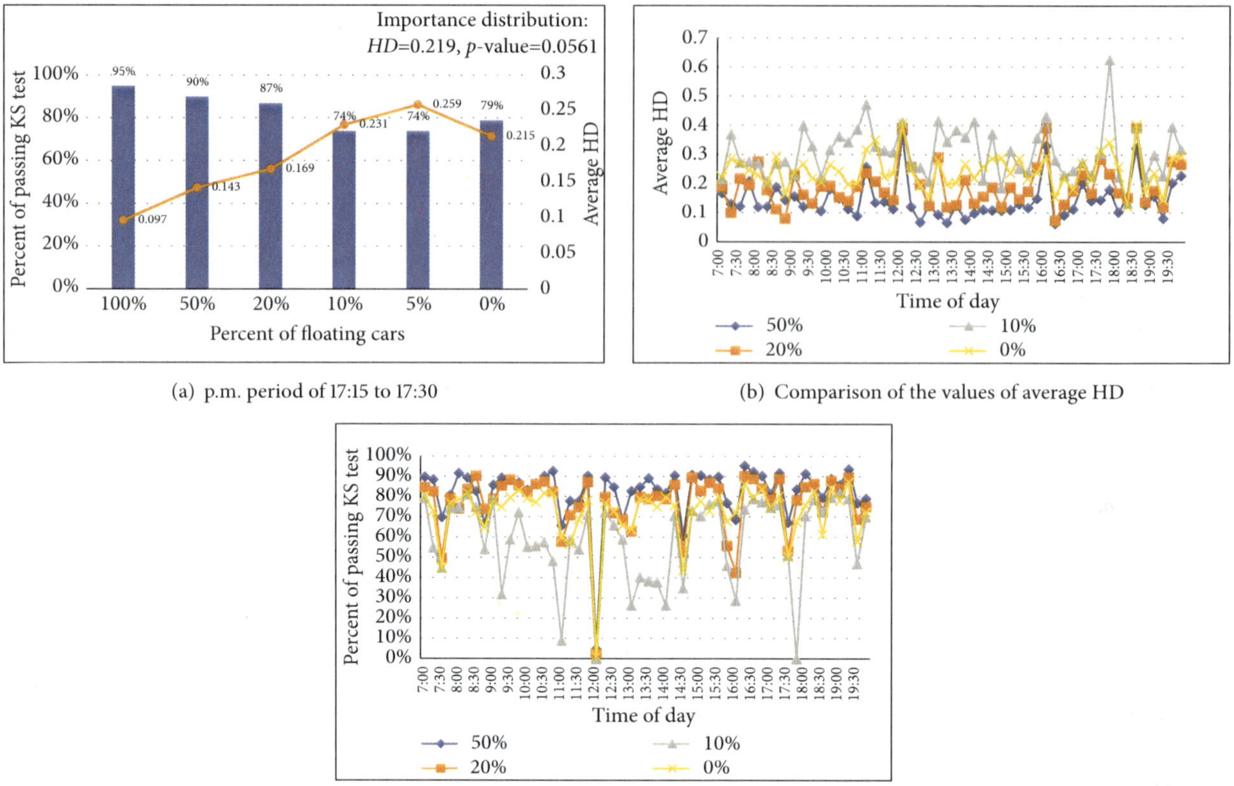

(a) p.m. period of 17:15 to 17:30

(b) Comparison of the values of average HD

(c) Comparison of the percents of passing KS test

FIGURE 7: The results comparisons for different percentages of floating cars. This figure is reproduced from a shorter version of this work presented at the 97th Annual Meeting of the Transportation Research Board [5].

distributions by using a sampling strategy. Finally, the result is a posterior distribution incorporating historical trends and the real-time travel time information as well.

The results of the case study demonstrate that the proposed approach produces the posterior distribution agreeing well with the empirical distribution. Besides, with different percentages of floating cars, our proposed approach performs equally well, even under condition with no data available. This suggests that, when the number of floating car sample size is less than five, the approach with the steady-state probabilities helps provide a promising result.

Nevertheless, this paper has some limitations. First, we used the historical observations as an a priori estimation. Information from estimated TTD in previous period should be integrated into the particle filter framework to update the parameters of the current importance distribution and provide a better a priori estimation. Second, the spatial TPMs of the Markov chain did not consider the temporal correlation between link travel times. A Markov process with a short memory taking into account the effect of travel time states at the preceding time interval on transition intensities may lead to better TPMs estimates. Finally, a density-based clustering method such as DBSCAN clustering or other DBSCAN variants should be developed to cluster link travel times for representing heterogeneous traffic states so that the ability of the proposed approach could be generalized to a large range of applications from urban signalized arterials to freeway continuous traffic instead of using traffic engineering rules. In addition, the effectiveness of the proposed approach still needs a further verification by conducting more case studies in other road links with different sources of trajectory data (e.g., smartphone and the ride service platform of Uber or DiDi Chuxing). These will be our future works. Last but not least, we should state that the work from this paper is an extended version of the earlier work presented at the 97th Annual Meeting of the TRB, Washington, DC, January 2018 [5]. For more details, it can be found in [5].

Conflicts of Interest

The authors declare that there are no conflicts of interest regarding the publication of this paper.

Acknowledgments

The authors would like to thank the anonymous reviewers for their help on a shorter version of the paper presented at the 97th Annual Meeting of the TRB, Washington, DC, January 2018. The research is supported by the National Natural Science Foundation of China under Grant no. 51238008.

References

[1] Y.-X. Ji and H. Zhang, "Travel Time Distributions on Urban Streets: Their Estimation with a Hierarchical Bayesian Mixture

Model and Application to Traffic Analysis using High - Resolution Bus Probe Data," in *93th Annual Meeting of the Transportation Research Board*, D.C, Washington, 2014.

[2] M. A. P. Taylor and S. Somenahalli, "Travel time reliability and the bimodal travel time distribution for an arterial road," *Road & Transport Research*, vol. 19, no. 4, pp. 37–50, 2010.

[3] F. Yang, M.-P. Yun, and X.-G. Yang, "Travel time distribution under interrupted flow and application to travel time reliability," *Transportation Research Record*, vol. 2466, pp. 114–124, 2014.

[4] M.-P. Yun, W.-W. Qin, Y. Liu, and J. Yu, "Analysis of Required Minimum Sample Size of Floating Cars for Estimating Urban Road Link Travel Time Considering Bimodal Distribution and Estimation Error," in *96th Annual Meeting of the Transportation Research Board*, D.C, Washington, 2017.

[5] W.-W. Qin and M.-P. Yun, "Estimation of Urban Link Travel Time Distribution Using Markov Chains and Bayesian Approaches," in *97th Annual Meeting of the Transportation Research Board*, D.C, Washington, 2018.

[6] D. Dailey and F. Cathey, "AVL-Equipped Vehicles as Traffic Probe Sensors," Technical Report No. WA-RD 534.1, Washington State Transportation Center (TRAC), 2002.

[7] Y.-H. Feng, J. Hourdos, and G. A. Davis, "Probe vehicle based real-time traffic monitoring on urban roadways," *Transportation Research Part C: Emerging Technologies*, vol. 40, pp. 160–178, 2014.

[8] T. Hunter, R. Herring, and A. Bayen, "Path and Travel Time Inference from GPS Probe Vehicle Data," in *Proceedings of the In Neural Information Processing Systems Foundation (NIPS) Conference*, Vancouver, Canada, 2009.

[9] R. Herring, A. Hofleitner, P. Abbeel, and A. Bayen, "Estimating arterial traffic conditions using sparse probe data," in *Proceedings of the 13th International IEEE Conference on Intelligent Transportation Systems, ITSC 2010*, pp. 929–936, Portugal, September 2010.

[10] B. S. Westgate, D. B. Woodard, D. . Matteson, and S. G. Henderson, "Travel time estimation for ambulances using Bayesian data augmentation," *The Annals of Applied Statistics*, vol. 7, no. 2, pp. 1139–1161, 2013.

[11] F. Guo, Q. Li, and H. Rakha, "Multistate travel time reliability models with skewed component distributions," *Transportation Research Record*, vol. 2315, pp. 47–53, 2012.

[12] S. Park, H. Rakha, and F. Guo, "Multi-state travel time reliability model: Impact of incidents on travel time reliability," in *Proceedings of the 14th IEEE International Intelligent Transportation Systems Conference, ITSC 2011*, pp. 2106–2111, USA, October 2011.

[13] A. Hofleitner, R. Herring, and A. Bayen, "Arterial travel time forecast with streaming data: A hybrid approach of flow modeling and machine learning," *Transportation Research Part B: Methodological*, vol. 46, no. 9, pp. 1097–1122, 2012.

[14] Z.-G. Li, C. Cai, A. K. Menon, Y. Xu, and F. Chen, "Estimation of link speed distribution from probe vehicle data," *Transportation Research Record*, vol. 2595, pp. 98–107, 2016.

[15] M. Ramezani and N. Geroliminis, "On the estimation of arterial route travel time distribution with Markov chains," *Transportation Research Part B: Methodological*, vol. 46, no. 10, pp. 1576–1590, 2012.

[16] Y. Lou, C. Zhang, Y. Zheng, X. Xie, W. Wang, and Y. Huang, "Map-matching for low-sampling-rate GPS trajectories," in *Proceedings of the 17th ACM SIGSPATIAL International Conference on Advances in Geographic Information Systems*, pp. 352–361, ACM, November 2009.

[17] I. Sanaullah, M. Quddus, and M. Enoch, "Developing travel time estimation methods using sparse GPS data," *Journal of Intelligent Transportation Systems: Technology, Planning, and Operations*, vol. 20, no. 6, pp. 532–544, 2016.

[18] Z.-L. Ma, L. Ferreira, M. Mesbah, and S. Zhu, "Modeling distributions of travel time variability for bus operations," *Journal of Advanced Transportation*, vol. 50, no. 1, pp. 6–24, 2016.

[19] D. J. Sun, X. Liu, A. Ni, and C. Peng, "Traffic congestion evaluation method for urban arterials: Case study of Changzhou, China," *Transportation Research Record*, vol. 2461, pp. 9–15, 2014.

[20] F. Yang, *Research on urban network traffic flow evolution based on experimental traffic theory and mathematical analysis [Ph.D. thesis]*, College of transportation engineering, Tongji University, China, 2015.

[21] Transportation Research Board, Transportation Research Board. Highway Capacity Manual.

[22] A. Doucet, N. de Freitas, and N. Gordon, *Sequential Monte Carlo Methods in Practice*, Springer, New York, NY, USA, 2001.

[23] M. S. Arulampalam, S. Maskell, N. Gordon, and T. Clapp, "A tutorial on particle filters for online nonlinear/non-Gaussian Bayesian tracking," *IEEE Transactions on Signal Processing*, vol. 50, no. 2, pp. 174–188, 2002.

[24] B.-J. Park, Y. Zhang, and D. Lord, "Bayesian mixture modeling approach to account for heterogeneity in speed data," *Transportation Research Part B: Methodological*, vol. 44, no. 5, pp. 662–673, 2010.

[25] E. Kidando, R. Moses, E. E. Ozguven, and T. Sando, "Bayesian nonparametric model for estimating multistate travel time distribution," *Journal of Advanced Transportation*, vol. 2017, Article ID 5069824, 9 pages, 2017.

[26] D. J. Spiegelhalter, N. G. Best, B. P. Carlin, and A. van der Linde, "Bayesian measures of model complexity and fit," *Journal of the Royal Statistical Society: Series B (Statistical Methodology)*, vol. 64, no. 4, pp. 583–639, 2002.

[27] J. Yeon, L. Elefteriadou, and S. Lawphongpanich, "Travel time estimation on a freeway using Discrete Time Markov Chains," *Transportation Research Part B: Methodological*, vol. 42, no. 4, pp. 325–338, 2008.

6

Study on the Equilibrium Discriminant Model of Urban Agglomeration Transport Supply and Demand Structure

Z. Y. Liu, C. B. Li, and M. Y. Jian

Transportation Institute, Inner Mongolia University, Hohhot 010021, China

Correspondence should be addressed to C. B. Li; bingbingnihao2008@126.com

Academic Editor: Jaeyoung Lee

In order to study the adaptability of urban agglomeration transport supply and demand structure, and to provide the basis for decision-makers when optimizing the structure of transport supply and demand in urban agglomeration, this paper proposed and verified a structure equilibrium theory on the basis of entropy theory. Firstly, based on the analysis of the influencing factors, an urban agglomeration transport demand structure model was established using entropy theory. Secondly, according to the actual passenger and freight turnover of urban agglomeration, a transport supply structure model was proposed with entropy theory. Then, by comparing the two entropy models, the equilibrium state of an urban agglomeration transport structure was analyzed. Finally, HuBaoOr urban agglomeration was taken as an example to verify the science and effectiveness of this discriminant model. The research in this paper lays a theoretical foundation for the achievement of urban agglomeration transport equilibrium structure and for realizing the best economic benefits and social benefits.

1. Introduction

With the development of information technology and city economy, the phenomenon of urban agglomeration increases, and the urban agglomeration will be an important trend of the urban development in the future. During the development process of urban agglomeration, transportation structure including transport supply structure and transport demand structure plays an important role as a base installation. Therefore, transportation structure should reach a state of harmony with the development of urban agglomeration. In an advanced and modern transport system of urban agglomeration, a relatively stable transport supply structure and transport demand structure formed under certain economic and demographic conditions should match each other. This state will be a dynamic equilibrium state and the utilization of any transport supply resource achieves the best outcome. However, due to the lack of comprehensive planning and effective coordination mechanism in our country, there are often disequilibrium phenomena in the transport supply structure and demand structure, which will hinder the economy development of urban agglomeration in turn.

Lots of literatures [1, 2] have focused on the equilibrium structure of transport supply and demand and proposed evaluating methods about supply and demand balance state of urban traffic. Due to serious traffic problem, the RACV (Royal Automobile Club of Victoria, Ltd) management strategy [3] was put forward to balance the traffic supply and demand. Yuan [4] studied the equilibrium problems of transport demand structure and supply structure by considering the relationship between transportation and social economic development. Taking adaptation of the regional freight supply and demand structure as the research object, Tang et al. [5] analyzed the corresponding influencing factors, built an evaluation model for the adaptation, and put forward the adaptability evaluation standard using the information entropy theory. Lu [6] discussed the main influencing factors and put forward the research content of transport supply strategy and demand management measures. Wu et al. [7] quantitatively analyzed the causal relationship between China's industrial structure and transportation structure through the test on industrial structure and transportation structure entropy based on the qualitative analysis of interaction mechanism between transport supply and demand. Wei Jun et al. [8]

and Daniel Hörcher [9] proposed the equilibrium conditions of transport demand and supply using system dynamics principle and considered the temporal, spatial, and directional demand imbalances. The supply and demand relationship with economy and transportation structure equilibrium problem considered by the equilibrium concept was also studied by Wu wenzheng [10] and Carey.G.W [11]. Ge Yu et al. [12] formed transport investment structure and transport demand coordination model of Beijing by comparing the transport investment structure and transport development characteristics between international metropolis and the city of Beijing. Pei Wu [13] studied the road network characteristics of urban agglomeration with some quantitative methods, such as the traffic assignment algorithm based on the path and the design algorithm of the integrated hybrid road network. Liu Zhijie [14] studied the hierarchy of groups and impact of urban agglomeration transportation systems from a macro point of view. Gao Yan [15] discussed the evaluation methodology for the intercity transportation system development of urban agglomeration, mainly on rail transit, based on some basic characteristics and associated theories of urban agglomeration.

In conclusion, there are many researches about transport supply and demand structure of a single city road network. The key problem they want to solve is traffic congestion and the proposed measures always include limiting the travel of cars on the demand side and increasing the capacity of the road network on the supply side within the city. There is very little research about an urban agglomeration, and if there is one, it often focuses on the macro development characteristic of urban agglomeration transport system and specific plans still within one administrative region lacking in-depth analysis of the internal relationship between transport supply and demand structure of urban agglomeration. The key problem to be solved in an urban agglomeration transport structure is to adjust supply to boost demand and promote the economic development of urban agglomeration through the coordination of the supply structure and demand structure subjected to the existing supply resources. Of course, the supply resources will change with the economic development of an urban agglomeration. So, we try to realize a dynamic equilibrium state through the adjustment and coordination measures which include the proportion of modes of transport and the spatial hierarchical structure of transport system between cities.

Since the formation and function of urban agglomeration cannot be separated from the improvement of the service capacity of transportation system, an equilibrium transport supply and demand structure has become an inevitable requirement for the development of urban agglomeration. Therefore, the equilibrium problem of transport supply and demand structure is one of the urgent problems to be solved at home and abroad.

Based on this, by studying the relationship between the transport supply structure and transport demand structure of urban agglomeration, the discriminant model of transport supply-demand equilibrium of urban agglomeration is built in this paper which can quantitatively evaluate the development degree and the equilibrium state of urban agglomeration transport supply-demand structure. This method can provide a theoretical foundation for realizing reasonable configuration and equilibrium structure of urban agglomeration transport resource.

2. Description of the Equilibrium Structure of Transport Supply and Demand

Multicenter group is the main spatial form of urban agglomeration. In this form, an urban agglomeration transport supply structure and demand structure include internal transport supply and demand of a city inside urban agglomeration, intercity transport supply, and demand between two center cities within urban agglomeration and intercity transport supply and demand between a center city and another non-center city within urban agglomeration. With respect to an urban agglomeration, the passenger and freight transport not only consider the traffic hub and traffic system in a city, but also consider the optimum transport path of the whole urban agglomeration in accordance with the minimum cost and the maximum return principle. Therefore, the internal and external transport supply and demand of an urban agglomeration should be unified planning including center city and non-center city.

In the rapid development of urban agglomeration in our country, there are many problems in construction of transportation system, which are mainly manifested in two ways, the coexistence of poor supply and unused portion of transport capacity and the coexistence of backward transport infrastructure and low utilization rate of part advanced infrastructures [16]. The transport supply-demand equilibrium structure of urban agglomeration researched in this paper is a relatively dynamic equilibrium structure at a certain time rather than an absolute equilibrium state with completely equal total supply and demand amount. It is common knowledge that the change of transportation supply always lags behind the change of transportation demand, so the relatively equilibrium state researched in this paper is a kind of dynamic equilibrium structure under the state of being short of supply. With the development of social economy of urban agglomeration, a new transport demand structure in the short term can be formed, while the transport supply structure cannot adapt to changed transport demand in the short term and this will result in a short term transport supply-demand disequilibrium structure of urban agglomeration. By this time, transportation governors and producers should adjust the transport supply structure to meet the changed demand structure. Before reaching the absolute equilibrium state by changing transport supply total amount, the transport supply and demand structure should be adjusted to achieve a relatively equilibrium state. The transition of supply-demand structure state is shown in Figure 1.

3. Analysis of the Factors Affecting the Equilibrium of Traffic Demand-Supply Structure of Urban Agglomeration

(1) Industrial Structure Adjustment of Urban Agglomeration. The change of industrial structure is the root of the change

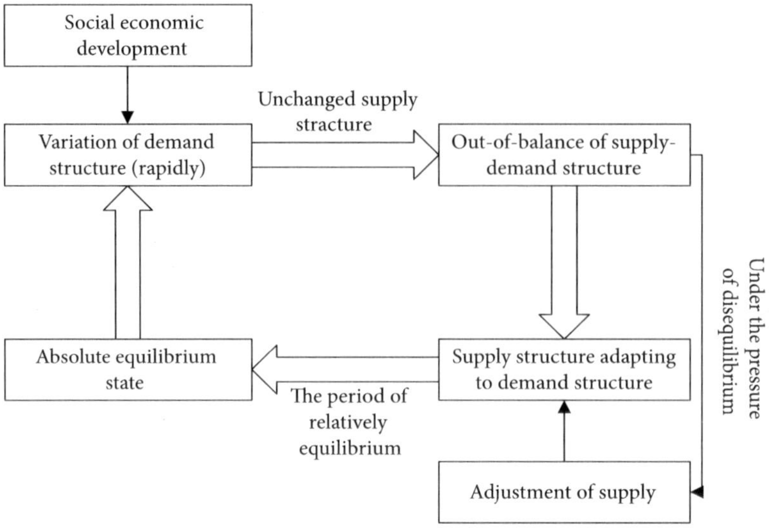

Figure 1: The transition of supply and demand structure state.

of the transport structure. Due to the adjustment in the ratio of various industries, type, flow, and direction of cargoes are all going to be different, and then the transport demand of urban agglomeration will change affecting the adaptability of transport supply and demand structure.

(2) Resident Consumption Structure of Urban Agglomeration. Income and consumption level of residents determine their preference to various modes of transport, thus forming a different passenger demand structure at different economic development level and stage. With the development of city economy, residents' consumption level continues to improve, and then the transport demand will also increase and show a trend of diversity, which will cause the change of urban agglomeration transport demand structure.

(3) Economic Relationship between Cities in Urban Agglomeration. The development of one city can well impact others' in urban agglomeration, while the economic relationship between different cities is different. The different degree of economic relationship between different cities leads to the different transport demand structure in an urban agglomeration. Here, we take the recessive economic gravity model [13] to represent the economic relationship between cities.

(4) Population Quantity and Size of Urban Agglomeration. The change of supply-demand equilibrium structure of passenger transport is closely related to population quantity and trip frequency. Increase of urban agglomeration population will cause the increase of city passenger transport demand, which will have an impact on the transport demand structure. Urban agglomeration population increases with the expansion of urban agglomeration scale, so the size of urban agglomeration can be reflected by the population in turn.

(5) Geographical Environment of Urban Agglomeration. Different urban agglomeration has different geographical environment, including the inland and coastal terrain differences, topography differences, and border distance differences. The different geographical environments need different transport modes, thus forming a unique transport demand structure of urban agglomeration.

(6) Other Influencing Factors. The energy structure, economic development strategy, transportation technology, and economic policy of urban agglomeration will also affect the transport demand structure. But, these factors are relatively stable in a certain period of time, almost no change, and their influence on urban agglomeration transport demand structure is also relatively small. So these factors will be temporarily ignored in the following construction of the transport demand structure function.

4. Building Equilibrium Discriminant Model of Urban Agglomeration Transport Supply-Demand Structure

4.1. Basic Idea of Modeling. We have to determine the research area before description of equilibrium problems of urban agglomeration transport supply and demand structure. The research area consists of two parts: all center cities and other closely neighboring cities (called non-center cities) involved in the urban agglomeration. $S = \{s_q \mid 1 \le q \le n\}$ denote the set of cities in the research area. $S' = \{s_p \mid 1 \le p \le m\}$ denote the set of center cities of selected urban agglomeration. It is obvious that $S' \subset S$.

(1) If $i = j$ and $1 \le i \le m$, $1 \le j \le m$, then s_i, s_j denote the same city in the urban agglomeration. At this point, the research object becomes a transport supply and demand relation of a single city in the urban agglomeration.

(2) If $i \ne j$ and $1 \le i \le j \le m$, the research object is intercity relationship between center cities in the urban agglomeration.

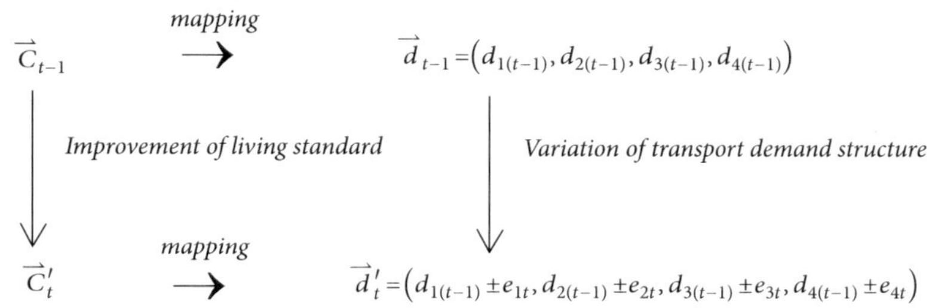

Figure 2: Mapping process from residents' consumption structure to transport demand.

(3) For any $s_i \in S$ and $s_i \notin S'$, $s_j \in S'$, the research object will be intercity relationship between a center city and a noncenter city in the urban agglomeration.

In order to describe equilibrium relationship between transport supply-demand structures of an urban agglomeration, firstly we need to establish urban agglomeration demand model and urban agglomeration supply model, respectively. Then we can determine the equilibrium state by comparing the above two models.

In this paper, the entropy theory is used to model urban agglomeration transport demand structure to quantitatively describe the spatial hierarchical transport demand structure between cities in an urban agglomeration based on the factors influencing the transport demand structure. As mentioned above, the relatively equilibrium state researched in this paper is a kind of dynamic equilibrium structure under the state of being short of supply, which means the total demand is greater than the total supply. So, in the modeling process of supply structure, actual transport volume of each traffic mode between cities of different hierarchical structures in an urban agglomeration is treated as transport supply, and the supply entropy is calculated. The discriminant model of transport supply-demand equilibrium state of urban agglomeration is built with the transport demand entropy and transport supply entropy which can reflect the equilibrium degree of the supply-demand structure intuitively.

4.2. Modeling Urban Agglomeration Transport Demand Structure. In the model of urban agglomeration transport demand structure, factors affecting demand are function variables. The function expression is as follows:

$$D(\vec{d}) = f(\vec{I}, F, \vec{C}, N, L, \varepsilon), \tag{1}$$

where $\vec{d}$ is transport demand vector, $\vec{I}$ is industrial structure vector, F is economic gravity between cities, $\vec{C}$ is consumption structure vector, N is population size, L is geographical environment, and ε denotes other influencing factors.

City transport demand structure vector and its influencing factors belong to different dimensions, and influencing factors include number attribute and nonnumber attribute which makes it difficult to quantify all the factors unified. So, entropy theory is introduced in this paper to quantitatively describe the relationship between the factors and urban agglomeration supply-demand structure.

(1) Industrial Structure Entropy of Urban Agglomeration. According to the definition of the structure entropy, we can calculate the corresponding industrial structure entropy in different periods and use the industrial structure entropy to reflect the changes of the industry in different economic development stages of urban agglomeration.

If $H_I(t)$ denotes the industrial structure entropy at period t, then

$$H_I(t) = -\sum_{j=1}^{3} P_{jt} \ln P_{jt}, \tag{2}$$

where P_{jt} is percentage of industry j at period t, and

$$P_{jt} = \frac{I_{jt}}{\sum_{j=1}^{3} I_{jt}} \tag{3}$$

$$I_{jt} = \sum_{i=1}^{n} I_{ijt}. \tag{4}$$

In this equation, I_{jt} is production value of industry j at period t and I_{ijt} is production value of industry j at period t of city s_i. j denotes the industry type, such as primary industry, secondary industry, and tertiary industry.

(2) Residents' Consumption Structure Entropy of Urban Agglomeration. Because the residents' consumption structure vector is hard to quantify description, the residents' consumption structure is mapped to transport demand of four kinds of transport modes. Then the change of residents' consumption structure corresponds to the change of transport demand. The mapping process can be described through Figure 2.

In Figure 2,

$$\vec{C}(t-1) = \sum_{i=1}^{n} \vec{C}_i(t-1) \tag{5}$$

$$d_j(t-1) = \sum_{i=1}^{n} d_{ij}(t-1) \quad (6)$$

$$e_{jt} = \sum_{i=1}^{n} e_{ijt}. \quad (7)$$

If $H_c(t)$ denotes the consumption structure entropy at period t, then

$$H_c(t) = -\sum_{j=1}^{4} Q_{j(t)} \ln Q_j(t) \quad (8)$$

$$Q_{j(t)} = \frac{\left(d_{j(t-1)} \pm e_{jt}\right)}{\sum_{j=1}^{4}\left(d_{j(t-1)} \pm e_{jt}\right)}, \quad (9)$$

where

$\overrightarrow{C}_{t-1}$ is residents' consumption structure vector at time $t-1$

$\overrightarrow{C}'_t$ is residents' consumption structure vector at time t

$\overrightarrow{d}_{t-1}$ is transport demand vector corresponding to residents' consumption structure at time $t-1$

$\overrightarrow{d}'_t$ is transport demand vector corresponding to residents' consumption structure at time t

$d_{j(t-1)}$ is transport volume of mode j (including roads, railways, aviation, and water transport mode) of urban agglomeration corresponding to residents' consumption structure at time $t-1$

$d_{ij(t-1)}$ is transport volume of mode j of city s_i corresponding to residents' consumption structure at time $t-1$

e_{jt} is transport demand variation of mode j corresponding to change of residents' consumption from time $t-1$ to time t

e_{ijt} is transport demand variation of mode j of city s_i corresponding to change of residents' consumption from time $t-1$ to time t

(3) Entropy about Economic Relationship between Cities of Urban Agglomeration. Firstly, recessive economic gravity model is used to measure the economic relationship between cities. Recessive economic gravity model can clearly express the development ability of cities and more accurately calculate the economic attractiveness between cities. The calculation formula of the model is as follows:

$$F_{ij}(t) = k_{ij} \bullet \frac{\left(P_i^1 + P_i^2\right)_{V_i} * \left(P_j^1 + P_j^2\right)_{V_j}}{r_{ij}^2}. \quad (10)$$

In this paper we just consider the relationship between cities at time t and there is no need to consider the potential economic demand of cities. So the above formula can be simplified to formula (11).

$$F_{ij}(t)' = k_{ij} \bullet \frac{P^1_{i\,V_i} * P^1_{j\,V_j}}{r_{ij}^2}, \quad (11)$$

where

F_{ij} is index of recessive economic gravity

P_i^1, P_j^1 are economic index of cities s_i and s_j (GDP, billion yuan)

P_i^2, P_j^2 are potential economic demand index of cities s_i and s_j (billion yuan)

V_i, V_j are population index of cities s_i and s_j (billion people)

r_{ij} is the minimum distance between cities s_i and s_j (hours)

k_{ij} is adjustment coefficient of recessive economic gravity, $k_{ij}=1$

If $H_F(t)$ denotes entropy of economic relationship between cities at period t, then

$$H_F(t) = -\sum_{i=1}^{n}\sum_{j=1}^{n} R_{ij} \ln R_{ij} \quad (12)$$

$$R_{ij} = \frac{F_{ij}(t)'}{\sum_{i=1}^{n}\sum_{j=1}^{n} F_{ij}(t)'}. \quad (13)$$

(4) Entropy of Urban Agglomeration Population Size. The urban agglomeration population size is mapped to transport demand of four kinds of transport modes as residents' consumption structure. The mapping process can be described through Figure 3.

In Figure 3,

$$N(t-1) = \sum_{i=1}^{n} N_i(t-1) \quad (14)$$

$$d_j(t-1) = \sum_{i=1}^{n} d_{ij(t-1)} \quad (15)$$

$$g_{jt} = \sum_{i=1}^{n} g_{ijt}. \quad (16)$$

If $H_N(t)$ denotes entropy of urban agglomeration population size at period t, then

$$H_N(t) = -\sum_{j=1}^{4} S_{jt} \ln S_{jt} \quad (17)$$

$$S_{jt} = \frac{\left(d_{j(t-1)} + g_{jt}\right)}{\sum_{j=1}^{4}\left(d_{j(t-1)} + g_{jt}\right)} \quad (18)$$

and

N_{t-1} is population size of urban agglomeration at time $t-1$

N'_t is population size of urban agglomeration at time t

$\overrightarrow{d}_{t-1}$ is transport demand vector corresponding to population size at time $t-1$

$\overrightarrow{d}'_t$ is transport demand vector corresponding to population size at time t

$d_j(t-1)$ is transport volume of mode j (including roads, railways, aviation, and water transport mode) of urban agglomeration corresponding to population size at time $t-1$

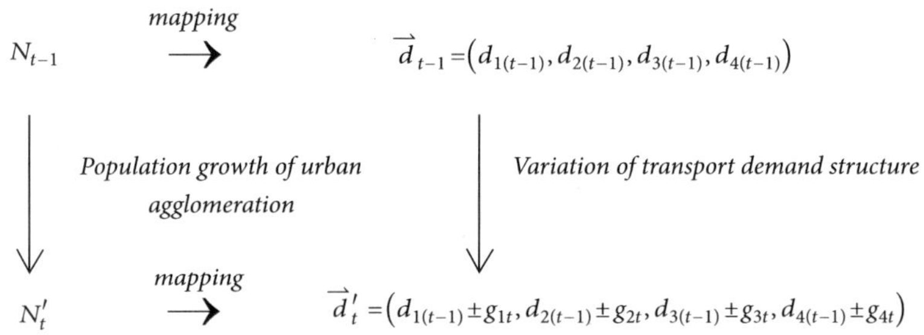

FIGURE 3: Mapping process from population size to transport demand.

$d_{ij}(t-1)$ is transport volume of mode j of city s_i corresponding to population size at time $t-1$

g_{jt} is transport demand variation of mode j corresponding to change of population size from time $t-1$ to time t

g_{ijt} is transport demand variation of mode j of city s_i corresponding to change of population size from time $t-1$ to time t

(5) Entropy of Geographical Environment of Urban Agglomeration. The geographical environment of urban agglomeration is a fixed factor which does not change with time. Therefore, the vector of the transport demand structure corresponding to the geographical location of the urban agglomeration is fixed, as formula (19).

$$\vec{d}_L = (d_{L1}, d_{L2}, d_{L3}, d_{L4}). \tag{19}$$

If H_L denotes entropy of geographical environment, then

$$H_L = -\sum_{i=1}^{4} L_i \ln L_i \tag{20}$$

$$L_i = \frac{d_{Li}}{\sum_{i=1}^{4} d_{Li}} \tag{21}$$

and d_{Li} is transport supply quantity of mode j corresponding to geographical environment of urban agglomeration.

According to the above entropy of the influencing factors the ultimate transport demand structure function is drawn by the method of undetermined coefficients. The transport demand structure entropy model is as formula (22).

$$\begin{aligned} H_D(t,d) &= aH_I(t) + bH_C(t) + cH_F(t) + dH_N(t) \\ &\quad + eH_L + \varepsilon(t). \end{aligned} \tag{22}$$

a, b, c, d, e are undetermined coefficients, which represent the weights of the five factors in part 2, and the sum of the weights is 1. The weight needs to be determined according to the specific development stage of urban agglomeration. $\varepsilon(t)$ is residual variable.

4.3. Modeling Urban Agglomeration Transport Supply Structure. The actual volume of the four transport modes is treated as urban agglomeration transport supply volume. If x_1, x_2, x_3 denote the weights of three types of supply-demand structure, respectively, which are a transport supply and demand relation of a single city in the urban agglomeration, intercity relationship between center cities, and intercity relationship between a center city and a non-center city in the urban agglomeration, then transport supply structure vector at time t is expressed as formula (23) and (24).

$$\vec{G}_t = (g_{1t}, g_{2t}, g_{3t}, g_{4t}) \tag{23}$$

$$\begin{aligned} g_{it} &= \sum_{j=1}^{m} g_{ijt} \\ &= x_1 \sum_{j=k=1}^{m} g_{ijt(j,k)} + x_2 \sum_{j=1}^{m} \sum_{k=i+1}^{m} g_{ijt(j,k)} \\ &\quad + x_3 \sum_{j=1}^{m} \sum_{k=m+1}^{n} g_{ijt(j,k)}, \end{aligned} \tag{24}$$

where g_{it} is the transport volume of mode i in urban agglomeration and $g_{ijt}(j,k)$ is the transport volume of mode i between the city s_j and city s_k at time t.

Entropy of transport supply structure at time t can be calculated through formula (25).

$$H_G(t,g) = -\sum_{i=1}^{4} T_{it} \ln T_{it}. \tag{25}$$

And

$$T_{\text{it}} = \frac{g_{\text{it}}}{\sum_{i=1}^{4} g_{\text{it}}}. \tag{26}$$

4.4. Equilibrium Discriminating of Urban Agglomeration Transport Supply and Demand Structure. To describe the disequilibrium degree of urban agglomeration transport supply-demand structure, transport supply structure entropy and transport demand structure entropy are compared in the model. If the ratio of supply entropy and demand entropy is closer to 1, the two entropies are closer, and the transport supply and demand structure is more balanced. The disequilibrium degree at time t is calculated as follows:

$$G(t) = \left| 1 - \frac{H_G(t,g)}{H_D(t,g)} \right|. \tag{27}$$

If $H_D(t,d) = H_G(t,g)$, $G(t)$ will be zero which indicates an absolute equilibrium of the supply-demand structure.

According to the definition of the concept of equilibrium and the division of transport supply and demand structure equilibrium level given by [4], this paper divides the equilibrium degree of urban transport demand and supply structure into categories as shown in Table 1.

Table 1: The division of transport supply and demand structure equilibrium level.

types	equilibrium level	$G(t)$
equilibrium structure	high quality equilibrium	$0 < G(t) \leq 0.1$
	marginal equilibrium	$0.1 < G(t) \leq 0.2$
interim stage	just equilibrium	$0.2 < G(t) \leq 0.3$
	near disequilibrium	$0.3 < G(t) \leq 0.4$
disequilibrium structure	mild disequilibrium	$0.4 < G(t) \leq 0.6$
	severe disequilibrium	$0.6 < G(t) \leq 1$

5. Case Analysis

In order to confirm the above method, HuBaoOr urban agglomeration is taken as an example to analyze the equilibrium state of transport supply and demand structure.

HuBaoOr urban agglomeration includes three center cities, Hohhot, Baotou, and Ordos, and three non-center cities, Ulanqab, Byannur, and Wuhai.

In the calculation process, some data are difficult to access, so only two factors, the industrial structure adjustment of urban agglomeration and city economic relationship between cities, are used in modeling transport demand structure and the two factors' weights are equal, $a = c = 0.5$.

(1) Entropy of Urban Agglomeration Transport Demand Structure. Firstly, industrial structure entropy of urban agglomeration will be calculated using formula (2). Production value of three main industries in HuBaoOr urban agglomeration is shown in Table 2.

$$I_{1t} = \sum_{i=1}^{6} I_{i1t} = 629.01\,(billion)$$

$$I_{2t} = \sum_{i=1}^{6} I_{i2t} = 6359.89\,(billion)$$

$$I_{3t} = \sum_{i=1}^{6} I_{i3t} = 6520.68\,(billion) \tag{28}$$

$$P_{1t} = \frac{I_{1t}}{\sum_{i=1}^{3} I_{it}} = 0.046560293$$

$$P_{2t} = \frac{I_{2t}}{\sum_{i=1}^{3} I_{it}} = 0.470768891$$

$$P_{3t} = \frac{I_{3t}}{\sum_{i=1}^{3} I_{it}} = 0.482670816.$$

Then industrial structure entropy H_I is 0.8495.

$$H_I = -\sum_{j=1}^{3} P_{jt} \ln P_{jt} = 0.8495. \tag{29}$$

Secondly, entropy about economic relationship between cities of urban agglomeration will be calculated using formula (12). Center cities are considered in the model to simplify the process. Index of GDP, population, and distance needed in the calculation process are shown in Tables 3, 4, and 5.

Table 2: Production value of three main industries in HuBaoOr urban agglomeration.

city	primary industry (billion yuan)	secondary industry (billion yuan)	Tertiary industry (billion yuan)
Hohhot	126.23	867.08	2097.21
Baotou	101.05	1830.64	1850.24
Ordos	98.96	2400.01	1727.15
Ulanqab	132.39	443.84	337.53
Byannur	165.64	450.55	271.24
Wuhai	4.74	367.77	237.31

Table 3: GDP of each city in 2015(billion yuan).

Hohhot	3090.52
Baotou	3781.93
Ordos	4426.13
Ulanqab	913.77
Byannur	887.43
Wuhai	609.82

Recessive economic gravity between center cities is calculated by formula (11).

$$Hohhot-Baotou:\ F_1' = k \bullet \frac{P_{Hu}^1 V_{Hu} * P_{Bao}^1 V_{Bao}}{r_{Hu-Bao}^2} = 335.85$$

$$Hohhot-Ordos:\ F_2' = k \bullet \frac{P_{Hu}^1 V_{Hu} * P_{Or}^1 V_{Or}}{r_{Hu-Or}^2} = 158.33$$

$$Baotou-Ordos:\ F_3' = k \bullet \frac{P_{Bao}^1 V_{Bao} * P_{Or}^1 V_{Or}}{r_{Ba0-Or}^2} = 175.40 \tag{30}$$

$$R_{1t} = \frac{F_1'}{\sum_{j=1}^{3} F_j'} = 0.50$$

$$R_{2t} = \frac{F_2'}{\sum_{j=1}^{3} F_j'} = 0.24$$

$$R_{3t} = \frac{F_3'}{\sum_{j=1}^{3} F_j'} = 0.26.$$

Then entropy about economic relationship between cities of urban agglomeration H_F is 1.0393.

$$H_F = -\sum_{j=1}^{3} R_{jt} \ln R_{jt} = 1.0393. \tag{31}$$

Table 4: Population of each city in 2015(billion people).

Hohhot	0.0287
Baotou	0.0265
Ordos	0.0194
Ulanqab	0.0214
Byannur	0.0167
Wuhai	0.0053

Table 5: Distance between center cities (hour).

	Transport mode	Distance(km)	Time(h)
Hohhot-Baotou	High-speed rail	165km	0.75h
Hohhot-Ordos	High-speed rail	275km	1.05h
Baotou-Ordos	railway	147km	1.03h

Finally, the demand structure entropy $H_D(t)$ of HuBaoOr urban agglomeration is 0.9444.

$$H_D(t) = aH_I(t) + cH_F(t) = 0.9444. \tag{32}$$

(2) Entropy of Transport Supply Structure. Transport supply of urban agglomeration includes a transport supply within a single city in the urban agglomeration, intercity supply between center cities in the urban agglomeration, and intercity supply between the center city and a non-center city, three types. Entropy of transport supply structure can be calculated with formula (25) and relevant data are showed in Table 6. The transport volume $g_{it}$of mode i in urban agglomeration including two parts, passenger volume $g_{1t}(passenger)$and freight volume $g_{1t}(freight)$.

$$\begin{aligned} g_{1t}(passenger) &= \sum_{j=1}^{3} g_{1jt}(passenger) = 4963.35 \\ g_{2t}(passenger) &= \sum_{j=1}^{3} g_{2jt}(passenger) = 4808.32 \\ g_{3t}(passenger) &= \sum_{j=1}^{3} g_{3jt}(passenger) = 922.02. \end{aligned} \tag{33}$$

According to China's statistic system, the conversion ratio of passenger turnover and freight turnover [17] is shown in Table 7.

$$\begin{aligned} g_{1t} &= g_{1t}(passenger) + 10g_{1t}(freight) = 18067.4 \\ g_{2t} &= g_{2t}(passenger) + g_{2t}(freight) = 1997.6 \\ g_{3t} &= g_{3t}(passenger) + 13.7g_{3t}(freight) = 1708.6 \\ T_{1t} &= \frac{g_{1t}}{\sum_{i=1}^{3} g_{it}} = 0.8297 \\ T_{2t} &= \frac{g_{2t}}{\sum_{i=1}^{3} g_{it}} = 0.0917 \\ T_{3t}' &= \frac{g_{3t}}{\sum_{i=1}^{3} g_{it}} = 0.0784 \end{aligned} \tag{34}$$

So, entropy of transport supply structure is 0.5736.

Then, the equilibrium degree $G(t)$ at time t can be acquired.

$$G(t) = \left|1 - \frac{H_G(t)}{H_D(t)}\right| = 0.392. \tag{35}$$

Comparing the value $G(t)$ of HuBaoOr urban agglomeration with Table 1, the supply-demand structure is at "near disequilibrium state," which means there is a slightly mismatch between transport supply structure and transport demand structure. If the later development is not planned in advance, the disequilibrium will be significant which may result in a series of transport problems and waste of resources.

In the current state, we can adjust the transport supply of each mode in real time according to the transport demand structure changed with the development of urban agglomeration in case of disequilibrium. We try to realize a dynamic equilibrium state within a certain variation range through the adjustment and coordination measures which include the proportion of modes of transport and the spatial hierarchical structure of transport system between cities.

At present, the development of HuBaoOr urban agglomeration is in its initial stage and there is a slight mismatch between transport supply structure and transport demand structure. Later, in order to promote the economic development of urban agglomeration, it is necessary to study it in detail from the point of view of supply structure, and the focal points of construction in different regions should be different. Investment should be focused on the supply modes with higher demand in the region.

6. Conclusion

In this paper, we propose a transport supply-demand equilibrium problem of the urban agglomeration. Entropy theory is used to build transport demand structure entropy model, supply structure entropy model, and discriminant model of transport supply-demand equilibrium state to calculate the equilibrium degree of the supply-demand structure in an urban agglomeration. Following this, we have provided a case analysis of the proposed model using data of HuBaoOr urban agglomeration.

The influence factors of transport demand used to reflect the demand structure of urban agglomeration can better reflect the potential transport demand of the urban agglomeration, which can really reflect the influence of social and economic development of urban agglomeration on transport demand structure. The discriminant model of transport supply-demand equilibrium structure of urban agglomeration can be applied to most of the general urban agglomeration to judge the transport supply and demand structure equilibrium state and the appropriate measures

Table 6: Turnover of passenger and freight of HuBaoOr urban agglomeration in 2015.

		turnover	roads	railways	aviation
Baotou	internal transport supply	passenger	13.35	1.53	0
		freight	49.35	9.56	0
	intercity transport supply with center cities	passenger	42.54	30.53	17.89
		freight	437.64	421.54	27.33
	intercity transport supply with non-center cities	passenger	4.62	58.25	21.45
		freight	62.25	84.63	19.34
Ordos	internal transport supply	passenger	14.74	7.88	0
		freight	56.25	9.24	0
	intercity transport supply with center cities	passenger	57.38	68.32	32.15
		freight	478.25	474.21	20.08
	intercity transport supply with non-center cities	passenger	8.69	28.43	32.54
		freight	51.36	95.36	15.36
Hohhot	internal transport supply	passenger	14.32	3.09	0
		freight	48.56	5.85	0
	intercity transport supply with center cities	passenger	72.38	43.25	37.68
		freight	552.79	517.83	18.26
	intercity transport supply with non-center cities	passenger	7.68	52.75	32.22
		freight	46.72	85.35	11.65

Table 7: Conversion ratio of passenger turnover and freight turnover.

	Roads transport	Railways transport	Aviation transport	Water transport
Conversion ratio of passenger turnover and freight turnover	10:1	1:1	13.7:1	2:1

should be taken to adjust supply and demand of various transport modes promoting the achievement of dynamic equilibrium state.

There are also some research shortcomings in this paper. This paper does not consider the effect of transportation objects on supply-demand structure. Because the transport modes preferred by passenger transport and freight transport are different, and the proportion of urban passenger and freight demand is obviously different owing to different industry characteristics of cities, it will result in different demand structure. Therefore, the supply structure of different cities in the urban agglomeration should be treated differently in future research. There are also many undetermined parameters of the model in this paper. The values of some parameters are assumed by authors rather than derived by actual data. This paper only gives a kind of research thought and methods of analyzing problems. The appropriate setting of the parameters' value and the effect of parameter values on the results are worth investigating in the future and important real-word constraints to consider in future research.

Conflicts of Interest

The authors declare that they have no conflicts of interest.

Acknowledgments

This work was financially supported by the Chinese National Natural Science Foundation (Project No. 51668048), the Inner Mongolia Natural Science Projects (Project No. 2014BS0501), and the Higher Educational Scientific Research Projects of Inner Mongolia Autonomous Region (Project No. NJZY17012).

References

[1] A. Zedgenizov and D. Burkov, "Methods for the Traffic Demand Assessment Based on the Quantitative Characteristics of Urban Areas Functioning," pp. 724–730.

[2] S. Chandra, M. Braughton, L. D. Galicia, A. Sanchez, M. Medina, and R. Aldrete, "A Multi-modal Transportation Score to Evaluate Infrastructure Supply-demand for Commuters," in *Proceedings of the International Conference on Sustainable Design, Engineering and Construction, ICSDEC 2016*, pp. 304–311, USA, May 2016.

[3] T. J. Sanderson and I. B. Greenwood, "Demand management: a motoring organisation's perspective," in *International Transport Conference*, vol. 91, pp. 103–108, 1991.

[4] Y. Jing, *Study on the structure equilibrium of transportation supply and demand*, Chang'an university, 2007.

[5] T. Jianqiao and W. Ciguang, "Adaptability between supply structure and demand structure of regional freight based on information entropy theory," *Journal of highway and transportation research and development*, vol. 30, no. 4, pp. 148–154, 2013.

[6] Lu. Huapu, "Strategies of supply of urban transport infrastructure and travel demand management," *Urban transport of China*, vol. 10, no. 3, pp. 1–6, 2012.

[7] Wu. Feng and Shi. Qizhou, "Entopy theory based study on relationship between transportation and industry systems," *Journal of transportation systems engineering and information technology*, vol. 6, no. 1, pp. 71–74, 2006.

[8] J. Wei and L. Li, "Model of dynamic relationship between traffic supply and demand," *Chang'an Daxue Xuebao (Ziran Kexue Ban)/Journal of Chang'an University (Natural Science Edition)*, vol. 30, no. 6, pp. 86–98, 2010.

[9] D. Hörcher and D. J. Graham, "Demand imbalances and multi-period public transport supply," *Transportation Research Part B: Methodological*, vol. 108, pp. 106–126, 2018.

[10] Wu. Wenzheng, *Study on equilibrium theory of transportation supply and demand*, Changan university, 2005.

[11] G. W. Accessibility, "Urbanization and Transportation: Towards an operational model transportation policy and urban rank: A model for the optimization of transport infrastructure. The transport Geography study group at the department of geography," *university of salford*, 2005.

[12] Yu. Ge, Z. Xiuyuan, and L. Wei, "Study on the adaptability of urban traffic infrastructural investment and traffic demand," *Mathematics in practice and theory*, vol. 12, no. 39, pp. 41–47, 2009.

[13] P. Wu, *The study on mixed highway network design for urban agglomeration*, Changsha university of science and technology, 2008.

[14] Z. Liu, *A study on transportation system planning for urban agglomerations*, Changan university, 2011.

[15] Gao. Yan, *Study on the evaluation for the inter-city transportation system development of urban agglomeration*, Changan university, 2009.

[16] Li. Yanyan, *Research on disequilibrium characteristics of urban traffic supply and demand*, Jilin University, 2011.

[17] *Dictionary of Communication Additions Series*, Shanghai Science and Technology Literature Press, 2008.

7

The Relative Influence of Proximity to Fast Road Infrastructure, Accessibility, and Deprivation on Crime

Margaret O'Mahony

Trinity Centre for Transport Research, Department of Civil, Structural & Environmental Engineering, Trinity College Dublin, Dublin 2, Ireland

Correspondence should be addressed to Margaret O'Mahony; margaret.omahony@tcd.ie

Academic Editor: Juan C. Cano

Irish rural areas saw an increase in the number of burglaries in the period following the completion of a network of high speed motorway interurban routes. To test the hypothesis that the new fast infrastructure was providing quick exits by urban based criminal gangs, Irish crime statistics along with census data were combined in a logistic regression analysis to assess the relative influence of accessibility, deprivation, population, and proximity to fast road infrastructure on the number of burglaries in areas. It was found that accessibility had a higher influence on crime than area deprivation and that proximity to fast road infrastructure was not significant.

1. Introduction

In relation to the association between socioeconomic status and criminal offending, the literature suggests that poverty has an effect on serious crime [1]. Relative deprivation is more important than absolute poverty [2] and others suggest that the educational level of parents is key [1]. Differences in spatial distribution of crime have been found [3]; for instance, crime is typically higher in urban areas [4] and in neighborhoods with high population density [5].

Drawve and Barnum [6] noted that crime is influenced by physical place features such as alcohol outlets, restaurants, and public transportation stops. They developed a more discerning method of analysis where different police divisions in a single city were viewed as their own environmental contexts to produce localized spatial risk assessments. Other work noted that it is the clustering of disadvantaged neighborhoods in close proximity that is particularly problematic [7].

Deprivation indices are practical measures that can be used to identify areas characterized by socioeconomic marginalization [8]. Stucky et al. [9] found that, consistent with previous research, lower levels of income were associated with higher violent and property crime counts and that local and nearby income inequality affects crime. When developing deprivation indices for Ireland, Haase and Pratschke [10] used scores that can enable comparisons to be made over time. This approach was first applied in an analysis of data from the 1991, 1996, and 2002 census [10] and is used in the current Pobal HP Deprivation Index for Small Areas [11]. The index focuses on three dimensions: demographic profile, social class composition, and the labour market situation. The index has gained recognition for being the first census-based index that has enabled comparability of a multivariate measure of deprivation over successive census periods. The work presented here uses the HP Deprivation Index as an indication of the socioeconomic characteristics of an area as a means of capturing the potential influence that may have on crime in the area.

In relation to the connection between transport and crime, others [12–14] have focused on the links between public transport and crime and in particular the impact that crime has on public transport use and personal safety while using transport facilities. Little research has been done on the influence of proximity to fast road networks on crime to date and the research presented here attempts to fill that gap.

Ireland is an island with a population of 4.7 million situated on the west of Europe. Its capital city is Dublin, on the east coast, and four other cities, Cork, Galway, Limerick, and Waterford, are key urban areas in a predominantly rural country. Faced with the challenges that the provision of

TABLE 1: Proximity to service centre scoring system [15].

Population	Level	Time from ED				
		<10 mins	10–20 mins	20–30 mins	30–60 mins	>60 mins
Towns 1,500–5,000	1	4	3	2	1	0
Towns 5,000–10,000	2	5	4	3	2	0
Towns 10,000–50,000 (excl 5 cities)	3	6	5	4	3	0
Towns > 50,000 (i.e., 5 cities)*	4	7	6	5	4	0

*Dublin, Cork, Galway, Limerick, Waterford.

transport infrastructure and public transport services to a large rural area pose, the Irish Government developed a transport accessibility index, made up of two components (described in more detail later): (1) proximity to public transport and (2) proximity to service centres (in other words, towns, or large metropolitan centres) [15]. There was a deliberate choice to only include transport related factors in the index and not to include measures of socioeconomic factors which were included in a separate deprivation index.

The key research question addressed here is to what extent proximity to motorways may be contributing to the high incidence of burglaries in rural areas in Ireland and to assess its importance relative to the influence of other variables such as accessibility, deprivation, and the area's population. In addressing the research question presented above, a multinomial logistic regression analysis is conducted using the number of burglaries reported in 2015 in 398 areas in Ireland.

2. Methods

The dependent variable, the number of burglaries in an area, was obtained from the crime statistics for 2015 [16]. It should be noted that some potential underreporting of crimes in the database is a possibility, as indicated in a CSO review [17], with a potential error of 10% noted for burglaries. One of the independent variables, the number of households in an area, was obtained from the 2011 census data [18], the most recent available at the time. Another independent variable, the deprivation index, was obtained from the Central Statistics Office (CSO) databases from the 2011 census [18] and the accessibility index was calculated on the basis of the 2006 census data [15], the most recent available. Proximity to a motorway was measured using Google maps [19]. More detail on how the accessibility and deprivation indices and the proximity to a motorway were obtained is presented below.

2.1. Accessibility Index. The accessibility index was measured by [15] at the level of electoral district (ED) and a score was developed for each. The first part of the score related to access to public transport. For bus stops, a 1 km zone was created around every bus stop captured and the proportion of the population in the ED that fell within this zone was calculated. The ED received a score on the following basis:

(i) 0% of ED's population within 1 km of a bus stop: Score 0

(ii) 1–50% of ED's population within 1 km of bus stop: Score 1

(iii) 50 to 99% of ED's population within 1 km of a bus stop: Score 2

(iv) 100% of ED's population within 1 km of a bus stop: Score 3 [15].

For train stations, a 10-minute off-peak drive time zone was created around every train station captured. A similar scoring system using the same percentages of the population as above was applied; for instance, a score of 0 was given to EDs where 0% of the population had a drive time to the nearest train station of less than 10 mins and, at the other end of the scale, a score of 3 was given to an ED if 100% of its population had a drive time to the nearest train station of <10 mins. After scoring the EDs for both the bus stop and train station parts of the index above, the ED was given the maximum (not combined) score it received. The maximum an ED could score therefore for proximity to both bus and train services was 3.

The second part of the accessibility index was the distance to service centre score. Table 1 indicates the scoring system as defined by [15]. Travel times were calculated using average off-peak road speeds. In terms of interpreting the scoring system, if an ED was within 30 mins of a city and also within 10 minutes of a town of between 1,500 and 5,000 residents, a score of 5 was assigned, it being the higher of the two scores allocated. The final stage in creating the accessibility index was to add the two maximum scores from the public transport and service centre analyses, resulting in a range from zero (most inaccessible) to 10 (most accessible).

2.2. Deprivation Index. The deprivation index was calculated by [11] using a number of demographic, social class, and labour market factors from census data to estimate the relative affluence or disadvantage of an area. The deprivation index ranges from −35 (most disadvantaged) to +35 (most affluent) and the frequency distribution for the areas are presented in Figure 1.

2.3. Proximity to Motorway. Motorways are the highest class of road infrastructure and mostly (but not all) operate at speed limits of 120 km/hr (74.5 miles per hour) and have at least two lanes in each direction. Roads of lower class can have two lanes in each direction (dual carriageway standard with speed limits of 100 km/hr (62 miles/hr)) for short stretches but typically they have one lane in each direction and operate

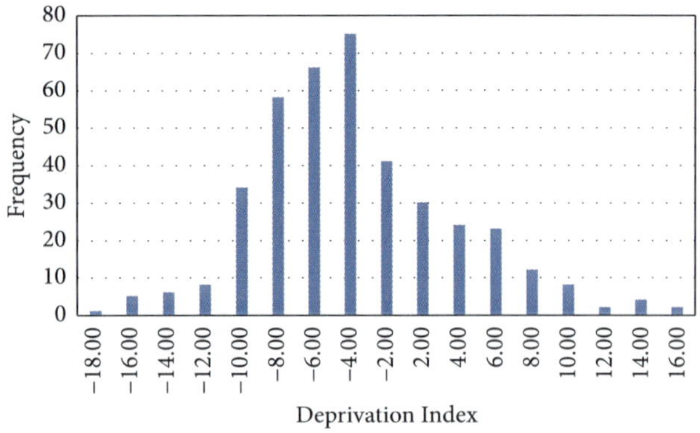

Figure 1: Frequency distribution of deprivation index.

at a speed limit of 100 km/hr. Large parts of the country are not close to the motorway network so it was considered that they would provide adequate controls against which the crime levels in areas close to motorways could be tested.

As the exact location of each burglary was not available, a proxy for the proximity to a motorway was used by measuring the time it took in traffic conditions in minutes from the nearest police station (to which the burglary was reported) to the nearest motorway using Google maps [19]. The results were then binned into the categories used in the analysis presented later.

3. Results and Discussion

After testing the correlation level between the number of burglaries from crime statistics for 398 regions within Ireland and deprivation index and finding it to be low at $R^2 = 0.26$, other independent variables were added to the analysis. A multinomial logistic regression was then performed with number of households, accessibility index, deprivation index, and proximity to a motorway (expressway) selected as independent variables. Table 2 includes variable definition and frequencies of the independent variables.

Before including variables in the model, a test for collinearity between them was completed, the results of which are shown in Table 3. It can be seen that each predictor has most of its variance loading on a different dimension indicating no multicollinearity [20]. An overall check on the relationship between the dependent and independent variables is the significance for the final model chi-squared value, after the independent variables have been added, the results of which are shown in Table 4, where significance is evident. The reduction in −2 LL value for the final model with the independent variables included compared with the model without them can also be seen in Table 4, another indicator that the addition of the independent variables improves the model.

The overall Nagelkerke statistic value was 0.72 representing relatively decent sized effects [20]. It can be seen in Table 4 that all independent variables are significant and therefore add to the model. The larger the chi-square value, the greater the loss of the model fit if that variable was

Table 2: Variable definition and frequencies.

Variable definition	Value	*N*	%
Number of burglaries	0–10	158	39.6
	11–20	85	21.3
	21–40	61	15.3
	41–60	19	4.8
	61–80	18	4.5
	81–100	8	2.0
	101–200	25	6.3
	201–1000	24	6.3
Number of households	0–250	61	15.3
	251–500	94	23.6
	501–750	68	17.0
	751–1200	51	12.8
	1201–3000	56	14.0
	3001–5000	27	6.8
	5001–7250	18	4.5
	7251–25000	24	6.0
Accessibility index	1.00	2	.5
	3.00	4	1.0
	4.00	26	6.5
	4.50	5	1.3
	5.00	46	11.5
	5.20	1	.3
	5.50	13	3.3
	6.00	96	24.1
	6.50	12	3.0
	6.70	1	.3
	7.00	84	21.3
	7.50	2	.5
	7.70	1	.3
	8.00	46	11.5
	8.30	1	.3
	8.40	1	.3
	8.50	4	1.0
	8.60	2	.5
	8.70	3	.8
	8.80	2	.5
	9.00	32	8.0
	9.10	1	.3
	9.30	1	.3
	9.60	1	.3
	9.70	1	.3
	10.00	10	2.5
Deprivation index	−18.00	1	.3
	−16.00	5	1.3
	−14.00	6	1.5
	−12.00	8	2.0
	−10.00	34	8.5
	−8.00	58	14.5
	−6.00	66	16.5
	−4.00	74	18.8
	−2.00	41	10.3

Table 2: Continued.

Variable definition	Value	N	%
	2.00	30	7.5
	4.00	24	6.0
	6.00	23	5.8
	8.00	12	3.0
	10.00	8	2.0
	12.00	2	.5
	14.00	4	1.0
	16.00	2	.5
	4.00	12	3.0
	5.00	25	6.3
	8.00	42	10.5
	9.00	34	8.5
	12.00	12	3.0
	13.00	76	19.3
	15.00	20	5.0
Proximity to motorway (mins)	21.00	32	8.0
	28.00	15	3.8
	31.00	27	6.8
	44.00	12	3.0
	47.00	20	5.0
	85.00	21	5.3
	92.00	32	8.0
	94.00	18	4.5

dropped. In this case, dropping number of households would result in the greatest loss of model fit, followed by accessibility index, proximity to motorway, and deprivation index, in that order. What is of interest here is the relatively low influence deprivation index is having on the model and this would not appear to support findings of other work [1]. Of further interest is the somewhat higher influence in the model of the accessibility index. Interpretation of this finding is complicated in that it combines proximity to public transport stations as well as an indication of how close the area is to a town or urban centre. The results support the results of others [4, 5] in suggesting that higher crime levels are evident in higher density or urban areas.

Classification accuracy compares the predicted group membership based on the logistic model with the actual. The benchmark is that the model is considered useful if it shows a 25% improvement over the rate of accuracy achievable by chance alone. The proportional by chance accuracy was computed by calculating the proportion of cases for each group based on the number of cases in each group for the dependent variable in Table 2 and then squaring and summing the proportion of cases in each group, the result of which was calculated to be 0.27. The proportional by chance accuracy criteria therefore $= 1.25 * 0.27 = 34\%$. The classification accuracy rate from the model was calculated to be 54.3% from Table 5 indicating that the model can be characterized as useful.

The model parameter estimates are presented in Table 6. The reference category refers to areas with more than 200 burglaries recorded and against which all other categories are referenced. Working down through each variable and examining the B values, we first look at the deprivation index. The B values are significant for the categories of 21–40, 41–60, and 101–200 burglaries which suggests that deprivation index is significant in distinguishing between geographical areas in those categories and areas in the reference category of >200 burglaries. The value of exp(B) of 0.847 for deprivation index in the 41–60-burglary category suggests that for every one unit increase in deprivation index the odds of an area with 41–60 reported burglaries decreased by 19% ($0.814 - 1 = -0.186$). The exp(B) values for the 21–40- and 101–200-burglary categories were similar at 0.847 and 0.895 suggesting similar relationships, all measured against the reference category, and when all other variables are controlled.

In examining the parameter estimates for the accessibility index, it can be noted first of all that this variable is significant in distinguishing the first three categories from the reference category but not for the others. Looking at the exp(B) values for the 0–10-burglary category, increasing the accessibility index by one unit would decrease the probability of an area being in the 0–10-burglary reported burglary category by 90%. Decreases in probability are also evident for the second and third categories, by 84% for the 11–20-burglary category and by 80% for the 21–40-reported-burglary category.

The number of households in an area is significant for almost all categories to some level except for one and the exp(B) values are all similar at ~0.999 or 1 meaning that for a unit increase in the number of households there is little change in the odds of an area being in those categories relative to the reference category. The last variable examined was the proximity of an area to a motorway. While the addition of this variable to the model was indicated as significant in Table 4, that is, it improved the model by its inclusion, it was not significant in distinguishing the categories from the reference category. To further demonstrate that proximity of an area to a motorway does not appear to influence the number of burglaries in that area, analysis of variance tests was completed indicating there was a significant difference between the proximity to motorway groups, $F(6, 391) = 45$, $P = 0.001$, but Games Howell post hoc tests revealed the differences were not between the lower and higher proximity distance groups as can be seen in Figure 2 where the means and confidence intervals (CI) are plotted. Overlap between the CI for the 20, 30, 90, and 100 km groups in particular suggests similarity in numbers of burglaries in these groups. Overlap is also evident for the 10 and 50 km proximity to motorway groupings.

The findings in relation to the influence of the number of households in an area and accessibility index tend to concur with the research of others [4, 5] who have suggested crime is higher in urban areas and in neighborhoods with high population density. Other research [7, 21–23] found a strong association between inequality and violent and property crime rates and this is somewhat aligned with the findings here in relation to the influence of deprivation where it was a useful indicator in some cases.

Offenders are known to commute a greater distance to commit a crime in rural areas [24], and in a rural context, the danger posed by outsiders has traditionally been seen as a

TABLE 3: Collinearity diagnostics.

Model	Eigenvalue	Condition index	Variance proportions (Constant)	Deprivation index	Accessibility index	Number of households	Proximity to motorway (mins)
1	3.153	1.000	.00	.02	.00	.02	.03
2	.973	1.801	.00	.43	.00	.20	.01
3	.515	2.475	.00	.45	.00	.28	.35
4	.342	3.037	.02	.04	.02	.28	.54
5	.017	13.485	.98	.06	.98	.22	.06

Collinearity diagnostics[a]

a. Dependent variable: number of burglaries.

TABLE 4: Model fitting and statistical significance of independent variables.

(a) Model fitting information

Model	Model fitting criteria −2 log likelihood	Likelihood ratio tests Chi-square	df	Sig.
Intercept only	1345.955			
Final	870.816	475.139	28	.000
N	398			
Nagelkerke	0.72			

(b) Likelihood ratio tests

Effect	Model fitting criteria −2 log likelihood of reduced model	Likelihood ratio tests Chi-square	df	Sig.
Intercept	979.601	108.785	7	.000
Deprivation index	892.174	21.358	7	.003
Accessibility index	913.287	42.471	7	.000
Number of households	1051.345	180.530	7	.000
Proximity to motorway (mins)	892.953	22.137	7	.002

TABLE 5: Classification table.

Observed	Classification Predicted 0–10	11–20	21–40	41–60	61–80	81–100	101–200	201–1000	Percent correct (%)
1.00	137	14	7	0	0	0	0	0	86.7%
2.00	53	27	4	0	0	0	1	0	31.8%
3.00	17	19	21	0	0	0	4	0	34.4%
4.00	2	2	11	1	1	0	2	0	5.3%
5.00	0	0	8	2	3	0	4	1	16.7%
6.00	0	0	3	0	1	0	2	2	0.0%
7.00	1	0	4	1	2	0	11	6	44.0%
8.00	0	0	0	0	2	0	6	16	66.7%
Overall%	52.8%	15.6%	14.6%	1.0%	2.3%	0.0%	7.5%	6.3%	54.3%

threat to rural communities [25, 26]. In the work presented here, proximity to fast road infrastructure was found to have limited influence on the number of burglaries in rural areas compared with the other variables tested. However, research in Australia found that for farm crime properties that are easily accessible by roads may be targeted [27]. Others have noted the common phenomenon of criminals from nearby cities travelling to a rural area to commit a crime and returning home [28].

3.1. Strengths and Limitations of the Research. The main strength of the research was the strong predictive performance of the model with a Nagelkerke R^2 value of 0.72 indicating that the set of independent variables used are

TABLE 6: Multinomial logistic regression model parameter estimate results.

Number of reported burglaries[a]	Parameter estimates							
	B	Std. error	Wald	df	Sig.	exp(B)	95% confidence interval for exp(B) Lower bound	Upper bound
0–10								
Intercept	26.325	5.665	21.596	1	.000			
Deprivation index	−.042	.074	.328	1	.567	.958	.829	1.108
Accessibility index	−2.301	.626	13.527	1	.000	.100	.029	.341
Number of households	−.004	.000	70.315	1	.000	.996	.995	.997
Proximity to motorway (mins)	.024	.018	1.672	1	.196	1.024	.988	1.061
11–20								
Intercept	22.388	5.645	15.730	1	.000			
Deprivation index	−.122	.073	2.805	1	.094	.885	.768	1.021
Accessibility index	−1.836	.620	8.762	1	.003	.159	.047	.538
Number of households	−.003	.000	52.478	1	.000	.997	.997	.998
Proximity to motorway (mins)	.010	.018	.288	1	.591	1.010	.974	1.046
21–40								
Intercept	19.450	5.597	12.077	1	.001			
Deprivation index	−.166	.069	5.773	1	.016	.847	.740	.970
Accessibility index	−1.617	.613	6.963	1	.008	.198	.060	.660
Number of households	−.001	.000	31.079	1	.000	.999	.998	.999
Proximity to motorway (mins)	.002	.018	.017	1	.897	1.002	.968	1.038
41–60								
Intercept	15.748	5.681	7.686	1	.006			
Deprivation index	−.206	.075	7.607	1	.006	.814	.703	.942
Accessibility index	−1.327	.629	4.449	1	.035	.265	.077	.910
Number of households	−.001	.000	15.493	1	.000	.999	.999	1.000
Proximity to motorway (mins)	−.019	.021	.815	1	.367	.981	.940	1.023
61–80								
Intercept	11.668	5.632	4.293	1	.038			
Deprivation index	−.118	.066	3.204	1	.073	.888	.780	1.011
Accessibility index	−.880	.618	2.031	1	.154	.415	.124	1.391
Number of households	−.001	.000	14.526	1	.000	.999	.999	1.000
Proximity to motorway (mins)	−.008	.018	.214	1	.643	.992	.957	1.028
81–100								
Intercept	13.015	6.073	4.592	1	.032			
Deprivation index	−.061	.076	.644	1	.422	.941	.811	1.092
Accessibility index	−1.275	.690	3.410	1	.065	.279	.072	1.081
Number of households	.000	.000	4.131	1	.042	1.000	.999	1.000
Proximity to motorway (mins)	−.017	.024	.460	1	.497	.984	.938	1.032
101–200								
Intercept	8.953	5.490	2.659	1	.103			
Deprivation index	−.111	.056	3.975	1	.046	.895	.802	.998
Accessibility index	−.714	.600	1.415	1	.234	.490	.151	1.587
Number of households	.000	.000	8.168	1	.004	1.000	.999	1.000
Proximity to motorway (mins)	.005	.014	.110	1	.741	1.005	.977	1.033

a. The reference category is areas with >200 burglaries.

useful in predicting the number of burglaries in an area. Most of this predictive power is driven by the population of the area and would be expected, but the work also highlights the importance of accessibility and deprivation, the latter having less influence than suggested by other research [1]. Accessibility is grounded in proximity to public transport and/or to a town or city and, as the second most influential variable in the model, suggests that burglary levels are higher when accessibility values are higher, another intuitive result. Although proximity to motorways is indicated as a

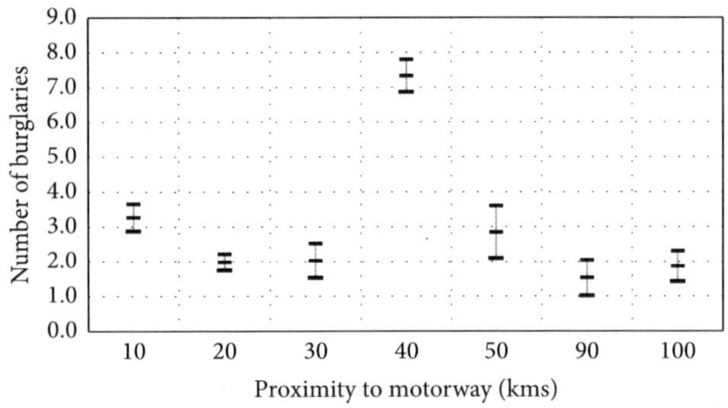

FIGURE 2: Means and confidence intervals of proximity to motorway groups.

statistically significant variable in the model, it is not useful in distinguishing areas by burglary rate which means it is not as useful a predictor.

A limitation of the work is that the data for all variables were not available for the same year. The accessibility index was derived by previous research for 2006 but was not repeated for later census data [15]. However, as it focuses on the proximity of a household to a bus stop and/or to a town or urban area, changes from year to year would be marginal and most influenced by a change in a bus route which in rural areas is subject to limited change. In terms of the methodological implications, an assessment was made of the potential impact of using the 2006 accessibility index with data from two other time points. The conclusion of the assessment was that while this is a limitation of the analysis, the potential error is likely to be small given the typically fixed nature of rural public transport routes year on year. Having said that, there is potential for some error but it is difficult to quantify. At the time the research was conducted, the 2011 census was the most recent and was used for obtaining the number of households and deprivation index. In terms of methodological implications, an assessment was made of the potential impact of using data from 2011 and combining it with data from two other time points. The alternative was to use census data from an earlier census, closer in time to 2006. On balance, and again due to the limited change in accessibility expected, year on year, it was decided that using the 2011 census data would be useful in capturing more recent household and deprivation index data. It is unlikely that there was a significant change in the number of households and deprivation between 2011 and 2015 as the country was in deep recession during those years with limited house building and an unchanging economic climate over that period, but some change cannot be discounted.

In relation to the choice of 2015 for the crime statistics data, this was necessary because data was not available in sufficiently accurate form for the year 2011 [17]. Improvements were made to the data collection and recording process by the police authority after 2011 following recommendations by the CSO [17] with a further review of the 2015 data demonstrating improvements, although a potential 10% underreporting in relation to burglaries is still likely in the 2015 data used here [17]. To assess the potential impact of the error, the model was rerun, after an average correction to allow for the underreporting was applied across the data. The observed predictive power for all independent variables improved marginally except for the accessibility index along with a slightly improved Nagelkerke R^2 value of 0.76. Again, the implications of using a different year for the crime data were assessed and the advantages and disadvantages were considered. Having assessed the information provided in the review by the CSO [17] of both the 2011 and 2015 datasets, it was considered prudent to use the 2015 dataset [17].

Another weakness was the reporting of crime statistics by police station at which they are reported rather than by the precise location of the crime. This was less than ideal and an assessment was made to consider the methodological implications and resulting limitations to the work. Individuals report crimes to the police station based in their area. The 398 police stations outside of Dublin county, and the focus of the analysis here, are located across the country, the geographical area of which is 70,273 km^2. At county level, some examples demonstrate the extent of the variation. Galway county has an area of 6,149 km^2 and 46 police stations cover that area meaning each covers an area of 133 km^2 on average giving an average radius from the police station to the edge of the area it covers of 6.5 km, assuming the police station is based at the centre of a circular area. Another large county has 27 police stations to cover 4,861 km^2 meaning each covers 180 km^2 with an average radius of 7.6 km, assuming a circular area of coverage (and assuming the road runs on a radial of the circle). An extrapolation of this to estimate the implications on the travel time to the nearest motorway would be perhaps overly simplistic and fraught with error but, for instance, a 10 km difference assuming a worst case scenario of the criminals travelling at the national speed limit of 80 km/hr would result in an error of 7.5 minutes. If they were using excessive speed, as might be expected when departing the scene of a crime, then this would be lower (at 100 km/hr it would be 6 mins, at 120 km/hr it would be 5 mins). It is difficult to assess how this error affects the model outcome but it is important to note there is an error associated with this variable.

4. Conclusions

The paper examines the influence of a number of area descriptive factors on crime rates, measured for the purposes of the work presented here, as number of burglaries reported in 2015. In particular, it examined the number of households in an area, accessibility index, deprivation index, and the proximity to fast road infrastructure. The logistic regression modelling exercise found that the number of households in an area had most influence on crime levels, followed by the accessibility index, proximity to motorway, and deprivation index in that order. Interesting results include the marginally higher relative importance of accessibility index compared with deprivation index suggesting that greater accessibility, influenced in part by proximity to a town or urban centre, has somewhat more influence on crime rates than the deprivation level of the area. While the proximity of an area to a motorway is shown in the model to be significant and improved the model marginally, its influence is not one that distinguishes areas of high and low distance from motorways in terms of the number of burglaries. This seems to refute the hypothesis that rural crime levels in areas close to fast road infrastructure were higher because of the efficient escape routes it provides for city based gangs.

Conflicts of Interest

The author declares that there are no conflicts of interest.

References

[1] J. Savolainen, R. Paananen, M. Merikukka, M. Aaltonen, and M. Gissler, "Material deprivation or minimal education? Social class and crime in an egalitarian welfare state," *Advances in Life Course Research*, vol. 18, no. 3, pp. 175–184, 2013.

[2] J. G. Bernburg, T. Thorlindsson, and I. D. Sigfusdottir, "Relative deprivation and adolescent outcomes in Iceland: A multilevel test," *Social Forces*, vol. 87, no. 3, pp. 1223–1250, 2009.

[3] R. Hudson, "Thinking through the relationships between legal and illegal activities and economies: spaces, flows and pathways," *Journal of Economic Geography*, vol. 14, no. 4, pp. 775–795, 2014.

[4] S. L. Myers Jr., "Crime in urban areas: New evidence and results," *Journal of Urban Economics*, vol. 11, no. 2, pp. 148–158, 1982.

[5] I. Hoch, "Factors in urban crime," *Journal of Urban Economics*, vol. 1, no. 2, pp. 184–229, 1974.

[6] G. Drawve and J. D. Barnum, "Place-based risk factors for aggravated assault across police divisions in Little Rock, Arkansas," *Journal of Crime and Justice*, vol. 41, no. 2, pp. 173–192, 2018.

[7] A. W. Chamberlain and J. R. Hipp, "It's all relative: Concentrated disadvantage within and across neighborhoods and communities, and the consequences for neighborhood crime," *Journal of Criminal Justice*, vol. 43, no. 6, pp. 431–443, 2015.

[8] P. Cabrera-Barona, C. Wei, and M. Hagenlocher, "Multiscale evaluation of an urban deprivation index: Implications for quality of life and healthcare accessibility planning," *Applied Geography*, vol. 70, pp. 1–10, 2016.

[9] T. D. Stucky, S. B. Payton, and J. R. Ottensmann, "Intra- and inter-neighborhood income inequality and crime," *Journal of Crime and Justice*, vol. 39, no. 3, pp. 345–362, 2016.

[10] T. Haase and J. Pratschke, Deprivation and its Spatial Articulation in the Republic of Ireland, 2005, http://trutzhaase.eu/wp/wp-content/uploads/R-2005-Deprivation-and-its-Spatial-Articulation.pdf.

[11] T. Haase, J. Pratschke, and J. Gleeson, "All-island deprivation index: towards the development of consistent deprivation measures for the island of Ireland," *Borderlands: The Journal of Spatial Planning in Ireland*, pp. 21–37, 2012.

[12] V. Spicer and J. Song, "The impact of transit growth on the perception of crime," *Journal of Environmental Psychology*, vol. 54, pp. 151–159, 2017.

[13] P. Cozens, R. Neale, J. Whitaker, and D. Hillier, "Managing crime and the fear of crime at railway stations—a case study in South Wales (UK)," *International Journal of Transport Management*, vol. 1, no. 3, pp. 121–132, 2003.

[14] O. T. McCarthy, B. Caulfield, and M. O'Mahony, "How transport users perceive personal safety apps," *Transportation Research Part F: Traffic Psychology and Behaviour*, vol. 43, pp. 166–182, 2016.

[15] Gamma and T. Haase, "Towards the development of a transport accessibility index. An analysis of geographical remoteness to support the Rural Transport Programme in the Republic of Ireland," http://trutzhaase.eu/publications/towards-the-development-of-a-transport-accessibility-index/.

[16] CSO Crime Statistics, 2015, http://www.cso.ie/en/statistics/crimeandjustice/.

[17] CSO, "Review of quality of crime statistics 2016," http://www.cso.ie/en/media/csoie/releasespublications/documents/crime-justice/2016/reviewofcrime.pdf.

[18] CSO, *Census 2011 Place of Work—Census of Anonymised Records (POWSCAR)*, Central Statistics Office, Cork, Ireland, 2012.

[19] Google maps https://maps.google.com/, 2017.

[20] A. Field, *Discovering Statistics Using IBM SPSS Statistics*, Sage Publishing, 4th edition, 2012.

[21] P. Buonanno and J. F. Vargas, "Inequality, crime, and the long run legacy of slavery," *Journal of Economic Behavior & Organization*, 2017.

[22] T. C. Pratt and F. T. Cullen, "Assessing macro-level predictors and theories of crime: a meta-analysis," *Crime and Justice*, vol. 32, pp. 373–450, 2005.

[23] R. J. Sampson, J. D. Morenoff, and T. Gannon-Rowley, "Assessing "neighborhood effects": social processes and new directions in research," *Annual Review of Sociology*, vol. 28, pp. 443–478, 2002.

[24] P. Wiles and A. Costello, *The "Road to Nowhere": The Evidence for Travelling Criminals*, HO Research Study no. 207, Home Office, London, UK, 2000.

[25] D. Gilling, "Governing crime in rural UK: risk and representation," in *Constable Countryside: Policing, Rurality and Governance*, R. I. Mawby and R. Yardley, Eds., pp. 69–79, Ashgate, Aldershot, UK, 2011.

[26] K. Halfacree, "Still 'out of place in the country'? Travellers and the post productivist rural," in *Constable Countryside: Policing, Rurality and Governance*, R. I. Mawby and R. Yardley, Eds., pp. 123–135, Ashgate, Aldershot, UK, 2011.

[27] E. Barclay and J. F. Donnermeyer, "Crime and security on agricultural operations," *Security Journal*, vol. 24, no. 1, pp. 1–18, 2011.

[28] R. I. Mawby, "Exploring the relationship between crime and place in the countryside," *Journal of Rural Studies*, vol. 39, pp. 262–270, 2015.

8

Design and Profit Allocation in Two-Echelon Heterogeneous Cooperative Logistics Network Optimization

Yong Wang,[1,2] Yingying Yuan,[1] Kevin Assogba,[3] Ke Gong,[1] Haizhong Wang,[3] Maozeng Xu,[1] and Yinhai Wang[4,5]

[1] *School of Economics and Management, Chongqing Jiaotong University, Chongqing 400074, China*
[2] *School of Management and Economics, University of Electronic Science and Technology, Chengdu 610054, China*
[3] *School of Civil and Construction Engineering, Oregon State University, Corvallis, OR 97330, USA*
[4] *Department of Civil and Environmental Engineering, University of Washington, Seattle, WA 98195-2700, USA*
[5] *Transportation Data Science Research Center, College of Transportation Engineering, Tongji University, Shanghai 201804, China*

Correspondence should be addressed to Yong Wang; yongwx6@gmail.com, Kevin Assogba; kevinassogba@gmail.com, and Ke Gong; gks_cn@163.com

Academic Editor: Ángel Ibeas

In modern supply chain, logistics companies usually operate individually and optimization researches often concentrate on solving problems related to separate networks. Consequences like the complexity of urban transportation networks and long distance deliveries or pickups and pollution are leading problems to more expenses and more complaints from environment protection organizations. A solution approach to these issues is proposed in this article and consists in the adoption of two-echelon heterogeneous cooperative logistics networks (THCLN). The optimization methodology includes the formation of cooperative coalitions, the reallocation of customers to appropriate logistics facilities, and the determination of the best profit allocation scheme. First, a mixed integer linear programing model is introduced to minimize the total operating cost of nonempty coalitions. Thus, the Genetic Algorithm (GA) and the Particle Swarm Optimization (PSO) algorithm are hybridized to propose GA-PSO heuristics. GA-PSO is employed to provide good solutions to customer clustering units' reallocation problem. In addition, a negotiation process is established based on logistics centers as coordinators. The case study of Chongqing city is conducted to verify the feasibility of THCLN in practice. The grand coalition and two heterogeneous subcoalitions are designed, and the collective profit is distributed based on cooperative game theory. The Minimum Cost Remaining Savings (MCRS) model is used to determine good allocation schemes and strictly monotonic path principles are considered to evaluate and decide the most appropriate coalition sequence. Comparisons proved the combination of GA-PSO and MCRS better as results are found closest to the core center. Therefore, the proposed approach can be implemented in real world environment, increase the reliability of urban logistics network, and allow decision makers to improve service efficiency.

1. Introduction

Cooperative game theory consists in the formation of coalitions and the effective allocation of the collective payoff to each participant. In modern supply chain, coalitions can be made up of facilities from either the same level or different echelons. Therefore, in a network comprising logistics centers (LCs), Distribution Centers (DCs), Pickup Centers (PCs), and customers, coalitions are generally formed vertically with LCs, DCs, and PCs or horizontally among DCs or PCs. A Two-echelon Heterogeneous Cooperative Logistics Network (THCLN) is a collaborative network with vertical synergies among LCs, DCs, and PCs, homogeneous horizontal synergies among DCs or PCs, and heterogeneous synergies among LCs and DCs or PCs. Properly optimizing THCLN can not only enhance the operational efficiency and avoid waste of resources, but also maximize players' profits. In addition, the implementation of a coordinated transportation system can contribute to achieve more stable coalitions [1]. However, well-developed negotiation mechanisms and a fair allocation of payoffs are necessary to increase players' willingness to participate [2].

The optimization of THCLN is a complex process since multiple customers, transportation resources, and profit allocations problems are combined to enhance practicability. Previous researches have focused on establishing a mixed integer programming model and designing efficient algorithms for similar optimization purposes. But due to the number of customers, evolutionary algorithms are usually confronted to problems modeling the amount of available data. Thus, a commonly used approach consists in clustering customers into small units before the application of intelligent algorithms [3, 4]. Clustering can also facilitate the management of companies' activities and allow decision makers to obtain more effective results by assigning customer clustering units to suitable suppliers [5, 6].

Solving customer clustering units' assignment problem is a basic network optimization problem and first appeared as the Quadratic Assignment Problem (QAP) in Dantzig et al. (1959). The standard assignment problem is a traditional binary programming problem usually solved by exact methods. However, as the number of tasks had increased, QAPs became more complicated to solve. Therefore, researchers like Li et al. [7] concentrated on applying heuristic algorithms to assess large-scale location and layout problems. Recently, Sghir et al. [8] proposed an optimization method based on intelligent multiple agents to solve QAPs and provided the algorithm with the capability of learning and dynamically selecting agents. The efficiency of QAP solutions facilitates the optimization of vehicle routing problems or any logistics transportation-related subject. In other words, the reassignment of customers units not only does have significant impacts on the configuration of the network but also influences future decisions.

With the rising need of intelligent approaches for logistics network optimization, heuristic algorithms like the Particle Swarm Optimization (PSO) and the Genetic Algorithm (GA) have been successfully applied in many domains [9, 10]. Shimizu and Miura [11] proposed a multipopulation based discrete PSO algorithm in order to handle binary decision variables and optimize large-scale logistics networks. Cupić and Teodorović [12] proposed an integer programing model to maximize companies' total profits and service quality and employed the Genetic Algorithm to solve a multiobjective hub location-allocation problem. Haddar et al. [13] proposed a 0-1 mixed interlinear programming model and combined the quantum PSO with an iterative heuristic algorithm to solve the Knapsack sharing problem. Guo et al. [14] used GA and PSO to study forward and reverse flows of fresh food supply chains in Shanghai and verified the validity of the model under low carbon emission environments. Wang et al. [15, 16] combined an improved ant colony algorithm with GA to optimize networks through the reassignment of customers to proper logistics facilities. Alshamsi and Diabat [17] introduced a GA capable of solving complex problems with thousands continuous and binary variables, constraints, and multiple integer variables. The proposed algorithm was applied and proved many times faster than GAMS. Atashpendar et al. [18] extended the speed-constrained multiobjective PSO algorithm to increase computation speed, convergence, and solution quality.

In logistics network cooperation, game theoretic methods are usually applied to model participants' behaviors in different scenarios and evaluate their contribution to the coalition as well as the proportion of allocated profits [19, 20]. Hernández and Peeta [21] evaluated the influences of collaboration mechanism on carriers and explored the differences with noncollaborative environments. Findings expressed high degree collaboration as beneficial to increase participants' profits. Guajardo and Rönnqvist [22] addressed the collaborative logistics network optimization problem and studied coalitions' layouts under core stability and equilibrium constraints. Applications in the forest transportation domain and for oil operations have been used to verify the relevance of the proposed optimization model. Wang et al. [15, 16] introduced collaboration as means of optimizing a single echelon multiple centers vehicle routing problem and applied an improved version of the Shapley value model to allocate profit.

Extensive articles have solved variants of network optimization problems, but there are still some shortcomings in the literature. (a) Researchers had significant achievements in the optimization of multicenter logistics network. However, few articles studied networks where PCs and DCs can independently collaborate with LCs. (b) The design of cooperative networks with different types of logistics facilities has not been studied particularly from aspects like coalitions' homogeneity or heterogeneity. (c) Collaboration among facilities of the same coalitions should incorporate sharing of transportation resources in order to yield more cost savings.

To alleviate computational complexity, our paper will first examine similar characteristics of customers and cluster them into groups. Therefore, large logistics areas will be divided into smaller and easily manageable units to be further assigned to distribution or pickup centers. For the allocation of customer clustering units and vehicles, we combine Genetic Algorithm (GA) and Particle Swarm Optimization (PSO) algorithm to increase solution efficiency. In order to achieve stability, the consultative process among the participants is coordinated by logistics centers owing to their capability at performing distribution and collection activities. In such a position, LCs could be more influential and better coordinate DCs and PCs. Furthermore, we compute the collective payoff of each coalition and propose a profit-sharing mechanism using the Minimum Cost Remaining Savings (MCRS) model to ensure a fair distribution. Before the formation of a coalition, DCs and PCs used to independently operate at the second echelon of the logistics network and caused the unreasonable exploitation of resources, half loaded vehicles, and the increase of traffic density. By including the sharing transportation fleets in collaboration's contracts, companies would significantly save cost and rationally use available resources. In addition, an effective customer clustering units' assignment procedure should reduce the complexity of the logistics network and make the optimization process conducive to obtaining more efficient customer assignment results and fair profit distribution.

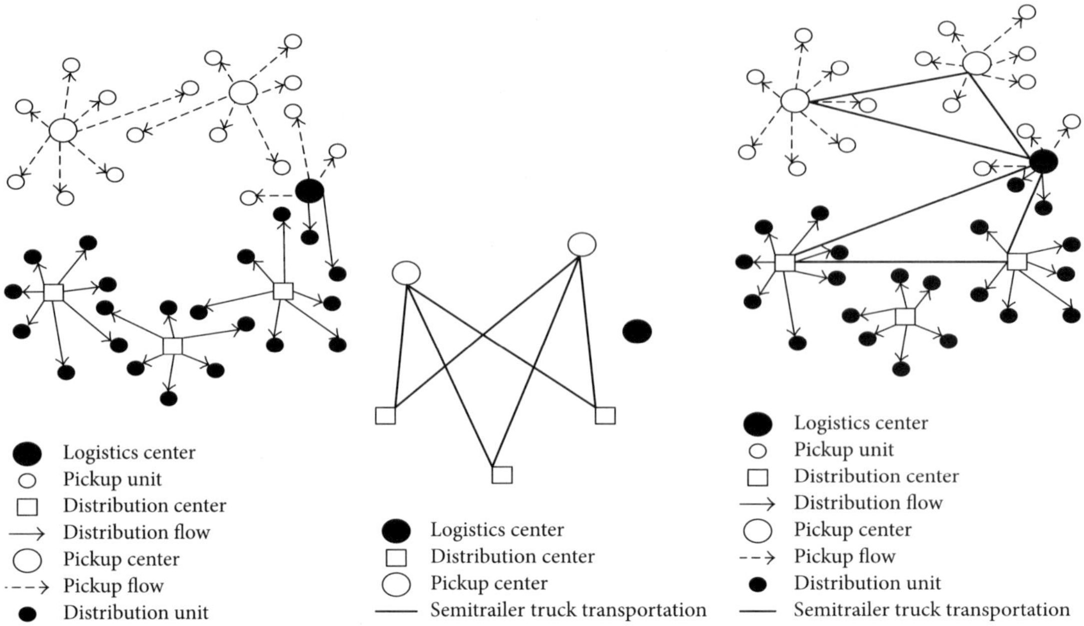

(a) Pickup and delivery before logistics network optimization (b) Product flow among facilities before logistics network optimization (c) After logistics network optimization

FIGURE 1: Logistics network diagram.

In summary, our research mainly contributes to the domain in the following directions: (1) a mixed integer programming model is established to interpret the two-echelon heterogeneous cooperative logistics network (THCLN) and minimize the total cost of nonempty coalitions; (2) variations in facilities transportation needs, routes, and required fleet size are considered as a result of cooperation, and a model is built to calculate payoffs based on members participation in the collective profit; (3) this paper combines the Genetic Algorithm and the Particle Swarm Optimization heuristics to propose a GA-PSO hybrid algorithm and find the most appropriated logistics facility for each customer clustering unit. (4) A practical study is adopted to support the effectiveness of the implemented approach in real world.

2. Problem Statement and Related Definition

2.1. Problem Description. THCLN increases the sustainability of logistics networks through the partitioning of customers, the sharing of transportation resources, and the reduction of greenhouse gas emissions. Figure 1 shows the network's structural changes which occur as logistics facilities agree to cooperate. Owing to their comprehensive functions, this article assigns the coordination of coalitions to LCs. For the synergy purposes, customer units can be redistributed among LCs, DCs, and PCs in accordance with the required type of service and their geographical locations. Therefore, distribution units can only be reassigned to LCs or DCs, whereas pickup units should be allocated to LCs or PCs. Transportation at the first echelon is assured by semitrailer trucks, and small vehicles are used to serve customers.

From a global point of view, Figure 1(a) illustrates a logistics networks characterized by the complexity of both distribution and pickup flows. In fact, companies usually invest in marketing events like advertisements and offer discounts in order to attract more customers to buy their products or require their services. Owing to these events, several customers overlook factors like distance or road conditions and make decisions which might prove irrational over a long period of time. Consequently, with the development of city, the increase in population and the number of logistics facilities, the entire network gains in complexity and gives rise to phenomena like traffic congestion and pollution. Despite the existence of cases where customers change suppliers as new facilities are established closer to them, others maintain partnerships due to loyalty entertained for years. As our proposed initial network presents a two-echelon transportation pattern, we also take into account semitrailer truck trips between PCs and DCs in order to highlight the existence of prior partnerships among centers. Those semitrailer truck transportations in Figure 1(b) can be interpreted as follows: PCs usually collect semifinished products from customers and process them before supplying DCs.

2.2. Logistics Network Optimization Architecture. Solving the THCLN consists in encouraging logistics facilities to synergize and share customers, semitrailer trucks, and vehicles. In addition, agreements should be achieved between PCs and DCs to use LCs as upper-level facility and redirect initial product flows through logistics centers. After the optimization, the network would be more organized, as shown in Figure 1(c), and long distance distributions or

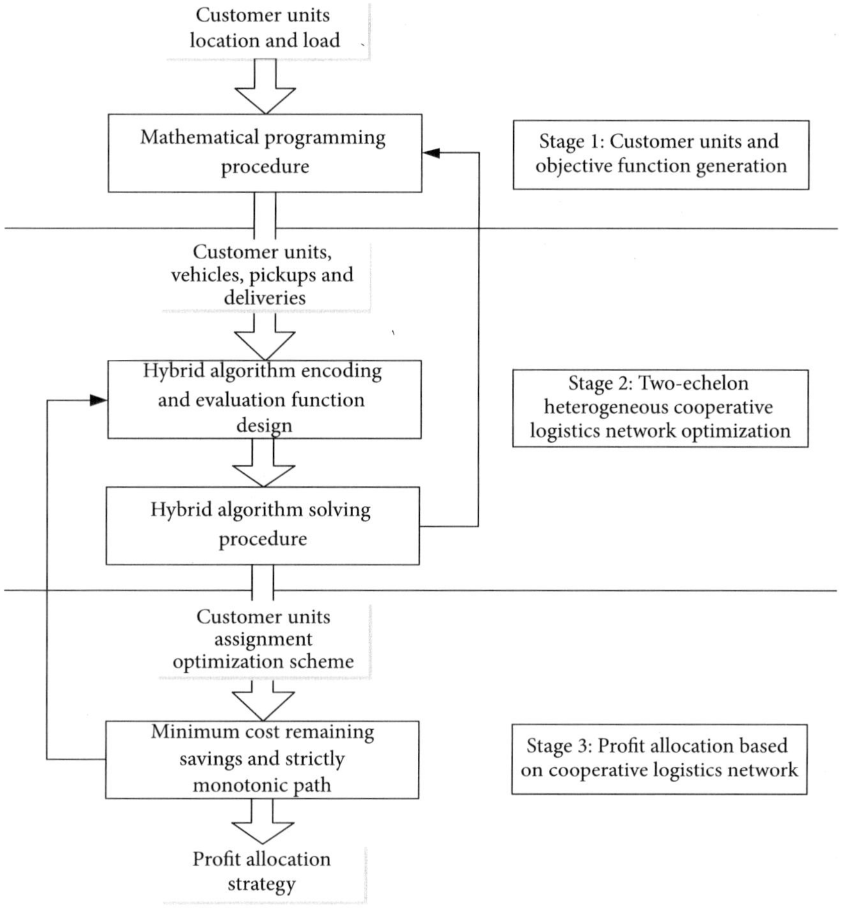

FIGURE 2: Logistics network optimization architecture.

pickups can be avoided within the same city. Products at the first echelon are still transported by means of semitrailer trucks and distribution or pickup operations carried out with a homogeneous fleet of vehicle.

The logistics network optimization architecture includes a three-phase solution approach shown in Figure 2. First, customer units' location and related demand data are collected and clustered, and a mixed integer linear mathematical model is proposed to interpret real world operations. Then, considering the available type of vehicle fleet as well as their capacities, GA and PSO are integrated to address the two-echelon heterogeneous logistics network optimization problem. Results obtained after running GA-PSO are used to compute the gap between objective function values before and after collaboration. Further, the Minimum Cost Remaining Savings method and the strictly monotonic path selection principles are consecutively applied. Their objectives are to distribute the cost savings among logistics facilities and evaluate the stability of one or multiple coalitions in the network.

2.3. Related Definition. In order to interpret the proposed problem into a mathematical model for further computations, some related variables are defined as follows:

$I\{i \mid i = 1, 2, 3, \ldots, m\}$ denotes the set of PCs, and m is the total number of PCs.

$J\{j \mid j = 1, 2, 3, \ldots, n\}$ denotes the set of DCs, and n is the total number of DCs.

$U\{u \mid u = 1, 2, 3, \ldots, a\}$ denotes the set of pickup units in the two-echelon heterogeneous logistics network, and a is the total number of pickup units.

$U'\{|u'|u' = 1, 2, 3, \ldots, b\}$ denotes the set of distribution units in the two-echelon heterogeneous logistics network, and b is the total number of distribution units.

$F\{f \mid f = m + 1\}$ expresses that LCs only apply their pickup capacities in the two-echelon heterogeneous logistics network.

$F'\{f' \mid f' = n+1\}$ expresses that LCs only apply their delivery capacities in the two-echelon heterogeneous logistics network.

$Q_{u,i}$ expresses the pickup quantity from the pickup unit u to LC or PC with in a working period, $i \in I \cup F$.

$Z_{j,u'}$ denotes the delivery demand transported from LC or DC to the distribution unit u' within a working period, $j \in J \cup F'$.

$S_{u,i}$ denotes the distance from the pickup unit u to LC or PC, $i \in I \cup F$.

$S'_{j,u'}$ denotes the distance from LC or DC to the distribution unit u', $j \in J \cup F'$.

$p_{i,h}$ expresses the transport quantity from pickup facility i to h within a working period, $i, h \in I \cup F$, and $h \neq i$.

$d_{j,h'}$ expresses the transport quantity from distribution facility j to h' within a working period, $j, h' \in J \cup F'$ and $h' \neq j$.

p_i expresses the pickup capacity of the ith PC or LC within a working period, where $i \in I \cup F$.

d_j expresses the delivery capacity of the jth DC or LC within a working period, where $j \in J \cup F'$.

T_p denotes the total pickup capacity of LC and DCs and can be expressed as $P = \sum_i p_i$, where $i \in I \cup F$.

T_d denotes the total delivery capacity of LC or PCs and can be expressed as $D = \sum_j d_j$, where $j \in J \cup F'$.

P_u expresses the pickup quantity of the pickup unit u within one working period.

$D_{u'}$ expresses the delivery demand of the distribution unit u' within one working period.

T expresses the total pickup quantity from all pickup units within one working period and can be expressed as $T = \sum_u q_u$, where $u \in U$.

T' expresses the total delivery demand of all distribution units within one working period, and can be expressed as $T' = \sum_{u'} q_{u'}$, where $u' \in U$;

$R\{r_{i,u} \mid r_{i,u} = 0, 1, \ i \in I \cup U, \ U \in U\}$ expresses the pickup relation matrix, where $r_{i,u} = 1$ indicates that LC or PC collects goods from pickup unit u, and $r_{i,u} = 0$ implies that LC or PC does not collect goods from the pickup unit.

$R'\{r'_{j,u'} \mid r'_{j,u'} = 0, 1, \ j \in J \cup G', \ u' \in U'\}$ expresses the distribution relation matrix, where $r'_{j,u'} = 1$ indicates that LC or DC delivers goods to distribution unit u' and $r'_{j,u'} = 0$ implies that LC or DC does not delivery goods to the distribution unit u'.

L_s denotes the loading capacity of a semitrailer truck.

L_v denotes the loading capacity of a vehicle.

f_{sr} denotes the freight rate of a semitrailer truck (dollar/pallet·kilometer).

f_{vr} denotes the freight rate of a vehicle (dollar/pallet·kilometer).

M_s denotes the average annual maintenance cost of a semitrailer truck (dollar/pallet·kilometer).

M_v denotes the average annual maintenance cost of a vehicle (dollar/pallet·kilometer).

γ expresses the variable cost coefficient of LCs

γ_i expresses the variable cost coefficient of DCs.

γ_j expresses the variable cost coefficient of PCs.

τ expresses the number of working periods.

$L_{\max}$ expresses the maximum delivery distance.

C expresses the cost savings provided by the coordinator when logistics facilities agree to cooperate.

G expresses the government incentive provided to logistics facilities in case of cooperation.

$L_{i,h}$ denotes the distance from the LC or PC i to h, $i, h \in I \cup F, i \neq h$.

$L'_{j,h'}$ denotes the distance from the LC or DC j to h', $j, h' \in J \cup F', j \neq h'$.

N_t denotes the number of trips of the vehicle within one working period.

N'_t denotes the number of trips of the semitrailer truck within one working period.

X_i denotes the cooperation relation between PCs and the coordinator; if PC i agrees to cooperate with LC, set $X_i = 1$; otherwise set $X_i = 0$.

X_j denotes the cooperation relation between DCs and the coordinator; if DC agrees to cooperate with LC, set $X_j = 1$; otherwise set $X_j = 0$.

φ denotes the fixed cost of LCs.

y_i denotes the fixed cost of PC i.

y_j denotes the fixed cost of DC j.

$N = \{0, 1, 2, \ldots, n\}$ denotes the sequence set of LC and DCs, where

$$\begin{Bmatrix} 0 & 1 & 2 & \cdots & n & 0 \\ \text{LC} & \text{DC}_1 & \text{DC}_2 & \cdots & \text{DC}_n & \text{LC} \end{Bmatrix} \tag{1}$$

with 0 representing LC. Notice that any DC can be the first node in coalitions without LC.

$q_{\beta,\alpha}$ denotes the variable transportation quantity between LCs and DCs, $\alpha, \beta \in N, \alpha < \beta$. For example,

$$\begin{Bmatrix} 0 & 1 & 2 & 0 \\ \text{LC} & \text{DC}_1 & \text{DC}_2 & \text{LC} \end{Bmatrix} \tag{2}$$

denotes that the semitrailer truck departs from LC, visits DC_1 and DC_2, and returns to LC; $q_{1,2}$ expresses the variable transport quantity from DC_1 to DC_2.

3. Model Formulation

The THCLN optimization problem is interpreted into a mixed integer linear programing model to minimize the total cost. The objective function is the aggregation of logistics facilities fixed and variable operations costs as well as transportation and maintenance costs of semitrailer trucks and vehicles. W_1 is the total transportation and maintenance cost among LCs and PCs (if cooperation occurs) within a working period; W_2 is the total transportation and maintenance cost among LCs and DCs (if cooperation occurs) within a working period; W_3 is the total transportation and maintenance cost of vehicles from LC or PC to each pickup unit and LC or DC to each distribution unit within a working period; and W_4 is the result of subtracting cooperation discounts from the sum of fixed and variable costs within a working period. The objective function can be expressed as follows:

$$\min \quad W = W_1 + W_2 + W_3 + W_4, \tag{3}$$

with

$$\begin{aligned}
W_1 &= \sum_{i,h\in I\cup F, i\neq h} \left(p_{i,h} \times f_{sv} \times L_{i,h} \times 2 \times N_t' + \frac{p_{i,h}}{L_s} \times \frac{M_s}{T} \right) \\
W_2 &= \sum_{j,h'\in J\cup F', j\neq h'} d_{j,h'} \times f_{sr} \times L'_{j,h'} \times N_t' + \sum_{\alpha,\beta\in J} q_{\beta,\alpha} \times f_{sr} \times L'_{j,h'} \times N_t' \\
&\quad + \frac{\sum_{j,h'\in J\cup F', j\neq h'} d_{j,h'} + \sum_{\alpha,\beta\in J} q_{\beta,\alpha}}{L_s} \times \frac{M_s}{T} \\
W_3 &= \sum_{u\in U} \sum_{i\in I\cup F} \left(Q_{u,i} \times r_{i,j} \times S_{u,i} \times 2 \times f_{vr} \times N_t \right) \\
&\quad + \frac{\sum_{i\in I\cup F, u\in U} \left(Q_{u,i} \times r_{i,u} \right)}{L_v} \times \frac{M_v}{T} \\
&\quad + \sum_{u'\in U'} \sum_{j\in J\cup F'} \left(Z_{j,u'} \times r'_{j,u'} \times S'_{j,u'} \times 2 \times f_{vr} \times N_t \right) \\
&\quad + \frac{\sum_{j\in J\cup F', u'\in U'} \left(Z_{j,u'} \times r'_{j,u'} \right)}{L_v} \times \frac{M_v}{T} \\
W_4 &= \sum_{i\in I} y_i + \sum_{j\in J} y_j + \varphi + \sum_{i\in I} \sum_{u\in U} \left(Q_{u,i} \times r_{i,u} \times y_i \right) \\
&\quad + \sum_{j\in J} \sum_{u'\in U'} \left(Z_{j,u'} \times r'_{j,u'} \times y_j \right) + \left[\sum_{i\in F} \sum_{u\in U} \left(Q_{u,i} \times r_{i,u} \right) \right. \\
&\quad \left. + \sum_{j\in F'} \sum_{u'\in U'} \left(Z_{j,u'} \times r'_{j,u'} \times y_j \right) \right] \times \gamma - \left(\sum_{i\in I\cup F} X_i \right. \\
&\quad \left. + \sum_{j\in J\cup F'} X_j \right) (C + G),
\end{aligned} \tag{4}$$

subject to

$$\sum_{i\in I\cup G} r_{i,u} = 1, \quad \forall u \in U \tag{5}$$

$$\sum_{j\in J\cup G'} r_{j,u'} = 1, \quad \forall u' \in U' \tag{6}$$

$$\begin{aligned}
r_{i,u} &= \{0, 1\}, \\
r_{j,u'} &= \{0, 1\} \\
&\forall i \in I \cup F,\ j \in J \cup F',\ u \in U,\ u' \in U'
\end{aligned} \tag{7}$$

$$p_i = \sum_{u\in U} r_{i,u} \times Q_{u,i}, \quad \forall i \in I \cup F \tag{8}$$

$$d_j = \sum_{u'\in U'} \left(r'_{j,u'} \times Z_{j,u'} \right), \quad \forall j \in J \cup F \tag{9}$$

$$P = \sum_{i\in I\cup F} r_{i,u} \times Q_{u,i}, \quad \forall u \in U \tag{10}$$

$$D = \sum_{j\in J\cup F'} \left(r'_{j,u'} \times Z_{j,u'} \right), \quad \forall u' \in U' \tag{11}$$

$$\begin{aligned}
S_{u,i} &> L_{\max}, \\
r_{i,u} &= 0, \\
Q_{u,i} &= 0, \\
&i \in I \cup F,\ \forall u \in U
\end{aligned} \tag{12}$$

$$\begin{aligned}
S'_{j,u'} &> L_{\max}, \\
r'_{j,u'} &= 0, \\
Z_{j,u'} &= 0, \\
&j \in J \cup F',\ \forall u' \in U'
\end{aligned} \tag{13}$$

$$\begin{aligned}
&Q_{u,i}, Z_{j,u'} \geq 0, \\
&\forall i \in I \cup F,\ \forall j \in J \cup F',\ u \in U,\ u' \in U'.
\end{aligned} \tag{14}$$

The objective function value is calculated using (3). Equations (4) are the subpart of the cost function and represent transportation, maintenance, and fixed and variable costs. Constraints (5)-(6) ensure that a customer unit is only served by a center. Constraint (7) introduces the binary relation variable between DCs, PCs, and their corresponding customers. Constraints (8)-(9) guarantee that delivery and pickup centers, respectively, meet their customers' needs. Constraint (10) stipulates that pickup activities can either be performed by LCs or PCs, whereas constraint (11) allows only LCs and DCs to deliver products to customer units. Constraints (12)-(13) regulate the maximum travelling distance of pickup and delivery vehicles. Constraint (14) defines nonnegative variables.

TABLE 1: Two-dimensional chromosome encoding table.

Pickup/distribution unit	1	2	⋯	i	⋯	L'	$L'+1$	⋯	j	⋯	L
Logistics facility number	$f_{k,1}$	$f_{k,2}$	⋯	$f_{k,i}$	⋯	$f_{k,L'}$	$f_{k,L'+1}$	⋯	$f_{k,j}$	⋯	$f_{k,L}$

4. The Hybrid Algorithm Solving Procedure

The Genetic Algorithm (GA) and Particle Swarm Optimization (PSO) algorithm are combined to design a GA-PSO hybrid algorithm used to address the two-echelon heterogeneous logistics network optimization model. GA is an evolutionary computing approach used to mimic the natural selection procedure and study combinatorial optimization problems [17, 23]. PSO is one of the swarm intelligence stochastic evolutionary metaheuristic approaches, proposed by Kennedy and Eberhart [24]. PSO is inspired by observations of the social behavior of bird flocking, fish schooling, and so on, based on swarm theory, and can be used to optimize the local and global locations according to the particle's position and velocity. Traditionally, these two algorithms have the merits of being population-based heuristics with the ability to evaluate the fitness of each possible solution. Through local and global search capabilities, GA and PSO have been successfully applied to solve problems in management, engineering, and pure science domains. Considering their individual abilities, we propose GA combined with PSO in a hybrid algorithm, which inherits the merits of both approaches and increases the probability of obtaining optimal solutions.

As a hybrid algorithm, GA-PSO is applied to meet the requirements of logistics network optimization for more complex algorithms. Existing traditional heuristics have already been proved efficient, but the increasing complexity of logistics networks limits their capacity at finding near-optimal solutions. In consequence, properly integrating different methods into a hybrid solution approach like GA-PSO can effectively improve the quality of optimization results. Other hybrid heuristics exist but some of them display weaknesses compared to GA-PSO in terms of performance. For example, Chen et al. [25] embedded a local search heuristic in the Ant Colony Optimization (ACO) algorithm to improve the solution search ability of ACO and address a task assignment problem. The proposed methodology had a satisfying global search capability but the local search presented insufficiencies. Combining GA and PSO on the contrary ensures good performances at both local and optimum search level. This integration can improve the optimization procedure and reduce the number of necessary iterations. The relevant operations and processes of the proposed hybrid algorithm are illustrated in sections below.

4.1. Hybrid Algorithm Encoding Scheme and Evaluation Function Design. Hybrid algorithm encoding and objective function designing are the key issues in the optimization operation processes [26]. In the two-echelon heterogeneous logistics distribution network, the pickup/delivery capacity of each logistics facility and the transportation cost among logistics facilities should be taken into account. Therefore, a two-dimensional chromosome encoding is proposed in this study. The first dimension of the chromosomes is denoted as a sequence of natural numbers: $1, 2, 3, \ldots, i, \ldots, L', L'+1, \ldots, j, \ldots, L$, and L' is the total number of pickup units, and $(L - L')$ is the total number of distribution units. The second dimension is the chromosome (particle) position. The chromosome position can be defined as the sequence number of logistics facilities including LC, DC, and PC that are assigned to serve each corresponding pickup unit and distribution unit. $f_{k,i}$ indicates the logistics facility including logistics center or pickup center assigned to the ith pickup unit in the kth chromosome, while $f_{k,j}$ represents the logistics facility including logistics center or distribution center assigned to the jth distribution unit in the kth chromosome. The two-dimensional chromosome encoding table is shown in Table 1. For example, if we set $f_{k,i} = 0, 1, 2$, $f_{k,j} = 0, 3, 4, 5$, and $i = 1, 2, 3, \ldots, L'$, $j = L' + 1, \ldots, L$, $f_{2,3} = 2$ denotes that the PC 2 is assigned to the third pickup unit in the second chromosome, if $L' = 30$, then $f_{2,35} = 3$ indicates that the DC 3 is assigned to the 35th distribution unit in the second chromosome.

The chromosome encoding method shown in Table 1 makes sure that each pickup/distribution unit can be served by a certain logistics facility, in other words, the pickup/distribution service can be correspondingly completed by LC, PC or DC. In addition, using the chromosome encoding method to initialize populations, the computational complexity can effectively be simplified. For example, assume there are four pickup units, five distribution units, one LC, two PCs, and three DCs in the two-echelon heterogeneous logistics distribution network. The first chromosome can be generated randomly as follows: 0 1 2 0 3 5 4 0 5, where pickup units 1, 4 and distribution unit 8 are served by LC, pickup units 2 and 3 are served by PC 1 and PC 2, respectively, and distribution units 5 and 7 are served by DC 3 and DC 5, respectively. The initial fitness function value can be computed based on the objective function. However, some randomly generated chromosomes in the initial population may not satisfy requirements like maximum delivery distance limit, missing the PCs or DCs, and so on. Therefore, new chromosomes need to be generated in order to substitute the previous unqualified chromosomes. To evaluate the effectiveness of our hybrid algorithm, the fitness function needs to be properly defined, and the unqualified chromosomes should be penalized. Set the unqualified chromosome number as $\gamma_k = \gamma_k + 1$, with the initial value of γ_k being equal to 0. The objective function of the kth chromosome is TC_k, and the fitness function can be formulated as follows:

$$F_k = \frac{1}{TC_k + \gamma_k \times M}, \tag{15}$$

where M is the penalty weight for each unqualified chromosome (particle), and it can be defined relatively to the value of the objective function.

4.2. GA Updates and the Corresponding PSO Operations. GA updates mechanism including the corresponding PSO operations is presented in the hybrid algorithm procedure. The fitness function value and a cumulative probability are firstly calculated for the Genetic Algorithm. Afterwards, the selection, crossover, mutation, and PSO operations are conducted sequentially before the offspring is released into the new population. In the selection process, the roulette wheel selection is proposed based on the elevated chromosomes fitness function value. In the crossover process, a random vector including 0 and 1 is presented to choose the genes corresponding to 0 from the primary parent and the ones corresponding to 1 from the second parent; the offspring can be created by combinations [27]. For example, consider a network where there are four pickup units, five distribution units, one LC, two PCs, and three DCs, the parents can be generated as PA1 = [0 1 2 0 5 3 4 0 4], PA2 = [1 0 1 2 3 0 5 4 0], and the random binary vector as EX1 = [1 0 0 0 1 0 1 0 0]. We can get as one of the offspring: offsprl = [1 1 2 0 3 3 5 0 4]. Nevertheless, the offspring chromosomes may not satisfy all the constraints. Therefore, the binary vectors need to be regenerated until the new chromosomes are qualified. The process of mutation occurs after the crossover: two particular orders of genes are reversed for pickups and deliveries in a chromosome. The inversed chromosomes including genes only reserved for pickups, those only reserved for deliveries, and the ones reserved for both pickups and deliveries are compared with the original chromosome, and the optimal one is selected using the mutation operation.

The PSO operations update both location and velocity information in our proposed algorithm. The position of each particle (chromosome) is presented as x_k^t where $k = 1, 2, \ldots, N$ (N is population size) and t is the iteration number. The corresponding velocity of particle is v_k^t. The position is evaluated in each iteration, and if the fitness function value is better than that of previous iterations, the position is updated as the "pbest". Meanwhile, the global best position of current iteration is updated as "gbest". The initial individual best position "pbest" and the global best position "gbest" are determined based on the initial population after GA operations. The related formulas to update both the velocity and position are presented as follows:

$$v_k^{t+1} = \begin{cases} w \times v_k^t + c_1 \times \text{rand}(t) \times (\text{pbest}_k^t - x_k^t) + c_2 \times \text{rand}(t) \times (\text{gbest}_k^t - x_k^t) & -V_p \le v_k^{t+1} \le V_p \\ -V_p + 2 \times V_p \times \text{rand}(t) & \text{others} \end{cases} \tag{16}$$

$$x_k^{t+1} = \begin{cases} \text{fix}\left(x_k^t + v_k^{t+1}\right) & -V_p \le x_k^{t+1} \le V_p \\ \text{rand int}\left[0, T_p'\right] & \text{others} \end{cases} \tag{17}$$

$$v_k^{t+1} = \begin{cases} w \times v_k^t + c_1 \times \text{rand}(t) \times (\text{pbest}_k^t - x_k^t) + c_2 \times \text{rand}(t) \times (\text{gbest}_k^t - x_k^t) & -V_d \le v_k^{t+1} \le V_d \\ -V_d + 2 \times V_d \times \text{rand}(t) & \text{others} \end{cases} \tag{18}$$

$$x_k^{t+1} = \begin{cases} \text{fix}\left(x_k^t + v_k^{t+1}\right) & -V_d \le x_k^{t+1} \le V_d \\ \text{rand int}\left[0, T_d'\right] & \text{others,} \end{cases} \tag{19}$$

where formulas (16) and (17) are used to update both the velocity and position for pickup services and formulas (18) and (19) are used to update both the velocity and position for delivery services. c_1 and c_2 are two acceleration coefficients. rand$(\cdot)$ represents a random fraction between 0 and 1. pbest_k^t denotes the individual best position of particle k at the tth iteration, and gbest_k^t is the global best position of particle k at the tth iteration. fix$(\cdot)$ assures that the position of each particle is an integer. V_p and V_d represent allowable maximum velocity for pickup and delivery services, respectively. rand int$[0, T_p']$ denotes an integer between 0 and T_p', and rand int$[0, T_d']$ denotes an integer between 0 and T_d', where T_p' and T_d' represent the numbers of logistics facility set for pickup service and delivery service, respectively. w is the inertia weight and can be formulated as follows:

$$w = \frac{(w_{\text{int}} - w_{\text{end}}) \times (S_{\max} - t)}{S_{\max}} + w_{\text{end}}, \tag{20}$$

where $S_{\max}$ indicates the maximum number of iterations, w_{int} denotes the initial inertia weight, and w_{end} denotes the inertia weight for the maximum number of iterations.

4.3. Hybrid Algorithm Procedure. Based on the features described above, the hybrid algorithm procedure is described in Figure 3 and the main steps can be detailed as follows.

Step 1 (algorithm initialization). In the hybrid algorithm procedure, 0 denotes that the logistics center, which can be used for pickup service and delivery service. An integer within $[0, T_p']$ or $[0, T_d']$ is randomly chosen and assigned to the gene position in each chromosome, and check whether

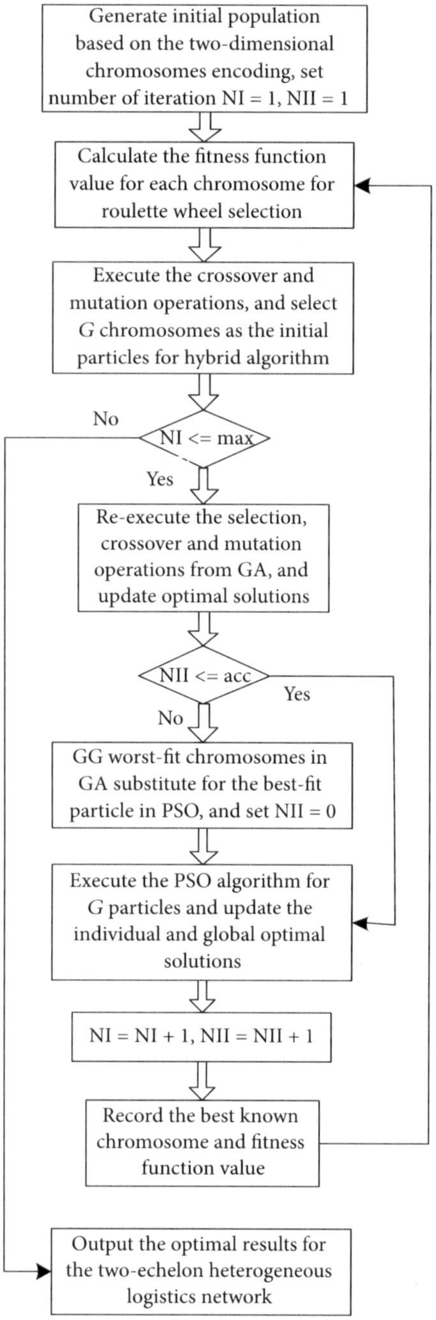

FIGURE 3: The flowchart of hybrid algorithm procedure.

the chromosomes fulfill the requirement or not. In case the condition is not respected, new chromosomes are generated until requirements are fulfilled, and calculate the fitness function value F_k based on formula (15).

Step 2. Execute the roulette wheel selection based on the elevated chromosomes fitness function value, execute crossover and mutation operations based on the crossover probability pc and the mutation probability pm, respectively, and update the optimal solutions and new chromosomes.

Step 3. Select the G chromosomes as the initial particles in the population, conduct the PSO operations as part of the hybrid algorithm, and calculate the individual optimal solution pbest_k^t and the global optimal solution gbest_k^t.

Step 4. Repeat the following steps until the maximum number of iterations is reached or the termination conditions are satisfied.

Step 4.1. Reexecute the selection, crossover, and mutation operations from GA; if the current fitness function value from the new offspring chromosomes is better than with the parents, update the optimal solutions and new chromosomes.

Step 4.2. Determine whether the number of iterations has achieved the periodic *acc* or not; if yes, replace the total *GG* of worst-fit chromosomes from GA with the corresponding number of best-fit particles from PSO algorithm; continue PSO with G particles, and update pbest_k^t and gbest_k^t; otherwise, the exchange procedure will not be conducted; execute the PSO algorithm on G particles and update the individual and global optimal solutions.

Step 5. If the number of iterations achieves the maximum number of iterations, the loop procedure will be finished. The current best known particle (chromosome) and fitness function value will be chosen as the optimal solution; otherwise, return to Step 2 to continue.

Step 6. Calculate and select the optimal solution from all feasible chromosomes. The optimal solution will be the final result for the two-echelon heterogeneous logistics network.

In the above hybrid algorithm procedure, the GA operations and PSO operations are reasonably combined by exchanging worst-fit solutions and best-fit solutions between GA and PSO algorithms. Therefore, it enhances the optimization capability of the hybrid algorithm and provides a more robust local and global search capability.

5. Profit Allocation Strategy

5.1. Minimum Cost Remaining Savings Model. The MCRS model is a game theoretic method used to solve cost or profit allocation problems [28]. The selection of MCRS to calculate logistics facilities profits is due to its well-known merits summarized as follows: (1) the model is simple to compute and each to implement; (2) results are generally fairly balanced and presented in a single vector, easy to visualize; (3) MCRS guarantees both individual and collective profit rationality.

The method proceeds to a two-step distribution, where a portion of the collective profit is distributed before the allocation of the remaining part in order to measure members' contributions. The upper and lower bound vectors $A_{\min} = \{A_{1\min}, \ldots, A_{j\min}, \ldots, A_{n\min}\}$ and $A_{\max} = \{A_{1\max}, \ldots, A_{j\max}, \ldots, A_{n\max}\}$ of the allocation scheme are first determined according to highest and lowest possible profit of each participant. $A_{j\max}$ denotes the maximum profit of logistics facility j and $A_{j\min}$ denotes its minimum value. Assume that A_j^* represents the actual profit of logistics facility j, $v(s)$ being the collective profit of alliance S and $v(z)$ the total benefit of the grand coalition. The value of the facility j's profit is determined by the following:

$$A_j^* = A_{j\min} + \frac{\left(A_{j\max} - A_{j\min}\right)}{\sum_{j\in N}\left(A_{j\max} - A_{j\min}\right)} \times \left(V(N) - \sum_{j\in N} A_{\min}\right), \tag{21}$$

where

$$\lambda = \frac{\left(A_{j\max} - A_{j\min}\right)}{\sum_{j=1}\left(A_{j\max} - A_{j\min}\right)} \tag{22}$$

$$\sum_{i=1}^{n} A_i = V(N),$$

where $V(N)$ refers to the hyperplane and λ is a coefficient used to balance the results. The bound vector's variables $A_{j\max}$ and $A_{j\min}$ can be solved by the following linear programming model:

$$\begin{aligned} \max\quad & A_j, \\ \min\quad & A_j \\ \text{s.t.}\quad & \sum_{j\in s} A_j \geq v(s) \\ & \sum_{j\in z} A_j = v(z). \end{aligned} \tag{23}$$

5.2. Strictly Monotonic Path Principles. The determination of good coalitions is subject to the verification of strictly monotonic path (SMP) principles. The multitude of coalition possibilities generally provides a wide range of profit allocation schemes. A sequence can be adopted only if the cost reduction percentage of each participant increases whenever a new facility joins. If π is a sequence of the coalition, $\pi(i)$ can represent the position of facility i in sequence π. $\eta(i, \pi, u)$ represents the cost reduction rate when facility i joins the coalition as the μth member and can be calculated as follows:

$$\eta(i, \pi, u) = \frac{\phi_i\left(\bigcup_{\pi(\mu)\leq u, u\in s} \mu, v\right)}{C_0 i}. \tag{24}$$

The process of selecting the SMP sequence is as follows.

Step 1. Select all sequence coalitions where the cost reduction percentage conforms to the principles of SMP, that is, monotonically increasing, and find the diagonal values in the matrix.

Step 2. Find the minimum value of the diagonal line and move to the next step or choose the second smallest value if the percentages are equal. Continue until all members are searched.

Step 3. The coalition sequence yielded from Step 2 is a candidate for the best profit distribution scheme. The process should be repeated for every sequence to ensure that all the possibilities have been evaluated.

6. Empirical Analyses

6.1. Data Description. We conduct a practical experimentation of the proposed network optimization methodology in Chongqing city, China, to evaluate the effects of collaboration and the applicability of our approach in real situations. Real locations of one LC, two PCs, three DCs, 50 pickup units, and 70 distribution units have been selected to illustrate irrational customer allocations observed in the urban logistics network. In the cooperative network, LC plays the roles of participant and coordinator. In order to clearly define the operations of the logistics center, we represent its pickup function by LC_P and the distribution function by LC_D. Figure 4 shows the logistics network before optimization. In order to facilitate the calculation, each distribution and pickup unit's demand is converted to the standard roll pallet quantity. The characteristics of all the logistics facilities are summarized in Table 2.

6.2. Improved ASO Parameter Setting and Optimization Results. In this section, we introduce the different parameters initialized for optimization purposes. They are separated in two categories such as objective function calculation parameters and algorithmic parameters and displayed as follows:

(1) Objective function parameters: $L_S = 2000$, $L_V = 200$, $f_{sr} = 0.000625$, $f_{vr} = 0.001875$, $M_S = 5000$, $M_v = 1700$, $T = 52$, $L = 50$, $N_t = 10$, $Nt' = 5$, $\xi_{p0} = 1500$, $\xi_{p1} = 1800$, $\xi_{p2} = 1050$, $\xi_{d0} = 1200$, $\xi_{d1} = 1200$, $\xi_{d2} = 600$, $\xi_{d3} = 900$. To encourage logistics facilities to cooperate, the local government provides the following as incentive: $G_{PC} = 1089$, $G_{PC1} = 1144$, $G_{PC2} = 923$, $G_{DC} = 996$, $G_{DC1} = 744$, $G_{DC2} = 526$, $G_{DC3} = 531$. The total cost savings provided if all the facilities form the grand alliance is $C = 6465$.

(2) Algorithm parameters: $N = 100$ is the popular size used to increase the diversity of initial chromosomes; $V_p = 2$ expresses the maximum velocity for pickup service; $V_d = 3$ expresses the maximum velocity for delivery service; $p_c = 0.8$ and $p_m = 0.02$ indicate the crossover and mutation probability, respectively; $G = 50$ indicates chromosomes as the initial particles in the population; $GG = 20$ expresses the number of chromosomes for exchanges between GA and PSO; $acc = 50$ denotes the number of iterations used for replacement between GA and PSO; $T_p' = 2$ and $T_d' = 3$ represent the numbers of logistics facility for pickup and delivery, respectively; $M = 100000$ is the penalty weight for each unqualified chromosome or particle; $S_{\max} = 1000$ is the maximum number of iterations; $\omega_{\text{int}} = 0.8$ is the initial inertia weight and $\omega_{\text{end}} = 0.3$ is the inertia weight of the maximum evolution generation; $c_1 = c_2 = 2$ are coefficients used for PSO speed calculation.

Table 2: Characteristics of six logistics facilities.

	Number of allocated customer units	Total demand (roll pallets)	Symbol
LC			
LC_P	19	9800	
LC_D	26	8260	
PC1	14	7800	
PC2	17	10380	
DC1	15	5800	
DC2	15	5180	
DC3	14	4500	

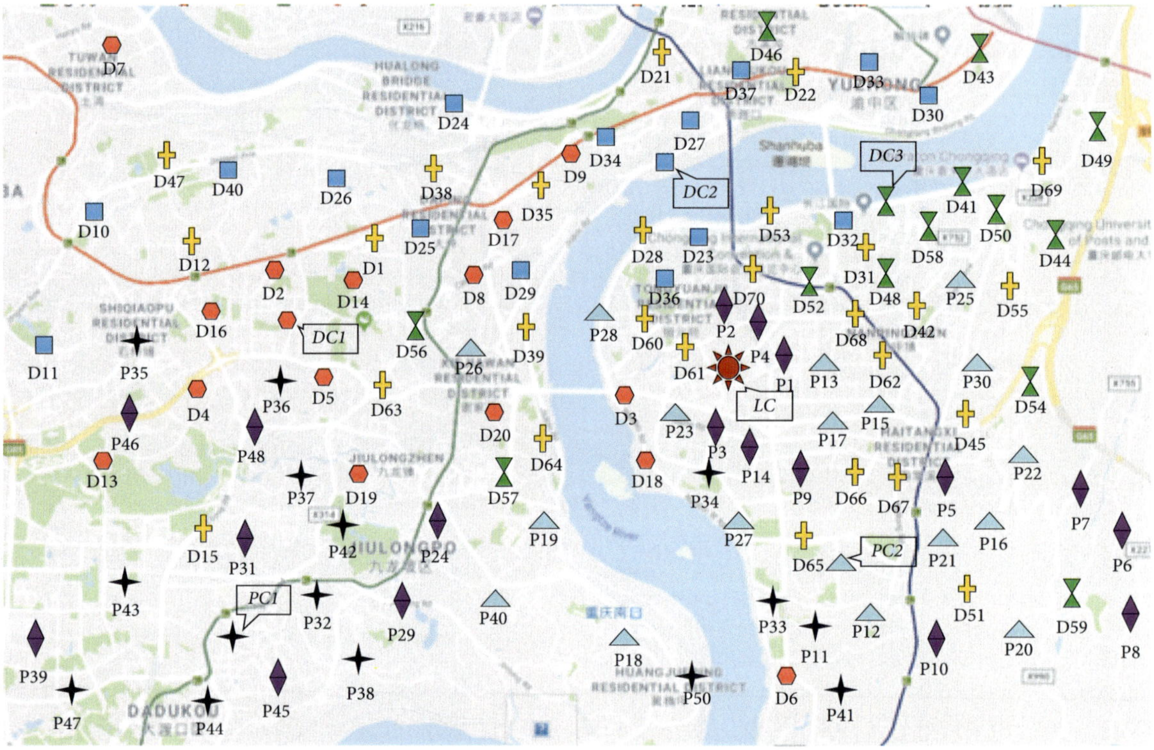

Figure 4: Pickup and distribution centers and units distribution diagram.

In this study, a working period consists of five working days. GA-PSO algorithm is used to assign customer units to corresponding logistics facilities and to compute the total cost in one working period. Cost savings are redistributed to participants of every coalition by applying the MCRS model. Optimization results are listed in Table 3, the initial customer units' assignment results are displayed in Table 4, and the best customer allocation in the grand alliance is summarized in the Table 5.

Compared with the initial customer allocation presented in Table 4, the cooperative network illustrated in Table 5 shows that the number of assigned customers varies from a logistics facility to another. While some centers have observed an increase, others have been allocated less units. For example, DC2's customers units decreased from 15 to 9, whereas PC1's customer units increased from 14 to 22. As a consequence, the total distance travelled by each facility's vehicles will decrease and savings will increase. Besides, the sharing of customers units allows LC, DCs, and PCs to reduce the number of unreasonable delivery and pickup trips, as well as the complexity of their corresponding transportation network.

For comparison purposes, we implement and test the proposed GA-PSO hybrid algorithm, CPTS (Cooperative Parallel Tabu Search), HGA (Hybrid Genetic Algorithm), and TLBO (Teaching–Learning-Based Optimization) with same data. CPTS utilizes the information interchange to accomplish the cooperation and boost both intensification and strategic diversification (James et al. 2009). HGA incorporates the elitism genetic operator and some improvement

TABLE 3: Comparison between initial and optimized network over one working period.

S	Initial		Optimized		$V(t)$
	Cost	Demand	Cost	Demand	
LC_P	5446	9800	4015	6820	648
LC_D	4980	10380	3558	8520	1422
PC1	5721	8260	3976	12000	1158
PC2	4613	7800	3338	7040	688
DC1	3320	5800	2081	9900	1239
DC2	2629	5180	1861	4860	769
DC3	2655	4500	1895	4580	760
LC_P, PC1	11166	18060	8813	18060	2353
LC_P, PC2	10058	17600	7262	17600	2796
LC_D, DC1	8300	16180	5250	16180	3050
LC_D, DC2	7609	15560	4786	15560	2824
LC_D, DC3	7635	14880	4750	14880	2886
LC_P, PC1, PC2	15779	25860	9412	25860	4411
LC_D, DC1, DC2	10929	16180	7621	16180	3308
LC_D, DC1, DC3	10955	15560	6054	15560	4901
LC_D, DC2, DC3	10265	14880	4955	14880	5310
LC_P, PC1, PC2, LC_D, DC1	24079	16180	16434	16180	7644
LC_P, PC1, PC2, LC_D, DC2	23388	15560	15984	15560	7404
LC_P, PC1, PC2, LC_D, DC3	23414	14880	15945	14880	7470
LC_P, PC1, PC2, LC_D, DC1, DC2	26708	21360	18890	21360	7819
LC_P, PC1, PC2, LC_D, DC1, DC3	26734	20680	17225	20680	9509
LC_P, PC1, PC2, LC_D, DC2, DC3	26044	20060	16163	20060	9881
LC_P, PC1, PC2, LC_D, DC1, DC2, DC3	29363	25860	16971	25860	12392

TABLE 4: Initial customer units' assignment.

	Customer units allocation
LC	
LC_P	P1 P2 P3 P4 P5 P6 P7 P8 P9 P10 P14 P24 P29 P31 P39 P45 P46 P48 P49
LC_D	D1 D12 D15 D21 D22 D28 D31 D35 D38 D39 D42 D45 D47 D53 D55 D60 D61 D62 D63 D64 D65 D66 D67 D68 D69 D70
PC1	P11 P32 P33 P34 P35 P36 P37 P38 P40 P42 P43 P44 P47 P50
PC2	P12 P13 P15 P16 P17 P18 P19 P20 P21 P22 P23 P25 P26 P27 P28 P30 P41
DC1	D2 D3 D4 D5 D6 D7 D8 D9 D13 D14 D16 D17 D18 D19 D20
DC2	D10 D11 D23 D24 D25 D26 D27 D29 D30 D32 D33 D34 D36 D37 D40
DC3	D41 D43 D44 D46 D48 D49 D50 D51 D52 D54 D56 D57 D58 D59

TABLE 5: Customer units' assignment in the grand coalition.

	Customer units allocation
LC	
LC_P	P1 P2 P3 P4 P6 P13 P14 P17 P22 P23 P25 P28 P30 P34
LC_D	D3 D6 D18 D21 D22 D23 D28 D36 D39 D45 D51 D52 D54 D59 D60 D61 D62 D65 D66 D67 D68 D70
PC1	P18 P19 P24 P26 P29 P31 P32 P35 P36 P37 P38 P39 P40 P42 P43 P44 P45 P46 P47 P48 P49 P50
PC2	P5 P7 P8 P9 P10 P11 P12 P15 P16 P20 P21 P27 P33 P41
DC1	D1 D2 D4 D5 D7 D10 D11 D12 D13 D14 D15 D16 D19 D20 D25 D26 D38 D40 D56 D57 D58 D63 D64 D47
DC2	D8 D9 D24 D27 D29 D33 D34 D35 D37
DC3	D17 D30 D31 D32 D41 D42 D43 D44 D46 D48 D49 D50 D53 D55 D69

TABLE 6: Algorithms optimization results comparison.

Sequence	Total cost (dollar)				Number of iterations			
	GA-PSO	CPTS	HGA	TLBO	GA-PSO	CPTS	HGA	TLBO
1	16971	17632	18639	17451	487	535	495	572
2	17235	17715	18334	17237	451	541	516	593
3	17158	17543	17865	17382	463	559	531	604
4	16971	17687	18026	17237	504	528	519	615
5	16971	17819	17865	17469	476	556	534	577
6	17023	17543	18223	17520	479	545	522	565
7	16971	17726	18409	17334	495	537	503	599
8	16971	17678	17987	17237	510	533	535	611
9	16971	17799	18172	17256	503	529	511	600
10	17196	17841	18464	17321	473	545	496	587
11	16971	17543	18326	17292	492	552	507	591
12	17304	17764	17865	17438	455	527	543	587
13	16971	17815	18079	17513	507	560	529	572
14	17189	17686	17897	17237	481	521	535	607
15	16971	17752	18363	17543	497	519	518	556
16	16971	17543	17865	17364	481	548	550	597
17	17105	17690	18144	17419	473	525	532	592
18	16971	17778	18251	17292	499	536	517	621
19	16971	17639	17962	17237	512	552	523	606
20	17113	17617	17994	17351	477	527	510	575
Average	17049	17691	18137	17357	486	539	521	591
t-test		−20.26	−17.79	−8.67		−10.73	−6.34	−21.31
p value		$2.5E-14$	$2.7E-13$	$5E-08$		$1.7E-9$	$4.4E-6$	$1E-14$

heuristic methods [9]. TLBO considers the solutions proposed by individuals' search, two-phase optimization process including teacher and learner phases to obtain the global best fitness values (Dokeroglu, 2015). Each algorithm is executed 20 times, and the best solution is selected as the optimal cost for convergence. The optimal total cost and the number of iterations can be used to measure the effectiveness of four algorithms shown in Table 6.

As shown in Table 6, the t-test and p value indicate that the total cost and the number of iterations are significantly different with the three algorithms. In addition, the number of times GA-PSO hybrid algorithm finds out the optimal solution is higher. For example, the best solution appears 12 out of 20 times with the GA-PSO, while it only appears four or five times with CPTS, HGA, and TLBO.

Compared with other three algorithms, the average optimization cost and number of iterations from our proposed algorithm are of better quality. Thus, we can conclude that GA-PSO has the following merits: (1) exchanging best-fit chromosomes and worst-fit particles between the GA and the PSO algorithm enhances the GA-PSO's capability to obtain better solutions. For example, the GA-PSO algorithm has a higher probability to obtain the best solution. (2) The proposed algorithm combines GA and PSO algorithm's global and local search capability. As illustration, the average cost from the proposed hybrid algorithm is lower than that of CPTS, HGA, and TLBO algorithms by 3.8%, 6.4%, and 1.8%, respectively. (3) The termination and periodic iteration rules are reasonably designed in the GA-PSO hybrid algorithm procedure, which improves the efficiency of the hybrid algorithm. Therefore, the average number of iterations from GA-PSO is significantly less than that of the CPTS, HGA, and TLBO algorithms by 53, 35, and 105 iterations, respectively. These findings indicate that our hybrid algorithm has better capability to obtain the optimal solution than that of the other three algorithms.

6.3. MCRS Model Application and Coalition Sequence Selection. Previous studies have already shown the efficiency of MCRS at profit distribution in multiple players' games [28]. Indeed, this method can be used to incorporate the coordinator's payoff into the design of the collective profit allocation plan and guarantee the stability of coalitions. Based on the fact that our case study's coordinator (LC) is also part of the network and receives the highest incentive from the government, we assume there is no extra payoff and set the synergy requirement $\sigma = 0$. In this study, we first consider the formation of a grand coalition before splitting the group into two subcoalitions and evaluate the influences of each of these scenarios. The combined form of all nonempty alliances is shown in Table 7.

The sum of the revenue of LC_P and LC_D constitute the total profit of LC. Profit allocation results display differences in facilities' benefits as they appear in different coalitions. For example, PC1's initial profit while being alone is 1158, but, after

TABLE 7: Profit allocation in two-echelon logistics distribution network.

S	$V(t)$	$\varphi(s,v)$
LC_P	648	(648; . . . ; . . . ; . . . ; . . . ; . . . ; . . .)
LC_D	1422	(1422; . . . ; . . . ; . . . ; . . . ; . . . ; . . .)
PC1	1158	(. . . ; 1158; . . . ; . . . ; . . . ; . . . ; . . .)
PC2	688	(. . . ; . . . ; 688; . . . ; . . . ; . . . ; . . .)
DC1	1239	(. . . ; . . . ; . . . ; . . . ; 1239; . . . ; . . .)
DC2	769	(. . . ; . . . ; . . . ; . . . ; . . . ; 769; . . .)
DC3	760	(. . . ; . . . ; . . . ; . . . ; . . . ; . . . ; 760)
LC_P, PC1	2353	(922; 1431; . . . ; . . . ; . . . ; . . . ; . . .)
LC_P, PC2	2796	(1378; . . . ; 1418; . . . ; . . . ; . . . ; . . .)
LC_D, DC1	3050	(. . . ; . . . ; . . . ; 1617; 1433; . . . ; . . .)
LC_D, DC2	2824	(. . . ; . . . ; . . . ; 1738; . . . ; 1086; . . .)
LC_D, DC3	2886	(. . . ; . . . ; . . . ; 1774; . . . ; . . . ; 1112)
LC_P, PC1, PC2	4411	(1377; 1616; 1418; . . . ; . . . ; . . . ; . . .)
LC_D, DC1, DC2	3308	(. . . ; . . . ; . . . ; 1372; 1212; 724; . . .)
LC_D, DC1, DC3	4901	(. . . ; . . . ; . . . ; 2121; 1564; . . . ; 1216)
LC_D, DC2, DC3	5310	(. . . ; . . . ; . . . ; 2203; . . . ; 1769; . . . ; 1338)
LC_P, PC1, PC2, LC_D, DC1	7644	(1438; 1677; 1478; 1617; 1434; . . . ; . . .)
LC_P, PC1, PC2, LC_D, DC2	7404	(1434; 1673; 1474; 1738; . . . ; 1085; . . .)
LC_P, PC1, PC2, LC_D, DC3	7470	(1435; 1674; 1475; 1774; . . . ; . . . ; 1112)
LC_P, PC1, PC2, LC_D, DC1, DC2	7819	(1418; 1653; 1439; 1372; 1212; 724; . . .)
LC_P, PC1, PC2, LC_D, DC1, DC3	9509	(1440; 1687; 1481; 2121; 1564; . . . ; 1216)
LC_P, PC1, PC2, LC_D, DC2, DC3	9881	(1444; 1669; 1457; 2204; . . . ; 1769; 1338)
LC_P, PC1, PC2, LC_D, DC1, DC2, DC3	12392	(1455; 1694; 1495; 2527; 1662; 1502; 2057)

TABLE 8: SMP based grand coalition sequence.

$\pi_1 = \{LC_P, PC1, PC2, LC_D, DC1, DC3, DC2\}$							
Player i	LC_P	PC1	PC2	LC_D	DC1	DC3	DC2
$\eta(i,\pi,1)$	11.9%						
$\eta(i,\pi,2)$	16.9%	25.0%					
$\eta(i,\pi,3)$	26.3%	29.2%	32.0%				
$\eta(i,\pi,4)$	26.3%	29.2%	32.0%	28.5%			
$\eta(i,\pi,5)$	26.4%	29.4%	32.1%	32.5%	43.2%		
$\eta(i,\pi,6)$	26.5%	29.5%	32.2%	42.6%	47.1%	45.8%	
$\eta(i,\pi,7)$	26.7%	29.6%	32.4%	51.0%	51.1%	50.1%	77.5%

sharing customers and transportation resources with LC_P, PC1 could save 1432. However, in certain circumstances, the profit can decrease as another participant enters the coalition. DC2 can be considered as an example to the extent that its initial profit (769) decreases after joining LC_D and DC1 and becomes 724. In fact, rational decision making requires every participant to give priority to coalitions where the increase of benefits is guaranteed. Therefore, we can conclude that the stability of any coalition depends on the profit of the members before and after new members adhere. In the process to form the grand coalition, cost reduction percentages have been calculated, presented in Figure 5, and illustrated in Figure 6.

The logistics center is denoted by LC_P and LC_D in Figure 5, while the legend in Figure 6 refers to it as LCP and LCD. We can notice at the fourth step in Figure 6 the simultaneous entrance of LCD and DC1 to emphasize that the delivery capacities of LC should not be applicable unless there is a DC in the coalition.

The examination of coalition sequences is critical to the profit distribution strategy and to participants' willingness to become member. In other words, the order in which members are added to the coalition affects the distribution of profits and the satisfaction of SMP principles. Nevertheless, following the examination of every possible combination, the SMP based coalition sequence is $\pi_1 = \{LC_P, PC1, PC2, LC_D, DC1, DC3, DC2\}$. Table 8 shows the cost reduction matrix representing members' behaviors as a new facility enters π_1.

In the design of the grand coalition, we have considered LC_P joining first the coalition followed by PC1 and PC2 in

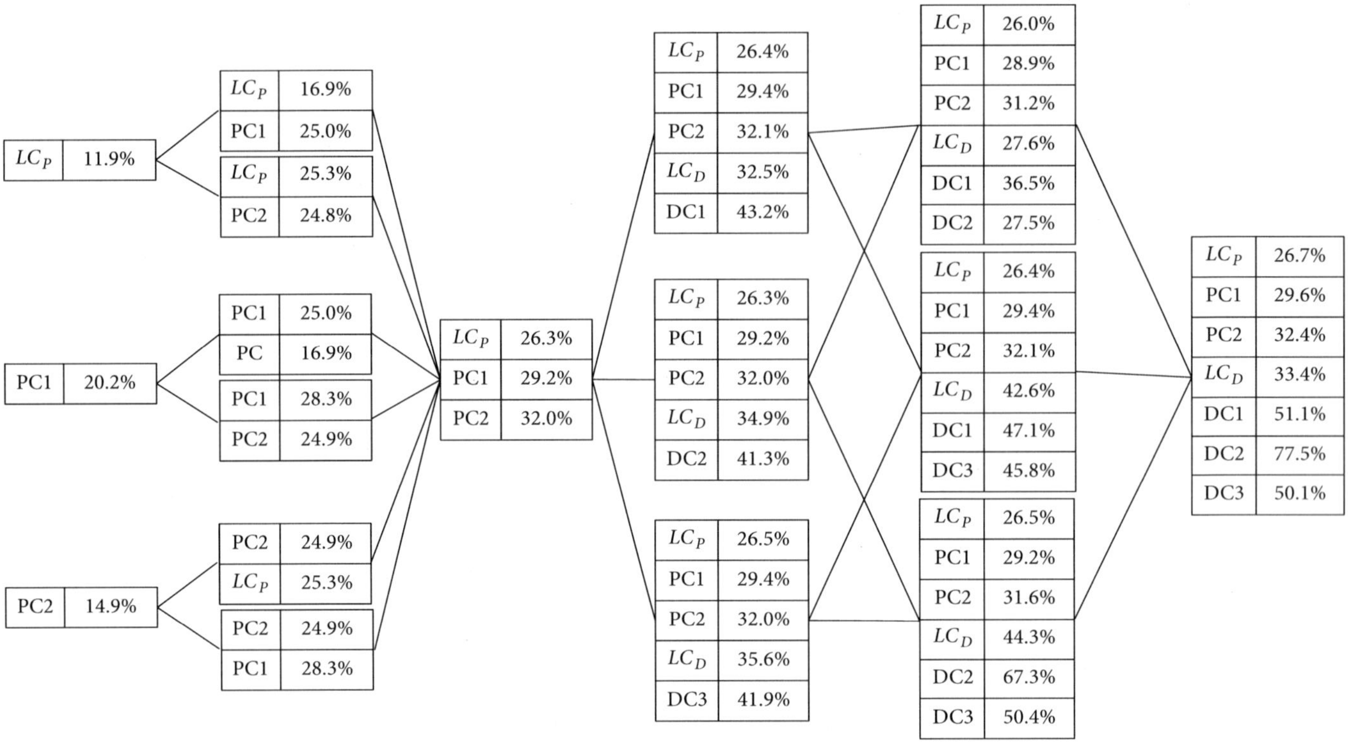

FIGURE 5: Grand alliance formation process.

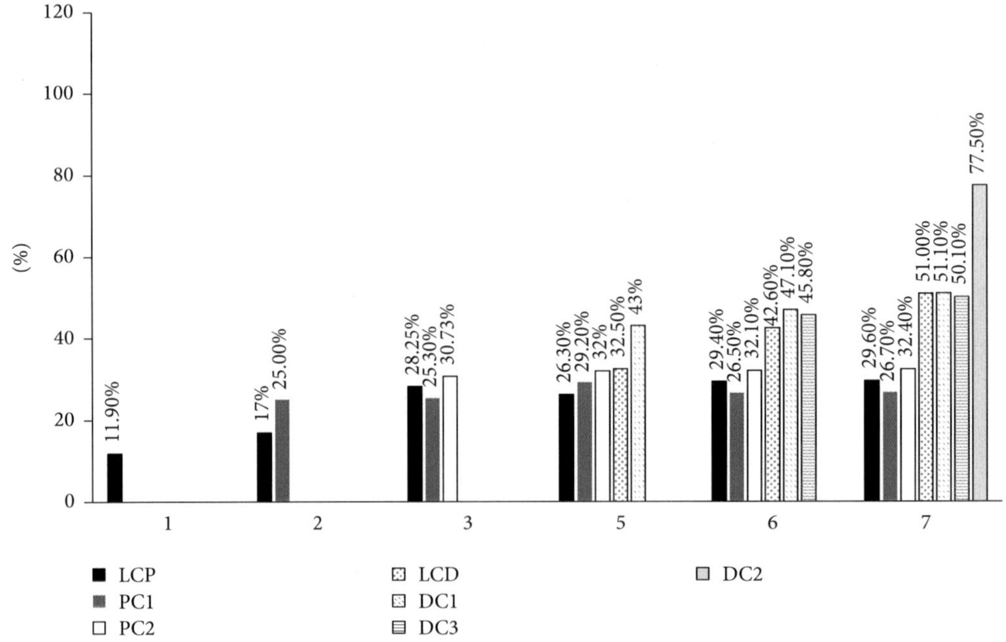

FIGURE 6: Cost reduction percentage diagram for the grand coalition.

order to assure that pickup activities are completed before deliveries and to conform to the proposed THCLN. The percentages of total operating costs reduced by LC, PC1, and PC2, at the level of unique pickup operations in the coalition, are, respectively, 26.3%, 29.2%, and 32.0%. After the adhesion of distribution centers in the order {DC1, DC3, DC2}, the final sequence for the grand coalition {LCP, PC1, PC2, LCD, DC1, DC3, DC2} yields as cost reduction percentage sequence {26.7%, 29.7%, 32.4%, 33.4%, 51.1%, 50.1%, 77.5%}, respectively. In the formation process, the profits of PCs have increased with the addition of DCs. In brief, we can conclude that LC as coordinator and the upper-level facility of PCs

TABLE 9: Profit allocations according to MCRS, Shapley, CGA, and GQP.

	MCRS	Shapley	CGA	GQP
LC_P	1377	1316	1344	1344
PC1	1616	1554	1312	1312
PC2	1417	1541	1755	1755
LCD	2527	2691	2923	2923
DC1	1662	1572	805	805
DC2	1502	1472	1214	1214
DC3	2058	2013	2807	2807

TABLE 10: Core center and distance comparison table.

Method	Profit allocation	Core center	Distance
MCRS	(1454.81, 1693.83, 1494.81, 2527.37, 1661.90, 1501.91, 2057.65)	(1188.91, 1695.29, 1529.81, 2328.50, 1581.68, 1455.18, 2130.29)	354.09
Shapley	(1315.98, 1554.43, 1540.91, 2691.34, 1572.2, 1472.27, 2013.04)		446.64
CGA	(1421.65, 1389.64, 1832.17, 2923.34, 804.72, 1214.17, 2806.61)		1308.43
GQP	(1421.65, 1389.64, 1832.17, 2923.34, 804.72, 1214.17, 2806.61)		1308.43

and DCs can better optimize the entire logistics network. In this article, we required LC to be first member to enter the coalition. However in other circumstances, LC may not consider the value of its profit and allow another facility to make the first move for the sake of a more stable coalition.

6.4. Alliance Stability. In this section, we examine the accuracy of the MCRS method at providing the best profit distribution schemes in the investigated case. In this regard, three different methods are used to calculate appropriate profits of each facility [29, 30]: the Shapley value model, the Cost Gap Allocation (CGA) model, and the Game Quadratic Programming (GQP) method. Moreover, the Euclidean distance between each distribution scheme and core center of the grand coalition will be evaluated to verify each method's performance. In accordance with the snowball theory [31], the best strategy should be the closest to the center. The vector at the core is determined using formula (25) where the collective profit of the grand coalition is $V(Z)$, β represents a member and α is an analytic expression to control the scope of the core. In addition, Table 9 presents the results of the four methods, and Table 10 gives the core center and corresponding distances.

$$\frac{v(Z)-v(Z-\{\beta\})}{v(Z)}\times\alpha+\sum_{k\epsilon Z}^{k\neq i} y_k = v(Z-\{\beta\}). \quad (25)$$

Despite the allocation of lowest profits to PC2, LCD, and DC2, as shown in Table 9, MCRS yields the lowest distance (354) to the core center in Table 10. Therefore, we can notice that MCRS is the closest to the center and confirms our profit distribution scheme the most appropriate. Besides, findings allow us to conclude that, in a cooperative coalition, individual profit should not be the most important criteria. Decision makers and coordinators have to observe the collective impacts of cooperation on the network.

6.5. Analysis of Two Heterogeneous Subcoalitions' Network. Segmentation is a strategic management approach used to control complex networks and consists in dividing the components into small parts. Some researchers have studied cooperation in logistics network optimization but concentrated on the design of multiple coalitions. Thus, in addition to the grand coalition, this paper has also considered other cases where the network would consist in two heterogeneous coalitions. In other words, PCs can independently cooperate with LC and a DC can form a coalition with LC and other DCs. It is important to notify that possibilities of constituting groups with different types of centers are numerous, but this article only evaluates circumstances where the same type of facility joins with LC in a single coalition. Our purpose is to examine the influence of multiple and single heterogeneous coalition in logistics network. Therefore, based on our solution methodology, the cost reduction percentages of possible sequences under two subcoalitions scenarios are shown in Tables 11 and 12.

Different from DCs coalitions where the best sequence can be directly selected, PCs coalition sequences should first be subject to the constraint of LC first entering. Indeed as Table 11 depicts, the cost reduction of $\pi_1 = \{PC1, LC_P, PC2\}$ is higher than the second option. However, according to the assumption that LC makes the first move, we are constrained to select sequence $\pi_1 = \{LC_P, PC1, PC2\}$. From Table 12, we can observe that cost reduction trend increases more rapidly under $\pi_2 = \{LC_D, DC1, DC3, DC2\}$. Therefore, the logistics network with two heterogeneous subcoalitions should be designed such that the first subcoalition follows the order $\{LC_P, PC1, PC2\}$, whereas the second respects the sequence

TABLE 11: Pickup coalition sequences based on SMP.

$\pi_1 = \{LC_P, PC1, PC2\}$			
Player i	LC_P	PC1	PC2
$\eta(i, \pi, 1)$	11.9%		
$\eta(i, \pi, 2)$	16.9%	25.0%	
$\eta(i, \pi, 3)$	26.3%	29.2%	32.0%
$\pi_1 = \{PC1, LC_P, PC2\}$			
Player i	PC1	LC_P	PC2
$\eta(i, \pi, 1)$	20.2%		
$\eta(i, \pi, 2)$	25.0%	16.9%	
$\eta(i, \pi, 3)$	29.2%	26.3%	32.0%

TABLE 12: Distribution coalition sequences based on SMP.

$\pi_2 = \{LC_D, DC1, DC3, DC2\}$				
Player i	LC_D	DC1	DC3	DC2
$\eta(i, \pi, 1)$	28.5%			
$\eta(i, \pi, 2)$	32.5%	43.2%		
$\eta(i, \pi, 3)$	42.6%	47.1%	45.8%	
$\eta(i, \pi, 4)$	51.0%	51.1%	50.1%	77.5%
$\pi_2 = \{LC_D, DC3, DC1, DC2\}$				
Player i	LC_D	DC3	DC1	DC2
$\eta(i, \pi, 1)$	28.5%			
$\eta(i, \pi, 2)$	35.6%	41.9%		
$\eta(i, \pi, 3)$	42.6%	47.1%	45.8%	
$\eta(i, \pi, 4)$	51.0%	51.1%	50.1%	77.5%

$\{LC_D, DC1, DC3, DC2\}$. The cost reduction percentage diagram of two subcoalitions is shown in Figure 7.

The horizontal axes on the graphical illustrations of Figures 7(a) and 7(b) represent the increase in participants' number, while the vertical axes mark the percentage of each facility. The amount of reduced cost monotonically increases after new members enter, thereby proving that the solutions bind the SMP rules. In the final appearance of the two subcoalitions, LC_P first joins the pickup coalition with 11.9% cost reduction and then PC1 joins the coalition for LC_P and PC1 to reduce up to 16.9%, 25% of their expenses in one working period; finally, PC2 joins, and the cost reductions of LC_P, PC1, and PC2 become 26.3%, 29.2%, and 32%, respectively. As for the delivery coalition, LC_D first joins the coalition with 28.5% cost reduction, then DC1 joins the coalition, and the cost reductions of LC_D and DC1 become 32.5% and 43.2% respectively; next, DC3 joins, and LC_D, DC1, and DC3 can reduce up to 42.6%, 47.1%, and 45.8% of their expenses in one working period; finally, DC2 enters to yield as cost reductions of LC_D, DC1, DC2, and DC3: 51.0%, 51.1%, 50.1%, and 77.5% respectively.

6.6. *Discussion.* The analysis of the profit allocation plans for the grand coalition and under two heterogeneous subcoalitions is a critical phase in the coordinator's decision making process. It offers the opportunity to evaluate existing alternatives and guarantees the reliability of the final decision as well as the sustainability of the cooperative logistics network. Each player's payoff is compared so as to estimate their best choices and conclude the most stable network. Table 13 presents the allocated profits of the five logistics facilities considered in the current case study.

TABLE 13: Comparison of different network scenarios.

	Two subcoalitions	The grand coalition
LCP	1377	1455
PC1	1616	1694
PC2	1417	1495
LCD	2527	2527
DC1	1662	1662
DC2	1502	1502
DC3	2058	2058
Total	12159	12393

As shown in Table 13, comparisons show that the profit received by each pickup centers is higher when they join the grand coalition. For example, each pickup facility can observe an increase of $78 in profit if they opt for the formation of the grand coalition. Subsequently, LCP, PC1, and PC2 will, respectively, reduce their profit by 5.4%, 4.6%, and 5.2% while moving from the grand coalition to form subcoalitions. Meanwhile, delivery centers would not constitute major problems to the final decision because their profits remain the same either in heterogeneous subcoalition or in the grand

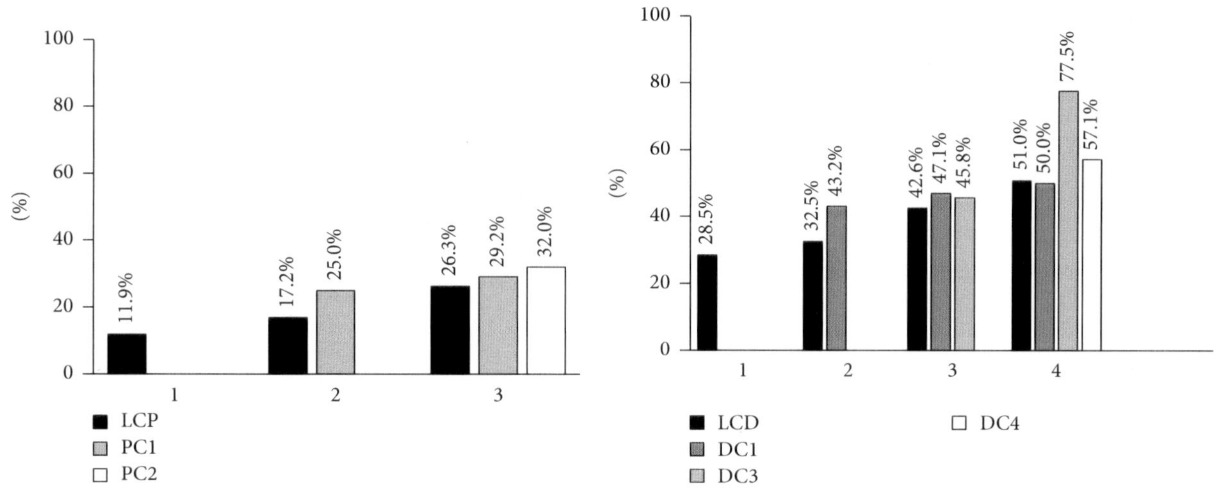

(a) First coalition formation

(b) Second coalition formation

FIGURE 7: Cost reduction percentage diagram for two subcoalitions.

coalition. Moreover, the collective profit in the logistics network increases when all the facilities join together to share resources. Though delivery centers' profits are identical, the difference generated by pickup activities in the collective network can enhance the superiority of the grand coalition. Therefore, findings indicate that forming a grand coalition rather than two subcoalitions is the best strategy to select both for participants and the coordinators.

In recent years, cooperation between logistics facilities has been playing an important role in the optimization of companies supply chains. Further incorporating transportation resources sharing can provide more cost savings. In addition, the existence of local governments' policies to encourage cooperation is also a sign of authorities will to achieve the sustainable development of their administrated areas. As one of the main development factors, transportation activities can be more organized with the reduction of the number of delivery and pickup vehicles. Therefore, encouraging the formation of a grand coalition is a relevant approach which would benefit not only logistics companies, but the whole society. However, forming a grand coalition may impose some managerial challenges. For example, the coalition's satisfaction of the strictly monotonic path principles may remain stable for a short period. Owing to the dynamics of modern logistics services and the variation of operations costs over time, participants can meet severe challenges resulting in internal financial and operational crisis. Consequently, the profit allocation will be influenced and the stability of the grand coalition threatened. As measure of prevention, different scenarios should be evaluated before coordinators launch negotiations. A possible means of handling such problems consists in splitting the grand coalition into several groups and evaluating potential risks related to individual facilities in the network before any final decision. Based on the proposed methodology, coordinators can also input estimated data for a decision period in order to find out the most appropriate type of coalition.

7. Conclusions

This paper studies the influence of cooperation among facilities on two echelons logistics networks with delivery and pickup operations. The grand and two heterogeneous coalitions have been studied to evaluate the preference of each participant willing to minimize expenses and maximize profit. The proposed THCLN is designed to interpret facilities possibilities of either cooperating with a similar center or participating in a group with different type of facilities. In other words, a pickup center can join another pickup center to form a subcoalition or join the grand coalition with distribution centers. The problem has been interpreted through a mixed integer linear model to optimize the total cost of nonempty coalitions. The solution methodology combines a GA-PSO hybrid algorithm to assign customers to the most suitable logistics facility and the MCRS method to properly allocate profits. Since different cooperative sequences have their corresponding cost reduction percentages, the selection of the best sequence is achieved using the SMP theory.

To test the effectiveness of THCLN as applicable network in real life, empirical analyses on a logistics network in Chongqing city, China, are conducted. As coordinator of the coalition formation process, we have selected the logistics center for its size and ability to collect and deliver goods. Comparisons have been done to prove that GA-PSO and MCRS are, respectively, proficient over other well-known methods. Computations have been carried out with CPTS, HGA, and TLBO algorithms, and GA-PSO has outperformed on the solution quality and the number of iterations. As for MCRS, its profit allocation scheme is closer to the core center than the Shapley value model, CGA, and GQP. Furthermore, we base on the assumption that the synergy requirement is equal to zero; that is, the coordinator should not receive an extra salary and share the collective profit to each member. The analysis of the results from the grand coalition and two heterogeneous subcoalitions concludes that forming a grand coalition is the most suitable.

In the future, this research can be directed to optimize logistics networks of more than two echelons in order to expand the cooperative mechanism to a larger environment. Moreover, as customer assignment is a basic network optimization strategy, searching vehicle routing solutions can be another direction to implement the proposed approach. The impact of information sharing on the stability of the grand coalition can also be modeled and the variations in a coalition's lifetime can be simulated. Besides, the study of cooperative fresh food and medical supply chains can be done to experiment the relevance of our optimization methodology in other domains.

Conflicts of Interest

The authors declare that there are no conflicts of interest regarding the publication of this paper.

Acknowledgments

This research is supported by the National Natural Science Foundation of China (Projects nos. 71402011, 71301180, 71471024, 41401255, and 71432003), China Postdoctoral Science Foundation (Projects nos. 2017T100692 and 2016M600735), National Social Science Foundation of Chongqing of China (no. 2017YBGL133), Natural Science Foundation of Chongqing of China (no. cstc2015jcyjA30012), and Postdoctoral Science Foundation of Sichuan Province, China (2017-22), and this research is partly supported by the Key Project of Human Social Science of Chongqing Municipal Education Commission (17SKG067 and 16SKGH067). In addition, the research is also partly supported by the Fundamental Research Funds for Central Universities (YWF-17-BJ-Y-04).

References

[1] O. Ö. Özener and Ö. Ergun, "Allocating costs in a collaborative transportation procurement network," *Transportation Science*, vol. 42, no. 2, pp. 146–165, 2008.

[2] Q. Wu, H. Ren, W. Gao, J. Ren, and C. Lao, "Profit allocation analysis among the distributed energy network participants based on Game-theory," *Energy*, vol. 118, pp. 783–794, 2017.

[3] H. Shi, Z. Li, W. Li, and Z. Yu, "Application of particle swarm optimization based on clustering analysis in logistics distribution," in *Proceedings of the 2009 International Conference on Management of E-Commerce and E-Government*, pp. 291–295, Nanchang, China, September 2009.

[4] R. Mesa-Arango and S. V. Ukkusuri, "Demand clustering in freight logistics networks," *Transportation Research Part E: Logistics and Transportation Review*, vol. 81, pp. 36–51, 2015.

[5] S. Baranowski, E. Busko, S. Shishlo et al., "Formation mechanism of logistics cluster in Belarus," *Agriculture and Agricultural Science Procedia*, vol. 7, pp. 12–20, 2015.

[6] I. Kumar, A. Zhalnin, A. Kim, and L. J. Beaulieu, "Transportation and logistics cluster competitive advantages in the U.S. regions: A cross-sectional and spatio-temporal analysis," *Research in Transportation Economics*, vol. 61, pp. 25–36, 2017.

[7] W.-J. Li and J. M. Smith, "An algorithm for quadratic assignment problems," *European Journal of Operational Research*, vol. 81, no. 1, pp. 205–216, 1995.

[8] I. Sghir, J.-K. Hao, I. B. Jaafar, and K. Ghédira, "A multi-agent based optimization method applied to the quadratic assignment problem," *Expert Systems with Applications*, vol. 42, no. 23, pp. 9252–9262, 2015.

[9] Z. Drezner, "Extensive experiments with hybrid genetic algorithms for the solution of the quadratic assignment problem," *Computers & Operations Research*, vol. 35, no. 3, pp. 717–736, 2008.

[10] C. Chen, M. Li, J. Sui, K. Wei, and Q. Pei, "A genetic algorithm-optimized fuzzy logic controller to avoid rear-end collisions," *Journal of Advanced Transportation*, vol. 50, no. 8, pp. 1735–1753, 2016.

[11] Y. Shimizu and T. Miura, "Effect of topology on parallel computing for optimizing large scale logistics through binary PSO," *Computer Aided Chemical Engineering*, vol. 30, pp. 1247–1251, 2012.

[12] A. Čupić and D. Teodorović, "A multi-objective approach to the parcel express service delivery problem," *Journal of Advanced Transportation*, vol. 48, no. 7, pp. 701–720, 2014.

[13] B. Haddar, M. Khemakhem, S. Hanafi, and C. Wilbaut, "A hybrid heuristic for the 0-1 Knapsack Sharing Problem," *Expert Systems with Applications*, vol. 42, no. 10, pp. 4653–4666, 2015.

[14] J. Guo, X. Wang, S. Fan, and M. Gen, "Forward and reverse logistics network and route planning under the environment of low-carbon emissions: A case study of Shanghai fresh food E-commerce enterprises," *Computers & Industrial Engineering*, vol. 106, pp. 351–360, 2017.

[15] Y. Wang, X. Ma, M. Liu et al., "Cooperation and profit allocation in two-echelon logistics joint distribution network optimization," *Applied Soft Computing*, vol. 56, pp. 143–157, 2017.

[16] Y. Wang, X. Ma, Z. Li, Y. Liu, M. Xu, and Y. Wang, "Profit distribution in collaborative multiple centers vehicle routing problem," *Journal of Cleaner Production*, vol. 144, pp. 203–219, 2017.

[17] A. Alshamsi and A. Diabat, "A genetic algorithm for reverse logistics network design: a case study from the GCC," *Journal of Cleaner Production*, vol. 151, pp. 652–669, 2017.

[18] A. Atashpendar, B. Dorronsoro, G. Danoy, and P. Bouvry, "A scalable parallel cooperative coevolutionary PSO algorithm for multi-objective optimization," *Journal of Parallel and Distributed Computing*, vol. 112, no. 2, pp. 111–125, 2018.

[19] B. Dai and H. Chen, "Profit allocation mechanisms for carrier collaboration in pickup and delivery service," *Computers & Industrial Engineering*, vol. 62, no. 2, pp. 633–643, 2012.

[20] S. R. Dabbagh and M. K. Sheikh-El-Eslami, "Risk-based profit allocation to DERs integrated with a virtual power plant using cooperative Game theory," *Electric Power Systems Research*, vol. 121, pp. 368–378, 2015.

[21] S. Hernández and S. Peeta, "A carrier collaboration problem for less-than-truckload carriers: characteristics and carrier collaboration model," *Transportmetrica A: Transport Science*, vol. 10, no. 4, pp. 327–349, 2014.

[22] M. Guajardo and M. Rönnqvist, "Operations research models for coalition structure in collaborative logistics," *European Journal of Operational Research*, vol. 240, no. 1, pp. 147–159, 2015.

[23] Y. Wang, X. Ma, and Y. Lao, "A fuzzy-based customer clustering approach with hierarchical structure for logistics network optimization," *Expert Systems with Applications*, vol. 41, no. 2, pp. 521–534, 2014.

[24] J. Kennedy and R. Eberhart, "Particle swarm optimization," in *Proceedings of the IEEE International Conference on Neural Networks*, pp. 1942–1948, Perth, Australia, December 1995.

[25] H. Chen, A. M. K. Cheng, and Y. Kuo, "Assigning real-time tasks to heterogeneous processors by applying ant colony optimization," *Journal of Parallel and Distributed Computing*, vol. 71, no. 1, pp. 132–142, 2011.

[26] T. J. Ai and V. Kachitvichyanukul, "A particle swarm optimization for the vehicle routing problem with simultaneous pickup and delivery," *Computers & Operations Research*, vol. 36, no. 5, pp. 1693–1702, 2009.

[27] M. A. Mohammed, M. K. Abd Ghani, R. I. Hamed, S. A. Mostafa, M. S. Ahmad, and D. A. Ibrahim, "Solving vehicle routing problem by using improved genetic algorithm for optimal solution," *Journal of Computational Science*, vol. 21, pp. 255–262, 2017.

[28] S. H. Tijs and T. S. Driessen, "Game theory and cost allocation problems," *Management Science*, vol. 32, no. 8, pp. 1015–1028, 1986.

[29] F. Cruijssen, P. Borm, H. Fleuren, and H. Hamers, "Supplier-initiated outsourcing: a methodology to exploit synergy in transportation," *European Journal of Operational Research*, vol. 207, no. 2, pp. 763–774, 2010.

[30] M. Frisk, M. Göthe-Lundgren, K. Jörnsten, and M. Rönnqvist, "Cost allocation in collaborative forest transportation," *European Journal of Operational Research*, vol. 205, no. 2, pp. 448–458, 2010.

[31] S. Lozano, P. Moreno, B. Adenso-Díaz, and E. Algaba, "Cooperative game theory approach to allocating benefits of horizontal cooperation," *European Journal of Operational Research*, vol. 229, no. 2, pp. 444–452, 2013.

9

Efficiency Assessment of Transit-Oriented Development by Data Envelopment Analysis: Case Study on the Den-en Toshi Line in Japan

Jing Guo,[1] **Fumihiko Nakamura**,[2] **Qiang Li**,[3] **and Yuan Zhou**[4]

[1]*Graduate School of Environmental Studies, Nagoya University, D2-1Furo-cho, Chikusa-ku, Nagoya 464-8603, Japan*
[2]*Institute of Urban Innovation, Yokohama National University, 79-5 Tokiwadai, Hodogaya, Yokohama 240-8501, Japan*
[3]*College of Resources Science and Technology, Beijing Normal University, 19 Xinjiekouwai Street, Beijing 100875, China*
[4]*Department of Geographical Sciences, University of Maryland-College Park, College Park, MD 20742, USA*

Correspondence should be addressed to Qiang Li; liqiang@bnu.edu.cn

Academic Editor: Taku Fujiyama

Transit-Oriented Development (TOD) is an urban planning approach that encourages a modal shift from private to public transportation. This shift can generate additional benefits from a sustainability perspective. This study aims to assess the efficiency of TOD by applying the data envelopment analysis (DEA) method. The ridership of public transportation is considered as the direct output characteristic of TOD efficiency, and nine indicators of ridership are selected as inputs on the basis of the core concepts of TOD. These concepts include density, diversity, and design (3Ds). The Tokyu Den-en Toshi Line in Japan is presented as a typical case of TOD because this line includes TOD and non-TOD stations. Assessing and comparing the results of all railway stations reveal that almost all indicator values of non-TOD stations are higher than those of TOD stations. The results suggest that TOD planning and programs are inefficient in terms of ridership generation. This implication, however, may be attributed to the inadequacy of the selected indicators for TOD assessment. The results obtained after adding operation year as input indicators and removing transfer station show that TOD stations perform efficiently and in accordance with expectations. This response indicates that the inclusion of influential factors is necessary for equitable TOD assessment. Therefore, other influential factors must be considered when evaluating the efficiency of TOD-based stations with different inherent attributes. In addition, the design input with the largest impact on all inefficient units was identified, suggesting that management of bus service and railway system should be well enhanced.

1. Introduction

Transportation infrastructure is one of the most important components of urban development. The efficient development and implementation of a transportation infrastructure plan are necessary to solve traffic congestion, greenhouse gas emissions, and other environmental consequences that accompany rapid urbanization and motorization [1]. From the perspective of sustainable and integrated urban development, the imbalance between land use and transportation systems is the most critical factor that affects the efficiency of transportation infrastructure. Transit-Oriented Development (TOD) strategy has been proposed as an efficient approach to solve this imbalance [2, 3].

TOD aims to encourage the use of public transportation [4] by increasing population density and mixed land use in areas surrounding transportation hubs and by promoting connectivity between stations and trip origins/destinations. Although the formal term of TOD was not introduced until 1993 [5], its underlying principle has been applied and implemented as a solution to urban growth challenges since the late 19th and early 20th centuries, mainly in the United States and Europe [6]. Later, in 1997, the density, diversity, and design (3D) framework of TOD was proposed and explained [7]. In several American cities, the 3D-based TOD plan has increased the ridership and modal shares of public transport and has decreased vehicle mileage by 3%–5% [8, 9]. Accordingly, the TOD concept has been applied with an emphasis on the 3D framework in urban and transport planning worldwide. However, its efficiency in practical applications may sometimes not meet expectations [10]. The low utilization of public transport cannot achieve the sustainable cycle of land–people–transport [11]. In other words, whether or not the TOD strategy can play its due role in urban planning and management is inconclusive. Therefore, evaluating the efficiency of TOD is a necessary challenge faced by urban administrators and policy makers who aim to successfully implement, enhance, and improve TOD.

To respond to such a demand, this study selects the appropriate indicators that reflect the 3D concept of TOD in practice. This study assesses the efficiency of a railway system that was developed on the basis of the 3D framework of TOD. Furthermore, it provides advice to improve the factors that most significantly influence the efficiency of TOD. In this study, data envelopment analysis (DEA), which was first proposed by Charnes et al. in 1978, is applied [12, 13]. Since its introduction, DEA has been widely used to measure the relative efficiency of various organizations, referred to as decision-making units (DMU), with multiple inputs and outputs [14–17]. DEA can identify the input with the largest impact on an inefficient unit. Such input can provide guidelines to policy makers. One line of the Japanese railway system is presented as a case study. In Japan, the integration of the transport system and land use has been practiced since the 1910s. This practice is consistent with the concept of TOD. After more than a century, the country has recognized advancements in railway operation and management. The efficiency evaluation of TOD cases in Japan can provide essential guidance or experiences for future transport and urban planning in other countries, especially in developing countries experiencing rapid urbanization.

2. Literature Review

The integration of transportation and land use dates back to the post-World War II era, when planners in Europe, most notably in Stockholm [18] and Copenhagen [19], led suburban development into satellite areas along transit corridors. Since the 1990s, a series of new concepts and approaches, such as the Smart Growth concept [20, 21] and New Urbanism approach [22], for the integration of transportation and land use have emerged to combat uncontrolled urban sprawl [6]. TOD is also gaining popularity as an approach to promote smart growth and sustainable development in the metropolitan areas of many developed and developing countries [23–26] where mass transit systems have been implemented to relieve traffic congestion.

A common definition or criterion remains to be established for TOD despite the substantial body of research that has discussed the concept. For example, Cervero et al. stressed the contributions of the 3Ds to the TOD concept; furthermore, they explained the 3D concept as density in the form of residence and jobs, diversity in the form of mixed land-use development, and design in the form of good street connectivity for pedestrians [25]. Other scholars interpreted TOD by borrowing words from its original explanation and providing new empirical additions. The various definitions for TOD generally state that TOD should promote compact and highly mixed-use development around transit hubs and should include accessible and walkable neighborhoods [18, 19].

Existing studies have demonstrated that TOD provides numerous advantages, such as facilitating polycentric cities and regions, mitigating urban sprawl, boosting public transport ridership, increasing bike usage and walkability, accommodating economic growth, and creating sustainable neighborhoods [6]. The most important function of TOD is to reduce traffic congestion and increase public transit use in areas experiencing accelerated urban development [25, 27, 28]. TOD has attracted increasing attention as a solution to the challenges of urban growth given its widely recognized benefits. TOD implementation, however, is not always successful. Thus, several studies have evaluated the efficiency of TOD (e.g., [9, 21]). Most existing research applied single and several indicators to measure the outcomes or benefits that TOD provides in terms of travel behavior, economics, and environment [28, 29]. Ridership is one of most representative indicators of TOD outcome [25]. Renne and Wells monitored the success of TOD through a survey that included and summarized the various indicators of the influences and the most useful indicators of TOD [3]. However, this study quantified TOD in the form of a single indicator and provided a limited explanation for the ridership efficiency of TOD. In addition, it lost meaningful information for the aspects that contribute to the success or failure of TOD. The multi-indicator evaluation method involves weight setting for each indicator. However, weight setting in existing studies is often empirical or less objective and could provide biased results.

In the past decade, DEA has been widely used to measure the efficiency of different models of transportation system operations or management [30]. The most significant advantage of DEA over traditional weight-based methods is that DEA does not require weight parameters to measure efficiency. Researchers have employed DEA to address two main research topics. One topic is the analysis and comparison of the efficiency of firms, networks, or nations in the field of economics and management science [31–33]. Only a small part of related works has focused on the evaluation of the operational performance and management of railways [30]. The other topic is the selection of the optimal solution for actual operations in vehicle dispatching and route planning [34, 35]. Up to now, DEA has not been applied in the assessment of transportation strategy, at least in terms of TOD.

2.1. DEA. As mentioned above, DEA identifies the frontier (envelopment surface), which is determined directly from data, and quantifies the relative efficiency of a set of comparable DMUs with several inputs and outputs. It solves a linear programming model to detect the efficient inputs and outputs that form the frontier and the inefficient inputs and outputs that form nonfrontier units. In addition to providing the efficiency level, DEA can identify the sources of inefficiency and the projection path to the frontier. The projection path to the envelope surface is determined by whether the model

TABLE 1: Indicators for DEA method.

		Indicators	Unit	Explanation
Density	x_1	Population density	Person/km^2	Population per square kilometer
	x_2	Land-use diversity	-	Shannon–Wiener Index
Diversity	x_3	Functional land area	km^2	Total residential, commercial, official, public service land area per catchment
	x_4	High-building area	km^2	Area with high buildings (over 4 floors) per catchment
	x_5	Bus stops	-	Bus stops per unit catchment area
	x_6	Daily bus operation length	km	Bus operation length per day in each catchment
Design	x_7	Bus routes	-	Number of routes to and from railway station
	x_8	Road length	km	Road length per catchment
	x_9	Number of station exits	-	Number of exits per railway station
	y	Ridership	Person	Average daily commuters per station

is output-oriented or input-oriented, the choice of which depends on the production process that characterizes the units (i.e., the process minimizes the use of inputs to produce a given output level or maximizes output level given an input level). A large output (i.e., ridership) is preferred in the evaluation of TOD efficiency given several certain inputs. Therefore, output-oriented DEA was selected in this study.

The CCR model, which was first proposed by Charnes et al. in 1978, is the basic DEA model. It functions under the assumption of constant returns to scale and reflects the fact that outputs change proportionally with inputs [12]. Assume that n DMUs ($j = 1, \ldots, n$), m inputs x_{ij} ($i = 1, \ldots, m$), and s outputs y_{rj} ($r = 1, \ldots, s$) exist for each DMU_j. The output-oriented CCR model can be formulated as follows:

$$\begin{aligned} \max \quad & \theta + \varepsilon \left(\sum_{i=1}^{m} s_i^- + \sum_{r=1}^{s} s_r^+ \right) \\ \text{s.t.} \quad & \sum_{j=1}^{n} \lambda_j x_{ij} + s_i^- = x_{i0} \\ & \sum_{j=1}^{n} \lambda_j y_{rj} - s_i^+ = \theta y_{r0} \\ & s_i^-, s_r^+ \geq 0 \\ & \lambda_j \geq 0, \end{aligned} \tag{1}$$

where θ is the efficiency score limited to less than 1, ε is the non-Archimedean infinitesimal, and x_{i0} and y_{r0} refer to the input and output of one certain unit DMU_0. s_i^- and s_r^+ are the input and output slack, which refers to the excess input or missing output that exists after the proportional change in the input or the output has reached the efficiency frontier. λ_j is the weight given to the DMU_j to achieve its optimal performance. It is calculated automatically during the model-solving process.

The improvement target (projection) of each inefficient unit is the result of respective slack values added to proportional reduction amounts (formula (2) and (3) for the output-oriented DEA model). The improvement target provides alternative ways to improve the performance of each inefficient unit. The difference between x_{i0} and x'_{i0}, y_{r0} and y'_{r0} should receive focus when changing the input and output.

$$x'_{i0} = x_{i0} - s_i^- \tag{2}$$

$$y'_{r0} = \theta y_{r0} + s_r^+. \tag{3}$$

2.2. Indicator Selection. As the output variable, ridership (y) is the number of passengers who ride a public transport system and directly reflects the efficiency of transportation policy and urban planning, such as TOD strategy. Populations living around transport hubs provide the trip demand and have the potential to use the transportation infrastructure; mixed land use in the catchment area helps generate trips by providing different services; and convenient design encourages high numbers of people to use public transportation. All of these effects promote modal shift and increase ridership. Therefore, from the core concept of TOD, that is, density, diversity, and design, nine indicators were selected as the input variables of ridership output (Table 1).

First, density is defined as the number of buildings or population per area. Here, population density (x_1) in a catchment, that is, population per unit area, is the primary indicator that directly determines trip demand and economic activity. The residential population tends to generate the boarding ridership, whereas the commercial and office population is related to the alighting ridership via commercial or employment attractiveness.

Second, diversity is reflected by land use and primarily represents the level of mixed land-use development. This indicator is obtained through the Shannon–Wiener index, which is widely applied to calculate species diversity in the field of ecology. It has been successfully applied in land-use research to estimate the number of land-use types and evenness [16]. Given that residential land, commercial land, official land, and public service land are the primary types of urban functional land areas, the functional land area will affect ridership by providing for the different needs or demands of passengers and other interested people. Furthermore, areas with high buildings (over 4 floors) surrounding transportation hubs affect trip demand generation and represent the basic characteristic of regional development level. Therefore, land-use diversity (x_2), functional land-use area

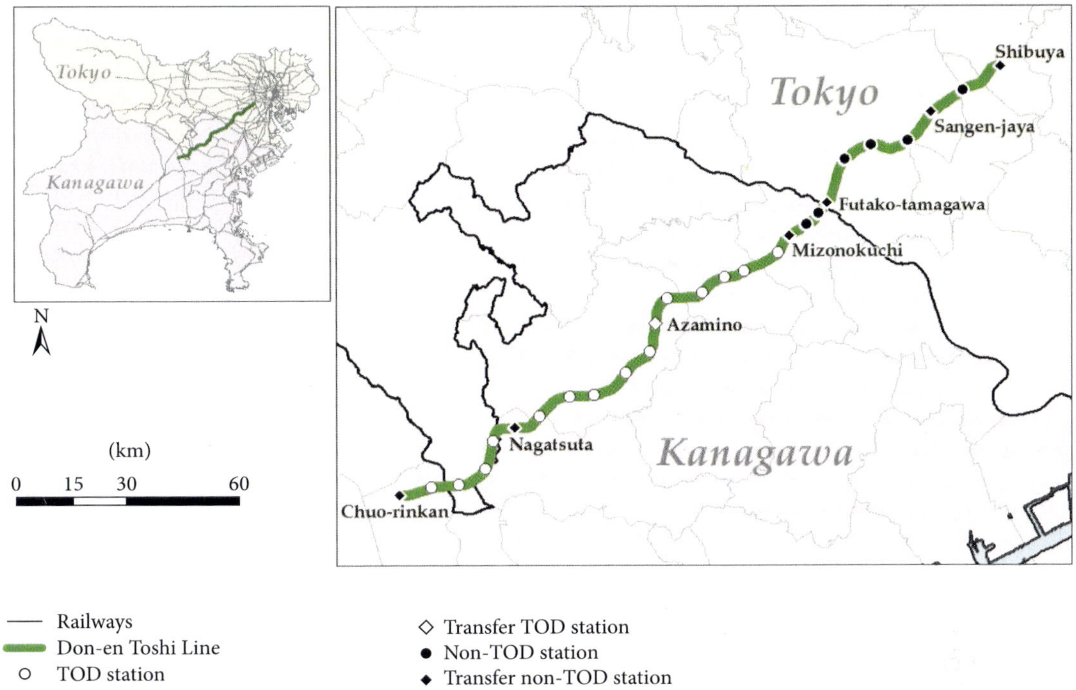

FIGURE 1: Tokyu Den-en Toshi Line.

(x_3), and high-building area (x_4) were selected as the three indicators of diversity input.

Third, design schemes can influence travel demand by increasing the accessibility of destinations by foot or by other public transportation systems [6]. It contains different implications from the micro to macro level, including the presence of shade trees and overhead lighting, design of on-site stores and services, and connections between work sites and worker residences. In this study, the design points to a basic transit infrastructure around rail stations that shortens the distance or time required to reach a destination. A highly advanced design would have the potential to facilitate public transit use and walking, as well as promote efficient energy use. Therefore, the number of bus stops (x_5), daily bus operation length (x_6), and bus routes (x_7) were selected to represent the connectivity between the rail station and passenger destinations. Connectivity eventually affects ridership choice and mode split. With respect to the total road length (x_8), we hypothesized that long roads in a certain area would increase road density and facilitate passenger access to railway stations. Finally, the number of railway station exits (x_9) also provides passengers with convenient access to aboveground roads or buses and underground stores or supermarkets.

3. Case Study

3.1. Tokyu Den-en Toshi Line. Japan's urban transportation system is famous for its excellent accessibility, efficiency, and punctuality. These characteristics resulted from advanced and reasonable transportation/urban planning. Moreover, in Japan, cities are commonly planned on the basis of TOD because the country has centered its land and city development along railroad construction for more than a century. *Ensenkaihatsu* (or "railway corridor development") was proposed by as early as 1910 and has a longer history than TOD. This urban development concept focused on residential district development along railways.

The Tokyu Den-en Toshi (DT) Line is a major commuter line that connects Shibuya in Tokyo and the Chuo-Rinkanin terminal in the neighboring Kanagawa Prefecture (Figure 1). The earliest interval, which is known as the Mizonokuchi Line, was opened in 1927 and connects the Futako-Tamagawa and Mizonokuchi Stations. In the mid-20th century, economic activity and employment rapidly concentrated in the central area of Tokyo, thus drastically increasing land and real estate prices and promoting ex-urban settlement. In response to this situation, the Japanese government adopted a new TOD plan. The DT Line was born at the right moment, and Tama New Town was subsequently developed. In 1966, the DT Line began operating from the Mizonokuchi Station to the Nagatsuta Station. An expansion was built as an access route to Tama Garden City [36]. The DT Line, which spans Shibuya Station to Futako-Tamagawa Station, was completed in 1977. In 1984, the line was extended from Nagatsuta Station to Chuo-Rinkan Station. Stations between the Mizonokuchi and Chuo-Rinkan Stations, except for Nagatsuta Station (Figure 1), were developed on the basis of TOD. To enable comparison in this study, the earlier/older stations are identified as non-TODs (12 stations), and the latter/newer stations are considered as TODs (15 stations).

The DT Line was selected as a case study in this work for two reasons. First, the DT Line and the Tama New Town that it traverses provide examples of transportation and land-use integration that are intended to stimulate population concentration and ridership growth. Second, this line has a relatively long history and was constructed under previous and new urban development policies (i.e., some of its stations were originally built with the birth of the line, whereas others are relatively new and were constructed through the TOD project). Therefore, earlier and later/newer stations can be compared to identify the factors that affect ridership.

3.2. Catchment Definition and Data Collection. Station catchment areas, which influence indicator values, need to be determined. In the transportation research field, the catchment area where local residents potentially use the railway is identified through (1) placing a buffer circle around a station (e.g., buffers with a radius of 400, 800, or 1200 m to define walkability to the public transportation facility) [37]; (2) network analysis [38] based on path networks, wherein a catchment area can be an area with a fixed walking or driving distance or travel time; (3) the boundary method, which is based on geo-code trip survey data (in this method, however, only 90% of the trip data are used within the study to remove the outliers) [39]; and (4) statistical modeling based on spatial theory, such as location and allocation method [40]. Among these methods, the boundary method is the most objective method that does not require intensive expert experience.

In this study, the Japanese Nationwide Person Trip Survey was used to identify station catchments through the boundary method. This survey is the most fundamental transportation questionnaire survey and is conducted every 10 years by the Ministry of Land, Infrastructure, Transport, and Tourism to investigate the actual travel behaviors of people in certain regions in Japan. In the Tokyo Metropolitan Area, approximately 730,000 individual questionnaires are collected. Survey items include individual trips, personal attributes, and travel behaviors. The origin collected for the survey was recorded by a small zone geo-code, and the survey results showed the station choices in each zone.

An origin–destination matrix from the zones of passenger residence (origin) to rail stations (destination) was constructed in accordance with the survey results and is shown in the flow map in Figure 2(a). Then, zones that generated few demands were removed on the basis of the rule that a catchment area should cover at least 90% of the ridership around the station (Figure 2(b)). A single catchment area nearly covers the buffer area with a 1,500 m radius around the railway. Given that individuals from the same small zone may select different rail stations, the catchment areas of two or more adjacent rail stations may overlap. Overlapped zones were divided into different stations by the number of station choices and distances to different stations (Figure 2(c)). Finally, the catchment area for each rail station was identified. Following the identification of catchment boundaries, all data for selected indicators (e.g., bus stop data, as an example shown in Figure 2(d)) were collected from the website of the Official Statistics of Japan and National Land Numerical Information. The collected data were then masked, reorganized, and standardized.

4. Results

4.1. Basic Statistical Characteristics. The initial analysis provided the basic statistical characteristics of all selected indicators for the 27 stations. All of the indicator values of non-TODs are higher than those of TOD stations. Therefore, TOD stations do not provide any obvious advantages. In detail, the mean population density of TOD stations is considerably lower than that of older stations. This result may be attributed to the different development histories of non-TOD and TOD stations. The higher land-use diversity of non-TODs than that of TOD stations can be attributed to the main function of the Tama New Town Station along the DT Line. The catchment of the Tama New Town Station mainly functions to relieve housing problems in Tokyo. Thus, this area contains vast tracts of residential land. By contrast, the catchment areas of older stations contain relatively more commercial and office areas than those of the Tama New Town Station. Non-TOD stations have higher functional land areas and numbers of high buildings per unit area than TOD stations, indicating that land use around old stations is better developed than that around new stations. TOD stations consistently have lower values for the five design indicators than non-TOD stations. This result suggested that infrastructure with less accessibility may not serve a high proportion of riders. Compared with non-TOD stations, TOD stations have lower average daily ridership, which is the direct output of transportation operation and management and land use. However, low ridership and indicator values do not necessarily indicate low efficiency and vice versa. Thus, an in-depth assessment is necessary to identify the efficiency of TOD stations.

4.2. TOD Efficiency. To explain whether different 3D patterns can generate correspondingly different ridership patterns, the DEA method was applied to calculate the efficiency of each station pattern. The results are shown in Table 3. Nine stations with E equal to 1 are considered as relatively efficient. However, seven of these stations are non-TOD stations, and only two are TOD stations. The mean efficiency score of TOD stations is lower than that of non-TOD stations, indicating that the 3D-based TOD strategy does not generate sufficient ridership as initially expected.

This finding can be partly explained by the individual attributes of each station. Station attributes include designation as a transfer station, opening date, and operating history. For example, the Shibuya Station is a transfer station with a long operating history in a catchment area with high population density. Accordingly, its corresponding land use is well developed, and its accessibility attracts high numbers of riders. Furthermore, non-TOD stations may achieve TOD goals given their specific attributes. For example, among non-TOD stations, the Azamino Station is highly efficient despite its lack of a long operating history because of its transfer function. The above result suggests that considering only 3D-related indicators cannot provide an adequately comprehensive assessment of TOD efficiency. The inherent attributes of each station cannot be ignored because they constrain the comparability of stations.

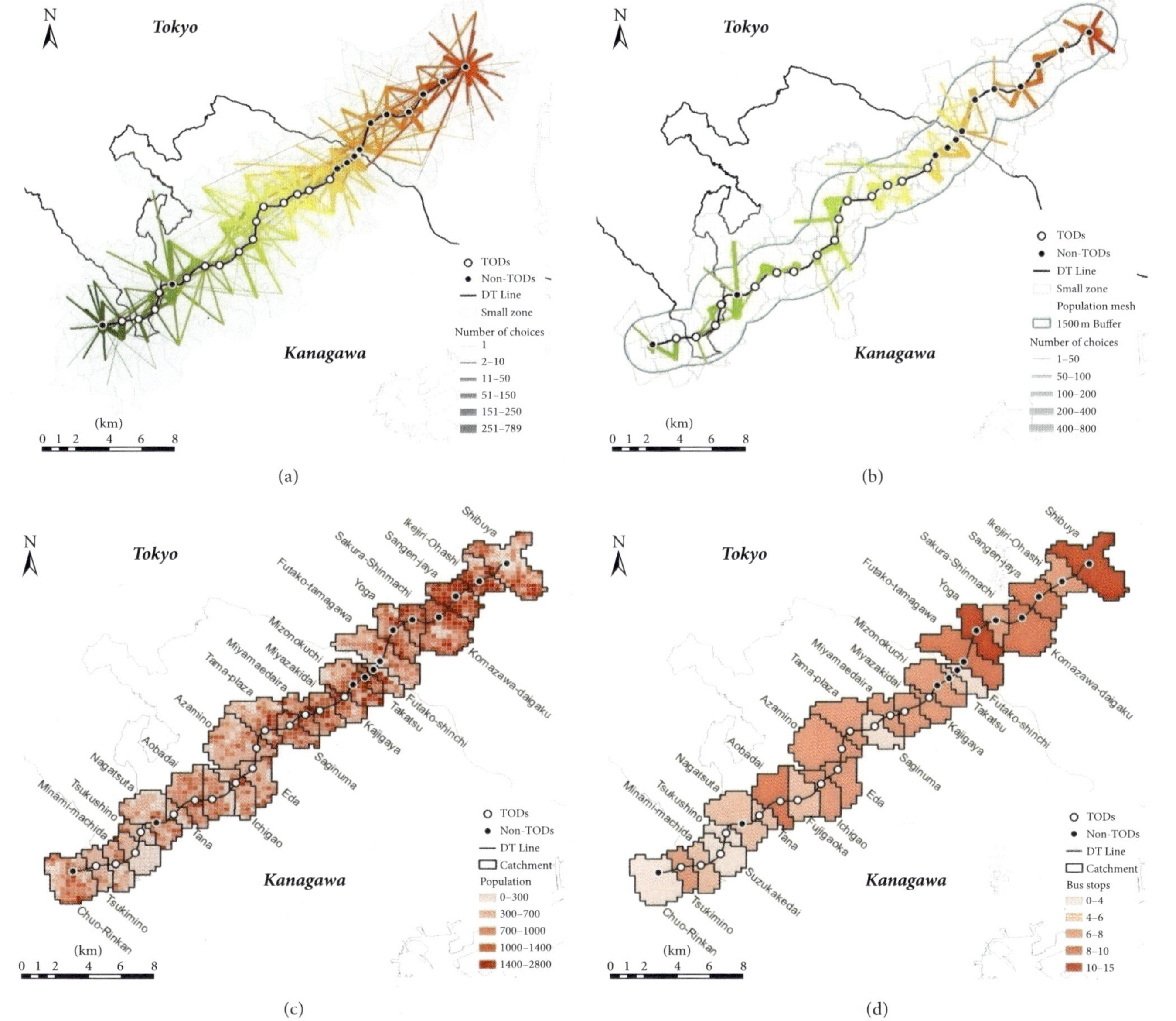

FIGURE 2: Generation of catchment area and distribution of a selected indicator for each catchment. (a) Original flow map; (b) deletion of extra zones and magnification; (c) finished catchment area; (d) bus stops for each catchment. *Note.* In (a) and (b), the flow color is used to distinguish the populations of different stations. The width of gray bar in legend represents the number of choices of separate stations.

4.3. TOD Efficiency considering Inherent Station Attributes. The inherent attributes of each station may affect its ridership proportion and efficiency. Inherent station attributes include the number of operation years, designation as a transfer station, number of trains running per day, and duration between trips. The former two attributes have been discussed above to account for the relatively inefficient performance of TOD stations. To avoid the biases resulting from operating history and transfer function, operation year was added to the DEA model. At the same time, seven transfer stations were removed from the DMU list to improve the comparability of stations. Hence, the DEA model was recalculated with 20 stations (6 non-TOD and 14 TOD stations) with 10 input indicators, including operation year.

After removing transfer station as an input (Table 4), the differences between the input indicator values of TOD and non-TOD stations have drastically decreased compared with those shown in Table 2. As shown in Table 2, the average ridership of non-TOD stations is almost three times larger than that of TOD stations. However, the average ridership of non-TODs is approximately 20% higher than that of TOD stations when only nontransfer stations are considered. Given that ridership is the output indicator of the DEA model, changes in ridership will inevitably lead to the different efficiency results shown in Table 5. Most TOD stations are relatively efficient, and the mean values of efficiency E for TODs and non-TODs are not considerably different. Yoga Station, a non-TOD station, performed relatively inefficiently

Table 2: Means of the indicators of non-TOD and TOD stations.

	Indicators	Non-TODs	TODs	Difference
Density	Population density	14.21	10.39	3.82
Diversity	Land-use diversity	0.97	0.64	0.33
	Functional land area	0.87	0.86	0.01
	High-building area	0.06	0.03	0.03
Design	Number of bus stops	7.37	6.05	1.32
	Daily bus operation length	575.44	423.51	151.93
	Bus routes	4.50	1.73	2.77
	Road length	19.36	21.37	−2.01
	Number of station exits	12.08	7.87	4.21
Daily ridership		130459	46747	83712

Table 3: Efficiency of each station.

Non-TOD stations			TOD stations		
Station	E	Operation years	Station	E	Operation years
			Kajigaya	0.824	51
Shibuya*	1	132	Miyazakidai	0.536	51
Ikejiri-Ohashi	0.364	40	Miyamaedaira	0.496	51
Sangen-Jaya*	0.739	110	Saginuma	1	51
Komazawa-Daigaku	0.760	40	Tama-Plaza	0.644	51
Sakura-Shimmachi	0.730	110	Azamino*	1	40
Yoga	0.487	110	Eda	0.895	51
Futako-Tamagawa*	1	110	Ichigao	0.507	51
Futako-Shinchi	1	90	Fujigaoka	0.513	51
Takatsu	1	90	Aobadai	0.905	51
Mizonokuchi*	1	90	Tana	0.809	51
Nagatsuta*	1	109	Tsukushino	0.286	49
Chuo-Rinkan*	1	88	Suzukakedai	0.274	45
			Minami-Machida	0.573	41
			Tsukimino	0.257	41
Mean	0.840	93.25	Mean	0.635	48.4

Note. * denotes a transfer station.

Table 4: Means of indicators of nontransfer stations.

	Indicators	Non-TODs	TODs	Difference
Density	Population density	15.74	10.48	5.26
Diversity	Land-use diversity	0.93	0.64	0.29
	Functional land area	0.87	0.86	0.01
	High-building area	0.03	0.02	0.01
Design	Number of bus stops	7.02	6.01	1.01
	Daily bus operation length	542.05	430.90	111.15
	Bus routes	2.83	1.71	1.12
	Road length	22.99	21.71	1.28
	Number of station exits	10.67	7.29	3.38
Operation year		80	49	31
Daily ridership		52478	40566	11912

given its 110-year operation history. The efficiency values of 12 TOD stations with 40–50 years of operation history, except for Ichigao and Tana Stations, are all higher than those of Yoga Station. This result suggested that TOD stations tend to reach their current efficiency more rapidly than non-TOD stations.

4.4. Potential for Improving Efficiency. The frontier project reflects the improvement target that the inputs and outputs of each inefficient station should reach. The distance (difference) between the real point and its projection indicates the volume required for improvement. Then, the percentage

Table 5: Efficiencies of nontransfer stations considering operation year.

Non-TOD stations		TOD stations	
Station	E	Station	E
		Kajigaya	1
		Miyazakidai	1
		Miyamaedaira	1
		Saginuma	1
Ikejiri-Ohashi	0.702	Tama-Plaza	1
Komazawa-Daigaku	1	Eda	1
Sakura-Shimmachi	1	Ichigao	0.708
Yoga	0.881	Fujigaoka	0.958
Futako-Shinchi	1	Aobadai	1
Takatsu	1	Tana	0.809
		Tsukushino	1
		Suzukakedai	0.894
		Minami-Machida	0.928
		Tsukimino	0.999
Mean	0.981	Mean	0.955

Table 6: Difference between the average projected and real values of inefficient stations (%).

	Indicators	Non-TODs	TODs
Density	Population density	6.33	17.89
Diversity	Land-use diversity	20.39	20.14
	Functional land area	6.04	8.67
	High building area	9.46	27.89
Design	Number of bus stops	28.68	26.32
	Daily bus operation length	38.31	46.32
	Bus routes	36.38	28.08
	Road length	19.27	23.17
	Number of station exits	6.34	0.62
Operation year		27.03	13.84

of difference relative to the original input can be calculated to discover the excess input indicator that contributes most significantly to inefficiency. The average difference percentage of each input for non-TOD stations and TOD stations (only including inefficient stations) is listed in Table 6.

All inefficient railway stations have higher numbers of redundant design indicators than the other two types of stations. The results indicated that bus location and services do not increase the corresponding ridership proportion of the railway. One possible reason is that passengers' daily transportation needs cannot be satisfied by the mutual independence of the railway and bus systems. Therefore, bus services and their feeder roles should be enhanced to improve the efficiency of the railway system. Moreover, diversity in the land use of areas surrounding inefficient stations does not generate adequate ridership, suggesting that the mixture of land-use types should be promoted despite requiring long-term effort. Population density is more redundant for inefficient TOD stations than for non-TOD stations. This result can be attributed to the number of serviced people and to age category (i.e., percentage of the elderly, adults, and children) and social level (i.e., percentage of unemployed, professionals, and managers) of the residents of a certain catchment area. For example, high-class workers and the elderly tend to use private transportation. Accordingly, increasing the density and optimizing the age and social category of the target population may help improve TOD efficiency. All aspects listed above should receive additional attention when planning future TOD stations.

5. Discussion and Conclusions

This study focused on measuring the efficiency of the TOD strategy implemented in the DT Line. TOD is a powerful urban planning approach and concept that is applied to resolve various urban problems. This study expounded on TOD assessment on the basis of the 3D framework. In this study, the assessment procedure included the identification of catchment areas and the selection of suitable indicators as input variables and ridership as output variables. The input variables are then subjected to the DEA method to validate the successful performance of the TOD strategy or program and to identify possible solutions to strengthen TOD implementation.

Existing TOD evaluations have failed to provide reliable conclusions regarding TOD performance. For example, a single-indicator assessment cannot adequately characterize the comprehensive efficiency level of a TOD strategy from the perspective of the 3D framework. Meanwhile, a simple weight-based method following the 3D concept is likely to be affected by weights with low objective values and may thus produce a biased result that may contradict reality. These situations suggest the necessity of designing a procedure for the comprehensive assessment of TOD strategies. By considering that density and diversity may generate transportation demand and that design must provide convenient service to satisfy such needs, we employed DEA to reflect the relative efficiency of TOD in accordance with 3D-based inputs and ridership output. Our DEA-derived results reflect the core principle of the TOD concept, which coordinates density, diversity, and design with ridership output.

Many TOD applications have demonstrated that TOD has a significant role in regional development. However, the present results suggested that TOD strategies cannot provide more positive effects than non-TOD strategies because, on the one hand, market power enables non-TOD strategies to achieve effects similar to those achieved by TOD strategies. More importantly, the current indicators selected for TOD assessment do not adequately reflect the 3D concept. In fact, the current indicators for TOD assessment only highlight aspects related to 3D, and some important factors that affect output are excluded from conventional TOD assessment. For example, operating history can also affect the output of TOD stations along the DT Line. Although TOD stations are comparable with non-TOD stations in general, development history results in the incomparable efficiency of new and old stations. Moreover, other factors, such as the age and social class of the catchment population, designation as transfer station, number of trains running per day, and duration between trips, have been overlooked. Adding operation year as an input indicator and removing transfer station revealed that the performance of TOD stations is relatively efficient and approaches expectations. This result indicated that the inclusion of other influential factors is necessary for equitable TOD assessment. Consequently, taking such factors into account is indispensable when evaluating the efficiency of TOD stations with different inherent attributes. In addition, the design input with the largest impact on all inefficient units was identified. Management of bus service and railway system should be well enhanced and encouragement for public transportation usage should be emphasized.

The applicability of the DEA-based TOD assessment approach is mainly dependent on indicator selection. The framework of the approach includes indicator selection from the perspective of 3Ds, catchment divisions, and the DEA model and is applicable to other public transportation systems (e.g., light rail, metro, bus, and bus rapid transit) and to other regions or countries. Nevertheless, the selected indicators may vary across different applications. Researchers and urban planners may have specific local or specialized priorities and knowledge that differ from those emphasized in the present case. Thus, other researchers should select indicators that are appropriate for their priorities and knowledge.

Nevertheless, the present study presents several critical issues that should be considered to improve the evaluation of TOD approaches. First, additional factors that affect the efficiency of TOD-based systems should be considered during the assessment and planning of a TOD project. These factors include the attributes of the target population (age structure, occupation, and income level) and the transportation system itself. Second, the design scheme should improve the convenience of accessing destinations by foot, bus, or bicycle. In this study, the selected indicators mainly characterized connectivity to railway stations by bus. However, such a characterization may be insufficient for design characterization. Other factors, such as plating strips, overhead lights, and bicycle path lengths, should be considered in future studies. Furthermore, the definition of the analytical unit is crucial for TOD assessment. In this study, the Nationwide Person Trip Survey in Japan was used to identify station catchment areas through the boundary method. When such data are unavailable, the effect of catchment size should be addressed by specifying a different buffer radius, travel distance, or travel time.

Conflicts of Interest

The authors declare that there are no conflicts of interest regarding the publication of this paper.

References

[1] P. Miller, A. G. de Barros, L. Kattan, and S. C. Wirasinghe, "Public transportation and sustainability: A review," *KSCE Journal of Civil Engineering*, vol. 20, no. 3, pp. 1076–1083, 2016.

[2] R. Cervero, "TCRP Report 102: Transit-Oriented Development in the United States: Experiences. Challenges and Prospects;" 2004.

[3] J. L. Renne and J. S. Wells, "Emerging European-style planning in the USA: Transit oriented development," *World Transport Policy & Practice*, vol. 10, no. 2, 2004.

[4] A. Nasri and L. Zhang, "The analysis of transit-oriented development (TOD) in Washington, D.C. and Baltimore metropolitan areas," *Transport Policy*, vol. 32, pp. 172–179, 2014.

[5] P. Calthorpe, *The Next American Metropolis: Ecology, Community, and the American Dream*, Princeton Architectural Press, New York, NY, USA, 1993.

[6] E. Papa and L. Bertolini, "Accessibility and Transit-Oriented Development in European metropolitan areas," *Journal of Transport Geography*, vol. 47, pp. 70–83, 2015.

[7] R. Cervero and K. Kockelman, "Travel demand and the 3Ds: density, diversity, and design," *Transportation Research Part D: Transport and Environment*, vol. 2, no. 3, pp. 199–219, 1997.

[8] R. Ewing and R. Cervero, "Travel and the built environment: a synthesis," *Transportation Research Record*, vol. 1780, pp. 87–114, 2001.

[9] Y. J. Singh, P. Fard, M. Zuidgeest, M. Brussel, and M. V. Maarseveen, "Measuring transit oriented development: A spatial multi criteria assessment approach for the City Region Arnhem and Nijmegen," *Journal of Transport Geography*, vol. 35, pp. 130–143, 2014.

[10] J. E. Evans and R. H. Pratt, "TCRP Report 95 – Chapter 17—Transit Oriented Development," 2007.

[11] Y. Hayashi and J. Roy, *Transport, Land-Use and the Environment*, Springer US, Land-Use and the Environment, 1996.

[12] A. Charnes, W. W. Cooper, and E. Rhodes, "Measuring the efficiency of decision making units," *European Journal of Operational Research*, vol. 2, no. 6, pp. 429–444, 1978.

[13] A. Charnes, W. Cooper, A. Y. Lewin, and L. M. Seiford, "Data Envelopment Analysis Theory, Methodology and Applications," *Journal of the Operational Research Society*, vol. 48, no. 3, pp. 332-333.

[14] S. F. Gomes Júnior, A. P. D. S. Rubem, J. C. C. B. Soares de Mello, and L. Angulo Meza, "Evaluation of Brazilian airlines nonradial efficiencies and targets using an alternative DEA approach," *International Transactions in Operational Research*, vol. 23, no. 4, pp. 669–689, 2016.

[15] M. Samà, C. Meloni, A. D'Ariano, and F. Corman, "A multicriteria decision support methodology for real-time train scheduling," *Journal of Rail Transport Planning and Management*, vol. 5, no. 3, pp. 146–162, 2015.

[16] T. Yoshida and K. Tanaka, "Land-use diversity index: a new means of detecting diversity at landscape level," *Landscape and Ecological Engineering*, vol. 1, no. 2, pp. 201–206, 2005.

[17] Q. He, J. Han, D. Guan, Z. Mi, H. Zhao, and Q. Zhang, "The comprehensive environmental efficiency of socioeconomic sectors in China: An analysis based on a non-separable bad output SBM," *Journal of Cleaner Production*, 2017.

[18] R. Cervero, "Sustainable new towns. Stockholm's rail-served satellites," *Cities*, vol. 12, no. 1, pp. 41–51, 1995.

[19] R. D. Knowles, "Transit Oriented Development in Copenhagen, Denmark: from the Finger Plan to Ørestad," *Journal of Transport Geography*, vol. 22, pp. 251–261, 2012.

[20] K. Behan, H. Maoh, and P. Kanaroglou, "Smart growth strategies, transportation and urban sprawl: Simulated futures for Hamilton, Ontario," *The Canadian Geographer*, vol. 52, no. 3, pp. 291–308, 2008.

[21] D. L. Winstead, "Smart growth, smart transportation: A new program to manage growth in Maryland," *Urban Lawyer*, vol. 30, no. 3, pp. 537–539, 1998.

[22] H. Dittmar and G. Ohland, *The New Transit Town: Best Practices In Transit-Oriented Development*, Island Press, 2003.

[23] L. Bertolini, C. Curtis, and J. Renne, "Station area projects in Europe and beyond: Towards transit oriented development?" *Built Environment*, vol. 38, no. 1, pp. 31–50, 2012.

[24] M. Givoni and D. Banister, *Integrated Transport: From Policy to Practice*, Routledge, New York, NY, USA, 2010.

[25] A. Galelo, A. Ribeiro, and L. M. Martinez, "Measuring and Evaluating the Impacts of TOD Measures – Searching for Evidence of TOD Characteristics in Azambuja Train Line," *Procedia - Social and Behavioral Sciences*, vol. 111, pp. 899–908, 2014.

[26] N. L. Widyahari and P. N. Indradjati, "The Potential of Transit-Oriented Development (TOD) and its Opportunity in Bandung Metropolitan Area," *Procedia Environmental Sciences*, vol. 28, pp. 474–482, 2015.

[27] B. P. Y. Loo, C. Chen, and E. T. H. Chan, "Rail-based transit-oriented development: lessons from New York City and Hong Kong," *Landscape and Urban Planning*, vol. 97, no. 3, pp. 202–212, 2010.

[28] C. Curtis, J. L. Renne, and L. Bertolini, "Transit Oriented Development. Making it Happen," *Journal of Transport Geography*, vol. 17, no. 6, p. 312, 2009.

[29] C. D. Higgins and P. S. Kanaroglou, "A latent class method for classifying and evaluating the performance of station area transit-oriented development in the Toronto region," *Journal of Transport Geography*, vol. 52, pp. 61–72, 2016.

[30] M. Kamruzzaman, D. Baker, S. Washington, and G. Turrell, "Advance transit oriented development typology: Case study in brisbane, australia," *Journal of Transport Geography*, vol. 34, pp. 54–70, 2014.

[31] D. S. Vale, "Transit-oriented development, integration of land use and transport, and pedestrian accessibility: Combining node-place model with pedestrian shed ratio to evaluate and classify station areas in Lisbon," *Journal of Transport Geography*, vol. 45, pp. 70–80, 2015.

[32] J. L. Renne and J. Wells, *Transit-Oriented Development: Developing a Strategy to Measure Success*, Washington, Wash, USA, 2005.

[33] L. Cavaignac and R. Petiot, "A quarter century of Data Envelopment Analysis applied to the transport sector: A bibliometric analysis," *Socio-Economic Planning Sciences*, vol. 57, pp. 84–96, 2017.

[34] X. Li, J. Yu, J. Shaw, and Y. Wang, "Route-Level Transit Operational-Efficiency Assessment with a Bootstrap Super-Data-Envelopment Analysis Model," *Journal of Urban Planning and Development*, vol. 143, no. 3, p. 04017007, 2017.

[35] J. Sanders, "Linking station node- and place functions to traffic flow: a case study of the Tokyu Den-En Toshi line in Tokyo, Japan," 2015.

[36] R. Cervero, A. Round, T. Goldman, and K.-L. Wu, *Rail Access Modes and Catchment Areas for the BART System*, Berkeley, 1995.

[37] J. Lohmann, E. Andersen, and A. Landex, "GIS-based Approaches to Catchment Area Analyses of Mass Transit," in *Esri International User Conference Proceedings*, pp. 1–13, 2009.

[38] T. Lin, R. Palmer, J. Xia, and D. A. Mcmeekin, "Automatic generation of station catchment areas: A comparison of Euclidean distance transform algorithm and location-allocation methods," in *Proceedings of the 2014 11th International Conference on Fuzzy Systems and Knowledge Discovery, FSKD 2014*, pp. 963–967, China, August 2014.

[39] O. Harrison and D. O. Connor, "Measuring Rail Station Catchment Areas in The Greater Dublin Area," in *Proceedings of the ITRN 2012*, 2012.

[40] D. Wheeler and A. Wang, "Catchment Area Analysis Using Generalized Additive Models," *Austin Biometrics Biostat*, vol. 2, no. 3, pp. 1–6, 2015.

10

Impact on Network Performance of Probe Vehicle Data Usage: An Experimental Design for Simulation Assessment

Lídia Montero[1,2] **Maria Paz Linares**[1,2] **Josep Casanovas**[1,2,3] **Esteve Codina**[1,2]
Gonzalo Recio[2] **Ester Lorente**[2] **and Juan Salmerón**[2]

[1] *Department of Statistics and Operations Research, Universitat Politècnica de Catalunya (UPC), C5-Carrer Jordi Girona 1-3, 08034 Barcelona, Spain*
[2] *inLab FIB, Universitat Politècnica de Catalunya (UPC), B6-Carrer Jordi Girona 1-3, 08034 Barcelona, Spain*
[3] *Barcelona Super Computing Center (BSC-CNS), Universitat Politècnica de Catalunya (UPC), Carrer Jordi Girona 29-31, 08034 Barcelona, Spain*

Correspondence should be addressed to Lídia Montero; lidia.montero@upc.edu

Academic Editor: Steven I. Chien

Probe-based technologies are proliferating as a means of inferring traffic states. Technological companies are interested in traffic data for computing the best routes in a traffic-aware manner and they also provide real-time traffic information with certain temporal accuracy. This paper analyses and evaluates how data provided by a fleet of probe cars can be used to develop a navigation service and how the penetration rate of this service affects a set of city-scale KPIs (Key Performance Indicators) and driver KPIs. The case study adopts a model-driven approach in which microscopic simulation emulates real-size fleets of probe vehicles that provide positions and speed data. What is noteworthy about the modelling behaviour is that drivers are segmented according to their knowledge of network conditions for selected trips: experts, regular drivers, and tourists. The paper presents and discusses the modelling approach and the results obtained from an experimental Barcelona CBD model designed to evaluate the penetration rates of probe vehicles and route guidance. An analysis of the simulation experiments reveals remarkable links among city-scale KPIs, which—from a multivariate point of view—is a novelty. A simulation-based framework for results analysis and visualization is also introduced in order to simplify the simulation results analysis and easily visualize OD paths for driver segments.

1. Introduction

Mobility is a key component in urban areas and should be addressed as part of a complex system, since it is a nonisolated component that strongly interacts with all the other components.

There is wide consensus about the brand new family of services that will be enabled by advances in intervehicular communications. From the early considerations of equipped vehicles as a network of mobile sensors [1] to more recent surveys [2], the capabilities for effectively monitoring traffic conditions have been studied. In urban areas, vehicles equipped with onboard sensors are expected to reach high concentrations in the near future. See [3] for a survey of works and applications related to traffic state monitoring.

In a parallel line, researchers have been motivated by advancements in vehicle positioning (onboard sensors) and in wireless communications that support Vehicle-to-Vehicle (V2V) and Vehicle-to-Infrastructure (V2I) applications. They have investigated how the collected data could be used to generate travel time information (see [4, 5]), and this research could be considered complementary to the estimation of links or path travel times from GPS probe vehicles. Summarily speaking, two complementary approaches are explored in the literature: [6] presents an overview of the statistical approaches; and this was later improved upon by [7] that combined GPS data with data from other travel time sources. A variant of these statistical models that exploit GPS data is analysed in [8] to identify the network paths whose travel times are estimated in [9]. An academic approach relying on traffic flow theory has been adopted by [3, 10] for the estimation of fundamental variables in arterials or urban motorways using Probe Vehicle Data (PVD) and Edie's definitions [11]. V2I applications under incident conditions have been developed and route guidance strategies evaluated by [12].

In the context of exploiting mobile data, a popular trend [13, 14] explores the use of handover information in cellular networks to estimate the traffic level of service, although they note important limitations on the overall performance.

The use of PVD has been investigated in some research projects such as Mobile Millennium or CarTel [1, 15, 16], which included a pilot traffic-monitoring system using the GPS in cellular phones to gather traffic information, process it, and distribute it back to phones in real time. Products and companies performing mobile crowdsourcing (Google Traffic, INRIX, and TomTom Traffic) allow for real-time data gathering. Machine learning applications for estimating travel time delays due to road work from GPS data have been proposed by [17].

Route guidance impact on travel time, safety, and environment have been intensively investigated [12, 18–21], usually in relation to their benefits under incident conditions and while simultaneously conducting quantitative assessments of the potential impacts of real-time routing guidance and advisory warning messages to guided vehicles. Some other authors have analysed different types of reactive [22] and proactive route guidance [23] policies using simulation, but the elaboration of traffic state estimation is simplistic.

The aim of this paper is to present a simulation-based platform that allows modelling several penetration rates for a fleet of PVD vehicles that feed travel time estimation between Points of Interest (POIs) and several penetration rates of route guidance for connected vehicles while considering driver behaviour and route choice models. Travel time estimates between POIs from PVD are critical inputs in Kalman filtering formulations, which some authors have addressed in the context of dynamic OD matrix estimation [24]: travel time availability and reliability from PVD guarantees a simplified linear formulation approach [25]. The resulting analysis platform can be easily adapted to any microscopic traffic simulator with the required extended functionalities, and it has been tested using a model of the Barcelona Business District developed and calibrated in the past for previous projects [3]. The conceptual framework is described first, and the section following that describes the simulation experiments, in which a large fleet of PVD vehicles are accounted for according to penetration rates (V2I) and additional factors considered in the experimental design. The next section provides an analysis of the results based on a set of network and driver KPIs, which are jointly considered by applying multivariate analysis techniques. The paper ends with our conclusions.

2. Simulation Testbed Framework

Figure 1 shows the simulation testbed framework, which is comprised of an Execution Controller, a Traffic Simulation Module, a Results Processing Module, and a Visual Analytics Module. The visualization and analytics tool has been implemented using the Shiny [26] web application framework for R, which simplifies the development of simulation results analyses by incorporating it into interactive web applications. The methodological framework can be implemented with any traffic simulation software by using the utilities that enable it to integrate—via API—the user-defined applications that implement the system's required functions, thus guaranteeing the transferability of the approach.

The Traffic Simulation Module was programmed using API extensions, and its components are as follows: emulation of PVD; auto demand split into vehicle classes according to driver type definition; estimation of lane and link travel times from data collected from PVD emulation; and customization of route choice models for guiding drivers according to the implemented navigation strategies. The assessment of navigation strategies (route guidance) is not the goal of the current work, although it is the aim of a follow-up project.

The simulation results analysis is performed by two fundamental and independent components: Results Processing and Visual Analytics modules. The Results Processing in Figure 1 covers all preprocessing of data automatically. In this way, the Visual Analytics Module can use the results of the traffic simulation environment directly without the need for a manual update.

2.1. Traffic Simulation Module. The traffic simulation component includes a microscopic traffic simulator and a set of custom modules and functions that were developed using API extensions. In addition to the functionalities already described at the beginning of this section, it also generates time-dependent system tables as well as link and lane traffic data by driver class.

In this work, an Aimsun [27] model was available from previous projects. Aimsun functional architecture and the interaction libraries (Aimsun API) support the extended modelling utilities that are required.

The exchange of information between the API applications and the microsimulator can be made at every simulation step (0.5 sec). The programming languages in which Aimsun provides its API are C++ and Python. While Python is used to easily collect some of the data, C++ is needed for emulating the probe vehicles due to performance reasons.

2.2. Probe Vehicle Data Emulation. This work assumes that the V2I and V2V technology is on board in probe cars. In a previous study by some of the authors, field test data from a fleet of 3 probe cars is discussed for Barcelona's CBD [3]. Collected and filtered data were used to calibrate the emulation of PVD by an API included in the Traffic Simulation Module (see [3]). Only basic vehicle sensors and no frontal camera data were used previously in the probe vehicles. The API-extension for emulating PVD depends on onboard sensors and technological specifications.

The aim of the current work is to emulate the "real-time data" of probe car data that is used in connected car guidance systems within different levels of probe car penetration. To this end, a reduced set of sensors for probe cars has been

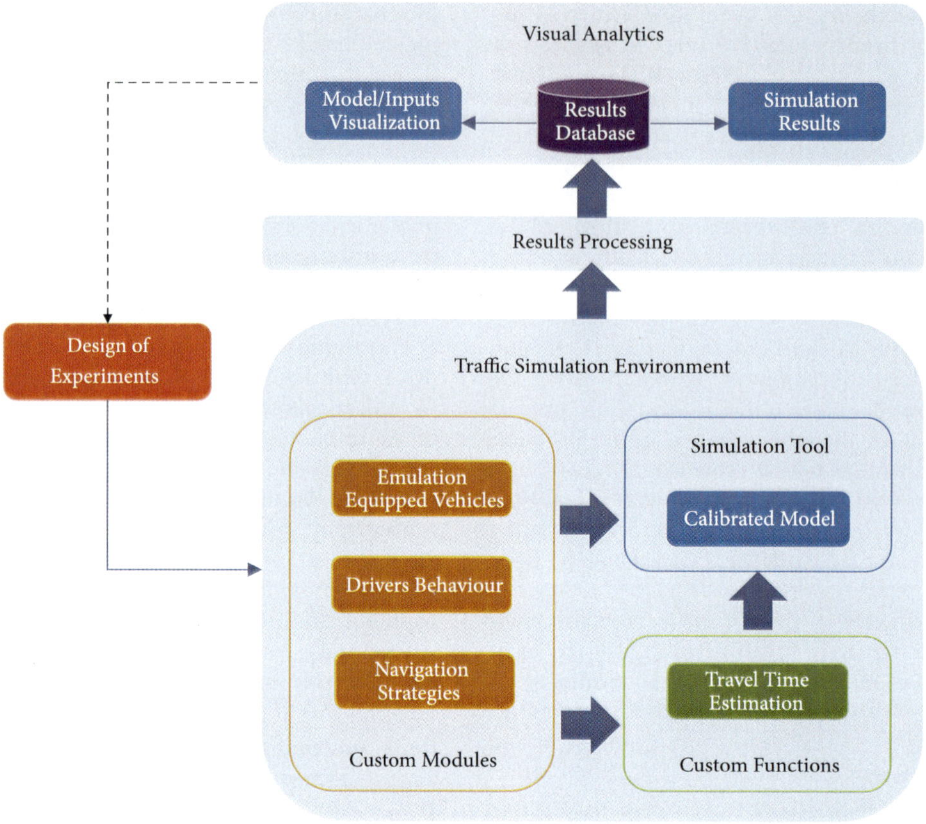

FIGURE 1: Simulation testbed architecture.

assumed, thus allowing data for vehicle position and speed at each simulation step.

2.3. Execution Controller. A simulation experiment consists of N replication executions that are launched from a controller, and each of them is set up by preprocessing and is then postprocessed at the end of the simulation in order to collect the results generated from each replication. KPIs are grouped into either network KPIs or driver type KPIs.

2.4. Results Processing and Visual Analytics Modules. The results of the simulation executions are stored on a server. The Results Processing Module (see Figure 1) contains the set of processes that are in charge of finding new execution results from recently performed simulation replicas and applying to them the postprocessing that the Visual Analytics Module needs.

The Visual Analytics Module in Figure 1 covers the visualization of input data, the model details, and the simulation results from any replication execution in the experimental design. It also consists of two implementations: that of the visualization application and the one that serves access to the web. The visualization is implemented with R-Shiny framework [26]. For performance reasons, C++ is used in some of the application processes.

3. Modelling Issues

Drivers are split into six groups according to guidance availability and their knowledge of network and traffic conditions. This work emphasizes driver behaviour modelling issues, which have not been considered in related papers in the literature [12, 18, 19, 22, 23, 28, 29].

3.1. Driver Behaviour. The first group is expert drivers, i.e., those who know the network and historic traffic conditions for the selected horizon of study. They are modelled with route choice selection and proportions by following experienced travel times where both satisfy dynamic user equilibrium (DUE) [30] and assume a historic demand pattern. DUE paths and proportions are loaded into the simulation environment from a precalculated binary file. The second group is regular drivers, i.e., those with knowledge of the network and historic traffic conditions for recurrent trips (50% randomly selected) but who use the main streets based on free-flow for nonrecurrent trips (50%). Expert and regular drivers exhibit driving characteristics related to the car-following model, such as reaction times, desired speed, and acceptance of speed limitations. According to the calibrated profile of Barcelona drivers, reaction time (1.0 s), reaction time at stop (1.35 s), reaction time at traffic light (1.35 s), speed acceptance, and Minimum Intervehicular Distance are assumed to be truncated normally distributed, with the former having mean 1.1, sd 0.1, min 0.9, and max 1.3 and the latter having mean 1.0 m, sd 0.3 m, min 0.5 m, and max 1.5 m. The third group is tourist drivers, who have limited knowledge of network and traffic conditions and use K-shortest paths based on free-flow conditions and main streets. Tourist drivers behave roughly with a 25% increment in reaction times, means, and limits (same standard deviation), and they strictly adhere to the speed limits, while Minimum Intervehicular Distance is truncated normally distributed with mean 1.25 m and sd 0.1 m, between 0.75 and 1.5 m. Finally, guided drivers constitute a design-dependent proportion for any expert, regular, or tourist driver class, and they are modelled with a 100% acceptance of navigation advice.

3.2. Travel Time Estimation. The selected approach is based on the first model presented in [5] for estimating lane travel times. The time-window concept, which can be viewed as a rolling horizon for updating lane/link travel time estimates, is critical for understanding the approach. The time-window interval is a design factor in the conducted simulation experiments because the penetration rate of PVD leads to lane- (link-) level vehicle data availability. Hence, as the time-window interval is increased, the percentage of lanes (links) with available data and data units also increases.

Position and speed data provided by the PVD can be emulated for each simulation step (0.5 sec) or at any multiple of the interval. Having PVD provided every 2 sec has been finally assumed for the experimental study described in this paper in order to limit the computational burden of executing replications. Thus, detection interval is a configurable parameter set to 2 sec for the current study. Many alternative possibilities could be considered to develop travel time estimators, but that is not the aim of this work and it is considered a topic for further research. The particular proposal implemented here considers three cases for every lane:

(i) Case 1: "No PVD in the last window": in this situation, the lane travel time value is the same as the lane travel time of the most recent time-window. If no data is available for this lane, then the lane travel time is set to free-flow travel time.

(ii) Case 2: "PVD from just one car in the last window": to compute the travel time in this case, consider the following:

(a) TT_1: it is estimated travel time for the fraction of the lane until the first detection of the vehicle. TT_1 is computed by dividing the length of this fraction by the probe vehicle's instantaneous speed at the first observation in this lane for the considered detection interval, which gives a time measure.

(b) TT_L: this value is the difference in seconds between the first and last detection intervals.

(c) TT_2: this is computed in the same way as TT_1. We divide the distance from the last observation at the beginning of the next section in the trajectory by the probe vehicle's instantaneous speed in the last observation (detection interval) of the time-window.

Finally, the travel time of the considered lane is $TT_1 + TT_L + TT_2$.

iii. Case 3: "PVD available from more than one car in the last time-window": in this case, the weighted travel time of the lane is $(1/n)\sum_{v=1}^{n} w_v TT_v$, where TT_v is the travel time of a car computed in the same way as in Case 2, and w_v is the fraction of the section length in $[0, 1]$ where v probe car is detected in the lane.

Hence, the weights w_v handle the lane changing behaviour of vehicles. In other words, if a vehicle travels in the slower lane at first and then switches to the faster lane and travels for some time before switching back to the slower lane and then exiting the segment, then 3 independent PVD detection types are assumed, which thus prevents underestimation of the travel time for the slower lane and overestimation of the travel time for the faster lane.

The estimation of link travel times from PVD is obtained as the mean of travel times for streams. Lane travel time estimates are combined into stream travel times according to the turning movements that are allowed.

3.3. Navigation Strategies. A navigation application is modelled as being available to a common percentage in driver classes, since PVD can be used to estimate travel times in network links and thus travel times between OD POIs. Route guidance assessment is not the aim of this paper (see [20] for an interesting discussion about the topic); thus, three routing strategies are implemented according to the possibilities of the AIMSUN [27] platform:

(i) Free-flow K-shortest paths when no probe fleet is available

(ii) Stochastic Route Choice: lane-based, according to instantaneous traffic conditions (lane travel times) as inferred from data provided by PVD

(iii) Stochastic Route Choice: link-based, according to instantaneous traffic conditions (link travel times) as inferred by PVD.

Stochastic Route Choice (SRC) modelling accounts for the estimated instantaneous K-shortest travel time paths, as defined by either user-defined link costs (link travel time estimates from probe car data) or user-defined stream costs (lane travel time estimates are combined and together they constitute stream travel times). Also, 100% rerouting across the trip is enabled (every time-window interval 1.5, 3, or 6 min) in order to account for dynamic and instantaneous travel-time-based route choice. K is set to 3 in all the experiments. A proportional SRC is calibrated to determine the probability of selecting a route for a K-shortest travel time path that is calculated using instantaneous travel times. Thus, choice probability P_k of a given alternative path k is

$$P_k = \frac{\mathrm{CP}_k}{\sum_{l \in K} \mathrm{CP}_l}, \quad (1)$$

where CP_i is the cost of path i. The proportional parameter is set to $\alpha = 1.2$ after calibration in order to consider the probability as inversely proportional to path costs.

The key point is that a dynamic (real-time) routing strategy for connected cars (guided cars) is applied from PVD-derived travel times. Travel time estimates used in

K-shortest path calculations might depend on either lane-based travel time estimates or overall link-based travel time estimates, both of which rely on PVD sent to a centralized subsystem.

Travel times between POIs can be inferred from OD path travel times and route choice proportions. They can then feed Kalman filtering formulations to estimate dynamic OD matrices, as proposed by the authors.

3.4. Collected KPIs. Default statistics in traffic microsimulation platforms are usually very rich. Statistics have been collected every 90 s and stored in an SQLITE database for each replication. Driver KPIs collected for each expert, regular, and tourist driver type—either guided or nonguided—are as follows:

(i) Mean travel time to cover 1 km (s/km)

(ii) Mean speed per vehicle (km/h)

(iii) Mean delay while covering 1 km (s/km)

(iv) Mean travel distance per vehicle (km)

(v) Mean travel time per vehicle (min).

Network KPIs are global statistics (all driver classes, buses included) and, over the whole simulation horizon, they are as follows:

(i) Total travel distance (km), *ttdis*

(ii) Total travel time (h), *ttt.h*

(iii) Fuel consumption (l), *fuelc*

(iv) Total CO2 emissions (kg), *co2*

(v) Total NOx emissions (kg), *nox*

(vi) Mean travel time to cover 1 km (s/km), *mtt.s.km*

(vii) Mean delay time while covering 1 km (s/km), *mdelay.s.km*

(viii) Density (veh/km), *density*

(ix) Mean speed (km/h), *mspeed*

(x) Mean flow (veh/h) (throughput measure), *mflow*

(xi) Throughput rate (%) completed trips divided into total demand, *thrputrate.*

4. Design of Experiments

The selected scenario is the Barcelona CBD, known as "L'Eixample" (see [3] for details), which comprises 7.46 km^2 and 250,000 inhabitants. The horizon study is 1 h, accounting for 42,500 trips. Passenger car demand is modelled as 15 min time-sliced demand whose origin-destination pattern reproduces the 9-10 h morning period in L'Eixample. The model includes the description of the 50 bus-routes operating in the area and accounts for frequencies and stops for boarding/alighting.

4.1. Design Factors. Factors considered in the *design of the simulation experiments* are

(i) *driver type distribution* (TD factor) into expert-regular-tourist for

 (a) base level: 40-50-10,

 (b) alternative levels: 20-70-10, 40-40-20, and 60-20-20;

(ii) *guidance penetration* (GP factor): percentage of cars that are connected cars whose route choice decisions follow those advised by a Navigation Tool fed by PVD:

 (a) base level 0%,

 (b) alternative levels: 10%, 20%, 30%, 70%, 80%, 90%, and 100%;

(iii) *demand pattern* (DP factor) into 4 levels—0%, 10%, 20%, and 30%—referring to a perturbation of the historic demand pattern in OD pairs belonging to the fourth percentile trip distance (according to Manhattan distance): they account for 42,500, 44,600, 46,860, and 48,600 trips, respectively:

 (a) base level: 0% means historic demand pattern,

 (b) alternative levels: 10%, 20%, and 30% increments;

(iv) *probe vehicle penetration percentage (PVD factor)*, modelled as common to any driver type into 4 levels: 0%, 10%, 20%, and 30%: base level is 0%; it indicates route guidance based on free-flow travel times; an additional Ground Truth level consisting of travel time estimates based directly on "simulated Ground Truth" was also included in some initial experimentation;

(v) *time-window length* (TW factor) which is the rolling horizon interval considered for the estimation of traffic variables from PVD:

 (a) base length: 3 min,

 (b) alternative lengths: 1.5 min and 6 min (TW is not affected when 0% PVD is set);

(vi) *navigation strategy* (NS factor) models driving recommendations based on either lane-level or link-level PVD when PVD is available: base level is lane-level, but it is not affected when 0% PVD is set.

The critical KPI, from a driver satisfaction point of view, is considered to be mean travel time (min), also for researchers interested in the assessment of travel times between POIs inferred from PVD. A detailed analysis on the base scenario for all design factors found that when $N = 5$ replications, this facilitates a global 5% relative precision in mean travel time (min) at 95% confidence for any driver type, while the greatest absolute error was about 1/3 min (Tourist). Table 1 shows absolute/relative errors at different confidence levels for driver types when 5 replicas are considered. Thus, the base scenario is set at factor levels: TD factor 40-50-10, GP factor 0%, DP factor 0%, and PVD factor 0%. Therefore, TW and NS levels are irrelevant.

Running the full factorial design is unfeasible for computational reasons, since $3{,}072 \times 5 = 15{,}360$ replications would be needed. Therefore, the first set of experiments (first round)

TABLE 1: Relative (%) and absolute errors (min) for long trips (Manhattan distance origin to destination centroid over 3 km) when 5 replicas are considered in the base scenario at 90%, 95%, and 99% confidence levels.

Mean Travel Time	Number of Replicas, 5										
min	TG	TNG	TOURIST	RG	RNG	REGULAR	EG	ENG	EXPERT	GUIDED	NON-G
Mean	**9.54**	**9.88**	**9.80**	**9.45**	**9.74**	**9.67**	**9.53**	**9.98**	**9.88**	**9.49**	**9.85**
Abs error (min)											
90	0.28	0.22	0.22	0.11	0.17	0.15	0.11	0.17	0.14	0.11	0.17
95	**0.36**	**0.29**	**0.29**	**0.14**	**0.23**	**0.20**	**0.14**	**0.22**	**0.19**	**0.15**	**0.22**
99	0.60	0.48	0.47	0.23	0.38	0.33	0.23	0.37	0.31	0.25	0.37
Rel error (%)											
90	2.90%	2.23%	2.24%	1.12%	1.79%	1.57%	1.12%	1.71%	1.46%	1.20%	1.73%
95	**3.77%**	**2.91%**	**2.92%**	**1.45%**	**2.34%**	**2.04%**	**1.46%**	**2.23%**	**1.91%**	**1.56%**	**2.26%**
99	6.26%	4.82%	4.84%	2.41%	3.88%	3.38%	2.42%	3.69%	3.16%	2.59%	3.74%

TG: tourist guided; TNG: tourist nonguided; RG: regular guided; RNG: regular nonguided; EG: expert guided; E-NG: expert nonguided. Tourist, regular, and expert columns indicate driver classes (averaging guided/nonguided cars) and guided/non-G columns indicate mean over driver classes.

was constrained in order to identify nonaliased factor main effects according to the Fedorov algorithm [31] for optimal designs: 29 experiments were given (thus, 145 replications were executed, each one taking around 2 h on an Intel Core i7-4790 CPU (frequency of 3.6 GHz), 4 cores, 8 GB DDR3 memory, and Windows 8.1 (x64 system)).

A second round of simulation experiments for the base level 40-50-10 in TD factor and 3 min for time-window (TW) factor was launched to quantify the most significant factors found in the first round of simulations. In decreasing order of importance, demand pattern (DP), PVD penetration, guidance penetration (GP), and navigation strategy (NS) factors are found. The second round design consisted of 140 new replications.

5. Results and Discussion

Network KPIs are affected by design factors, when either gross effects or net effects are considered (i.e., elaborated from the linear model for each global network KPI in all design factors). A gross effect factor indicates the factor impact when it is considered alone, and this is the mean of KPIs at different levels of the selected factor. A net effect for a factor means the effects of all other factors—except for the selected factor—are taken into account before calculating the *net* effect for each level of the selected factor. A heatmap representation has been chosen to summarize the gross and net effects of design factors on network KPIs (see Figure 2). Clearly, demand pattern (DP) and guidance penetration (GP) strongly affect all network KPIs. The gross effects of driver class configuration (TD) impact total travel time and emission KPIs. The net effects of PVD, NS, TW, and TD factors decrease when the other factors are considered, while the net effects of the DP and GP factors remain the same or are even magnified for some KPIs, such as those related to emissions and total travel distance. There is a reduction in the net effects of DP on the mean travel time for covering one km (mtt.s.km) and the mean delay by km (mdelay.s.km) once the remaining factors are controlled for. Dendrograms of KPIs and design factors are grouped from bottom to top according to the most similar KPIs. They start by assigning each item to its own cluster and proceed to find the closest (most similar) pair of clusters and then merge them into a new single cluster. DP and GP effects are similar on the factor side, although on the KPI side similarities arise between emission KPIs, travel time (mtt.s.km), and delay (mdelay.s.km) for covering one km.

Network KPIs are numeric variables with different scales and internal correlation; thus a normalized PCA (principal component analysis) is applied, which explains almost 95% of the total inertia (variability in KPIs) in the first factorial plane. Each axis in the plane models a "hidden variable" that combines the original ones. The first and second axes explain, respectively, 56% and 39% of the total variability. On the other hand, the projection of the cloud of replicas and KPIs onto the first factorial plane reveals the meaning of hidden variables (see Figure 3(a)).

The first factorial axis (see Figure 3(a)) is a size axis that is positively associated with the most contributive variables, such as the total travel time (ttt.h), density, and the CO2 and NOx emissions, as well as DP factor (total demand). In the opposite direction of mean speed, this axis clearly represents quantity of trips (total demand), and it is dominated obviously by the DP factor. Once the total number of trips is controlled for, the network performance is represented on a second axis that is orthogonal to the first one. The network throughput flow (mflow) and the total travelled distance (ttdis) are positively correlated, indicating an increase in total distance when throughput flow increases. This total travelled distance is also more positively related to fuel consumption than to CO2 and NOx emissions. On the negative side of the second axis, there is also a positive correlation with density when we look at the mean delay (mdelay.s.km) and the mean travel time (mtt.s.km) needed to cover a km.

According to the overlapped representation of ellipses around the centres of demand pattern levels: low increments over historic demand are located to the left of the interpreted size axis 1; high increments go to the right; and the largest variability range changes from axis 1 to axis 2, thus increasing congestion. Clearly, the second axis is a hidden variable that

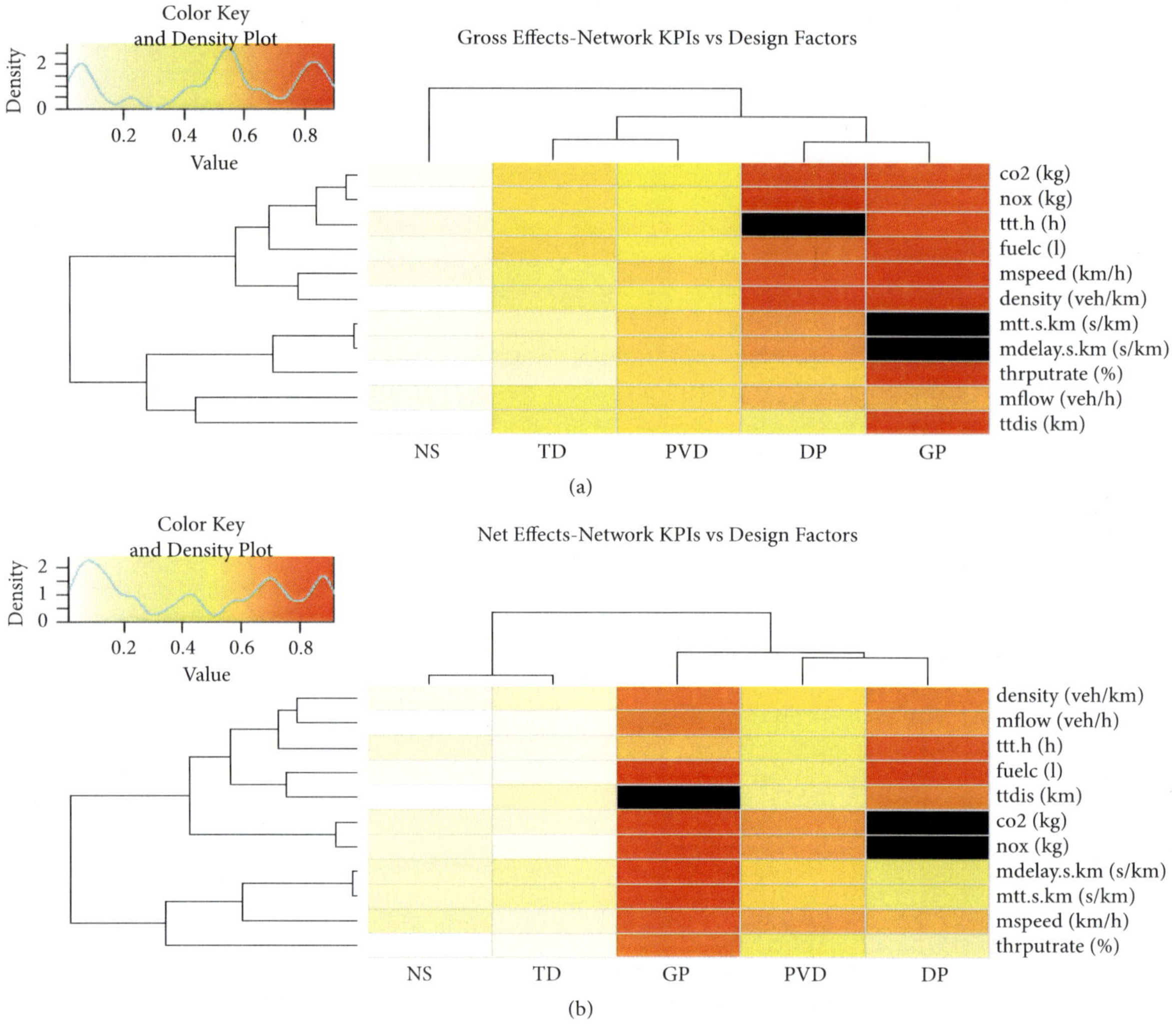

FIGURE 2: Gross (a) and net (b) effects of design factors on network KPIs. All replicas. Colour proxy developed from (partial) Spearman correlation. First round of simulations.

indicates quality of service in terms of congestion (better in the positive part of axis 2). Once the demand pattern factor DP has been clearly identified, the rest of the factor levels are located in the diagram to gain interpretability on the quantity-quality hidden variables plane.

In Figure 3(b), the meaning of the axis quantity-quality is reinforced for the second round of simulations, which has more homogeneous KPI results in terms of variability. Clearly, the second axis is a latent congestion-level axis where delay and mean time for covering one km (mdelay.s.km and mtt.s.km) are 180 degrees opposite of the throughput rate. Once the demand factor DP has been clearly identified, the rest of the factor levels are located in the diagram to gain interpretability from the quantity-quality latent plane (Figure 3(a)).

For the second round of simulations in the base level of driver type configuration (TD), the meaning of the quantity-quality latent plane is reinforced (Figure 3(b)). Blue and yellow arrows in Figure 3 from left to right join the gravity centres for DP level ellipses from low to high demands.

Marginal net effects of mean speed and fuel consumption have been analysed for design factor levels in Figure 4. It is noteworthy that a guidance penetration (GP factor) increment up to 30% benefits mean speed, but the mean speed tends to decrease for guidance levels over 30%, since guidance strategies are merely reactive. Nevertheless, as demand increases (DP factor), mean speed decreases almost linearly. However, the fuel consumption KPI tends to increase as either GP or DP increases.

Clearly, increasing the guidance penetration (GP) has a positive effect on performance in terms of mean speed (up to a 30% limit), but also on fuel consumption KPIs, as shown in Figure 4; and lane-level guidance outperforms the results of link-level guidance for the two selected KPIs in Figure 4.

A diagram in the style of a macro fundamental diagram [32] is presented in Figure 5(a). It considers flow, density axis, and contour curves for speed, and it concerns the first round of simulations, with replications grouped according to unsupervised clustering techniques based on factorial space distances. Five clusters are identified by hierarchical

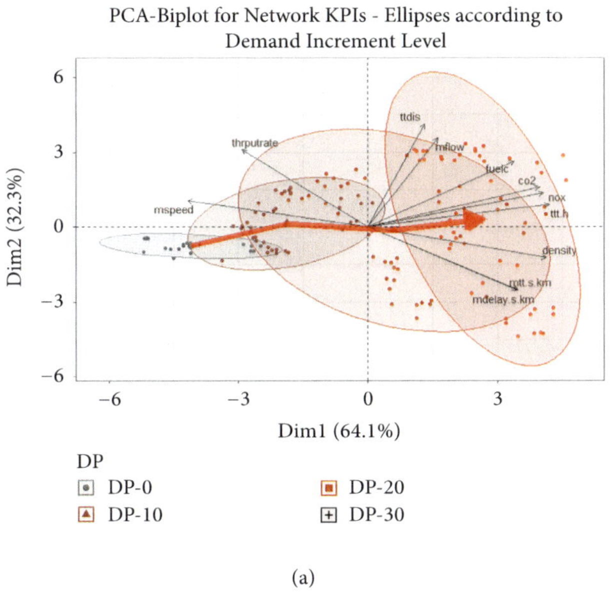

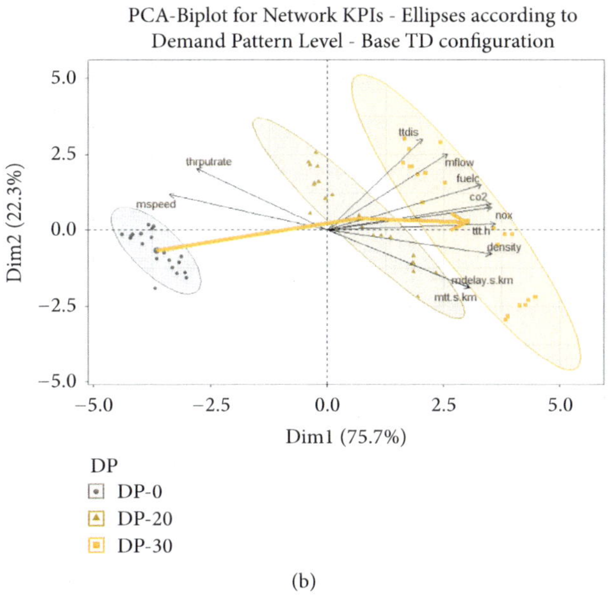

FIGURE 3: Biplot of principal component analysis of network KPIs. 90% confidence ellipses around demand levels. (a) All configurations-first round of simulations and (b) only base driver class. Second round of simulations.

classification rules. Note that there are two clusters (3 and 4) at the break point of congestion: one is over the curve indicating a large throughput (network performance, cluster 4); the second (cluster 3) is mostly under the curve, which indicates lower performance (throughput). The difference between these two groups relies on the 30% DP with 70%, 80%, and 90% GPs in cluster 4 over the curve, while cluster 3 points exhibit 20% DP and low GP levels. Cluster 5 belongs to 30% DP, with no situations in which serious damage to network performance occurs due to guidance being provided. In Figure 5(b), simulations in cluster 2 have 20% DP, so congestion is found for GP over 30%. Routes that are experienced according to recurrent conditions (historic demand) are not valid when an increasing demand appears; therefore, the reactive navigation strategies tend to maintain the level of service by increasing total travel distance and fuel consumption. Otherwise, the level of service decays and NOx and CO2 emissions increase.

Finally, Table 2(a) presents interesting figures for the means of speed and trip travel time (according to driver KPIs) in different scenarios for the second round of simulations. No guidance is the reference for each class of driver and DP level. Under base demand (0% DP), all drivers tend to increase their speed moderately when probe vehicle penetration grows (guidance is available for a variable part of the population according to experimental design), but nonguided expert and regular driver speeds are better than those for guided drivers (since recurrent congestion conditions are present). As more demand is injected into the network (DP 20% and 30%), guided versus nonguided expert differences increase in favour of guided ones, but even nonguided vehicles from any class benefit from incremental PVD penetration. In Table 2(b), the mean trip travel time for guided vehicles is always less than trip travel time for nonguided vehicles when demand increases (20% and 30% DP), when low PVD penetration is present (10% PVD) and the overall benefit within each driver class is enhanced as PVD increases. Nevertheless, as PVD increases, nonguided drivers benefit from shorter mean travel times. A detailed analysis of the results suggests that reactive navigation strategies should be considered in detail, because longer distance routes are produced for guided vehicles and, therefore, speed is higher for guided vehicles while their final trip travel times are not reduced since they follow longer routes.

6. Conclusions

The research carried out was based on an approach consisting of a general framework and simulation architecture for emulating and evaluating penetration rates of probe car fleets and navigation services for a subset of drivers. The contribution of this paper relies on a detailed simulation of driver classes. Reactive navigation strategies were calculated according to travel times estimated from PVD, and they have been shown to be advantageous for guided and nonguided cars up to a certain level of around 30% of guided vehicles for any PVD penetration. Nevertheless, reactive guidance increases the level of service at the expense of increasing guided trip lengths (the overall trip travel time decreases), thus increasing fuel consumption. This is unhelpful in terms of sustainability. Additionally, when demand increases and the level of service for drivers decreases, it is worth using route guidance for a fraction of the vehicles because this reduces NOx and CO2 emissions (despite the longer distances and consumption). We can definitively say that any assessment of mobility services must consider several KPIs, since a simple increment in driver speed or a reduction in travel time does not automatically revert positively in terms of

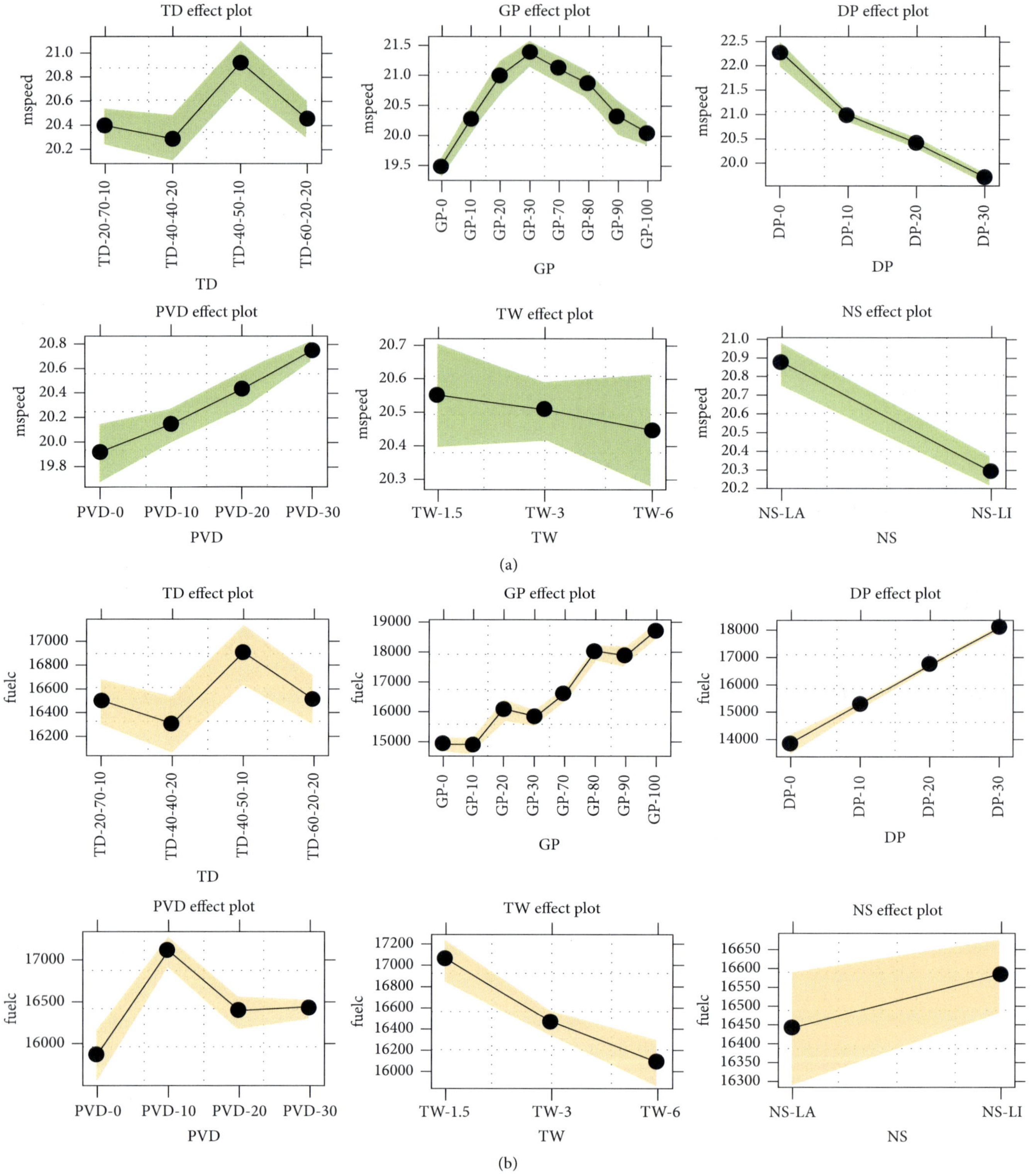

FIGURE 4: Speed (a) and fuel consumption (b). Network KPI net effects on design factor levels. All driver class configurations. First round of simulations.

sustainability (fuel consumption and emissions reduction). In recurrent traffic conditions, navigation devices are not suitable for expert drivers, but those drivers do benefit from their being used by the other cars. Furthermore, because travel times involved in guidance have been estimated from a fleet of PVD and even their lowest tested penetration rate is 10%, the overall effect on network KPIs is positive. Hence, after extensive analysis of the results obtained for several

TABLE 2: Mean speed (top) and mean travel time (bottom) for driver type 40-50-10 expert-regular-tourist (second round of simulations) according to DP and PVD factors.

(a)

Mean Speed (km/h)			PVD Factor - Probe Vehicle Data Penetration					
DP	Driver Type	Base-0%	10%		20%		30%	
0%	Expert-G		22.27	*1.28%*	22.50	*2.34%*	22.51	*2.39%*
	Expert-NG	21.99	22.24	*1.16%*	22.55	*2.54%*	22.70	*3.23%*
	Regular-G		22.06	*1.65%*	22.33	*2.88%*	22.37	*3.09%*
	Regular-NG	21.70	21.94	*1.07%*	22.26	*2.55%*	22.43	*3.35%*
	Tourist-G		21.09	*2.31%*	21.41	*3.82%*	21.56	*4.58%*
	Tourist-NG	20.62	20.86	*1.17%*	21.20	*2.84%*	21.41	*3.84%*
20%	Expert-G		20.91	*3.14%*	21.45	*5.79%*	21.46	*5.84%*
	Expert-NG	20.28	20.48	*0.98%*	21.22	*4.66%*	21.37	*5.40%*
	Regular-G		20.81	*4.36%*	21.29	*6.74%*	21.39	*7.27%*
	Regular-NG	19.94	20.15	*1.05%*	20.90	*4.82%*	21.08	*5.68%*
	Tourist-G		19.96	*4.94%*	20.50	*7.79%*	20.70	*8.83%*
	Tourist-NG	19.02	19.26	*1.26%*	19.99	*5.07%*	20.12	*5.79%*
30%	Expert-G		19.95	*5.39%*	20.48	*8.21%*	20.62	*8.93%*
	Expert-NG	18.93	19.38	*2.39%*	20.15	*6.43%*	20.34	*7.47%*
	Regular-G		19.95	*6.89%*	20.33	*8.96%*	20.52	*9.98%*
	Regular-NG	18.66	19.15	*2.61%*	19.83	*6.26%*	20.08	*7.60%*
	Tourist-G		19.34	*8.35%*	19.71	*10.43%*	19.92	*11.59%*
	Tourist-NG	17.85	18.30	*2.53%*	18.98	*6.33%*	19.22	*7.65%*

(b)

Mean Travel Time (min)			PVD Factor - Probe Vehicle Data Penetration					
DP	Driver Type	Base-0%	10%		20%		30%	
0%	Expert-G		5.50	*−0.47%*	5.37	*−2.80%*	5.34	*−3.22%*
	Expert-NG	5.52	5.37	*−2.82%*	5.14	*−6.96%*	5.24	*−5.13%*
	Regular-G		5.47	*−0.53%*	5.35	*−2.72%*	5.33	*−3.08%*
	Regular-NG	5.50	5.35	*−2.71%*	5.12	*−6.98%*	5.21	*−5.20%*
	Tourist-G		5.62	*−0.85%*	5.47	*−3.48%*	5.48	*−3.28%*
	Tourist-NG	5.66	5.51	*−2.77%*	5.26	*−7.17%*	5.36	*−5.43%*
20%	Expert-G		6.34	*−1.25%*	6.02	*−6.25%*	6.01	*−6.36%*
	Expert-NG	6.42	6.35	*−1.09%*	5.83	*−9.23%*	5.89	*−8.28%*
	Regular-G		6.31	*−2.18%*	6.04	*−6.48%*	6.03	*−6.60%*
	Regular-NG	6.46	6.33	*−1.88%*	5.83	*−9.75%*	5.91	*−8.41%*
	Tourist-G		6.43	*−2.12%*	6.14	*−6.52%*	6.16	*−6.33%*
	Tourist-NG	6.57	6.45	*−1.92%*	5.95	*−9.50%*	6.03	*−8.25%*
30%	Expert-G		6.70	*−2.65%*	6.40	*−6.99%*	6.42	*−6.68%*
	Expert-NG	6.88	6.79	*−1.24%*	6.32	*−8.18%*	6.33	*−8.02%*
	Regular-G		6.73	*−3.07%*	6.45	*−7.17%*	6.45	*−7.04%*
	Regular-NG	6.94	6.81	*−1.89%*	6.31	*−9.04%*	6.39	*−8.03%*
	Tourist-G		6.75	*−4.48%*	6.53	*−7.51%*	6.56	*−7.07%*
	Tourist-NG	7.06	6.97	*−1.37%*	6.45	*−8.69%*	6.53	*−7.54%*

KPIs, it can be concluded that OD travel times between POIs inferred from PVD appear to be reasonable inputs for simplifying dynamic OD matrix estimation formulations that are being developed by the authors in an ongoing work (also proposed in [24]). Finally, PVD penetration should represent a nonnegligible percentage of drivers, since 20–30% of the results are consistently better for any KPI at the network level than those obtained for a 10% PVD penetration rate (a fleet of 4,250 probe vehicles in the base scenario).

Conflicts of Interest

The authors declare that there are no conflicts of interest regarding the publication of this paper.

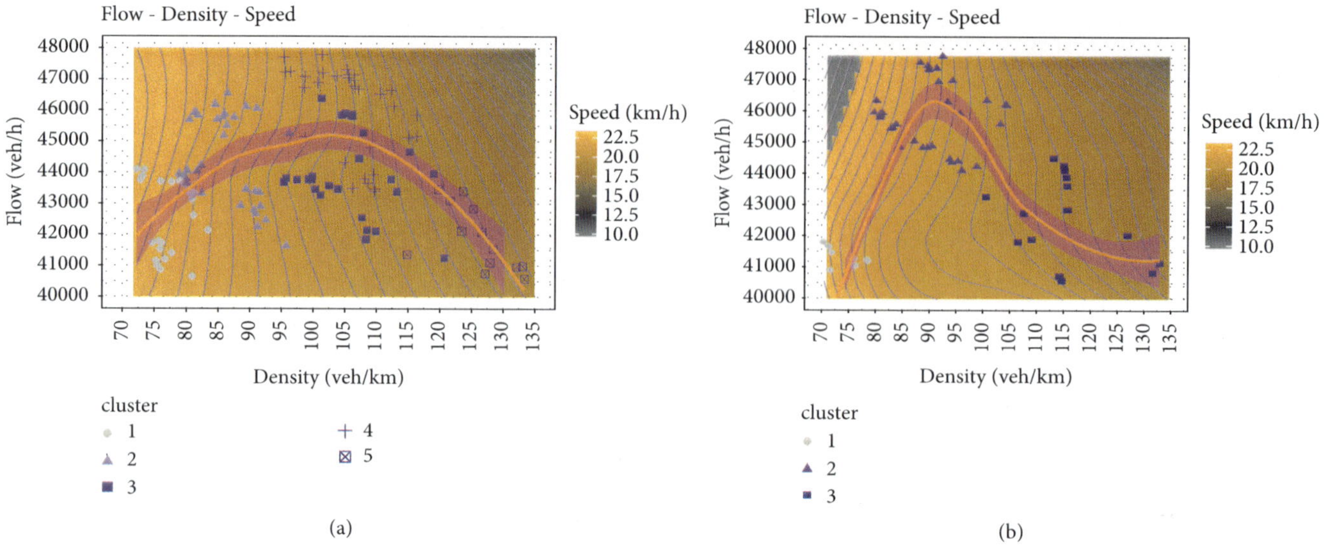

FIGURE 5: Flow-density-speed diagram with replications grouped by unsupervised clustering techniques. (a) First round of simulations. (b) Second round of simulations.

Acknowledgments

Throughout this work, the authors have benefited from the support of inLab FIB team at UPC and the suggestions of Jaume Barceló (Emeritus Professor at UPC). This research was funded by Spanish R+D Programs, TRA2016-76914-C3-1-P, and by Secretaria d'Universitats-i-Recerca-Generalitat de Catalunya, 2017-SGR-1749.

References

[1] B. Hull, V. Bychkovsky, Y. Zhang et al., "CarTel: a distributed mobile sensor computing system," in *Proceedings of the 4th International Conference on Embedded Networked Sensor Systems (SenSys '06)*, pp. 1–14, Boulder, Colorado, USA, November 2006.

[2] U. Lee and M. Gerla, "A survey of urban vehicular sensing platforms," *Computer Networks*, vol. 54, no. 4, pp. 527–544, 2010.

[3] L. Montero, M. Pacheco, J. Barceló, S. Homoceanu, and J. Casanovas, "Case study on cooperative car data for estimating traffic states in an urban network," *Transportation Research Record*, vol. 2594, pp. 127–137, 2016.

[4] H. Rim, C. Oh, K. Kang, and S. Kim, "Estimation of Lane-Level Travel Times (L2TT) in V2X-based Traffic Information System," in *Proceedings of 90th Transportation Research Board Annual Meeting - CD-ROM*, 2011.

[5] I. Sanaullah, M. Quddus, and M. Enoch, *Estimating Link Travel Time from Low-Frequency GPS Data*, 2013.

[6] E. Jenelius, M. Rahmani, and H. N. Koutsopoulos, "Travel time estimation for urban road networks using low frequency GPS probes," *Transportation Research Board 91st Annual Meeting Compendium of Papers. CD-ROM*, 2012.

[7] E. Jenelius and H. N. Koutsopoulos, "Travel time estimation for urban road networks using low frequency probe vehicle data," *Transportation Research Part B: Methodological*, vol. 53, pp. 64–81, 2013.

[8] M. Rahmani and H. N. Koutsopoulos, "Path inference from sparse floating car data for urban networks," *Transportation Research Part C: Emerging Technologies*, vol. 30, pp. 41–54, 2013.

[9] M. Rahmani, E. Jenelius, and H. N. Koutsopoulos, "Non-parametric estimation of route travel time distributions from low-frequency floating car data," *Transportation Research Part C: Emerging Technologies*, vol. 58, pp. 343–362, 2015.

[10] T. Seo, T. Kusakabe, and Y. Asakura, "Estimation of flow and density using probe vehicles with spacing measurement equipment," *Transportation Research Part C: Emerging Technologies*, vol. 53, pp. 134–150, 2015.

[11] L. C. Edie, "Discussion of traffic stream measurements and definitions," in *Proceedings of 2nd International Symposium on the Theory of road Traffic Flow*, 1963.

[12] J. Lee and B. Park, "Evaluation of route guidance strategies based on vehicle-infrastructure integration under incident conditions," *Transportation Research Record*, vol. 2086, pp. 107–114, 2008.

[13] M. G. Demissie, G. H. de Almeida Correia, and C. Bento, "Intelligent road traffic status detection system through cellular networks handover information: An exploratory study," *Transportation Research Part C: Emerging Technologies*, vol. 32, pp. 76–88, 2013.

[14] C. Chen, J. Ma, Y. Susilo, Y. Liu, and M. Wang, "The promises of big data and small data for travel behavior (aka human mobility) analysis," *Transportation Research Part C: Emerging Technologies*, vol. 68, pp. 285–299, 2016.

[15] U. C. Berkeley, *Mobile Millenium*, http://traffic.berkeley.edu/.

[16] R. J. Herring, *Real-Time Traffic Modeling and Estimation with Streaming Probe Data using Machine Learning*, UC Berkeley, 2010.

[17] B. Du, S. Chien, J. Lee, L. Spasovic, and K. Mouskos, "Artificial neural network model for estimating temporal and spatial freeway work zone delay using probe-vehicle data," *Transportation Research Record: Journal of the Transportation Research Board*, vol. 2573, pp. 164–171, 2016.

[18] E. Paikari, L. Kattan, S. Tahmasseby, and B. H. Far, "Modeling and simulation of advisory speed and re-routing strategies

in connected vehicles systems for crash risk and travel time reduction," in *Proceedings of the 26th IEEE Canadian Conference on Electrical and Computer Engineering (CCECE '13)*, Saskatchewan, Canada, May 2013.

[19] A. Olia, H. Abdelgawad, B. Abdulhai, and S. N. Razavi, "Assessing the Potential Impacts of Connected Vehicles: Mobility, Environmental, and Safety Perspectives," *Journal of Intelligent Transportation Systems: Technology, Planning, and Operations*, vol. 20, no. 3, pp. 229–243, 2016.

[20] J. Wang, J. Lv, C. Wang, and Z. Zhang, "Dynamic route choice prediction model based on connected vehicle guidance characteristics," *Journal of Advanced Transportation*, vol. 2017, Article ID 6905431, pp. 1–8, 2017.

[21] J.-S. Oh and R. Jayakrishnan, "Emergence of private advanced traveler information system providers and their effect on traffic network performance," *Transportation Research Record*, vol. 1783, pp. 167–177, 2002.

[22] F. P. Deflorio, "Evaluation of a reactive dynamic route guidance strategy," *Transportation Research Part C: Emerging Technologies*, vol. 11, no. 5, pp. 375–388, 2003.

[23] J. Pan, M. A. Khan, I. S. Popa, K. Zeitouni, and C. Borcea, "Proactive vehicle re-routing strategies for congestion avoidance," in *Proceedings of the 8th IEEE International Conference on Distributed Computing in Sensor Systems, DCOSS 2012*, Hangzhou, China, May 2012.

[24] X. Yang, Y. Lu, and W. Hao, "Origin-destination estimation using probe vehicle trajectory and link counts," *Journal of Advanced Transportation*, vol. 2017, Article ID 4341532, 2017.

[25] J. Barceló, L. Montero, M. Bullejos, O. Serch, and C. Carmona, "A kalman filter approach for exploiting bluetooth traffic data when estimating time-dependent od matrices," *Journal of Intelligent Transportation Systems: Technology, Planning, and Operations*, vol. 17, no. 2, pp. 123–141, 2013.

[26] RStudio Project, *R. Shiny Web Application framework for R*, https://www.rstudio.com/products/shiny/.

[27] Transport Simulation Systems, Aimsun Microscopic Simulator (v 8.0), 2013, http://www.aimsun.com.

[28] J. Wang, J. Lv, Q. Zhang, and X. Yan, "Impact of connected vehicle guidance information on network-wide average travel time," *Advances in Mechanical Engineering*, vol. 9, no. 1, 2016.

[29] S. Minelli, P. Izadpanah, and S. Razavi, "Evaluation of connected vehicle impact on mobility and mode choice," *Journal of Traffic and Transportation Engineering (English Edition)*, vol. 2, no. 5, pp. 301–312, 2015.

[30] *Dynamic Traffic Assignment: A Primer*, Transportation Research Board, Washington, D.C., USA, 2011.

[31] V. Federov, *Theory of Optimal Experiment Designs*, New York, NY, USA, 1972.

[32] C. F. Daganzo and N. Geroliminis, "An analytical approximation for the macroscopic fundamental diagram of urban traffic," *Transportation Research Part B: Methodological*, vol. 42, no. 9, pp. 771–781, 2008.

Development of a Representative EV Urban Driving Cycle based on a k-Means and SVM Hybrid Clustering Algorithm

Xuan Zhao, Qiang Yu, Jian Ma, Yan Wu, Man Yu, and Yiming Ye

School of Automobile, Chang'an University, Xi'an, China

Correspondence should be addressed to Xuan Zhao; zhaoxuan@chd.edu.cn

Academic Editor: Constantinos Antoniou

This paper proposes a scientific and systematic methodology for the development of a representative electric vehicle (EV) urban driving cycle. The methodology mainly includes three tasks: test route selection and data collection, data processing, and driving cycle construction. A test route is designed according to the overall topological structure of the urban roads and traffic flow survey results. The driving pattern data are collected using a hybrid method of on-board measurement method and chase car method. Principal component analysis (PCA) is used to reduce the dimensionality of the characteristic parameters. The driving segments are classified using a hybrid k-means and support vector machine (SVM) clustering algorithm. Scientific assessment criteria are studied to select the most representative driving cycle from multiple candidate driving cycles. Finally, the characteristic parameters of the Xi'an EV urban driving cycle, international standard driving cycles, and other city driving cycles are compared and analyzed. The results indicate that the Xi'an EV urban driving cycle reflects more aggressive driving characteristics than the other cycles.

1. Introduction

Fuel depletion, environmental disruption, and air pollution have contributed to the development of electric vehicles (EVs) [1]. The driving range calculation and state of charge estimation for EVs are generally performed based on the international standard driving cycles [2, 3]. In the study of energy consumption prediction and energy management optimization strategies for EVs, some scholars have also adopted the international standard driving cycles [4–6]. However, the data of these international standard driving cycles were collected from internal combustion engine vehicles (ICEVs). Because of the differences in the torque characteristics, power characteristics [7–11], and braking characteristics between EVs and ICEVs, the driving cycles of EVs are significantly different from those of ICEVs [12–14]. The research on driving range calculation, state of charge estimation, energy consumption prediction, and energy management optimization strategies for EV under the international standard driving cycles can produce large errors. Therefore, it is important to conduct research on the driving cycles of EVs [15, 16].

The driving cycle is a speed-time profile that represents a typical real-world driving pattern in a certain city or region [3, 17]. Driving cycles have a wide range of applications in the vehicle and transportation fields [18], and they are often used to simulate traffic conditions during laboratory chassis dynamometer bench tests and in automotive simulation research [3]. Moreover, these cycles serve as a standardized measurement procedure for the evaluation and certification of new vehicle models, the monitoring of exhaust emissions and fuel consumption for ICEVs, and the estimation of the driving range and energy consumption per unit mileage for EVs [3, 19]. Because of differences in city size, road topology, road types, traffic conditions, vehicle ownership, economics, cultures, geographical features, and vehicle types, the characteristics of driving cycles in different cities and regions are different [20, 21]. Hence, it is necessary to develop a unique driving cycle for each typical city and region. Moreover, many researchers have found that the energy consumption of an EV for a real-world driving cycle greatly differs from that based on the international standard driving cycles [19, 22–25].

Many researchers have also successfully developed real-world driving cycles, such as the Dublin driving cycle [3],

Hong Kong driving cycle [17], California driving cycle [26], Edinburgh driving cycle [27], Hamburg driving cycle [28], Winnipeg driving cycle [29], Mashhad driving cycle [30], Beijing driving cycle, Shanghai driving cycle, Chengdu driving cycle, Chongqing driving cycle, Ningbo driving cycle, Changchun driving cycle, Jilin driving cycle, Zitong driving cycle, Jiutai driving cycle, and Mianyang driving cycle [31]. To date, no studies have focused on the development of typical urban driving cycles in northwestern China. As the core city in Northwest China, Xi'an is characterized by the comprehensive economic, cultural, traffic, and population dynamics driven by the western development in China. Furthermore, Xi'an is a typical ancient capital of China. Notably, the center of the city is the bell tower, and the first ring road was built along the ancient city wall. Xi'an has a population of 9.61 million, and the number of vehicles totals more than 3 million. The total length of urban roads is 2562 km. Therefore, this paper takes Xi'an as an example to study the construction methodology of the EV urban driving cycle.

This paper is organized as follows. The test route selection and data collection method are presented in Section 2. The data processing procedure uses principal component analysis (PCA) to reduce the dimensionality of characteristic parameters, and the k-means and support vector machine (SVM) hybrid method is used for clustering calculations in Section 3. The driving cycle is constructed in Section 4. Section 5 compares the EV driving cycle constructed in this paper with other driving cycles. Finally, Section 6 presents the conclusions of this study.

2. Test Route Selection and Data Collection Method

The test routes must cover various urban road types, business and nonbusiness districts, densely populated areas and non-populated areas and consider the urban topological structure, driving speed, traffic flow, travel time, and origin-destination (O-D) pattern [30, 33–35]. First, the overall topological structure of urban roads in Xi'an was analyzed using ArcGIS software. The lengths and proportions of urban roads in Xi'an are shown in Figure 1. The total length of Xi'an urban roads is 2562 km, including 141 km of expressways, 597 km of main roads, 1001 km of secondary roads, and 823 km of branch roads. To investigate traffic flow, 11 traffic flow monitoring points were established on various types of roads: 2 expressways, 5 main roads, 4 secondary roads, and 1 branch road. Moreover, to avoid the differences between working days and nonworking days, traffic flow monitoring tests were conducted over 14 days. The traffic flow survey results obtained are shown in Figure 2. According to the traffic flow survey results, the peak times are 7:30-9:30 and 17:30-19:30, and the off-peak times are 12:00-13:30 and 19:30-21:00 [17, 20, 29–31]. Considering the abovementioned comprehensive factors, a circular road with a length of 38.4 km was finally selected as the test route. This route included the 11.3 km of expressways, 10.7 km of main roads, 9.4 km of secondary roads, and 7.0 km of branch roads, as shown in Figure 3. The details of the test route are shown in Table 1. Because Xi'an is located in a plain area, the slopes of the urban roads are mainly caused by the viaducts, and the influence of the road gradient on data acquisition is not considered in the test route selection process.

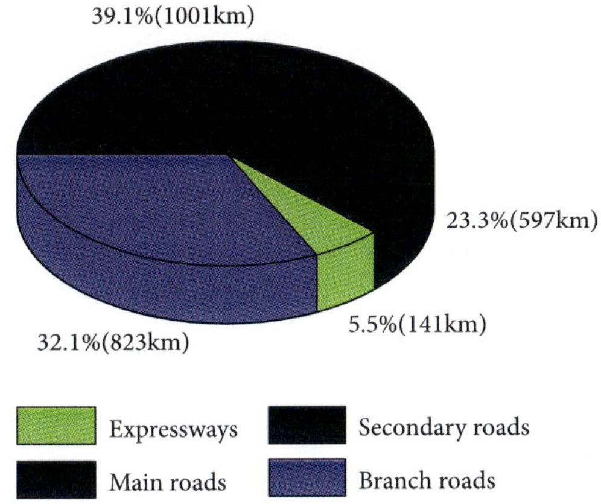

FIGURE 1: Lengths and proportions of urban road types.

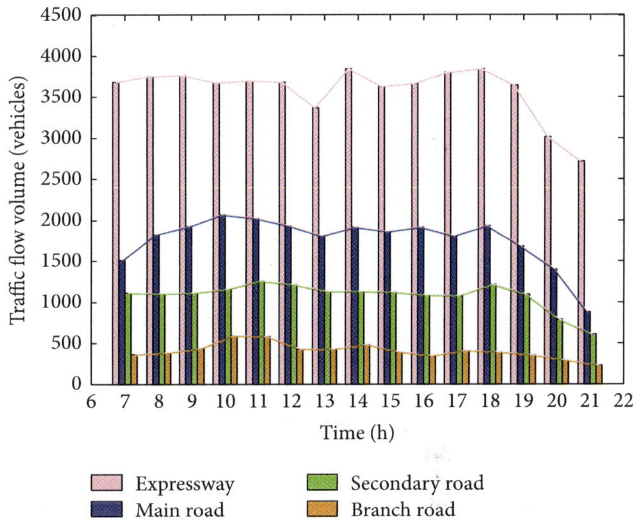

FIGURE 2: Traffic flow survey results.

In driving cycle construction, the common data acquisition methods include the on-board measurement method and chase car method. On-board measurement method uses on-board diagnostics (OBD) installed in a test vehicle to collect trip activity information. The advantage of this approach is that the collected data can accurately represent the actual driving cycle based on large-scale studies. However, the test costs to obtain a reasonable sample size can be extremely high, and the data processing workload is large [32, 34, 36]. Chase car method is widely used to collect speed-time data from real-world driving cycles. Such method involves randomly following a target vehicle and imitating its driving pattern in the traffic stream. When the previous target vehicle stops or is lost, another tracking target is selected [37]. A global position system (GPS) is installed in the test car to collect data, and a large number of driving characteristics are simulated and sampled through the car-following process. Chase car method has been commonly adopted due to its low cost requirements [17, 27, 33, 35].

Table 1: Details of the test route.

Number	Road name	Road type	Road length (km)	Number of traffic lights
1	Wenyi south road	Secondary road	1.2	3
2	Mingsheng road	Branch road	0.6	0
3	Jiandong road	Branch road	1.7	5
4	Leju road	Branch road	0.6	1
5	East Gate south road	Main road	1.1	1
6	East Gate road	Main road	0.4	1
7	Changle road	Branch road	1.5	0
8	Xingqing road	Main road	0.6	3
9	Changle west road	Main road	1.6	2
10	North section of the east first ring road	Express way	1.0	0
11	Taihua south road	Main road	0.8	2
12	Hanyuan road	Secondary road	2.0	0
13	East second ring road	Express way	2.0	0
14	North second ring road	Express way	1.8	0
15	Taihua south road	Main road	0.7	1
16	Xuanwu road	Secondary road	1.6	3
17	Weiyang road	Main road	0.7	0
18	Fangxin road	Branch road	1.6	2
19	Mingguang road	Secondary road	1.0	3
20	Gongnong road	Secondary road	0.5	3
21	Ziqiang west road	Secondary road	0.8	0
22	Xinghuo road	Main road	1.3	1
23	Daqing road	Main road	2.1	1
24	Taoyuan road	Secondary road	1.5	4
25	Fengqing road	Main road	1.4	2
26	West section of south second ring road	Express way	5.0	0
27	Chang'an road	Secondary road	0.5	0
28	Xingshan temple east road	Branch road	1.0	1
29	Cuihua road	Secondary road	0.3	1
30	South second ring road	Express way	1.5	0

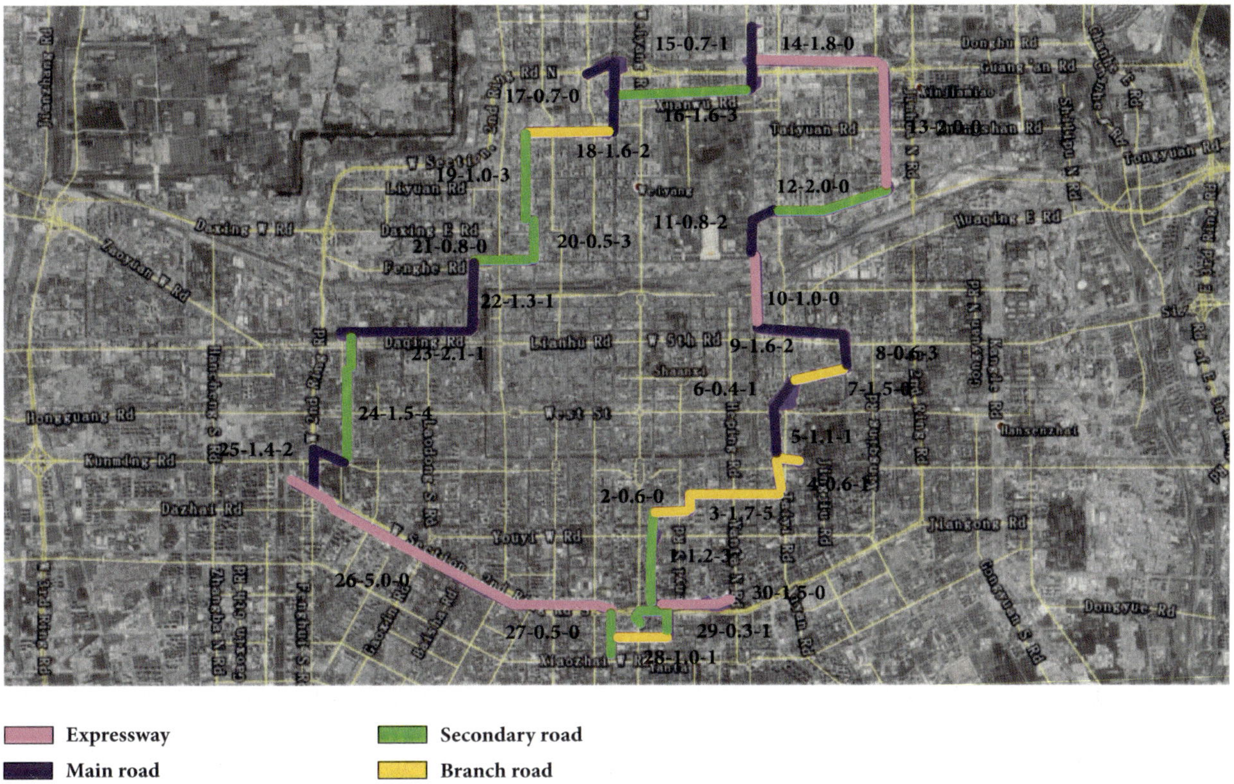

FIGURE 3: Test route. Note: the red data in the A-B-C format represents road information, where A represents the serial number of the roads, B represents the length of the roads, and C represents the number of traffic lights.

Considering on-board measurement method and chase car method has distinct merits and demerits, and this paper used a hybrid method of on-board measurement method and chase car method to collect test data. When an available EV was driving on the preset test route, the test vehicle chased it and imitated its driving pattern; when no target EV was available, the driving patterns of the test vehicle were collected. The GPS signal is easily influenced by the urban buildings, which can cause signal loss and data spikes; therefore, both GPS and OBD were used to collect driving pattern data. Speed-time data obtained by OBD were mainly used to supplement and improve abnormal GPS data. Moreover, 14 trained professional drivers were selected to minimize the influence of the driver on the collected data. In addition, a BYD E6 pure EV, which is the most commonly owned EV in the market, was selected as the test vehicle. The technical characteristics of the test EV are shown in Table 2. The experimental equipment installed in the test vehicle mainly included a GPS, OBD, a gyroscope, a driving recorder, a 12-V lead-acid battery, and a power inverter. The GPS recorded the speed, time, and distance of each trip. The OBD recorded the speed, time, distance, and battery state of charge. Additionally, the gyroscope recorded the acceleration signal and road gradient, and the driving recorder was used to collect traffic condition and chased vehicle information. Finally, 12-V lead-acid battery and power inverter were used to power the test equipment. The test equipment is shown in Figure 4. All data were recorded at a constant sampling frequency of 1 Hz [38–40]. Driving data was collected by test vehicles moving along preset routes under real-world traffic conditions. Moreover, to consider the effect of the traffic flow and travel time on the test results, a 14-day test was conducted. The daily test periods were the morning peak from 7:30 to 9:30, the midday off-peak from 12:00 to 13:30, the afternoon peak from 17:30 to 19:30, and the evening off-peak from 19:30 to 21:00. The tests were conducted under clear weather conditions, on dry roads, and at wind speeds less than 1 m/s. Data collection ceased in cases of abnormal traffic conditions [30]. Test data were collected for a total of 56 trips totaling approximately 428 hours.

3. Data Processing

In the data processing procedure, the raw data were first denoised and smoothed by using a wavelet decomposition and reconstruction method. Then, the preprocessed data were partitioned into 36388 kinematic segments according to (1). In addition, a PCA algorithm was adopted to reduce the dimensionality of the characteristic parameters. Finally, the k-means and SVM hybrid clustering algorithm was used to classify the kinematic segments that were used to construct the driving cycle.

$$
\begin{array}{cc}
\text{acceleration} & a \geq 0.15\text{m/s}^2 \\
\text{deceleration} & a \leq -0.15\text{m/s}^2 \\
\text{uniform} & v \geq 2\text{m/s} \cap -0.15\text{m/s}^2 < a < 0.15\text{m/s}^2 \\
\text{idling} & v < 2\text{m/s} \cap -0.15\text{m/s}^2 < a < 0.15\text{m/s}^2
\end{array} \tag{1}
$$

where a is acceleration and v is speed.

TABLE 2: Technical characteristics of the test EV.

Characteristic parameter	Value
Curb weight (kg)	2380
Mass (kg)	2755
Length (mm)	4560
Width (mm)	1822
Height (mm)	1630
Wheel based (mm)	2830
Front/rear track width (mm)	1585/1560
Centroid height (mm)	640
Maximum speed (km/h)	140
Maximum endurance mileage (km)	400
Rated voltage (V)	600
Maximum torque (Nm)	450
Maximum power (kW)	120

FIGURE 4: Test equipment.

3.1. Reduction of the Dimensionality of Characteristic Parameters Based on PCA. PCA was proposed and applied by statisticians K. Pearson and H. Hotelling in 1901 [41, 42]. PCA is a multivariate statistical method that transforms a large number of correlated variables into a few uncorrelated principal components through orthogonal transformation and can maintain the original variable information as much as possible. Generally, we use characteristic parameters to describe the vehicle motion state. If all the characteristic parameters are used for classification, it will increase both the computational complexity and the difficulty of analysis [41, 42]. Although each parameter characterizes different motion information, the parameters are not independent, and some parameters can be expressed by a combination of several other parameters. Therefore, we use the PCA algorithm to reduce the dimensionality of the characteristic parameters [41, 42].

According to the existing literature, eight characteristic parameters are selected: the maximum speed, minimum speed, average speed, standard deviation of speed, maximum acceleration, maximum deceleration, average acceleration, and standard deviation of acceleration [30, 33, 36].

We assume that the sample size is n, the number of observed parameters is p, and the sample observation matrix is $A = [a_{ij}]_{n\times p}$.

$$A_{n\times p} = \begin{bmatrix} a_{11} & a_{12} & \cdots & a_{1p} \\ a_{21} & a_{22} & \cdots & a_{2p} \\ \vdots & \vdots & & \vdots \\ a_{n1} & a_{n2} & \cdots & a_{np} \end{bmatrix} \quad (2)$$

where a_{ij} is the j-th characteristic parameter of the i-th kinematic segment.

The principal component expressions obtained after orthogonal transformation using the PCA are as follows:

$$\begin{aligned} y'_1 &= {e_1}^T X = e_{1\text{-}1}x_1 + e_{1\text{-}2}x_2 + \cdots + e_{1\text{-}p}x_p \\ y'_2 &= {e_2}^T X = e_{2\text{-}1}x_1 + e_{2\text{-}2}x_2 + \cdots + e_{2\text{-}p}x_p \\ &\vdots \\ y'_p &= {e_p}^T X = e_{p\text{-}1}x_1 + e_{p\text{-}2}x_2 + \cdots + e_{p\text{-}p}x_p \end{aligned} \quad (3)$$

where $y'_1, y'_2, \cdots, y'_p$ are independent of each other and called the first, second, ..., p-th principal components, respectively; $e_1, e_2, \cdots, e_p$ are eigenvectors corresponding to eigenvalues arranged in descending order, namely, $\lambda_1 \geq \lambda_2 \geq \cdots \geq \lambda_p \geq 0$.

The principal component variance contribution ratio P and cumulative variance contribution ratio CP are calculated as follows:

$$P = \frac{\lambda_i}{\sum_{k=1}^{p} \lambda_k} \times 100\%, \quad (i = 1, 2, \cdots, p) \quad (4)$$

$$CP = \frac{\sum_{k=1}^{i} \lambda_i}{\sum_{k=1}^{p} \lambda_k} \times 100\%, \quad (i = 1, 2, \cdots, p) \quad (5)$$

The principal components corresponding to the eigenvalues whose cumulative variance contributions ratio are greater than 90% are selected to form the comprehensive evaluation index that encompasses 90% of the entire data information.

The number of principal components with cumulative variance contribution ratio exceeding 90% is obtained, where $Y_i = e_i^T X,\ i = 1, 2, ..., k\ (k \leq p)$. Additionally, the principal component expression and the principal component score are calculated.

The calculated principal component expression is shown in (6).

$$\begin{bmatrix} Y_1 \\ Y_2 \\ Y_3 \\ Y_4 \\ Y_5 \\ Y_6 \\ Y_7 \\ Y_8 \end{bmatrix} = \begin{bmatrix} 0.567 & 0.499 & 0.562 & 0.158 & 0.203 & 0.074 & 0.149 & 0.143 \\ -0.151 & -0.046 & -0.096 & -0.196 & 0.443 & 0.599 & 0.579 & -0.193 \\ -0.007 & -0.341 & -0.167 & 0.574 & 0.395 & -0.117 & 0.149 & 0.579 \\ -0.220 & 0.239 & -0.006 & -0.650 & 0.139 & -0.140 & -0.001 & 0.658 \\ -0.006 & 0.043 & -0.041 & 0.155 & -0.484 & 0.734 & -0.210 & 0.393 \\ -0.017 & 0.019 & 0.000 & 0.054 & -0.589 & -0.245 & 0.759 & 0.114 \\ -0.212 & 0.750 & -0.535 & 0.304 & 0.067 & -0.056 & 0.000 & -0.074 \\ -0.749 & 0.106 & 0.600 & 0.256 & 0.019 & -0.007 & -0.020 & -0.036 \end{bmatrix} \times \begin{bmatrix} X_1 \\ X_2 \\ X_3 \\ X_4 \\ X_5 \\ X_6 \\ X_7 \\ X_8 \end{bmatrix} \quad (6)$$

where Y_1, $Y_2, \ldots,$ and Y_8 are the first, second, ..., and eighth principal components, respectively; X_1, $X_2, \ldots,$ and X_8 represent maximum speed, minimum speed, average speed, standard deviation of speed, maximum acceleration, maximum deceleration, average acceleration, and standard deviation of acceleration, respectively.

The PCA algorithm is used to reduce the dimensionality of the characteristic parameters. The principal component variance, variance contribution ratio, and the cumulative variance contribution ratio are shown in Table 3. Notably, from the first principal component to the eighth principal component, as the principal component variance decreases, the amount of information contained in the principal component gradually decreases. The first principal component Y_1 contains the most information, in this case, 36.15% of the original indicator information. The second principal component Y_2 contains 32.94% of the original indicator information, and the third principal component Y_3 contains 28.50% of the original indicator information. The cumulative contribution ratio of the first three principal components reaches 97.59%; therefore, they essentially encompass all the original indicator information. Thus, to simplify the problem and reduce the complexity of the analysis, the first three principal components are selected as the characteristic parameters for clustering analysis. The principal component score matrix is shown in Table 4. These scores are calculated for the first three principal components of 36388 kinematic segments. The component matrix is shown in Table 5, which

Table 3: PCA results.

Number of principal component	Principal component variance	Variance contribution rate /%	Cumulative variance contribution rate /%
1	2.892	36.15	36.15
2	2.635	32.94	69.09
3	2.280	28.50	97.59
4	0.182	2.27	99.86
5	0.005	0.06	99.92
6	0.004	0.05	99.97
7	9.9e-4	0.02	99.99
8	5.9e-4	0.01	100.00

Table 4: Principal component score matrix.

Segment number	Principal component 1	Principal component 2	Principal component 3
1	2.569	0.748	0.271
2	1.909	1.510	-0.645
3	1.998	0.898	-0.158
4	1.853	1.552	0.188
5	1.829	0.467	-0.316
⋮	⋮	⋮	⋮
10000	-0.459	1.424	-0.054
10001	0.026	0.102	1.151
10002	0.306	-1.141	0.380
10003	0.217	0.946	0.817
10004	-0.644	1.510	-0.258
⋮	⋮	⋮	⋮
36384	1.972	0.619	0.470
36385	0.865	1.777	-1.059
36386	1.472	0.200	0.742
36387	2.110	-1.821	-0.838
36388	2.670	0.368	0.373

Table 5: Component matrix.

Characteristic parameters	Principal component 1	Principal component 2	Principal component 3
maximum speed	0.965	-0.245	-0.010
minimum speed	0.848	-0.074	-0.514
average speed	0.955	-0.155	-0.252
standard deviation of speed	0.268	-0.317	0.866
maximum acceleration	0.346	0.720	0.597
maximum deceleration	0.125	0.973	-0.177
average acceleration	0.254	0.939	0.225
standard deviation of acceleration	0.243	-0.313	0.874

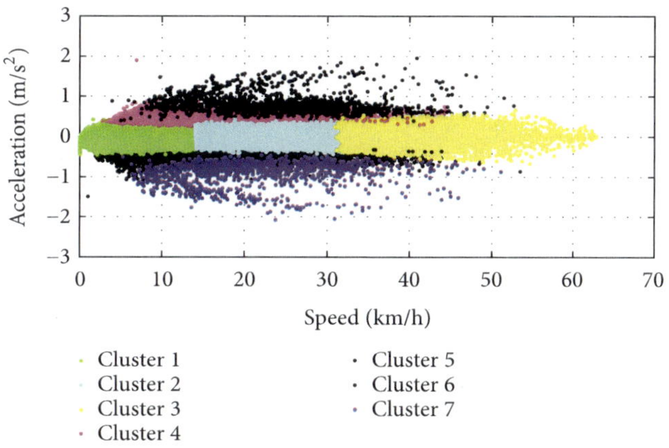

FIGURE 5: k-means clustering result.

displays the correlation coefficient between the principal component and the original characteristic parameter.

3.2. Clustering Based on the k-Means and SVM Hybrid Algorithm

3.2.1. Clustering Based on k-Means. The k-means algorithm has been widely used in cluster computing in the existing literature [3, 4, 19, 24, 43]. The k-means clustering algorithm uses the Euclidean distance between a sample and cluster center as a criterion for similarity determination, and it has the advantages of efficiency and simplicity [19, 43]. The Euclidean distance is calculated as shown in (7).

$$d_{ij} = \sqrt{\sum_{k=1}^{P} \left(x_{ik} - x_{jk}\right)^2} \tag{7}$$

where P is the number of characteristic parameters and x_{ik} and x_{jk} are the k-th characteristic parameters of the i-th sample and the j-th sample, respectively.

The number k of clusters in the k-means algorithm has an important influence on the clustering results, but generally, an accurate value of k is initially unknown [44, 45]. If the number of clusters is greater than the true value, the algorithm wrongly divides the same class of data into multiple classes, which will cause the clustering result boundaries to be blurred [46]. Conversely, merging different classes of data into one class will result in a decrease in the compactness of the cluster [47]. Therefore, clustering stability is generally used to determine the k value [48]. The concept of the approach is that if repeated clustering is performed for the same sample data, then a suitable k value should produce the same clustering results. In other words, stability is viewed as an indication of whether the k value fits the data. We repeated the clustering stability simulation test and finally determined that the number of clusters was seven. The classification results of the k-means clustering algorithm are shown in Figure 5.

The analysis of the k-means clustering effect shows that when the distances between the segment and other cluster centers are large enough, and the characteristic parameters of the segment are close to those of a certain class, the clustering effect is good. In contrast, when the distance between the segment and all cluster centers is similar and the characteristic parameters of the segment are significantly different from those of most of the segments in the class, the clustering effect is poor. There are two main reasons for these results. First, k-means is a hard clustering algorithm. When clustering involves multiple classes or the distance between cluster centers is small, the clustering effect is poor, and it is easy to reach local optima that cannot be incrementally clustered. Second, the convergence condition of the k-means clustering algorithm is iterated until the cluster center no longer changes. This approach may cause the data within the class to be very similar but does not fully consider the distance between classes; therefore, only local optima are guaranteed, and the global optimization is not achieved.

3.2.2. Clustering Based on the k-Means and SVM Hybrid Algorithm. The SVM method was first proposed by Cortes and Vapnik in 1995 and has significant advantages in solving nonlinear and high-dimensional pattern recognition problems [49, 50]. The SVM method is based on the Vapnik-Chervonenkis dimensional statistical learning theory and is an approximate implementation of minimizing structural risk. The basic principle of SVM is to establish an optimal classification hyperplane that can not only separate samples without errors, but also maximize the classification interval [50]. A schematic diagram of the optimal hyperplane is shown in Figure 6. SVM has been widely used for classification identification due to its many advantages, such as a high level of robustness and strong generalization ability.

We assume that the size of the sample set is l and that the sample set consists of two classes: $S = \{(x_i, y_i), i = 1, 2, \cdots, l, x_i \in R^d, y_i \in \{+1, -1\}\}$. If x_i belongs to the first class, $y_i = 1$; otherwise, $y_i = -1$. If the sample can be correctly divided into two classes using a classification hyperplane, then the sample is linearly separable and satisfies (8).

$$wx_i + b \geq 1,$$

$$y_i = 1$$

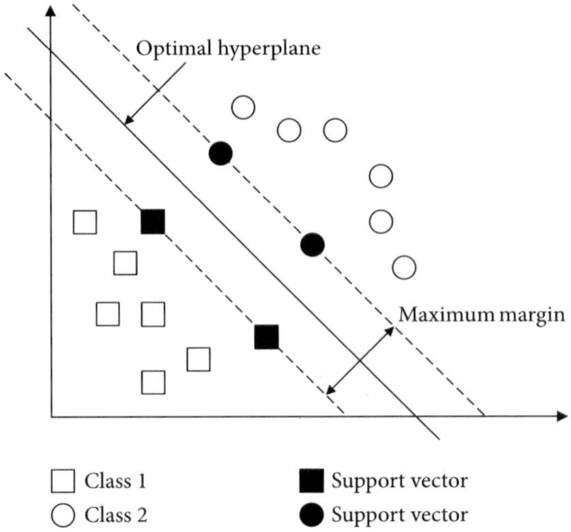

FIGURE 6: Schematic diagram of the optimal hyperplane.

$$wx_i + b \leq -1, \quad y_i = -1, \quad i = 1, 2, \cdots, l \tag{8}$$

where x_i is the sample data, y_i is a decision attribute, w is the normal vector of the hyperplane, and b is the offset of the classification straight line.

If the interval from the sample point to the classification hyperplane is $\varepsilon = |wx_i + b| = 1$, the distance between two types of samples is $2/\|w\|$. To select an optimal classification hyperplane from a large number of classification hyperplanes and separate the samples of different classes as much as possible, the distance from the sample set to the classification hyperplane is maximized. Therefore, the optimization goal is to find the optimal classification hyperplane under the constraint of (9).

$$\begin{aligned} &\min \quad \frac{\|w\|^2}{2} \\ &\text{s.t.} \quad y_i (wx_i + b) \geq 1, \quad i = 1, 2, \cdots, l \end{aligned} \tag{9}$$

The above problem can be transformed into a dual problem according to the Lagrange theory and then solved with a quadratic programming method. In the linear inseparability problem, the nonlinear mapping of $\Phi : R^d \longrightarrow H$ is used to map the original input space samples to the feature space H. Then, the optimal classification hyperplane is constructed in the high-dimensional feature space H. The dual problem after mapping to the high-dimensional feature space becomes (10).

$$\begin{aligned} &\max \quad Q(a) = \sum_{i=1}^{l} a_i - \frac{1}{2} \sum_{i=1}^{l} \sum_{j=1}^{l} a_i a_j y_i y_j K(x_i, x_j) \\ &\text{s.t.} \quad \begin{cases} \sum_{i=1}^{l} a_i y_i = 0 \\ 0 \leq a_i \leq C, \end{cases} \quad i = 1, 2, \cdots, l \end{aligned} \tag{10}$$

where $a_i > 0$, $i = 1, 2, \cdots, l$ is the Lagrange coefficient and $K(x_i, x_j)$ is the kernel function that satisfies the Mercer condition. C is a penalty factor used to control penalty degree for misclassified samples.

Assuming that $a^* = (a_1^*, a_2^*, \cdots, a_l^*)^T$ is the solution of (10), then the final optimal classification function can be obtained as shown in (11).

$$f(x) = \text{sgn}\left(\sum_{i=1}^{l} a_i^* y_i K(x_i, x) + b^*\right) \tag{11}$$

This paper builds a multiclassification model based on SVM to classify the driving segments. The specific classification processes include training set screening, kernel function parameter optimization, classification prediction, and clustering result evaluation.

The first step is training set screening. Because a SVM is a supervised learning algorithm, selecting an appropriate training set before classification is extremely important. To improve the classification accuracy and efficiency, the optimal segments were selected from the k-means clustering results as the SVM training set, and the remaining segments were used as the testing set. The selection principles of the optimal segments are as follows. We selected an appropriate number of training sets to avoid under-learning and over-learning issues. Moreover, we selected representative segments from the k-means clustering results, and these segments were as close to a cluster center as possible and as far from the other cluster centers as possible. According to the above principles, 2353 representative segments were selected as the training set of the SVM, and the remaining segments were used as the testing set. Additionally, all segments were normalized to eliminate the influence of the dimensions on the classification results.

The second step is kernel function parameter optimization. Kernel functions can map the sample data in the original low-dimensional space to a high-dimensional feature space and transform them into linearly separable data. Meanwhile, the optimal classification hyperplane in the high-dimensional

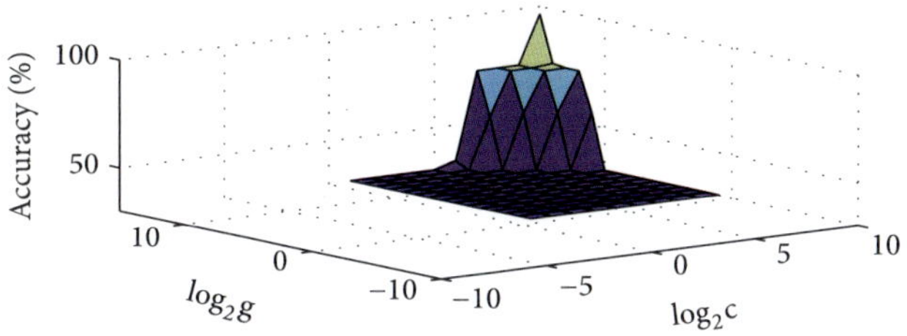

FIGURE 7: Optimization results.

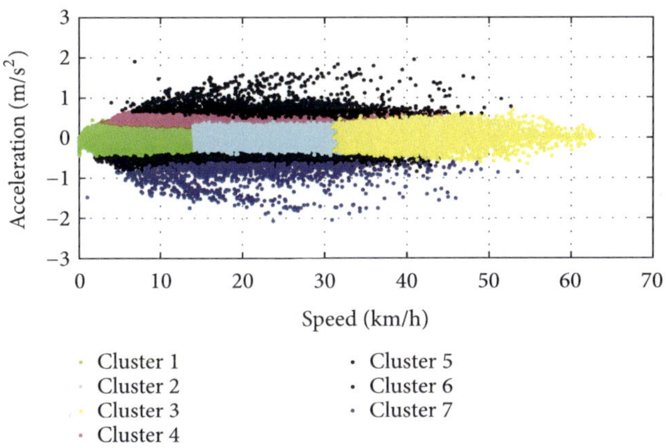

FIGURE 8: k-means and SVM hybrid model clustering results.

space can be obtained. Because the radial basis kernel function (RBF) can precisely express the features of the training set, accurately reflect the structure feature of the high-dimensional space, effectively control the dimension of the model solution set, and obtain the global optimal solution, this paper uses a RBF as the basic kernel function of the SVM.

$$K(x, x_i) = \exp\left(-\text{gamma}\,\|x_i - x\|^2\right) \quad (12)$$

where $\|x - x_i\|$ is the two-norm distance, x_i is the support vector, x is the predicted sample, and gamma is the kernel function parameter g.

Based on K-fold cross validation method, this paper adopts a grid search algorithm to obtain the optimal parameters c and g. The optimization results are shown in Figure 7. When c=0.25 and g=4.0, classification recognition accuracy reaches a maximum of 100%.

The third step in the proposed method involves classification prediction and clustering result evaluation. The SVM model was trained using the optimal parameters. The clustering results of driving segments using the k-means and SVM hybrid model are shown in Figure 8.

To evaluate the clustering results, this paper introduces two indicators: compactness and separation [47, 51].

Compactness (CP) is an internal cluster evaluation criterion that uses the norm distance between all data sets in each cluster and the cluster center to evaluate the compactness of the cluster. A small CP value indicates that the data within the cluster are similar or close to one another.

$$\overline{CP_i} = \frac{1}{|\Omega_i|}\sum_{x_i \in \Omega_i} \|x_i - w_i\| \quad (13)$$

where Ω_i is the ith cluster, w_i is the cluster center in Ω_i, x_i represents the data in Ω_i, and $|\Omega_i|$ represents the quantity of data in Ω_i. As a global measure of compactness, the average of all clusters is calculated as follows.

$$CP = \frac{1}{K}\sum_{k=1}^{K}\overline{CP_k} \quad (14)$$

where K is the number of clusters.

Separation (SP) is an external evaluation criterion between clusters that uses the average Euclidean distance between each two cluster centers to evaluate the degree of separation of the clusters. A high SP value indicates that the distance between clusters is large.

$$SP = \frac{2}{K^2 - K}\sum_{i=1}^{K}\sum_{j=i+1}^{K}\left\|w_i - w_j\right\|_2 \quad (15)$$

Table 6 gives the calculation results for CP and SP. A comparison of the values indicates that the k-means and SVM hybrid clustering results have small CP and large SP values; therefore, the clustering results with the k-means and SVM hybrid model are significantly better than those of the k-means alone.

The categorized driving segment characteristics are shown in Table 7 and include low constant speed driving segments, medium constant speed driving segments, high constant speed driving segments, weak acceleration driving segments, strong acceleration driving segments, weak deceleration driving segments, and strong deceleration driving segments.

4. Driving Cycle Construction

The driving cycle is constructed by connecting the most representative driving segments according to predetermined rules until the driving cycle duration is reached. Generally, the duration of the general international standard driving cycles and real-world representative driving cycles is between

TABLE 6: The calculation results of *CP* and *SP*.

Calculation results	k-means	k-means and SVM
CP	1.393	1.317
SP	4.954	5.242

TABLE 7: Driving segment characteristics.

Category number	The number of segments	Average speed	Average acceleration	Definition
1	10914	5.47	-0.008	low constant speed driving segments
2	9238	22.62	-0.006	medium constant speed driving segments
3	6503	40.54	0.007	high constant speed driving segments
4	3729	18.75	0.463	weak acceleration driving segments
5	1341	24.41	0.840	strong acceleration driving segments
6	3573	18.00	-0.489	weak deceleration driving segments
7	1090	21.38	-0.946	strong deceleration driving segments

TABLE 8: Time proportion and length of seven driving segments.

Driving segments	Time proportion (%)	Time length (s)
1	30.53	366
2	13.98	168
3	12.51	150
4	14.28	171
5	7.84	94
6	14.72	177
7	6.14	74

600 s and 1800 s. Therefore, according to the Hong Kong, Colombo, and Sydney driving cycles, the duration of the Xi'an EV urban driving cycle in this paper is set to 1200 s [17, 33, 36, 43]. Then, the duration of each driving segment class in the constructed Xi'an EV urban driving cycle is determined based on experimental data. First, this paper calculates the cumulative time of the seven driving segment classes in the collected experimental data and determines the time proportions of the seven driving segment classes. As shown in Table 8, the time proportions of the seven driving segment classes are 30.53%, 13.98%, 12.51%, 14.28%, 7.84%, 14.72%, and 6.14%, respectively. Then, according to the time proportions of the seven driving segment classes in the experimental data, the time length of the seven driving segment classes in the constructed Xi'an representative EV urban driving cycle is determined. Their time lengths are 366 s, 168 s, 150 s, 171 s, 94 s, 177 s, and 74 s. In the process of constructing the driving cycle, we first randomly select a driving segment of no more than 5 s as the initial segment. Then, according to the following principles, the most representative driving segments are selected. The selection principle of representative driving segments is as follows: (1) the selected driving segment should be as close as possible to the cluster center; (2) the difference between the final speed of the previous driving segment and the initial speed of the current driving segment is less than 1km/h; (3) the selection of the driving segment is without replacement. Finally, the selected current driving segment is linked to the previous driving segment until the time length of each driving segment class is reached. This approach yields a constructed candidate driving cycle.

According to the test data and classification results, the time proportion and length of the seven driving segments are shown in Table 8.

Considering the randomness in the construction process of the driving cycle, one construction result is difficult to fully represent the Xi'an EV urban driving cycle. Therefore, the construction process is repeated to generate a large number of candidate driving cycles. Moreover, the scientific assessment criteria are set, and the most representative driving cycle is selected from candidate driving cycles to form the Xi'an EV urban driving cycle.

The assessment criteria flow diagram is shown in Figure 9. First, the average speed, maximum speed, average acceleration, average deceleration, standard deviation of acceleration, proportion of acceleration, proportion of deceleration, proportion of uniform speed, and proportion of idling are selected as assessment parameters [17, 37]. Then, we calculate the relative error (RE) of the assessment parameters of the candidate driving cycles and the overall test data and select the driving cycles for which the RE of all the assessment parameters is less than 10%. Finally, the mean relative error (MRE) is calculated, and the candidate driving cycle with the minimum MRE is selected. If there is more than one candidate driving cycle with the same minimum MRE, the

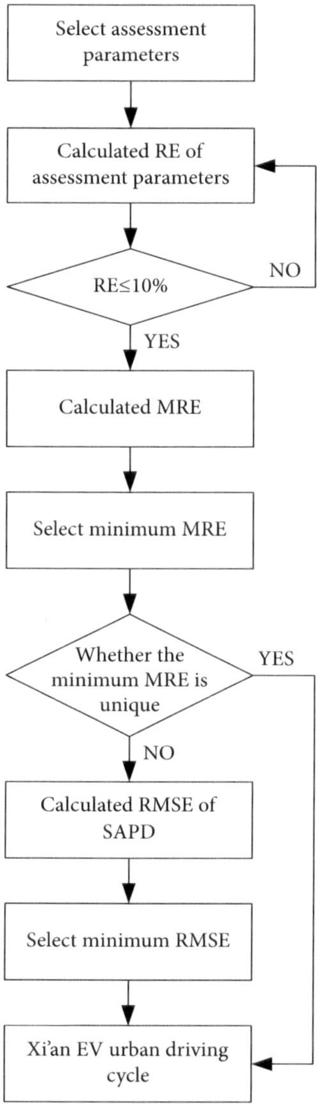

FIGURE 9: Assessment criteria flow diagram.

root mean square error (RMSE) of the speed-acceleration probability distribution (SAPD) is calculated. The driving cycle with the minimum RMSE is then selected as the most representative Xi'an EV urban driving cycle.

$$RE = \left(\frac{\alpha_c - \alpha_t}{\alpha_t}\right) \times 100\% \tag{16}$$

where α_c and α_t are the assessment parameters of the candidate driving cycles and test data, respectively.

$$MRE = \frac{1}{n}\sum_{i=1}^{n}\left(\frac{\alpha_c - \alpha_t}{\alpha_t}\right) \tag{17}$$

where n is the number of all assessment parameters.

$$RMSE = \sqrt{\frac{1}{M \times N}\sum_{j=1}^{M}\sum_{i=1}^{N}\left(x_{ij} - y_{ij}\right)^2} \tag{18}$$

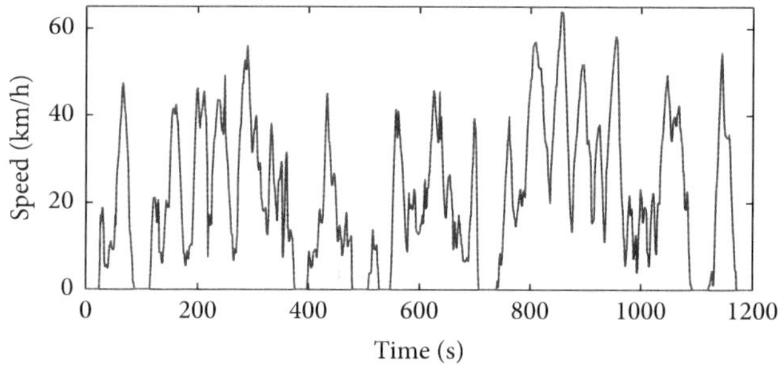

FIGURE 10: Xi'an EV urban driving cycle.

where N and M are the numbers of speed bins and acceleration bins, respectively, and x_{ij} and y_{ij} are the frequency values of the ijth bin of the candidate driving cycle and the test data, respectively.

The construction of Xi'an EV urban driving cycle is shown in Figure 10. The SAPDs of Xi'an EV urban driving cycle and test data are shown in Figures 11 and 12, respectively. A comparison of the Xi'an EV urban driving cycle and test data is shown in Table 9. According to the comparison results, the REs of all assessment parameters are less than 8.81%, the MRE is 4.03%, and the RMSE of the SAPD is 1.32%. The small MRE and RMSE values indicate that the constructed Xi'an EV urban driving cycle is very close to that based on experimental data from the real-world driving cycle. Therefore, the Xi'an EV urban driving cycle constructed in this paper can effectively represent the speed-time driving pattern of the real-world cycle in Xi'an.

5. Comparison of Driving Cycles

5.1. Comparison of the Xi'an and International Standard Driving Cycles. To study the differences between the Xi'an EV urban driving cycle constructed in this paper and the international standard driving cycles, eight characteristic parameters of seven general international standard driving cycles are compared as shown in Figure 13 and Table 10. The seven general international standard driving cycles are the European Economic Commission 15-mode test cycle (ECE 15), New European Driving Cycle (NEDC), Japanese Industrial Standards Committee 08 test cycle (JC08), Japan 10/15 mode test cycle (J10/15), Federal Test Procedures 72 and 75 (FTP72, FTP75), and Worldwide harmonized light duty test cycle (WLTC) [52], respectively. The comparison of results shows that the Xi'an EV urban driving cycle has greater average acceleration and deceleration values, larger proportions of acceleration and deceleration, and smaller proportions of uniform speed and idling compared to the other driving cycles. Therefore, compared with various general international standard driving cycles, the Xi'an EV urban driving cycle is characterized by more aggressive driving features.

5.2. Verification of the Construction Method. To verify the advanced and scientific nature of the construction method of the driving cycle proposed in the paper, the driving cycles

TABLE 9: Comparison of the Xi'an EV urban driving cycle and test data.

Characteristic parameters	Test data	Driving cycle	RE	MRE	RMSE
Average speed (km/h)	20.01	20.74	3.65%		
Maximum speed (km/h)	68.22	63.86	6.39%		
Average acceleration (m/s^2)	0.76	0.78	2.63%		
Average deceleration (m/s^2)	0.78	0.77	1.28%		
Standard deviation of acceleration (m/s^2)	0.84	0.79	5.95%	4.03%	1.32%
Proportion of acceleration(%)	34.5	32.9	4.64%		
Proportion of deceleration(%)	32.1	31.9	0.62%		
Proportion of uniform speed(%)	15.9	17.3	8.81%		
Proportion of idling(%)	17.5	17.9	2.29%		

Note: RE is the relative error of the assessment parameters of the candidate driving cycles and the overall test data; MRE is the mean relative error of all assessment parameters; and RMSE is the root mean square error of the SAPD.

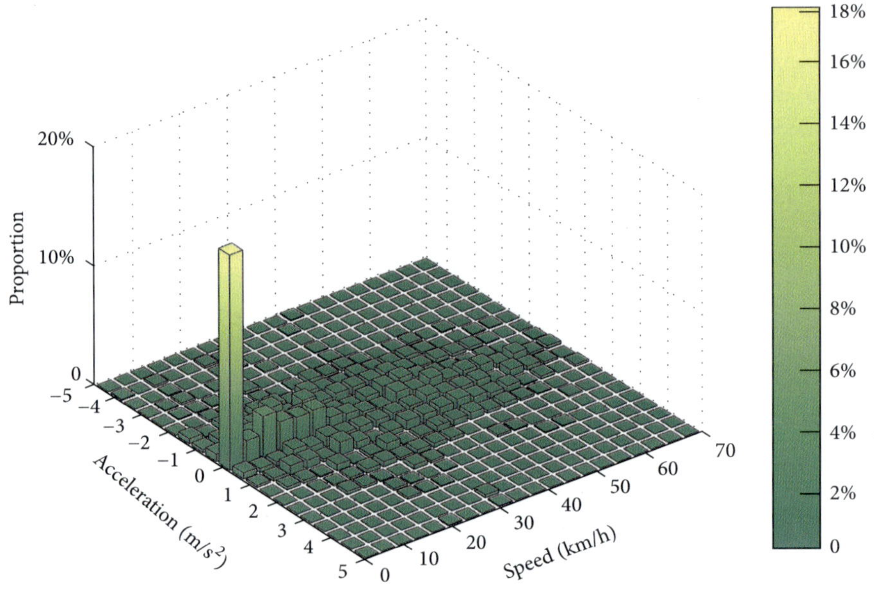

FIGURE 11: SAPD of the Xi'an EV urban driving cycle.

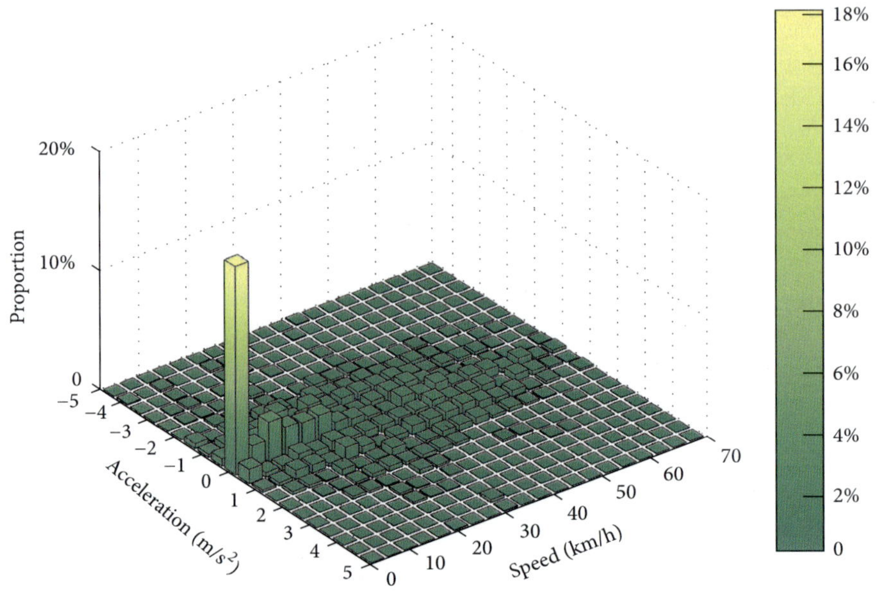

FIGURE 12: SAPD of the test data.

TABLE 10: Comparison of results for the Xi'an and international standard driving cycles.

Driving cycle	Xi'an	ECE15	NEDC	JC08	J10/15	FTP-72	FTP-75	WLTC
Vehicle type	EV	ICEV	ICEV	ICEV	ICEV	ICEV	ICEV	ICEV
Road type	Urban	Urban	Composite	Composite	Urban	Urban	Urban	Composite
Average speed(km/h)	20.74	18.4	33.6	24.4	17.7	31.5	34.1	46.5
Maximum speed(km/h)	63.86	50.0	120	82.0	70.0	91.3	91.3	131.3
Average acceleration(m/s^2)	0.78	0.64	0.51	0.42	0.63	0.59	0.60	0.42
Average deceleration(m/s^2)	0.77	0.75	0.71	0.44	0.62	0.69	0.70	0.44
Proportion of acceleration(%)	32.9	21.5	23.8	27.8	25.9	32.8	32.4	29.4
Proportion of deceleration(%)	31.9	18.5	17.6	25.9	26.4	28.3	28.2	27.8
Proportion of uniform speed(%)	17.3	29.2	34.8	17.4	22.2	20.9	21.2	30.3
Proportion of idle(%)	17.9	30.8	23.8	28.9	25.4	18.0	18.2	12.5

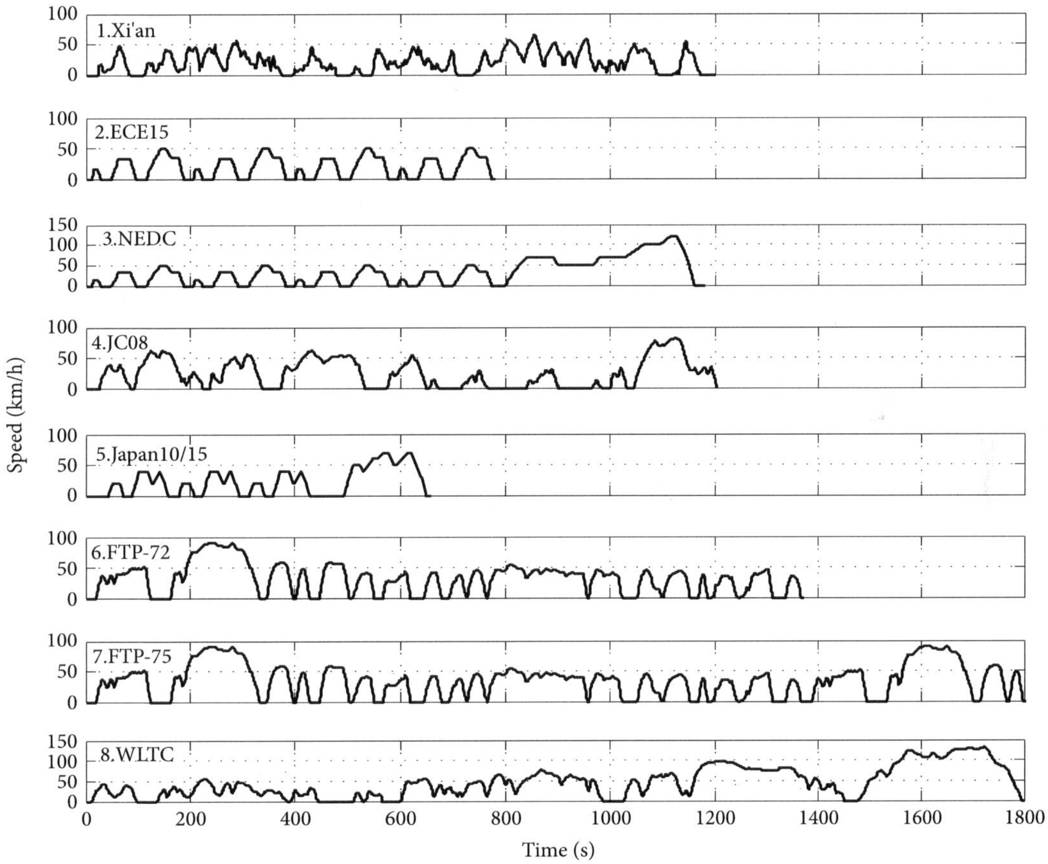

FIGURE 13: Driving cycles.

constructed by the k-means and SVM hybrid clustering algorithm, k-means clustering algorithm, and fuzzy c-means (FCM) clustering algorithm were compared. The driving cycles constructed by the k-means clustering algorithm and FCM clustering algorithm are shown in Figures 14 and 15, respectively. A comparison of the characteristic parameter results is shown in Table 11. The results indicate that the characteristic parameters of the driving cycle based on the k-means and SVM hybrid clustering algorithm are closest to those of real-world test data. Therefore, the driving cycle constructed in this paper accurately reflects the speed-time and speed-acceleration characteristics of the real-world.

5.3. Comparison of Typical City Driving Cycles. To study the differences between the Xi'an EV urban driving cycle and other typical city driving cycles, this paper introduces the Winnipeg, Dublin, Mashhad, Hongkong, Ningbo, and Tianjin driving cycles and studies the difference and similarity among driving cycles through comparing the characteristic parameters. A comparison of the results is shown in Table 12.

Table 11: Comparison of the characteristic parameter results.

Characteristic parameters	Test data	k-means and SVM	k-means	FCM
Average speed (km/h)	20.01	20.74	21.59	22.47
Maximum speed (km/h)	68.22	63.86	63.07	61.42
Average acceleration (m/s^2)	0.76	0.78	0.69	0.74
Average deceleration (m/s^2)	0.78	0.77	0.76	0.79
Standard deviation of acceleration (m/s^2)	0.84	0.79	0.76	0.8
Proportion of acceleration(%)	34.5	32.9	36.0	38.33
Proportion of deceleration(%)	32.1	31.9	32.6	30.92
Proportion of uniform speed(%)	15.9	17.3	14.3	14.01
Proportion of idling(%)	17.5	17.9	17.1	16.74
MRE (%)	-	4.03	6.11	6.88
RMSE (%)	-	1.32	1.89	2.04

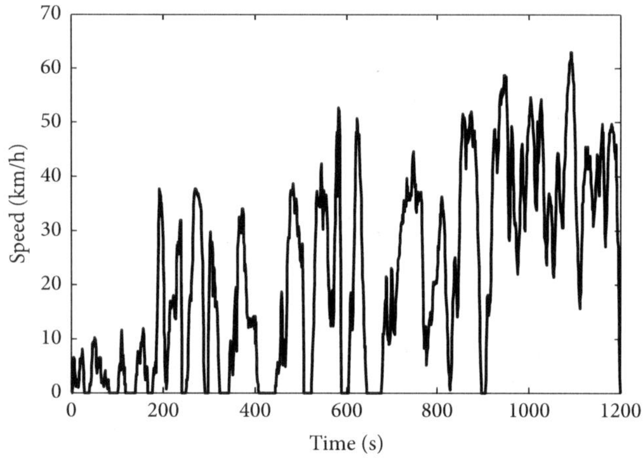

Figure 14: Driving cycle based on k-means clustering algorithm.

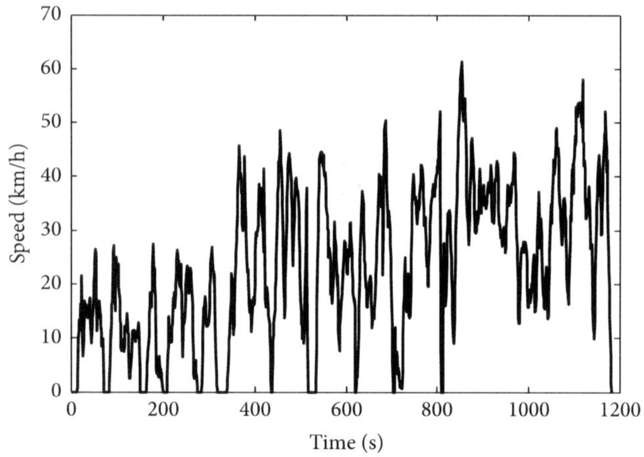

Figure 15: Driving cycle based on FCM clustering algorithm.

The average and maximum speeds of the Xi'an EV urban driving cycle are similar to those of the Mashhad and Ningbo driving cycles. Moreover, the average acceleration and average deceleration of Xi'an EV urban driving cycle are obviously greater than those of other typical city driving cycles, and the proportions of acceleration, deceleration, uniform speed, and idling are similar to those of the Winnipeg driving cycle. Additionally, the average acceleration and average deceleration of the EV driving cycles are obviously greater than those of the ICEV driving cycles, and the EV driving cycles display more aggressive driving characteristics.

According to an analysis of the results, the Xi'an EV urban driving cycle reflects a more aggressive driving pattern than other driving cycles. The reasons for this finding are threefold. First, car ownership and the urban population have increased rapidly in Xi'an. The city size and transportation infrastructure construction cannot keep up with the increase in car ownership, and urban traffic congestion is serious. Second, the urban driving cycle in this paper is based on EVs, but general international standard driving cycles are based on traditional ICEVs. The start-up acceleration of EVs is obviously higher than that of ICEVs due to the low-speed constant torque characteristics of the motor [53, 54]. In addition, because the motor provides a fast response and good robustness, slight changes in the accelerator pedal will cause significant speed changes, which is not satisfied by the engine. Therefore, the average acceleration of the EV driving cycle is greater than that of the ICEV driving cycles. Third, to increase the energy utilization rate and driving range of EVs, the motor regenerative braking system can convert kinetic energy into electric energy and recharge batteries during the braking process. Specifically, when the driver releases the accelerator pedal, the EV enters into the regenerative braking mode. However, the ICEV remains in a uniform speed gliding state; therefore, EV experiences greater deceleration.

6. Conclusions

This paper proposes a scientific and systematic methodology for the development of a representative EV urban driving cycle. The methodology mainly includes three tasks: test route selection and data collection, data processing, and driving cycle construction. For test route selection and data collection, the overall topological structure of urban roads and traffic flow monitoring results are used as the main factors for the selection of test routes. We combine the advantages of the on-board measurement and chase car methods to collect driving pattern data. In the data processing stage, the

Table 12: Comparison of the results for typical city driving cycles.

Driving cycle	Xi'an	Winnipeg	Dublin	Mashhad	Hongkong	Ningbo	Tianjin
Vehicle type	EV	PHEV	EV	ICEV	ICEV	ICEV	ICEV
Road type	Urban	Composite	Urban	Urban	Urban	Urban	Urban
Size of experiment	56 trips, 428 hours, 2158 km	76 volunteers, one year	1485 journals	450 km	29 hours	24 hours	25 hours
Average speed(km/h)	20.74	31.4	30.87	20.27	25.0	23.7	22.5
Maximum speed(km/h)	63.86	100	84.5	60.90	77.7	60.2	70.2
Average acceleration(m/s^2)	0.78	0.60	0.62	0.53	0.60	0.51	0.36
Average deceleration(m/s^2)	0.77	0.60	0.64	0.54	0.60	0.58	0.43
Proportion of acceleration(%)	32.9	31.4	26.7	37.67	34.5	37.0	36.0
Proportion of deceleration(%)	31.9	31.2	25.0	37.48	34.2	33.0	30.0
Proportion of uniform speed(%)	17.3	14.9	27.6	3.16	12.0	11.0	21.0
Proportion of idle(%)	17.9	21.8	20.7	21.70	17.8	20.0	12.0
Reference	-	[29]	[3]	[30]	[17]	[32]	[32]

PCA algorithm is used to reduce the dimensionality of the characteristic parameters. The driving segments are classified using a k-means and SVM hybrid clustering algorithm, and the classification results of the hybrid method are obviously superior to those of the k-means method. In the driving cycle construction stage, scientific assessment criteria are studied to select the most representative driving cycle from multiple candidate driving cycles. In addition, the SAPDs of the representative Xi'an EV urban driving cycle and real-world test data are compared, and the results show that the Xi'an EV urban driving cycle constructed in this paper effectively represents the speed-time driving pattern of the real-world cycle. Finally, the characteristic parameters of the Xi'an EV urban driving cycle, international standard driving cycles, and other typical city driving cycles are compared and analyzed. The results indicate that the Xi'an EV urban driving cycle reflects a more aggressive driving characteristic than other driving cycles, mainly due to the difference in the start-up acceleration and regenerative braking characteristics of EVs and ICEVs. Therefore, a representative EV driving cycle should be used for energy optimization, state of charge estimation, and driving mileage prediction for EVs, as well as the evaluation and certification of new EV models.

Conflicts of Interest

The authors declare that they have no conflicts of interest.

Acknowledgments

This research is funded by the National Key R&D Program of China (2017YFC0803904), National Natural Science Foundation of China (51507013), China Postdoctoral Science Foundation (2018T111006, 2017M613034), Postdoctoral Science Foundation of Shaanxi Province (2017BSHEDZZ36), Shaanxi Province Industrial Innovation Chain Project (2018ZDCXL-GY-05-03-01), Shaanxi Provincial Key Research and Development Plan Project (2018ZDXM-GY-082), and Shaanxi Innovative Talents Promotion Plan Project (2018KJXX-005).

References

[1] S. Zhang, R. Xiong, and C. Zhang, "Pontryagin's Minimum Principle-based power management of a dual-motor-driven electric bus," *Applied Energy*, vol. 159, pp. 370–380, 2015.

[2] S. Zhang and R. Xiong, "Adaptive energy management of a plug-in hybrid electric vehicle based on driving pattern recognition and dynamic programming," *Applied Energy*, vol. 155, pp. 68–78, 2015.

[3] J. Brady and M. O'Mahony, "Development of a driving cycle to evaluate the energy economy of electric vehicles in urban areas," *Applied Energy*, vol. 177, pp. 165–178, 2016.

[4] W. N. Lourido, L. E. Munoz, and J. E. Pereda, "A methodology to obtain a synthetic driving cycle through GPS data for energy analysis," in *Proceedings of the 12th IEEE Vehicle Power and Propulsion Conference, VPPC 2015*, Canada, October 2015.

[5] S. Xie, H. He, and J. Peng, "An energy management strategy based on stochastic model predictive control for plug-in hybrid electric buses," *Applied Energy*, vol. 196, pp. 279–288, 2017.

[6] Z. Chen, B. Xia, C. You, and C. C. Mi, "A novel energy management method for series plug-in hybrid electric vehicles," *Applied Energy*, vol. 145, pp. 172–179, 2015.

[7] L. Dang, N. Bernard, N. Bracikowski, and G. Berthiau, "Design optimization with flux weakening of high-speed PMSM for electrical vehicle considering the driving cycle," *IEEE Transactions on Industrial Electronics*, vol. 64, no. 12, pp. 9834–9843, 2017.

[8] Z. Chen, R. Xiong, K. Wang, and B. Jiao, "Optimal energy management strategy of a plug-in hybrid electric vehicle based on a particle swarm optimization algorithm," *Energies*, vol. 8, no. 5, pp. 3661–3678, 2015.

[9] Y. Wang, H. Fujimoto, and S. Hara, "Driving force distribution and control for EV with four in-wheel motors: a case study of acceleration on split-friction surfaces," *IEEE Transactions on Industrial Electronics*, vol. 64, no. 4, pp. 3380–3388, 2017.

[10] B. Gasbaoui and A. Nasri, "A novel 4WD electric vehicle control strategy based on direct torque control space vector modulation technique," *Intelligent Control Automation*, vol. 3, no. 03, pp. 236–242, 2012.

[11] K. Hartani, M. Bourahla, Y. Miloud, and M. Sekour, "Electronic differential with direct torque fuzzy control for vehicle propulsion system," *Turkish Journal of Electrical Engineering & Computer Sciences*, vol. 17, no. 1, pp. 21–38, 2014.

[12] S. A. Oleksowicz, K. J. Burnham, A. Southgate et al., "Regenerative braking strategies, vehicle safety and stability control systems: critical use-case proposals," *Vehicle System Dynamics*, vol. 51, no. 5, pp. 684–699, 2013.

[13] J. Nadeau, M. Boisvert, and P. Micheau, "Implementation of a cooperative strategy between a vehicle's mechanical and regenerative brake system," in *Proceedings of the 2014 IEEE Vehicle Power and Propulsion Conference, VPPC 2014*, Portugal, October 2014.

[14] Z. Zhang, L. Wang, J. Zhang, and R. Ma, "Study on requirements for load emulation of the vehicle with an electric braking system," *IEEE Transactions on Vehicular Technology*, vol. 66, no. 11, pp. 9638–9653, 2017.

[15] C. Hou, M. Ouyang, L. Xu, and H. Wang, "Approximate Pontryagin's minimum principle applied to the energy management of plug-in hybrid electric vehicles," *Applied Energy*, vol. 115, pp. 174–189, 2014.

[16] M. Helmbrecht, C. Olaverri-Monreal, K. Bengler, R. Vilimek, and A. Keinath, "How electric vehicles affect driving behavioral patterns," *IEEE Intelligent Transportation Systems Magazine*, vol. 6, no. 3, pp. 22–32, 2014.

[17] W. T. Hung, H. Y. Tong, C. P. Lee, K. Ha, and L. Y. Pao, "Development of a practical driving cycle construction methodology: A case study in Hong Kong," *Transportation Research Part D: Transport and Environment*, vol. 12, no. 2, pp. 115–128, 2007.

[18] S. Shi, N. Lin, Y. Zhang et al., "Research on Markov property analysis of driving cycles and its application," *Transportation Research Part D: Transport and Environment*, vol. 47, pp. 171–181, 2016.

[19] A. Fotouhi and M. Montazeri-Gh, "Tehran driving cycle development using the k-means clustering method," *Scientia Iranica*, vol. 20, no. 2, pp. 286–293, 2013.

[20] S. H. Kamble, T. V. Mathew, and G. K. Sharma, "Development of real-world driving cycle: case study of Pune, India," *Transportation Research Part D: Transport and Environment*, vol. 14, no. 2, pp. 132–140, 2009.

[21] M. Knez, T. Muneer, B. Jereb, and K. Cullinane, "The estimation of a driving cycle for Celje and a comparison to other European cities," *Sustainable Cities and Society*, vol. 11, pp. 56–60, 2014.

[22] M. Fellah, G. Singh, A. Rousseau, S. Pagerit, E. Nam, and G. Hoffman, "Impact of real-world drive cycles on PHEV battery requirements," *SAE Technical Papers*, 2009.

[23] E. Tate, M. O. Harpster, and P. J. Savagian, "The electrification of the automobile: from conventional hybrid, to plug-in hybrids, to extended-range electric vehicles," *Sae International Journal of Passenger Cars Electronic & Electrical Systems*, vol. 1, no. 1, pp. 156–166, 2009.

[24] L. Berzi, M. Delogu, and M. Pierini, "Development of driving cycles for electric vehicles in the context of the city of Florence," *Transportation Research Part D: Transport and Environment*, vol. 47, pp. 299–322, 2016.

[25] G. Wager, J. Whale, and T. Braunl, "Driving electric vehicles at highway speeds: The effect of higher driving speeds on energy consumption and driving range for electric vehicles in Australia," *Renewable & Sustainable Energy Reviews*, vol. 63, pp. 158–165, 2016.

[26] H. He, J. Guo, N. Zhou, C. Sun, and J. Peng, "Freeway driving cycle construction based on real-time traffic information and global optimal energy management for plug-in hybrid electric vehicles," *Energies*, vol. 10, no. 11, 2017.

[27] A. Esteves-Booth, T. Muneer, H. Kirby, J. Kubie, and J. Hunter, "The measurement of vehicular driving cycle within the city of Edinburgh," *Transportation Research Part D: Transport and Environment*, vol. 6, no. 3, pp. 209–220, 2001.

[28] R. Günther, T. Wenzel, M. Wegner, and R. Rettig, "Big data driven dynamic driving cycle development for busses in urban public transportation," *Transportation Research Part D Transport & Environment*, vol. 51, pp. 276–289, 2017, http://dx.doi.org/10.1016/j.trd.2017.01.009.

[29] R. Smith, S. Shahidinejad, D. Blair, and E. L. Bibeau, "Characterization of urban commuter driving profiles to optimize battery size in light-duty plug-in electric vehicles," *Transportation Research Part D: Transport and Environment*, vol. 16, no. 3, pp. 218–224, 2011.

[30] M. A. Pouresmaeili, I. Aghayan, and S. A. Taghizadeh, "Development of Mashhad driving cycle for passenger car to model vehicle exhaust emissions calibrated using on-board measurements," *Sustainable Cities and Society*, vol. 36, pp. 12–20, 2018.

[31] Q. Wang, H. Huo, K. He, Z. Yao, and Q. Zhang, "Characterization of vehicle driving patterns and development of driving cycles in Chinese cities," *Transportation Research Part D: Transport and Environment*, vol. 13, no. 5, pp. 289–297, 2008.

[32] H. Achour and A. G. Olabi, "Driving cycle developments and their impacts on energy consumption of transportation," in *Proceedings of the International Conference on Sustainable Energy & Environmental Protection*, vol. 112, pp. 1778–1788, 2012.

[33] U. Galgamuwa, L. Perera, and S. Bandara, "Development of a driving cycle for Colombo, Sri Lanka: an economical approach for developing countries," *Journal of Advanced Transportation*, vol. 50, no. 7, pp. 1520–1530, 2016.

[34] Z. Yu, S. Shi, S. Wang et al., "Statistical inference-based research on sampling time of vehicle driving cycle experiments," *Transportation Research Part D: Transport and Environment*, vol. 54, pp. 114–141, 2017.

[35] H. Y. Tong and W. T. Hung, "A framework for developing driving cycles with on-road driving data," *Transport Reviews*, vol. 30, no. 5, pp. 589–615, 2010.

[36] N. H. Arun, S. Mahesh, G. Ramadurai, and S. M. S. Nagendra, "Development of driving cycles for passenger cars and motorcycles in Chennai, India," *Sustainable Cities and Society*, vol. 32, pp. 508–512, 2017.

[37] S.-H. Ho, Y.-D. Wong, and V. W.-C. Chang, "Developing Singapore Driving Cycle for passenger cars to estimate fuel consumption and vehicular emissions," *Atmospheric Environment*, vol. 97, pp. 353–362, 2014.

[38] J. I. Huertas, J. Díaz, D. Cordero, and K. Cedillo, "A new methodology to determine typical driving cycles for the design

of vehicles power trains," *International Journal on Interactive Design and Manufacturing*, vol. 12, no. 1, pp. 319–326, 2018.

[39] J. D. K. Bishop, C. J. Axon, and M. D. McCulloch, "A robust, data-driven methodology for real-world driving cycle development," *Transportation Research Part D: Transport and Environment*, vol. 17, no. 5, pp. 389–397, 2012.

[40] P. Seers, G. Nachin, and M. Glaus, "Development of two driving cycles for utility vehicles," *Transportation Research Part D: Transport and Environment*, vol. 41, pp. 377–385, 2015.

[41] K. Pearson, "Principal components analysis," *The London, Edinburgh, and Dublin Philosophical Magazine and Journal of Science*, vol. 6, no. 2, p. 559, 1901.

[42] Y. Hua, H. Jiang, H. Tian, X. Xu, and L. Chen, "A comparative study of clustering analysis method for driver's steering intention classification and identification under different typical conditions," *Applied Sciences*, vol. 7, no. 10, p. 1014, 2017.

[43] P. Yuhui, Z. Yuan, and Y. Huibao, "Development of a representative driving cycle for urban buses based on the K-means cluster method," *Cluster Computing*, pp. 1–10, 2018.

[44] W. M. Rand, "Objective criteria for the evaluation of clustering methods," *Journal of the American Statistical Association*, vol. 66, no. 336, pp. 846–850, 1971.

[45] G. S. Davidson, B. N. Wylie, and K. W. Boyack, "Cluster stability and the use of noise in interpretation of clustering," in *Proceedings of the IEEE Symposium on Information Visualization 2001 (INFOVIS)*, pp. 23–30, USA, October 2001.

[46] S. Ben-David and U. Von Luxburg, "Relating clustering stability to properties of cluster boundaries," in *Proceedings of the 21st Annual Conference on Learning Theory, COLT 2008*, pp. 379–390, Finland, July 2008.

[47] U. Von Luxburg, "Clustering stability: An overview," *Foundations and Trends in Machine Learning*, vol. 2, no. 3, pp. 129–168, 2009.

[48] S. Ben-David, D. Pál, and H. U. Simon, "Stability of k-means clustering," in *Proceedings of the Conference on Learning Theory*, vol. 4539, pp. 20–34, 2007.

[49] C. Cortes and V. Vapnik, "Support-vector networks," *Machine Learning*, vol. 20, no. 3, pp. 273–297, 1995.

[50] V. N. Vapnik, *The Nature of Statistical Learning Theory*, Springer, 1995.

[51] A. Rakhlin and A. Caponnetto, "Stability of K-means clustering," in *Proceedings of the 20th Annual Conference on Neural Information Processing Systems, NIPS 2006*, pp. 1121–1128, Canada, December 2006.

[52] M. Tutuianu, P. Bonnel, B. Ciuffo et al., "Development of the World-wide harmonized Light duty Test Cycle (WLTC) and a possible pathway for its introduction in the European legislation," *Transportation Research Part D: Transport and Environment*, vol. 40, pp. 61–75, 2015.

[53] X. Zhao, Q. Yu, M. Yu, and Z. Tang, "Research on an equal power allocation electronic differential system for electric vehicle with dual-wheeled-motor front drive based on a wavelet controller," *Advances in Mechanical Engineering*, vol. 10, no. 2, 2018.

[54] X. Zhao, S. Xu, Y. Ye, M. Yu, and G. Wang, "Composite braking AMT shift strategy for extended-range heavy commercial electric vehicle based on LHMM/ANFIS braking intention identification," *Cluster Computing*, pp. 1–16, 2018.

12

Urban Traffic Noise Maps under 3D Complex Building Environments on a Supercomputer

Ming Cai ,[1,2,3] Yifan Yao ,[1,2,3] and Haibo Wang [1,2,3,4]

[1]*School of Intelligent Systems Engineering, Sun Yat-sen University, Shenzhen, China*
[2]*Guangdong Provincial Key Laboratory of Intelligent Transportation System, Guangzhou, China*
[3]*Guangdong Provincial Engineering Research Center for Traffic Environmental Monitoring and Control, Guangzhou, China*
[4]*School of Civil and Transportation Engineering, Hebei University of Technology, Tianjin, China*

Correspondence should be addressed to Haibo Wang; wanghb9@mail.sysu.edu.cn

Academic Editor: Konstantinos Vogiatzis

The complexity of the 3D buildings and road networks gives the simulation of urban noise difficulty and significance. To solve the problem of computing complexity, a systematic methodology for computing urban traffic noise maps under 3D complex building environments is presented on a supercomputer. A parallel algorithm focused on controlling the compute nodes of the supercomputer is designed. Moreover, a rendering method is provided to visualize the noise map. In addition, a strategy for obtaining a real-time dynamic noise map is elaborated. Two efficiency experiments are implemented. One experiment involves comparing the expansibility of the parallel algorithm with various numbers of compute nodes and various computing scales to determine the expansibility. With an increase in the number of compute nodes, the computing time increases linearly, and an increased computing scale leads to computing efficiency increases. The other experiment is a comparison of the computing speed between a supercomputer and a normal computer; the computing node of Tianhe-2 is found to be six times faster than that of a normal computer. Finally, the traffic noise suppression effect of buildings is analyzed. It is found that the building groups have obvious shielding effect on traffic noise.

1. Introduction

With rapid growth of population and the continued expansion of transportation systems, traffic noise pollution becomes quite a nuisance to urban residents. In Europe, the recent Environmental Noise Directive (END) revision has updated the current situation on the application of the END [1] and noise pollution continues to be a major health problem. The growing amount of noise pollution can lead to a number of serious diseases, including sleep disturbance [2, 3], stroke [4], male infertility [5], and learning impairment [6]. Moreover, long-term exposure to traffic noise can easily lead to high blood pressure [7] and affect the quality of life in the neighborhood [8].

Since END 's prescription for noise maps and action plans, there have been many efforts by the scientific community to conduct high-level study and propose new mitigation systems for the main sources of noise in urban areas: road traffic, railway traffic, airport, and industrial, ranging from regulations to operational approaches [9–11]. A noise map is an important tool for observing and controlling noise pollution and the basis for epidemiology study relating annoyance to noise and noise exposure in urban areas. In urban planning, a noise map is used to analyze the sound qualities of the soundscapes in a specific urban area to generate recommendations for the urban design of the soundscapes [12]. For example, Palma de Mallorca (Spain) conducted a study analyzing various noise mitigation measures, which consider not only the reduction of noise and the number of people who can benefit from these measures but also the net monetary benefits generated, by using a traffic noise map [13]. Another study proposed a method for sorting road stretches by priority by using a noise map. The method is based on the so-called "road stretch priority index", in which the index involves a number of variables that are weighted according to their influence on the road traffic noise problem, and road stretches of different priorities thus require different plan actions [14]. Similarly, Daniel Naish proposed a regional road traffic noise management strategy (RRTNMS); in this strategy, the road was ranked according to the predicted sound pressure level, and different control measures were implemented according to the results of the ranking [15]. In both studies, noise maps play a key role in the prediction of traffic noise and the display of final result. Furthermore, noise impact indicators and the mean of the expected individual annoyance scores in the population are calculated on the basis of a noise map. Those indicators include the percentages of people being highly annoyed, annoyed, and slightly annoyed and are used to compare the effect of noise reduction [16].

The research on noise maps started with the publication of END; since then, many scholars have researched noise maps, covering the calculating model and the update method for a noise map [17–23]. The calculation of a noise map is a complex process that involves a traffic noise prediction model and sound propagation attenuation. Nicolas Fortin et al. implemented a noise prediction method within the Orbis-GIS2 software [24]; the method can produce large noise maps on a personal computer in a few hours, but the noise map is two-dimensional rather than three-dimensional. A small-scale 3D noise map can be implemented well by combining the prediction model and measured data to obtain a high-precision noise map [25]. However, since the complexity of the noise map calculation is the square of the computing scale which includes receiver point's density and map size, a large-scale 3D noise map requires an excessive amount of time when using a normal computer with a CPU containing no more than 16 computing cores (the normal computer information used in the experiments in this article is described in Section 9). A study in 2009 considered computing a noise map on a supercomputer; in this effort, software called Noise Propagation Model is used to obtain the noise map in a supercomputer [26]; the parallel algorithm assigns tasks at one time rather than using dynamic assignment, and the noise map is two-dimensional. In addition, a traffic noise map was computed in the cloud computing environment [27]; however, in that study, the noise map produced was an existing service in the cloud computing system, and the computational framework was not explained.

In the paper, a method for computing a large-scale noise map in three dimensions on a supercomputer is presented to solve the problem of time-consuming calculation of dynamic three-dimensional noise maps. The supercomputer used in this study is Tianhe-2, and a computation strategy using Tianhe-2 is introduced and its performance is analyzed. The entire noise map calculation process includes a noise prediction model, a parallel calculation algorithm, and a visualization scheme of noise map. Then, combined with taxi trajectory data, a real-time noise map implementation method was proposed to dynamically update the noise map of the entire city. In addition, two experiments were carried out to analyze the efficiency of the parallel algorithm, one is about the expansibility analysis of the noise map parallel computing algorithm proposed in this paper, and the other is the analysis of the computing efficiency of Tianhe-2. Finally, we analyze the suppression effect on traffic noise of buildings from the changes in noise distribution on the ground and building surface caused by building groups.

2. Traffic Noise Prediction among Building Groups

Noise map is composed of noise receiver points and it reflects the noise distribution in an urban area. Vehicles running on the road generate noise that propagates across buildings in a city before reaching the receiver point. Therefore, a model for predicting traffic noise has three steps: the first step is to predict the acoustical power of road, the second step is to predict sound attenuation caused by the building group in an urban area, and the last step is to combine all of the road's acoustic power with the sound attenuation caused by the building group to calculate the noise value of the noise receiver point [28].

The factors affecting road noise emissions include traffic volume, road speed, and road surface materials. Among them, road surface materials are inherent attributes of roads, and good road surface materials can effectively reduce road noise emissions [29, 30].This paper is based on China's road environment. China's roads are basically asphalt pavements, so the impact of pavement materials as a fixed item is not reflected in the model parameters. As a result, the acoustical power of road is calculated using (1) and (2):

$$L_{eqi} = L_{oi} + 10\lg\frac{N_i}{TV_i} + 10\lg\left(\frac{r_0}{r}\right) + 10\lg\left(\frac{\psi}{\pi}\right) - 16 \quad (1)$$

$$L_{eq} = 10\lg\left(10^{0.1L_{eqs}} + 10^{0.1L_{eqm}} + 10^{0.1L_{eql}}\right) \quad (2)$$

where L_{eqi} is equivalent sound level of the ith type traffic flow to the receiver point without shelter, L_{eq} is Equivalent sound level of the road to the receiver point without shelter, L_{oi} is average sound level of the ith type traffic flow to the reference point, N_i is traffic flow volume of the ith type traffic flow, V_i is average speed of the ith type traffic flow, T is the prediction time, r_0 is distance between the reference point and the lane ($r_0 = 7.5m$), r is distance between the receiver point and the lane, and ψ is the field angle formed by the receiver point, the origin point of the road, and the end point of the road, with the peak of the angle being the receiver point.

In the model, the sound attenuation caused by the building group has two parts:

$$A_{building} = A_{building1} + A_{building2} \quad (3)$$

$A_{building1}$ denotes the immediate shelter effect on the propagation path of sound caused by building group, and the relevant prediction equation is

$$A_{building1} = 0.1Cd_b \quad (4)$$

where C is the density of the building group on the propagation path of sound and d_b is the length of the sound propagation path.

$A_{building2}$ is the shelter effect of sound diffusion caused by the building group, and the relevant prediction equation is

$$A_{building2} = -10\lg(1-p) \quad (5)$$

where p is the ratio of the projection length on the road of buildings near the road divided by the road length.

In total, the noise value of the receiver point should combine all roads and the attenuation of those roads:

$$L_A = 10 \lg \sum_k 10^{0.1(L_{eq}^k - A_{building}^k)} \quad (6)$$

where L_A is the equivalent sound level of receiver point A, L_{eq}^k is the equivalent sound level of road k to a receiver point without shelter, and $A_{building}^k$ is the sound attenuation caused by the building group between road k and the receiver point.

3. Parallel Algorithm in a Supercomputer

Urban areas contain a large number of buildings with complicated layouts and irregular outlines, which increases the computation complexity of a noise map, especially a three-dimensional one. However, simplifying the prediction model would cause the noise map to have low prediction accuracy. In this condition, a supercomputer is applied to compute a large-scale 3D noise map. In this study, Tianhe-2 is used to compute a noise map in three dimensions. Tianhe-2 was announced in June 2013, with Intel Xeon Phi accelerators contributing more than 85% of the 55pflops peak performance [31]. The strong computing ability of Tianhe-2 supports the computation of a large-scale 3D noise map on the basis of its hardware; thus, it is critical to have an appropriate and efficient parallel algorithm. The parallel algorithm presented here can be divided into two sections: one is the map tiling, which is the structural basis of parallel computation, and the other section is the coordination of the compute nodes, which implements the computation of noise map.

3.1. Map Tiling. When the distance between the sound source and the receiver point is sufficiently large, the impact of the sound source to the receiver point is so small that it can be ignored. Therefore, breaking up the noise map into a tiling map composed of many smaller rectangle blocks is a reasonable approach for calculating the noise.

The map is tiled into a set of rectangle blocks that are 200 meters in length and 200 meters in width. In each block, the distance between noise receiver points is always set as 4m, so there are always 2500 receiver points in a block. While computing the noise of a block, the roads and buildings in a wider area are brought into such a state that each receiver point in the block can receive all impacts of the roads near it. Range of the area is determined by the number of roads inside it. As shown in Figure 1, when computing block 0, the roads and buildings in the grey area are also brought into the computation. The initial size of the grey area is $600m \times 600m$, and it gradually expands until the number of roads inside it is larger than a threshold θ.

3.2. Coordination of the Compute Nodes. Because a supercomputer is composed of a large number of compute nodes and the map is tiled into a set of blocks, one compute node can compute one block at a time. As a result, N blocks can be computed simultaneously if N compute nodes are applied to compute the noise map. The critical point lies in coordinating those compute nodes. In the algorithm, one node is specifically used to control the computation of the other nodes. In the control node, a task list is running to assign tasks to the compute nodes. The control node sends the task number to the compute node, which represents the block index that the compute node is supposed to compute, and the compute node sends the end flag to the control node when the task is finished; the process is shown in Figure 2.

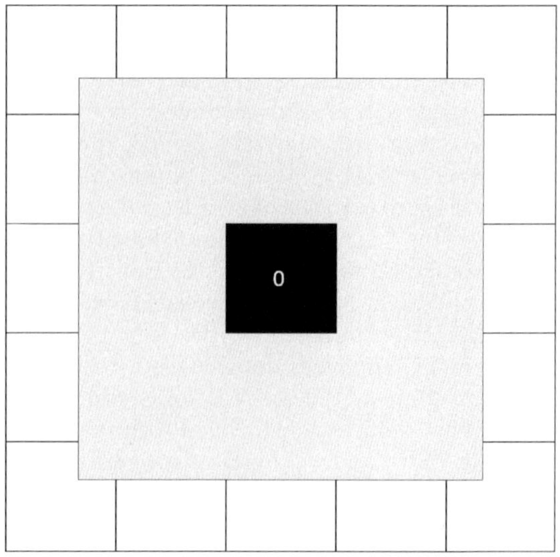

FIGURE 1: Take eight neighborhoods of block 0 into the computation.

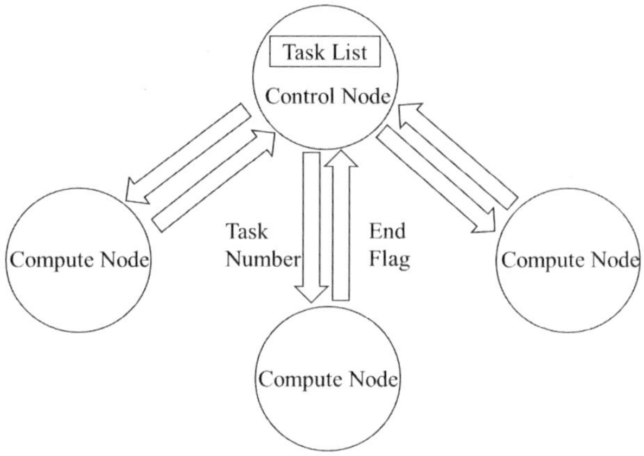

FIGURE 2: The process of parallel control on a supercomputer.

To maximize the computational efficiency and reduce the idle time of compute node, the control logic of control node is supposed to be dynamic. This process requires that whenever a compute node completes its task, the control node assigns a new task for it. The control logic is shown in Figure 3. First, the control node assigns the first task for each compute node, and the control node turns into the loop state. In the loop state, the control node checks whether any compute node has completed its computations and then sends an end flag to that compute node. Whenever the control node finds an idle compute node, it sends a new task to the idle compute node until all tasks are completed.

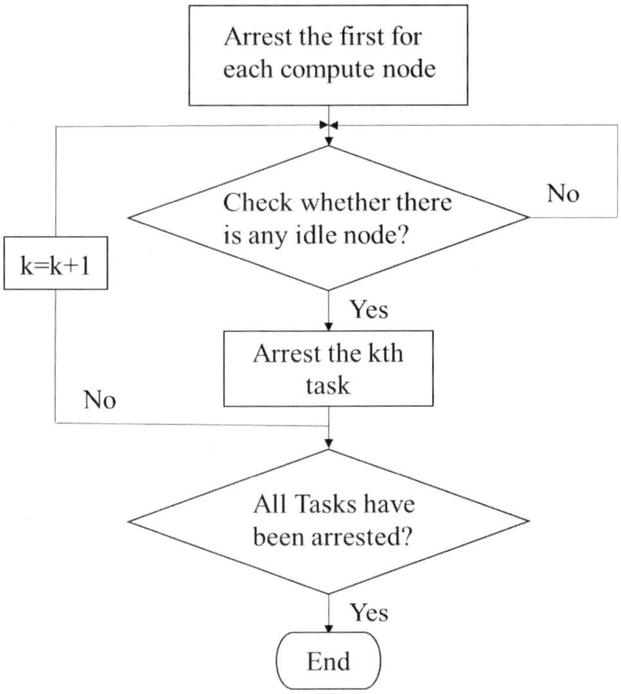

FIGURE 3: Control logic of the control node on a supercomputer.

4. Rendering a 3D Noise Map

After computing a noise map on the supercomputer, the noise data must be rendered to visualize the result. It is difficult to render a large-scale noise map without any simplification process. One of the most convenient and rapid methods of simplification is to partly render the noise map according to the current zoom level. The basis of rendering the noise map is the conversion of a noise value into a color value.

4.1. Render in Grading. The zoom level determines the range of noise map to be rendered; thus, the noise receiver points required to render the noise map can be selected at a certain distance, and not all of the receiver points must be used in the rendering if the zoom level is small and the rendering range is large. The distance depends on the current zoom level:

$$\delta = k\lambda \tag{7}$$

where δ is the distance between the noise receiver points selected to render the noise map, k is the proportionality coefficient, and λ is the current zoom level.

In addition, the receiver point includes the point of a building face; thus, the rendering of the noise distribution in buildings is required. The rendering of the noise distribution in buildings depends on the zoom level as well, and buildings are rendered when the zoom level is greater than a given threshold:

$$\lambda \geq \lambda_0 \tag{8}$$

where λ_0 is the threshold of the zoom level.

4.2. Noise Conversion to Color. The noise map is composed of a variety of receiver points. Each receiver point has a noise value after computing, and visualization requires the conversion from noise to color. Color is expressed by an RGB vector: a noise value larger than 80 dB corresponds to red, whose RGB vector is $(1, 0, 0)$, a noise value less than 40 dB corresponds to green, whose RGB vector is $(0, 1, 0)$, and a noise value between 40 dB and 80 dB is calculated as

$$RGB = \left(\frac{(Noise - 40)}{40}, \frac{(80 - Noise)}{40}, 0 \right) \tag{9}$$

where *Noise* is the noise value of receiver point.

4.3. Display of the Rendering of Noise Map. The traffic noise rendering result is displayed in Figure 4. Figure 4(a) is at a small zoom level that is not in excess of the threshold, so the building is not rendered. Figure 4(b) is at a zoom level that is in excess of the threshold, so the building is rendered. Figure 4(c) shows the noise distribution on the building face in detail. The green area indicates that the area is very quiet, the red area indicates that the level of noise pollution is serious, and the yellow area indicates that the area is at a normal noise level.

5. Strategy for Obtaining a Dynamic Noise Map

The dynamic noise map can reflect the real-time noise pollution and is an extremely effective tool for noise monitoring and implementing a short-term noise prevention plan. Dynamap project has put forward the aim for realizing a real-time noise map, it focuses on the research of a technical solution able to ease and reduce the cost of noise map through an automatic monitoring system [32]. Another project called Noise Tube proposes measuring and mapping noise pollution through public mobile phones [33]. This section proposes a method for drawing dynamic noise maps based on taxi trajectory data and supercomputers to effectively solve the problems of acquiring real-time traffic data and excessive time-consuming when computing noise maps.

5.1. Real-Time Traffic Data. There are many taxis in every city, the tracking information of which is owned by those taxi enterprises. This information is available in real-time. In this study, the traffic data used to predict noise are calculated from the tracking information of almost 20000 taxis in Guangzhou. The instantaneous speed of the taxi is inferred from the real-time location information of the taxi; then, the speed of each taxi on the same road over one hour is averaged to obtain the hourly average speed of the road. Studies have shown that road traffic volume and average road speed have different degrees of influence on traffic noise, and there is a clear correlation between them [34].To obtain the hourly traffic volume, large amounts of traffic monitor video are used to determine the pattern between the road speed and the road traffic volume for different road types, and the hour traffic volume is determined from the hourly average speed. Figure 5 shows the taxi tracking data in Guangzhou, as visualized by

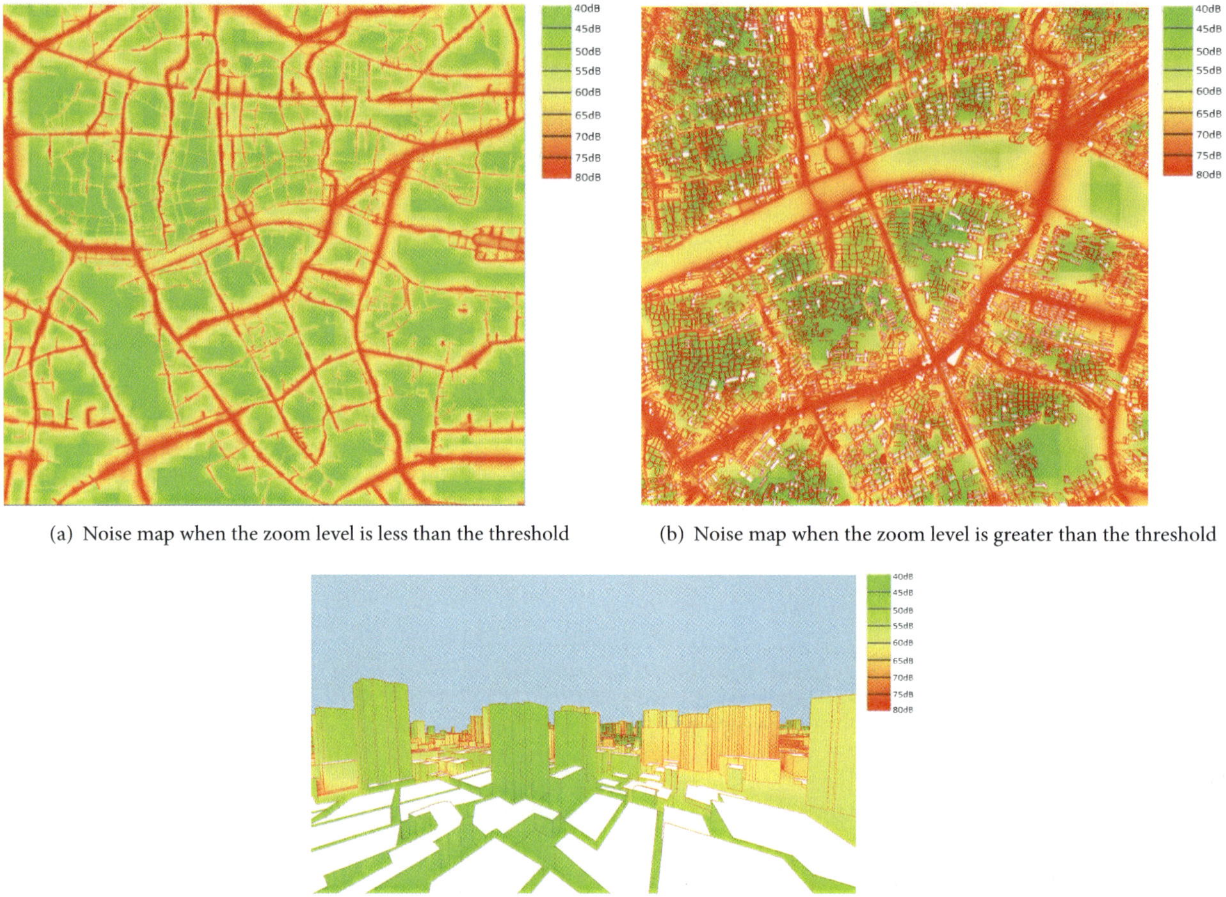

(a) Noise map when the zoom level is less than the threshold

(b) Noise map when the zoom level is greater than the threshold

(c) Noise distribution on the building faces

FIGURE 4: 3D noise map calculated by a supercomputer.

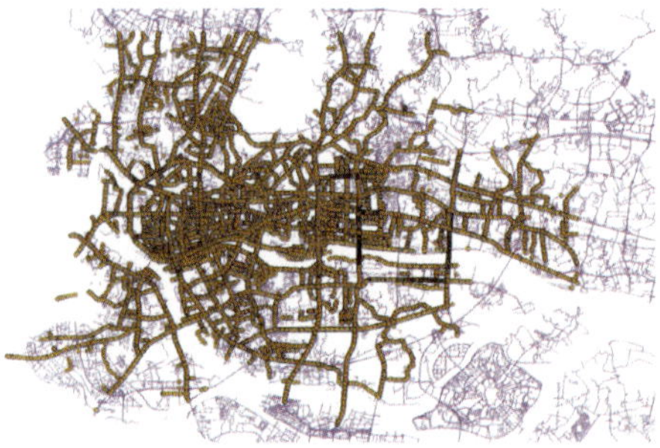

FIGURE 5: Real-time taxi tracking data in Guangzhou.

a GIS tool. Through this method, real-time traffic data does not need to be obtained through all-day traffic monitoring, and traffic data of the entire city can be obtained; accordingly, this method is not precise enough to obtain traffic data with high accuracy.

5.2. Offline Computing. Computing a traffic noise map for the main urban area of Guangzhou on Tianhe-2 while using 300 compute nodes still requires 22 hours, so an optimization measure is still required to increase the computational speed and obtain a dynamic traffic noise map. It is unreasonable to use more compute nodes to realize the dynamic noise map because doing so would be too expensive. Fortunately, considering the computing process, the sound attenuation caused by the building group $A_{building}$ is correlated only to the position of the road and buildings according to (4) and (5). Therefore, offline computing is possible for $A_{building}$ [35].

In the first computation of noise map, $A_{building}$ is calculated using (4) and (5), and the result is saved in the database. In the following calculation of the noise map, $A_{building}$ is read from the database rather than calculated using (4) and (5). The offline computing of the sound attenuation caused by the building group improves the computational efficiency vastly, requiring only approximately 222 seconds to compute the traffic noise map in the main urban area of Guangzhou on Tianhe-2 when using 300 compute nodes; thus, there is enough surplus time to generate traffic data from the taxi data.

6. Efficiency Analysis

The computation of a large-scale 3D traffic noise map is quite vast, and the computing time is determined by the computing efficiency. To evaluate whether the methodology in the paper is feasible, the efficiency is analyzed in this section. The efficiency analysis is divided into two parts: the algorithm and the hardware. The first part is the expansibility of the algorithm, and the second part is the computing efficiency of the supercomputer.

6.1. Expansibility of the Algorithm. The expansibility of algorithm refers to the variation of algorithm's efficiency with a change in the computing resources and the scale of computation. In the experiment, the efficiency of the algorithm is indicated by the computing time, the computing resource is indicated by the number of compute nodes, and the scale of computation is indicated by the step length of the receiver points (in the same map, the smaller the step length is, the greater the number of receiver points is).

The computing time of a noise map is composed of two parts: one is the time for initialization, which lapses at the beginning of the operation of each node, and the other part is the time for computing the blocks. Because the parallel algorithm allows the compute node to be running constantly until all tasks are completed, the computing time of a noise map is similar to the running time of each compute node. From the discussion above,

$$N_{node} \times t_{initial} + N_{block} \times t_{block} = T_{actual} \times N_{node} \quad (10)$$

After simplification,

$$t_{initial} + \left(\frac{N_{block}}{N_{node}}\right) \times t_{block} = T \quad (11)$$

where $t_{initial}$ is the time for initialization of each compute node, t_{block} is the time for computing each block, N_{node} is the number of compute nodes, N_{block} is the number of blocks, and T_{actual} is actual computing time of the noise map.

The noise map used to measure the expansibility of the algorithm contains 127 road segments and 102 buildings, and the area is $1332m \times 1785m$; thus, the noise map is tiled into 7×9 blocks. The number of compute nodes and the step lengths of receiver point are changed to measure the expansibility of the algorithm, as indicated in Table 1. In addition, the threshold θ in map tiling is set as $9 \times N_{block}/N_{node}$.

$t_{initial}$ and t_{block} in (11) are estimated from Table 1 using the least squares method, and when different numbers of compute nodes are used, $t_{initial}$ and t_{block} have different values. The fitting results using the least squares method are displayed in Figure 6, where *Ratio* represents N_{block}/N_{node}.

In Figure 6, the abscissa is N_{block}/N_{node}, and the ordinate is the computing time T_{actual}; the step length of the receiver point varies from 10 m to 2 m. In the five regression formulas, the slope represents time for computing each block t_{block}, and the intercept represents the time required for the initialization of each compute node $t_{initial}$.

Equation (11) reveals that the computing time is composed of two parts: the initial time $t_{initial}$ and the average computing time of each block t_{block}. A specific value between the times is correlated to the efficiency of the algorithm, as shown in Table 2 for a computation using 3 compute nodes.

The following three conclusions can be drawn from the experiment implemented above:

TABLE 1: Expansibility of algorithm.

Step length between receiver points /m	Number of compute nodes	Computing time /s
10	1	201
	2	109
	3	73
	4	61
	5	53
8	1	310
	2	160
	3	113
	4	94
	5	77
6	1	538
	2	286
	3	199
	4	169
	5	140
4	1	1204
	2	635
	3	444
	4	368
	5	312
2	1	4766
	2	2568
	3	1821
	4	1480
	5	1260

TABLE 2: Specific value between t_{block} and $t_{initial}$.

Step length	t_{block} /s	$t_{initial}$ /s	$t_{block}/t_{initial}$
10	2.9688	13.988	0.212
8	4.61	18.169	0.254
6	7.8935	39.305	0.201
4	17.755	81.783	0.217
2	69.654	375.05	0.186

(a) Equation (11) is fitted quite well when using the least squares method from Figure 5; all coefficients of the association approach 1 very well, indicating that the computing time is explicitly revealed by (11).

6.2. Computing Efficiency of the Supercomputer. Nearly 22 hours is required to calculate the noise map of the main city zone in Guangzhou on supercomputer when using 300 compute nodes. Thus, an experiment to determine the computing efficiency of a supercomputer is implemented to illustrate the necessity of using a supercomputer to compute a large-scale noise map.

A normal computer with a quad-core CPU and 16 GB random-access memory is used for comparison with the

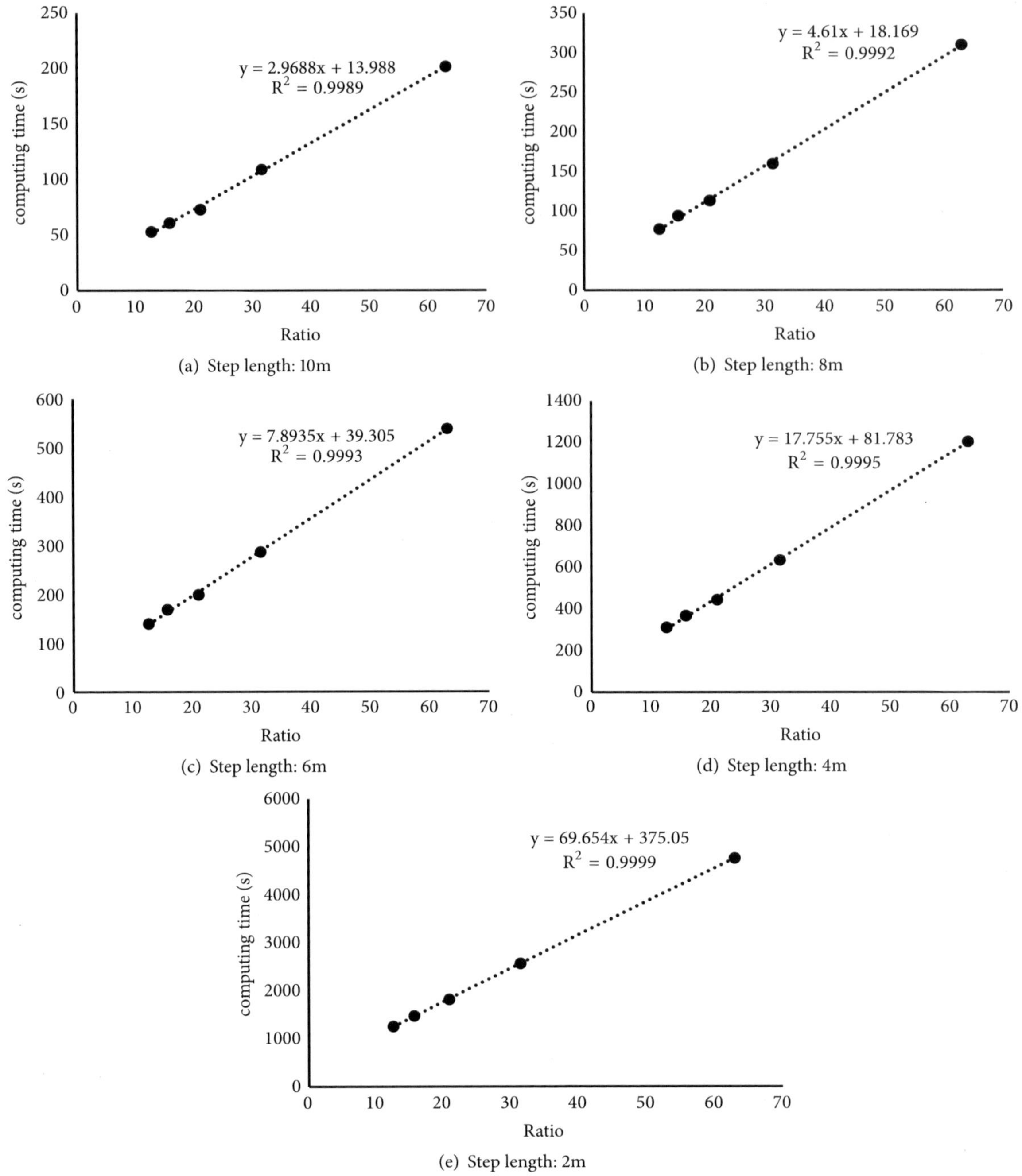

FIGURE 6: Fitting Result of N_{block}/N_{node} and T_{actual}.

supercomputer, and the map used in the experiment is the same map that was used in the expansibility experiment. The map is computed using one, two, and three compute nodes in the normal computer and in the supercomputer, and the result is shown in Table 3.

From Table 3, it is evident that the supercomputer is six times faster than a normal computer in terms of computation speed. As a result, it takes more than 132000 hours to calculate the same noise map of the main city zone in Guangzhou on a normal computer with a quad-core CPU because one core is used to control the other nodes. Thus, a supercomputer must be used to compute a large-scale noise map.

7. Inhibitory Effect of Building Groups on Traffic Noise

Building groups are the main barriers of traffic noise in urban area; thus, a scene is chosen to reflect the variation of sound field distribution causing by building groups. The scene is shown as in Figure 7 with 1034 road segments and 406 buildings.

In order to reveal the effect of building groups, a noise map without buildings of the scene is computed; then, a difference map between noise map with buildings and noise map without buildings is rendered as shown in Figure 8.

TABLE 3: Comparison between a normal computer and a supercomputer.

Number of compute nodes	Computing time /s	
	Normal computer	Supercomputer
1	1233	201
2	670	109
3	453	73

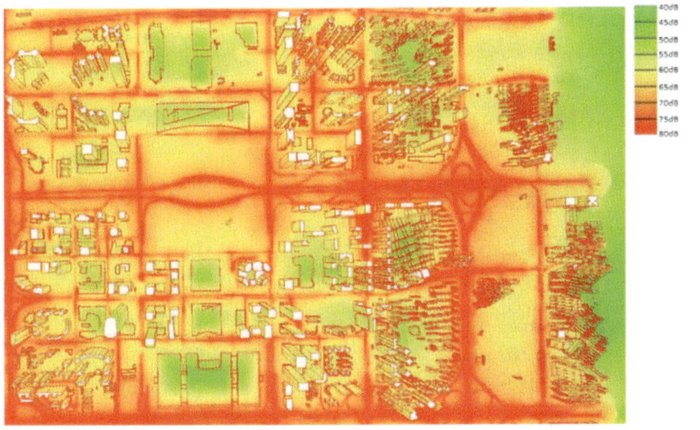

FIGURE 7: Selected scene for noise analysis.

TABLE 4: Building attenuation of traffic noise.

Point number	Noise attenuation /dB
1	26.2
2	36.5
3	32
4	33.6
5	25.6
6	32.8
7	26.1
8	35.5
9	37

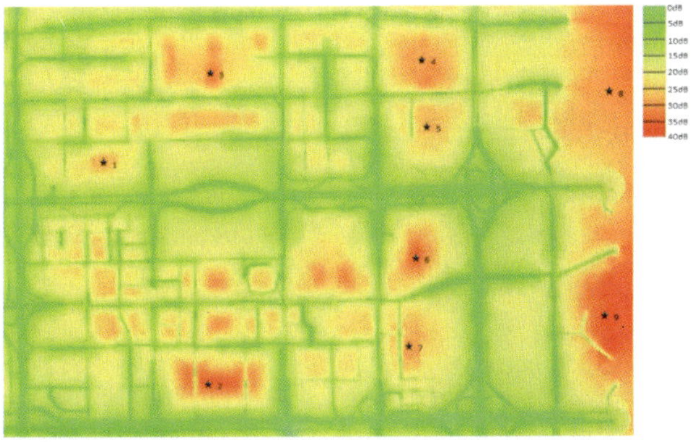

FIGURE 8: Difference map between noise map with or without buildings.

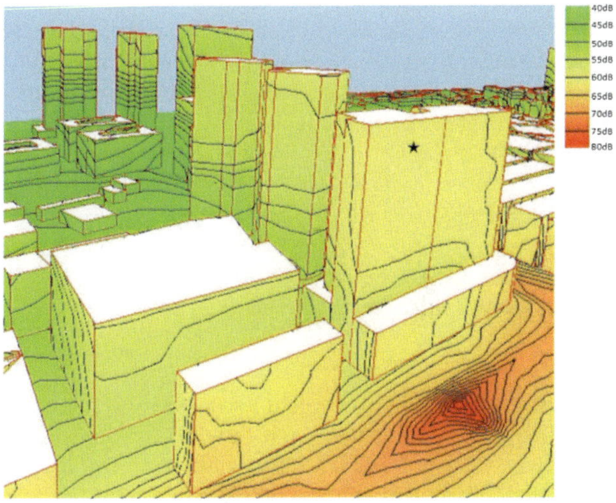

FIGURE 9: Noise distribution on the building surface.

9 points are selected to accurately show the inhibitory effect of building groups as Table 4. From Figure 8, it is obvious that building groups have significant inhibitory effect on traffic noise; the depression of noise pressure level ranges from 0dB to 40dB. In combination with Figure 7, it can be found that the depression of noise pressure level is determined by the density of building groups: the denser the buildings, the greater the noise attenuation value.

Building groups not only have an inhibitory effect on horizontal noise, but also have an inhibitory effect on high buildings. Figure 9 shows the noise distribution on the building surface and black lines are noise pressure level contours. According to the distance attenuation principle of sound, the noise level at the upper floor of a building should be less than the noise level at the lower floor; however, in Figure 9, the noise level at the upper floor is higher than that at the lower floor in many buildings; it is because the noise at lower floor of a building will be blocked by other buildings, and the noise at upper floor is not blocked.

The noise distribution on the surface of buildings marked with asterisks is shown in Table 5. Each column in Table 5 represents each floor of the building. It can be seen from Table 5 that the noise value increases rather than decreases as the height increases.

8. Conclusion

A systematic methodology for computing a large-scale 3D traffic noise map on a supercomputer was proposed in the paper. The method covers the noise prediction model and the parallel computation algorithm used on a supercomputer. It offers a highly efficient and feasible approach for computing

Table 5: Noise distribution of a building surface.

Height/m	0	7.5	15	22.5	30	37.5	45	52.5	60	67.5	75
Noise level of a floor/dB	53.36	55.73	54.02	54.94	56.31	57.18	57.47	57.53	57.55	57.59	58.35
	52.8	53.1	53.41	54.35	55.72	56.68	56.84	56.89	56.92	57.01	57.55
	52.73	52.97	53.21	54.18	55.51	56.25	56.65	56.68	56.7	56.81	57.46
	52.76	52.97	53.19	54.33	55.62	56.39	56.78	56.79	56.81	56.92	57.31
	53.08	53.19	53.45	54.77	56.06	56.81	57.3	57.29	57.31	57.4	57.68

a large-scale 3D noise map. Then, to visualize the computing result, a rendering method is developed to display the rendering result of the traffic noise pollution in an urban city. Moreover, the strategy of obtaining a dynamic noise map was presented and can generate a real-time noise map in theory.

Two experiments were implemented to analyze the efficiency. The first one considered the expansibility of the parallel algorithm. When the computing scale is fixed, the computing time is positively correlated with the ratio between the number of computing blocks and the number of compute nodes: with an increase in the number of compute nodes, the computing time decreases linearly. When the computing node is fixed, with an increase in the computing scale, the ratio between the computing time and the time required for initialization remains about the same; i.e., the algorithm achieves high usage of the compute nodes. The second experiment aimed to measure the computational speed of a supercomputer. The result showed that supercomputer Tianhe-2 is six times faster than a normal computer in terms of computing speed. Therefore, it is necessary to use a supercomputer to compute a large-scale noise map.

In addition, a scenario was chosen to analyze the impact of building groups on traffic noise. The results showed that buildings have a significant inhibitory effect on traffic noise, and due to the sheltering effect of buildings, the noise pressure level at the lower floor is lower than the upper floor in some buildings.

9. Software Availability

All software used in the paper can be requested from the corresponding author by email and will be provided free of charge. Program for the parallel computing of noise map is coded in C++ and applied on supercomputer or computer with multicore and application for visualization of the computing result is coded in C#. The supercomputer used in the paper is Tianhe-2, and a normal computer with a quad-core CPU (Inter Core i5. 3.30GHZ) and 16 GB of RAM is used for comparison with the supercomputer in Section 6.2. The dataset for testing the parallel algorithm can be provided in XML file and dataset for visualization noise map can be provided in text files. All files are available from the corresponding author too.

Conflicts of Interest

The authors declare that they have no conflicts of interest.

Acknowledgments

This work is supported by the National Natural Science Foundation of China (no. 11574407), the Science and Technology Planning Project of Guangzhou City, China (no. 201704020142), and the Joint Foundation for Super-Computing Science Applied of NSFC and Guangdong (no. 2015_515).

References

[1] E. U. Directive, “Directive 2002/49/EC of the European parliament and the Council of 25 June 2002 relating to the assessment and management of environmental noise,” *Journal of the European Communities, L*, vol. 189, no. 12, pp. 0012–0026, 2002.

[2] M. G. Smith, I. Croy, M. Ögren, O. Hammar, E. Lindberg, and K. Persson Waye, “Physiological effects of railway vibration and noise on sleep,” *The Journal of the Acoustical Society of America*, vol. 141, no. 5, pp. 3262–3269, 2017.

[3] K. V. Weyde, N. H. Krog, and B. Oftedal, “Nocturnal road traffic noise exposure and childrens sleep duration and sleep problems,” in *Proceedings of the Nocturnal road traffic noise exposure and children's sleep duration and sleep problems. International Journal of Environmental Research Public Health*, pp. 491–502, 2017.

[4] M. Sørensen, M. Hvidberg, Z. J. Andersen et al., “Road traffic noise and stroke: A prospective cohort study,” *European Heart Journal*, vol. 32, no. 6, pp. 737–744, 2011.

[5] K.-B. Min and J.-Y. Min, “Exposure to environmental noise and risk for male infertility: A population-based cohort study,” *Environmental Pollution*, vol. 226, pp. 118–124, 2017.

[6] P. H. T. Zannin and D. P. Z. Zwirtes, “Evaluation of the acoustic performance of classrooms in public schools,” *Applied Acoustics*, vol. 70, no. 4, pp. 626–635, 2009.

[7] W. Babisch, W. Swart, D. Houthuijs et al., “Exposure modifiers of the relationships of transportation noise with high blood pressure and noise annoyance,” *The Journal of the Acoustical Society of America*, vol. 132, no. 6, pp. 3788–3808, 2012.

[8] D. Botteldooren, L. Dekoninck, and D. Gillis, “The influence of traffic noise on appreciation of the living quality of a neighborhood,” *International Journal of Environmental Research and Public Health*, vol. 8, no. 3, pp. 777–798, 2011.

[9] R. Zhang, Z. He, H. Wang, F. You, and K. Li, “Study on selftuning tyre friction control for developing main-servo loop integrated chassis control system,” *IEEE Access*, vol. 5, pp. 6649–6660, 2017.

[10] Y. Liu, B. Xia, C. Cui, and M. Skitmore, "Community response to construction noise in three central cities of Zhejiang province, China," *Environmental Pollution*, vol. 230, pp. 1009–1017, 2017.

[11] P. Gagliardi, L. Fredianelli, D. Simonetti, and G. Licitra, "ADS-B system as a useful tool for testing and redrawing noise management strategies at Pisa Airport," *Acta Acustica united with Acustica*, vol. 103, no. 4, pp. 543–551, 2017.

[12] K. Vogiatzis and N. Remy, "Soundscape design guidelines through noise mapping methodologies: An application to medium urban agglomerations," *Noise Mapping*, vol. 4, no. 1, pp. 1–19, 2017.

[13] M. Ausejo, M. Tabacchi, M. Recuero, C. Asensio, R. Pagán, and I. Pavón, "Design of a noise action plan based on a road traffic noise map," *Acta Acustica united with Acustica*, vol. 97, no. 3, pp. 492–502, 2011.

[14] A. Ruiz-Padillo, A. J. Torija, Á. Ramos-Ridao, and D. P. Ruiz, "A methodology for classification by priority for action: Selecting road stretches for network noise action plans," *Transportation Research Part D: Transport and Environment*, vol. 29, pp. 66–78, 2014.

[15] D. Naish, "A method of developing regional road traffic noise management strategies," *Applied Acoustics*, vol. 71, no. 7, pp. 640–652, 2010.

[16] H. C. Borst and H. M. E. Miedema, "Comparison of noise impact indicators, calculated on the basis of noise maps of DENL," *Acta Acustica united with Acustica*, vol. 91, no. 2, pp. 378–385, 2005.

[17] M. Cai, J. Zou, J. Xie, and X. Ma, "Road traffic noise mapping in Guangzhou using GIS and GPS," *Applied Acoustics*, vol. 87, pp. 94–102, 2015.

[18] O. C. de Souza Neto and S. Paul, "Road traffic noise mapping in the Federal University of Santa Maria, Brazil," *The Journal of the Acoustical Society of America*, vol. 138, no. 3, pp. 1732-1732, 2015.

[19] M. Cai, Y. Yao, and H. Wang, "A traffic-noise-map update method based on monitoring data," *The Journal of the Acoustical Society of America*, vol. 141, no. 4, pp. 2604–2610, 2017.

[20] F. Alías and J. C. Socoró, "Description of anomalous noise events for reliable dynamic traffic noise mapping in real-life urban and suburban soundscapes," *Applied Sciences (Switzerland)*, vol. 7, no. 2, 2017.

[21] D. Xie, Y. Liu, and J. Chen, "Mapping Urban environmental noise: A land use regression method," *Environmental Science & Technology*, vol. 45, no. 17, pp. 7358–7364, 2011.

[22] H. Wang, H. Chen, and M. Cai, "Evaluation of an urban traffic Noise–Exposed population based on points of interest and noise maps: The case of Guangzhou," *Environmental Pollution*, vol. 239, pp. 741–750, 2018.

[23] J. Morel, C. Marquis-Favre, and L.-A. Gille, "Noise annoyance assessment of various urban road vehicle pass-by noises in isolation and combined with industrial noise: A laboratory study," *Applied Acoustics*, vol. 101, pp. 47–57, 2016.

[24] N. Fortin, E. Bocher, and J. Picaut, "An opensource tool to build urban noise maps in a GIS," in *Proceedings of the Open Source Geospatial Research and Education Symposium (OGRS*, pp. 226–235, 2012.

[25] P. J. Sequeira Goncalves, "3D noise map of an hospital and noise sources evaluation," *The Journal of the Acoustical Society of America*, vol. 123, no. 5, pp. 3263-3263, 2008.

[26] A. Czyzewski, M. Szczodrak, and A. Czyżewski, "Software for Calculation of Noise Maps Implemented on Supercomputer," *TASK Quarterly*, vol. 13, pp. 363–377, 2009.

[27] K. Marciniuk, M. Szczodrak, and B. Kostek, "Performance of noise map service working in cloud computing environment," *Archives of Acoustics*, vol. 41, no. 2, pp. 297–302, 2016.

[28] H. Wang, M. Cai, and Y. Yao, "A modified 3D algorithm for road traffic noise attenuation calculations in large urban areas," *Journal of Environmental Management*, vol. 196, pp. 614–626, 2017.

[29] G. Licitra, M. Cerchiai, L. Teti, E. Ascari, and L. Fredianelli, "Durability and variability of the acoustical performance of rubberized road surfaces," *Applied Acoustics*, vol. 94, pp. 20–28, 2015.

[30] K.-Y. Ho, W.-T. Hung, C.-F. Ng, Y.-K. Lam, R. Leung, and E. Kam, "The effects of road surface and tyre deterioration on tyre/road noise emission," *Applied Acoustics*, vol. 74, no. 7, pp. 921–925, 2013.

[31] TOPorg, "China's Tianhe-2 is World's Top Supercomputer," *Communications of the ACM*, 2013.

[32] P. Bellucci, L. Peruzzi, and G. Zambon, "LIFE DYNAMAP project: The case study of Rome," *Applied Acoustics*, vol. 117, pp. 193–206, 2017.

[33] N. Maisonneuve, M. Stevens, M. E. Niessen, and L. Steels, "NoiseTube: Measuring and mapping noise pollution with mobile phones," *Environmental Science and Engineering*, vol. 2, no. 6, pp. 215–228, 2009.

[34] K. Vogiatzis and P. Kopelias, "Benefits and limitations toward a sustainable road environment during the years of economic recession," *International Journal of Sustainable Development and Planning*, vol. 10, no. 5, pp. 701–712, 2015.

[35] M. Cai, Z. Zhang, X. Ma, and F. Li, "Rapid updating of 3D road traffic noise maps in large cities," *Noise Control Engineering Journal*, vol. 64, no. 3, pp. 335–341, 2016.

13

Exploring Travel Time Distribution and Variability Patterns using Probe Vehicle Data

Peng Chen,[1,2] Rui Tong,[1] Guangquan Lu,[1,2] and Yunpeng Wang[1,2]

[1]*Beijing Key Laboratory for Cooperative Vehicle Infrastructure Systems and Safety Control, School of Transportation Science and Engineering, Beihang University, Xue Yuan Road No. 37, Hai Dian District, Beijing 100191, China*
[2]*Beijing Advanced Innovation Center for Big Data and Brain Computing, Beihang University, Xue Yuan Road No. 37, Hai Dian District, Beijing 100191, China*

Correspondence should be addressed to Peng Chen; cpeng@buaa.edu.cn

Academic Editor: Yair Wiseman

Exploring travel time distribution and variability patterns is essential for reliable route choices and sophisticated traffic management and control. State-of-the-art studies tend to treat different types of roads equally, which fails to provide more detailed analysis of travel time characteristics for each specific road type. In this study, based on a vast amount of probe vehicle data, 200 links inside the Third Ring Road of Beijing, China, were investigated. Four types of roads were covered including urban expressways, auxiliary roads of urban expressways, major roads, and secondary roads. The day-of-week distributions of unit distance travel time were first analyzed. Kolmogorov-Smirnov test, Anderson-Darling test, and chi-squared test were employed to test the goodness-of-fit of different distributions and the results showed lognormal distribution was best-fitted for different time periods and road types compared with normal, gamma, and Weibull distribution. In addition, four reliability measures, that is, unit distance travel time, coefficient of variation, buffer time index, and punctuality rate, were used to explore the day-of-week travel time variability patterns. The results indicated that urban expressways, auxiliary roads of urban expressways, and major roads have regular and distinct morning and afternoon peaks on weekdays. It is noteworthy that in daytime the travel times on auxiliary roads of urban expressways and major roads share similar variability patterns and appear relatively stable and reliable, while urban expressways have most reliable travel times at night. The results of analysis help enable a better understanding of the volatile travel time characteristics of each road type in urban network.

1. Introduction

Nowadays, high traffic demand and limited road capacities make people spend much more time on their daily journeys. Travel time reliability (TTR), defined as the level of consistency of travel conditions over time [1], has been increasingly considered as an important measure of the performance of transportation systems as well as travelers' perceptions of such performance. In order to comprehensively characterize TTR, travel time distribution (TTD) and variability patterns need to be explored as a prior, which is essential for reliable route choices and sophisticated traffic management and control [2].

Thanks to advanced traffic sensing technologies, various travel time related information can be collected conveniently nowadays. The technologies essentially include station-based traffic state measurement (e.g., loop detector, video camera, and microwave sensor) and point to point travel time collection (e.g., automatic vehicle identification systems, license plate recognition systems, mobile, Bluetooth, and probe vehicles). The acquisition results of station-based devices strongly depend on the spatial layout and fixed position of traffic detectors. In contrast, probe vehicles equipped with the global positioning system (GPS) could travel all over the network and record the travel time and location information of vehicles at a certain interval. These data are known as probe vehicles data, representing the relatively complete operation conditions for urban traffic. With increasing amounts of data available, there has been a surge of literature devoted to the analysis of TTR and TTD in recent years.

Urban traffic times are essentially volatile due to various influencing factors, for example, weather, road types, roadway geometry, traffic control, accident, and varying traffic demands. Martchouk et al. [3] compared the travel time data collected using Bluetooth detection during good and bad weather on freeway segments, and found that there was an increase in travel time that continues throughout the snowstorm event. A number of studies considered the possible impact of roadway geometry on the travel time of urban arterials, such as number of lanes and left turn lane types [4, 5]. Chen et al. [6] investigated travel time data on one urban arterial incorporating the effects of traffic signal timing, and the results demonstrated that traffic control has a distinct effect on urban travel time. Several studies analyzed the relationships between travel time or speed and volume-to-capacity ratio of expressways [3, 7, 8] and found that the travel time variability is greatly affected by volume-to-capacity ratio.

It is noteworthy that most studies on travel time analysis employed travel time data from only one road type [3, 9, 10] or treated different types of roads equally [5, 11], which failed to provide more detailed analysis of travel time characteristics for each specific road type. In practice, travel times of different road types are supposed to have diverse variability patterns due to different road characteristics and capacities and attractiveness for travelers. Only limited study [12] adopted the travel time data from highway and urban roads to identify the relatively coarse distinction caused by two facility types. TTR needs to be investigated at fine grained levels of spatial and temporal aggregation. Thus, based on a vast amount of probe vehicle data, this study aims to explore the TTD and variability patterns for different road types (e.g., urban expressways, auxiliary roads of urban expressways, major roads, and secondary roads) in typical time periods (e.g., peak versus off-peak, weekday versus weekend).

The remainder of this paper is organized as follows. Section 2 provides a literature review of TTD and variability patterns, followed by a detailed description of the datasets used for analysis. In Section 3, we investigate and compare the goodness-of-fit of four typical distributions to travel times for different road types in different time periods. Then, travel time variability (TTV) patterns are identified and discussed in terms of four reliability measures in Section 4. Last, summary and conclusions are given in Section 5.

2. Literature Review

2.1. Travel Time Distribution. Most existing studies on TTV put significant effort into identifying the best statistical model for fitting travel times. The fitted distributions can be categorized into three main classes, that is, single-mode distributions [9, 13–16], multimode distributions [6, 17–19], and truncated distributions [4, 20, 21].

Single-mode distributions apply one kind of standard distributions to characterize travel time, like normal, lognormal, gamma, Weibull, burr, and other distributions that are commonly used. For example, Emam and Al-Deek [15] tested the lognormal, gamma, Weibull, and exponential distributions for representing travel time data of a freeway from weekdays and found that a lognormal distribution was best-fitted. Nie et al. [22] adopted gamma distribution to model travel time on arterial and local streets during morning peak, mid-of-day, and evening peak. Lei et al. [9] compared Generalized extreme, Weibull, burr, normal, gamma, lognormal, Generalized Pareto, and other distributions for modeling travel times on urban expressways with varying levels of service and concluded that generalized extreme and Generalized Pareto are preferable fits than others. Kieu et al. [16] selected burr, gamma, lognormal, normal, and Weibull to model public transport travel time and found that lognormal can be recommended as the descriptor because of its desirable performance and attractive mathematical characteristics. Single-mode distribution is simple and the parameter(s) of the distribution is easy to estimate. Such a fitting approach is usually adopted by traffic engineers with its attractive mathematical characteristics, though the goodness-of-fit may be not quite desirable due to substantially different free flow and congested conditions.

Multimode distribution appears more accurate because more than one component will be used to model travel time states. The components can come from one or multiple standard distributions. For example, Chen et al. [6] compared different regression mixture models including the model whose all components are Gaussian distributions and the model with mixture of Gaussian and lognormal distributions to estimate urban arterial travel time. Guo et al. [18] discussed multistate versions of normal, gamma, and lognormal distributions, and the results showed that the multistate lognormal model consistently outperforms other models, especially during peak hours. Benefiting from its multistate nature, multimode distribution is capable of adapting the different characteristics of travel time states in urban roads, for example, caused by the diversity of vehicle model (e.g., cars versus buses) and signal timings (e.g., green light versus red light). Nevertheless, the proportion and parameter(s) of each component are usually determined by extra estimation methods, such as Maximum Likelihood Estimation (MLE) and Expectation Maximization (EM) algorithm, which needs extensive computational expense.

Truncated distribution enables the travel time to be restricted within a certain limited range. Thus, excessively short or long travel times, which correspond to unreasonable values generated from the tail of standard distribution, can be excluded [4]. The upper and lower bounds can be defined by the maximum and minimum observed travel times, respectively, if the sample size is large enough [20]. As a result, the applicability of standard distributions can be improved if truncated distribution is adopted. However, the corresponding probability density function (PDF) of the truncated distribution should be modified in order to ensure that the integration of PDF in the specified range is equal to 1, which increases the computation load to some extent.

2.2. Variability Patterns. Three main types of TTV can be generally found in the literature, that is, vehicle-to-vehicle variability that characterizes the difference between travel times of different vehicles traveling the similar route at the same time, period-to-period variability that relates to vehicles traveling the similar route at different periods within a day, and day-to-day variability that represents the travel time variations between similar trips at the same time period on different days [16, 23, 24].

In order to identify the variability patterns of travel time, researchers have defined several reliability measures for quantitative analysis. TTR measures can be generally grouped into three categories, that is, the probability index, the statistical index, and the buffer time type index. The probability index mainly reflects the probability that the travel time can match the specified conditions in the form of probability distribution. The statistical index analyzes TTV based on historical data or real-time information, while the buffer time type index indicates the reservation time to ensure that the probability of reaching the destination is large enough from the travelers' point of view.

Moylan [25] and Rakha et al. [26] used standard deviation to measure the link TTV patterns. Yazici et al. [12] analyzed the differences in TTV patterns between urban roads and highways by calculating the coefficient of variation and found that higher travel times correspond to lower reliability on highways, yet correspond to higher reliability on urban roads. Kieu et al. [16] also employed the coefficient of variation to measure the variability of bus travel time on urban corridors. Van Lint and Van Zuylen [10] used width and skewness of the day-to-day TTD to predict freeway variability patterns. Pu [27] analytically examined a number of reliability measures and explored their mathematical relationships and interdependencies, including the 95th percentile travel time, standard deviation, coefficient of variation, travel time index, and planning time index. He concluded that the coefficient of variation is a good proxy for a range of reliability measures, and median-based buffer index or failure rate is recommended when TTDs are heavily skewed. Chase et al. [28] calculated thirteen reliability measures for 983 freeway segments from 15-minute space mean speed data and explained the measures as to what they represent, which may help decision makers to effectively prioritize traffic management and geometric improvements. Moylan [25] considered the buffer time, planning time index, failure rate, and frequency of congestion of freeway segments. Alvarez and Hadi [29] measured the failure rate and misery index of general purpose lanes and high-occupancy toll lanes. The Florida Department of Transportation [30] used the percentage of travel time less than the median travel time plus a certain acceptable additional time such as the percentage of 5%, 10%, 15%, and 20% above the expected travel time to estimate TTV. Table 1 presents the operational definitions of common reliability measures and the classes to which they belong.

To sum up, though TTD and variability patterns have been extensively investigated, limited studies shed light on the distinction at fine grained levels of spatial and temporal aggregation, for example, travel times for different road types during typical time periods. To fill this gap, this study conducts detailed analysis of travel time characteristics for each specific road type, for example, urban expressways, auxiliary roads of urban expressways, major roads, and secondary roads, based on a vast amount of proven vehicle data in the urban network of Beijing, China. Period-to-period and day-to-day variability are taken into comprehensive consideration. The results help reflect the travelers' preference to different road types and further enhance our understanding of traveling behavior in urban networks.

3. Dataset

In this study, the probe vehicle data collected in the urban network of Beijing, China, during one week from June 1st (Monday) to 7th (Sunday), 2015, were utilized. About every one minute, taxis equipped with GPS devices uploaded a set of instantaneous information, such as location, direction, speed, and being occupied or not occupied by passengers. Note that only the taxis occupied by passengers were considered, which are supposed to be closer to the behavior of regular vehicles. And enough data (at least five sets of traffic information) was uploaded within two minutes for each link. The data quality is good overall. The probe vehicle data were first preprocessed to remove erroneous information and a Kalman filtering process is utilized to achieve smoothing the vehicle trajectories. Then, by resorting to the map matching method, the travel speed data were further aggregated into every two minutes for each link within the network.

Note that in Beijing several two-way urban expressways, that is, Rings 2–5, enclose the urban area. The lengths of Rings 2–5 are 33, 48, 65 and 98 km, respectively. The urban expressways are surrounded by auxiliary roads, and the auxiliary roads and urban expressways are connected by various types of interchanges with paired entrances and exits for merging or diverging traffic. Besides, major and secondary roads are widely distributed in the urban network.

In order to explore the TTD and variability patterns of urban roads, a large area inside the Third Ring Road in Beijing was selected and in total 200 links covering four road types, that is, urban expressways, auxiliary roads of urban expressways, major roads, and secondary roads, were exploited. Each type of road includes 50 links that are evenly distributed in the road network, and hence TTVs for each type of road are supposed to be represented comprehensively. Figure 1 shows the spatial distribution of different types of roads of interest.

4. Day-of-Week Travel Time Distribution Analysis

In this section, the probability distributions of travel time during different time periods were analyzed to investigate the characteristics of travel time variation.

4.1. Calculation Procedure of Travel Time Distribution Analysis. As mentioned above, this study considered links from different road types. It is expected that various urban traffic states from different road types in different time periods can be described and compared in detail. For analyzing travel time distributions of each road type, we selected four typical periods of time, that is, peak hours on weekdays, off-peak hours on weekdays, peak hours on weekends, and off-peak hours on weekends. Peak hours represent the most congested

Table 1: Travel time reliability measures.

Category	Measure	Definition
Probability index	Variance	The expected value of the square of the deviations of a random variable from its mean value
	Standard deviation	The square root of the variance
	Coefficient of variation	Standard deviation divided by the average travel time
Statistical index	Kth percentile	Kth percentile of the travel time distribution
	Skewness statistic	The ratio of the difference between 90th and 50th percentile travel time to the difference between 50th and 10th percentile
	Width statistic	The ratio of the difference between 90th and 10th percentile travel time to the 50th percentile travel time
	Travel time index	The ratio of the average travel time to free-flow travel time
	Planning time index	The ratio of 50th percentile travel time to free-flow travel time
Buffer time type index	Buffer time index	The ratio of the difference between 95th percentile and the average travel time to the average travel time
	Failure/on-time performance	Percent of trips with travel time less than, for example, 1.1 $*$ median travel time, 1.25 $*$ median travel time, and so on
	Frequency of congestion	Percent of time that the travel time is larger than double the free-flow travel time
	Misery index	The average of the highest five percent of travel times divided by the free-flow travel time

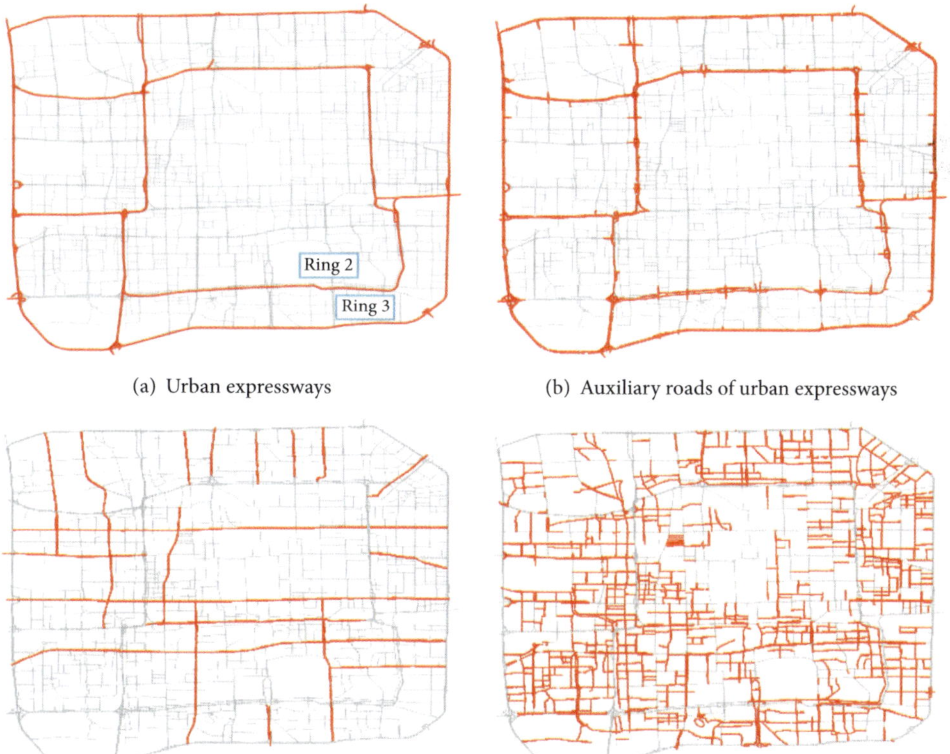

(a) Urban expressways

(b) Auxiliary roads of urban expressways

(c) Major roads

(d) Secondary roads

Figure 1: Spatial distribution of different types of roads.

periods during the day while off-peak hours represent the smooth periods. They were distinguished for each road type by referring to the average travel time of the day. During each time period, it is assumed that link travel time is independent and identically distributed. For providing recommendations to large-scale network computing, such as path finding problems, four common types of single-mode probability distributions including normal, lognormal, gamma, and Weibull were employed to test whether they can fit the travel times desirably. The parameters of each distribution were estimated by the fitting functions in Matlab, such as *normfit*, *lognfit*, *gamfit*, and *wblfit*, using the travel time data of each period. The goodness-of-fit of each distribution was calculated using standard statistical tests such as Kolmogorov-Smirnov (K-S), Anderson-Darling (A-D), and chi-squared (χ^2) test at the 5% significance level. If the p value is larger than the significance level, the distribution is supposed to fit the original data significantly. The χ^2 test is sensitive to the choice of the number of intervals. However, it seems that there is no optimal method to choose the width of the interval [4]. In this study, we set the initial number of intervals to be ten, which is also the default value set for χ^2 test in Matlab. The complete calculation procedure is described below.

Step 1. Transform original vehicle speed data collected from probe vehicles into unit distance travel time. In this study, the unit distance was set as one hundred meters and the calculation of travel time was made in the unit of seconds per 100 meters. This step is to eliminate the influence of different road length on travel time analysis.

Step 2. Choose peak and off-peak hours according to the unit distance travel time on weekdays and weekends. For simplicity, one hour was set as the length of the analysis period.

Step 3. Determine parameter(s) of different distribution types from unit distance travel time. Here, the fitting functions in Matlab were used to estimate the optimal values of the parameter(s).

Step 4. Check the goodness-of-fit of four probability distributions by K-S test, A-D test, and χ^2 test. Accordingly, whether each candidate distribution can be accepted by three statistical tests was examined.

Step 5. Calculate the average acceptance rates of each candidate distribution, and those for each time period and for each road type.

4.2. Analysis Results. Note that Step 4 in the above procedure aims to test whether the travel time data from each time period of each road type follow the four types of distributions using three statistical tests. Considering that 50 links are related for each road type, the unit distance travel times of one individual link in each period will be tested ($4\times3=$) 12 times. For a specific road type, candidate distribution, statistical test, and time period, the tests will be performed ($50\times5=$) 250 times for weekdays and ($50\times2=$) 100 times for weekends. The goodness-of-fit results of TTD for different road types during different periods are presented in Table 2.

The following presents the analysis of distribution fitting results investigated from three perspectives, that is, probability distribution types, time periods, and road types.

4.2.1. Distribution Types. First, the best-fitted distribution type was analyzed on the whole for altogether 16 time periods through comparing the average acceptance rates of each distribution. It was found that lognormal distribution has the highest average acceptance rates; that is, $14/16=87.5\%$. It is worth noting that the best-fitted distribution type for TTD of urban expressways during peak hours on weekends is Weibull distribution and for TTD of the auxiliary roads of urban expressways during off-peak hours on weekends is gamma distribution. However, even in such cases the difference between the acceptance rates of lognormal distribution and the best-fitted distribution is minimal and less than 1%. Then, the best-fitted distribution type was analyzed within each time period. The results showed that the average acceptance rates of lognormal, gamma, normal, and Weibull distribution are 83.7%, 81.9%, 78.8%, and 78.2%, respectively. Apparently, lognormal distribution seems superior compared with other three types of distribution.

4.2.2. Time Periods. In this study, the best-fitted time period was defined as the one whose average acceptance rate of all 12 statistical tests, that is, K-S, A-D, and χ^2 tests, is highest. Accordingly, three best-fitted time periods were identified as off-peak hours of major roads on weekdays, off-peak hours of major roads on weekends, and off-peak hours of auxiliary roads of urban expressways on weekends, with the average acceptance rates of 93.0%, 92.8%, and 92.7%, respectively. By comparison, three worst-fitted time periods are peak hours of secondary roads on weekends, peak hours of secondary roads on weekdays, and peak hours of urban expressways on weekdays, with the average acceptance rates of 68.4%, 67.7%, and 67.1%, respectively. It is interesting to notice that the average acceptance rates on weekdays were always less than on weekends for each road type, and those during peak hours were always less than during off-peak hours within each time period except for the weekends of urban expressway. It helps demonstrate that the typically different traffic states during peak and off-peak hours on weekdays and weekends make a difference of travel time characteristics.

4.2.3. Road Types. In practice, travel times of different road types are supposed to have distinct characteristics. When mixing weekdays and weekends for each road type, the average acceptance rates of urban expressways, auxiliary roads of urban expressways, major roads, and secondary roads are 73.6%, 89.0%, 89.6%, and 70.4%, respectively. It implies that the travel times on auxiliary roads of urban expressways and major roads fitted standard distributions better than the other types of road. The related TTV patterns of each road type will be analyzed in terms of reliability measures in the following.

Table 2: The goodness-of-fit test results of four road types during different periods.

Road types	Time periods		Distributions	Test methods: Accepted by K-S (%)	Test methods: Accepted by A-D (%)	Test methods: Accepted by χ^2 (%)	Average acceptance rates of: each distribution type (%)	Average acceptance rates of: each time period (%)	Average acceptance rates of: each road type (%)	Best-fitted distribution
Urban expressways	Weekdays	Peak hours 17:30–18:30	Normal	58.4	73.2	66.4	66.0	67.1	73.6	Lognormal
			Lognormal	60.4	75.2	69.6	68.4			
			Gamma	58.4	73.6	69.6	67.2			
			Weibull	56.8	70.4	73.2	66.8			
		Off-peak hours 12:30–13:30	Normal	72.8	80.4	77.2	76.8	76.6		Lognormal
			Lognormal	77.6	84.0	79.6	80.4			
			Gamma	73.6	82.0	76.0	77.2			
			Weibull	65.2	75.6	74.8	71.9			
	Weekends	Peak hours 15:00–16:00	Normal	70.0	83.0	76.0	76.3	76.9		Weibull
			Lognormal	72.0	84.0	75.0	77.0			
			Gamma	70.0	83.0	76.0	76.3			
			Weibull	75.0	80.0	79.0	78.0			
		Off-peak hours 12:00–13:00	Normal	67.0	78.0	80.0	75.0	73.8		Lognormal
			Lognormal	70.0	78.0	80.0	76.0			
			Gamma	70.0	78.0	79.0	75.7			
			Weibull	61.0	72.0	73.0	68.7			
Auxiliary roads of urban expressways	Weekdays	Peak hours 17:30–18:30	Normal	75.6	82.4	78.8	78.9	82.0	89.0	Lognormal
			Lognormal	83.2	88.8	87.2	86.4			
			Gamma	78.8	88.4	85.6	84.3			
			Weibull	73.6	80.8	80.8	78.4			
		Off-peak hours 12:30–13:30	Normal	89.2	92.8	89.6	90.5	90.8		Lognormal
			Lognormal	90.8	95.2	90.8	92.3			
			Gamma	91.2	93.6	90.0	91.6			
			Weibull	88.0	91.2	86.8	88.7			
	Weekends	Peak hours 10:00–11:00	Normal	90.0	92.0	88.0	90.0	90.4		Lognormal
			Lognormal	90.0	96.0	90.0	92.0			
			Gamma	90.0	95.0	90.0	91.7			
			Weibull	84.0	91.0	89.0	88.0			
		Off-peak hours 13:30–14:30	Normal	94.0	95.0	89.0	92.7	92.7		Gamma
			Lognormal	93.0	96.0	90.0	93.0			
			Gamma	94.0	96.0	91.0	93.7			
			Weibull	93.0	93.0	88.0	91.3			

Table 2: Continued.

Road types	Time periods		Distributions	Test methods: Accepted by K-S (%)	Test methods: Accepted by A-D (%)	Test methods: Accepted by χ^2 (%)	Average acceptance rates of: each distribution type (%)	Average acceptance rates of: each time period (%)	Average acceptance rates of: each road type (%)	Best-fitted distribution
Major roads	Weekdays	Peak hours 17:30–18:30	Normal	77.6	82.4	75.6	78.5	83.9	89.6	Lognormal
			Lognormal	87.6	92.0	86.4	88.7			
			Gamma	85.2	88.0	84.4	85.9			
			Weibull	82.8	85.2	79.2	82.4			
		Off-peak hours 12:30–13:30	Normal	91.2	94.8	92.8	92.9	93.0		Lognormal
			Lognormal	94.8	96.8	93.6	95.1			
			Gamma	92.8	95.6	92.0	93.5			
			Weibull	87.6	91.2	93.2	90.7			
	Weekends	Peak hours 15:30–16:30	Normal	87.0	89.0	85.0	87.0	88.8		Lognormal
			Lognormal	92.0	92.0	92.0	92.0			
			Gamma	89.0	92.0	88.0	89.7			
			Weibull	86.0	88.0	85.0	86.3			
		Off-peak hours 12:30–13:30	Normal	94.0	97.0	89.0	93.3	92.8		Lognormal
			Lognormal	94.0	96.0	92.0	94.0			
			Gamma	94.0	96.0	89.0	93.0			
			Weibull	90.0	92.0	91.0	91.0			
Secondary roads	Weekdays	Peak hours 17:00–18:00	Normal	52.4	63.2	64.4	60.0	66.7	70.4	Lognormal
			Lognormal	68.0	77.6	76.4	74.0			
			Gamma	62.8	72.8	71.6	69.1			
			Weibull	55.2	62.0	74.0	63.7			
		Off-peak hours 12:30–13:30	Normal	62.4	70.8	70.0	67.7	71.6		Lognormal
			Lognormal	71.2	80.4	78.0	76.5			
			Gamma	68.0	76.0	74.8	72.9			
			Weibull	62.0	69.6	76.0	69.2			
	Weekends	Peak hours 16:30–17:30	Normal	56.0	69.0	61.0	62.0	68.4		Lognormal
			Lognormal	68.0	78.0	75.0	73.7			
			Gamma	66.0	75.0	74.0	71.7			
			Weibull	59.0	67.0	73.0	66.3			
		Off-peak hours 12:30–13:30	Normal	69.0	75.0	74.0	72.7	74.8		Lognormal
			Lognormal	79.0	84.0	78.0	80.3			
			Gamma	76.0	81.0	72.0	76.3			
			Weibull	63.0	75.0	72.0	70.0			

5. Day-of-Week Variability Patterns of Different Road Types

The temporal variation of travel times for each road type was analyzed by exploring the variability patterns, which is beneficial for traffic management agencies to make sophisticated strategies and for travelers to make smart route choices. To this end, four reliability measures were employed, that is, unit distance travel time, coefficient of variation, buffer time index, and punctuality rate. The detailed analysis results are provided below.

5.1. Unit Distance Travel Time. As mentioned before, we transformed the original travel time data from probe vehicles into unit distance travel time. According to the state-of-the-art studies [1], travel times at different quantiles have been commonly used to present the variation tendency changes through the whole day. Thus, we selected the 15th percentile, average, and the 95th percentile travel time to study the variability patterns. The 15th percentile travel time was usually regarded as free flow travel time, and the 95th percentile travel time as the time budget within which travelers will arrive on time. The average travel time of each road type can be calculated as $\overline{\mathrm{TT}_{X,w}}$ in

$$\overline{\mathrm{TT}_{X,w}} = \frac{\sum_{n=1}^{N} \mathrm{TT}_{X,n,w}}{N \times m}, \tag{1}$$

where $\overline{\mathrm{TT}_{X,w}}$ is the average value of all links travel time for road type X within time window w during N days; $\mathrm{TT}_{X,n,w}$ is link travel time for road type X within time window w on day n; m is the number of links, that is, 50 in this study.

Figure 2 illustrates the variability of different percentile unit distance travel times for each road type through the day. The main findings are summarized below.

On weekdays, the morning and afternoon peaks, which look like two bumps (camel shape) and highly skewed distributions, can be obviously distinguished for each road type. Note that the duration of peak hours on urban expressways is comparatively longer than the other types of road, indicating that urban expressways bear higher traffic demand in accordance with its designed function. Moreover, the TTDs over the four hierarchical road types seem to follow a similar and predictable pattern over the day. On the other hand, on weekends all types of roads did not have obvious morning and evening peaks as on weekdays and the unit distance travel times were shortened on the whole to a large extent. It is also interesting to notice that secondary roads tend to have larger unit distance travel times in daytime on both weekdays and weekends. These events may be due to the limited traffic capacity of secondary roads, where it is easier to reach saturation than other road types.

More detailed analysis shows that the morning peak on weekdays appeared around 8:00–9:00 and afternoon peak around 18:00–19:00, which well agrees with the commuting time periods. On weekends, the peak hours through the day appeared around 11:00–12:00 or 16:00–17:00, indicating distinctly different traveling behavior and variability patterns compared with weekdays. Furthermore, during the night periods, for example, 0:00–6:00 and 20:00–0:00, the 15th percentile, average, and the 95th percentile travel times of all road types were short and close to each other. In particular, the 15th percentile travel times of all road types were almost unchanging through the day on both weekdays and weekends. While, during the daytime, for example, 6:00–10:00, the average and the 95th percentile travel times increased significantly. Especially during peak hours, the 95th percentile travel time was about two to three times of the average travel time on weekdays and weekends and as much as six to nine times of the 15th percentile travel time on weekdays and three to five times on weekends.

5.2. Coefficient of Variation. Although the standard deviation is one of the most commonly used variability measures, the absolute value of the standard deviation may yield questionable inferences about the variability [11]. Thus, variability patterns were measured using the coefficient of variation (CV) which normalizes the standard deviation based on the average travel time of the selected time periods. The CV of travel times for road type X within time window w during N days can be calculated as $\mathrm{CV}_{X,w}$ in

$$\mathrm{CV}_{X,w} = \frac{(1/(N \times m)) \sum_{n=1}^{N} \left(\mathrm{TT}_{X,n,w} - \overline{\mathrm{TT}_{X,w}}\right)^2}{\overline{\mathrm{TT}_{X,w}}}. \tag{2}$$

The CVs of travel times for each road type are shown in Figure 3. Overall, it is interesting to notice that though weekdays and weekends have distinctly different traffic characteristics, the CVs of travel times for each road type did not have such significant difference. It implies that TTR measures can help interpret the essential characteristics of variability patterns.

On both weekdays and weekends, the CVs for urban expressways in daytime were relatively smaller and had a narrower fluctuation range compared with major roads and secondary roads. It indicates relatively stable traffic states of uninterrupted flow and reliable travel times on urban expressways. It should be pointed out that, during the periods of 1:00–3:00, the CVs for urban expressway fluctuated significantly. One possible reason is that during such periods the vehicle speed data collected from probe vehicles were scarce and related to greater randomness of driving behavior.

Furthermore, on weekdays the maximum CVs of urban expressways and auxiliary roads of urban expressways through the day were around 1. The secondary roads had a large value of the maximum CV around 1.6 and the major roads around 1.8. On weekends, the maximum CV of urban expressways through the day was about 1. The values for auxiliary roads of urban expressways and secondary roads were around 1.4, and major roads still had the largest maximum CV of around 1.8. The interrupted nature of travel flows on major roads makes travel times more variable. Due to interactions between volatile traffic regimes and signal control strategies, the resulting travel times on major roads often present various distributions under different levels of congestion [6] and thus larger CVs in operation.

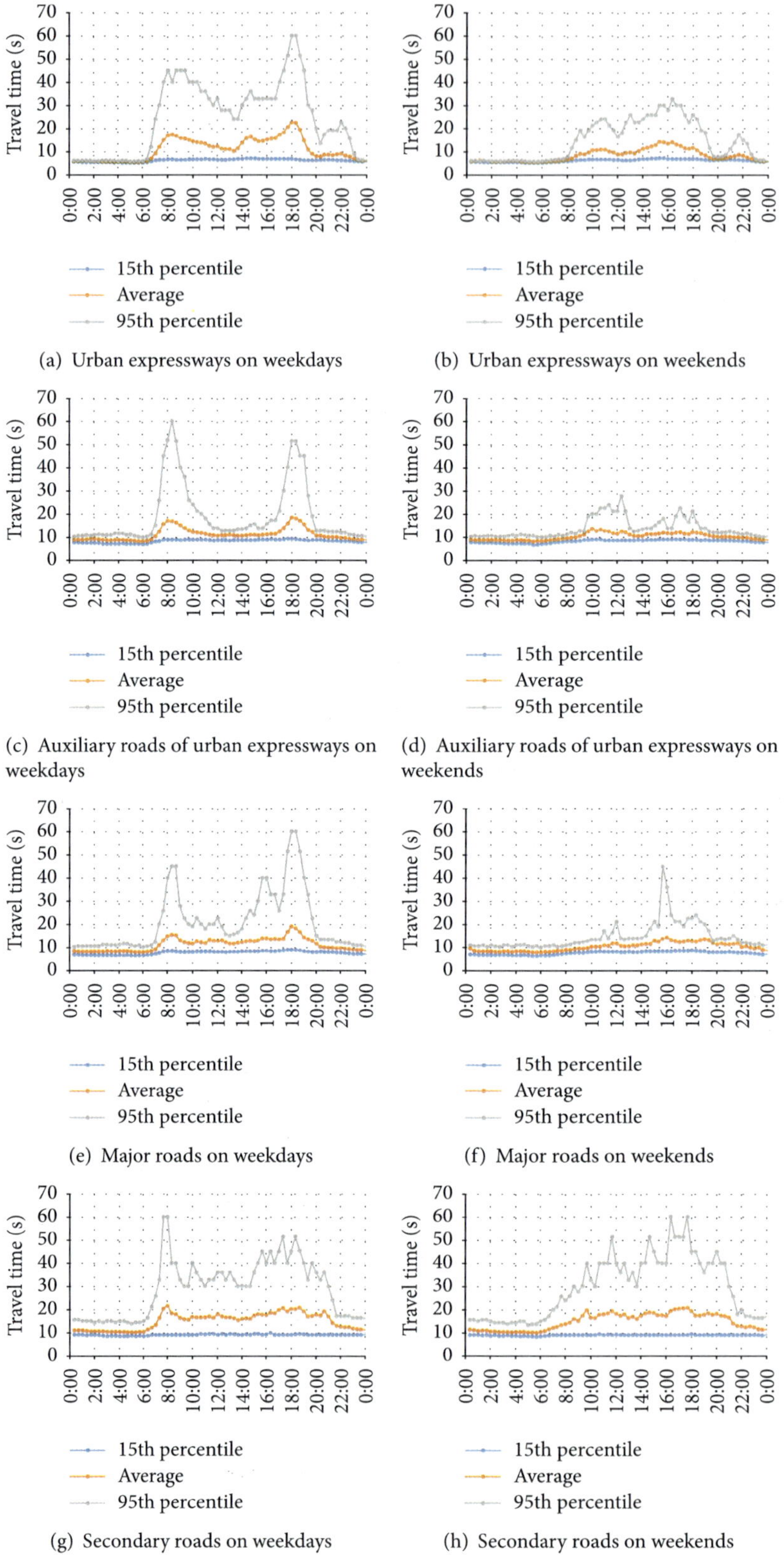

(a) Urban expressways on weekdays

(b) Urban expressways on weekends

(c) Auxiliary roads of urban expressways on weekdays

(d) Auxiliary roads of urban expressways on weekends

(e) Major roads on weekdays

(f) Major roads on weekends

(g) Secondary roads on weekdays

(h) Secondary roads on weekends

FIGURE 2: Unit distance travel time for different types of road through the day.

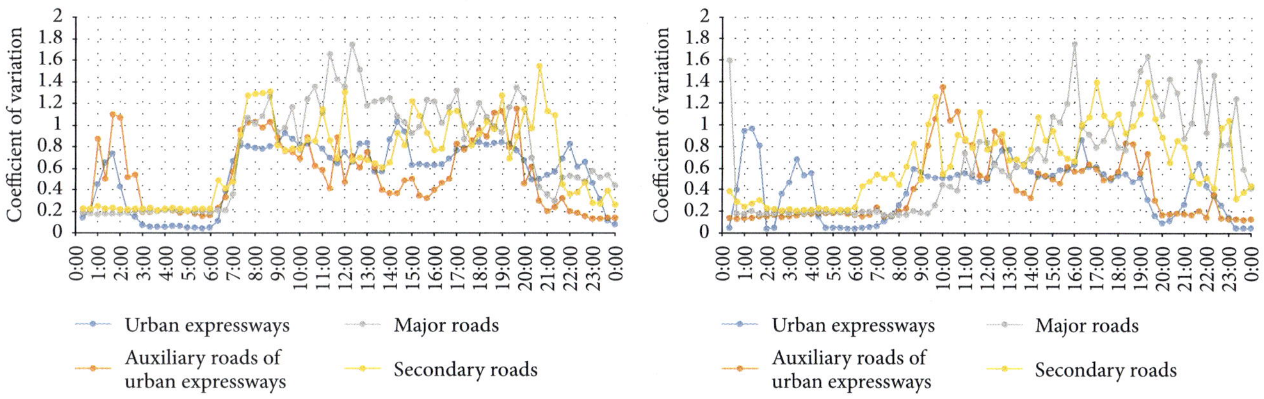

(a) On weekdays (b) On weekends

FIGURE 3: Coefficient of variation through the day.

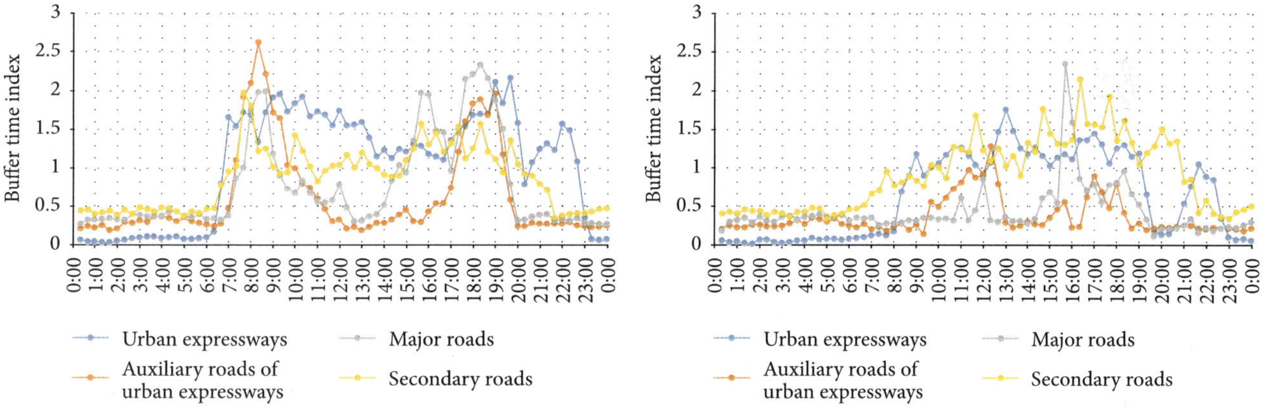

(a) On weekdays (b) On weekends

FIGURE 4: Buffer time index through the day.

5.3. Buffer Time Index. The buffer time index (BTI) represents the additional time that travelers have to spend on the basic of average travel time when they would like to arrive at their destination on time [31]. The BTI for road type X can be expressed as BTI_X in

$$\mathrm{BTI}_X = \frac{\mathrm{TT}_{X,w,95\mathrm{th}} - \overline{\mathrm{TT}_{X,w}}}{\overline{\mathrm{TT}_{X,w}}}, \quad (3)$$

where $\mathrm{TT}_{X,w,95\mathrm{th}}$ is the 95th percentile travel time for road type X within time window w during N days.

The BTIs of travel times for each road type are shown in Figure 4. It was found that the BTIs seem to have shapes very close to the 95th percentile travel times in Figure 2. In more detail, on weekdays BTIs had obvious peak values in the morning and afternoon, for example, around 8:00–9:00 and 18:00–20:00, for auxiliary roads of urban expressways and major roads. The BTIs for urban expressways were minimal until 6:00 and then maintained larger values ranging from 1 to 2, implying volatile and unreliable traffic states through the day. Though, with relatively lower values, the BTIs for secondary roads shared the similar changing tendency as urban expressways. Note that the maximum value of BTIs among four road types on both weekdays and weekends was more than 2, which implies that the drivers would spend three times as much as the average travel time in that period when planning trips to guarantee they would have 95 percent of the possibilities to arrive on time. The minimum value of BTI was close to 0 and related to urban expressways. For auxiliary roads of urban expressways and major roads the minimum value of BTI was around 0.25, while it was about 0.5 for secondary roads, implying usually unreliable travel times provided by secondary roads.

Pu [27] pointed out that some reliability metrics may be inconsistent in their depictions of reliability, such as the case of BTI that may remain constant for different values of CV. The same conclusions can be drawn in this study. It was obvious that on weekdays the CVs of urban expressways and auxiliary roads of urban expressways had a large fluctuation in early morning, for example, around 0:00–3:00; however, the BTIs of the same road types were stable during that period. Similar phenomena can also be recognized on weekends, for instance, for urban expressways in early morning, for example, around 0:00–4:00, as well as major roads

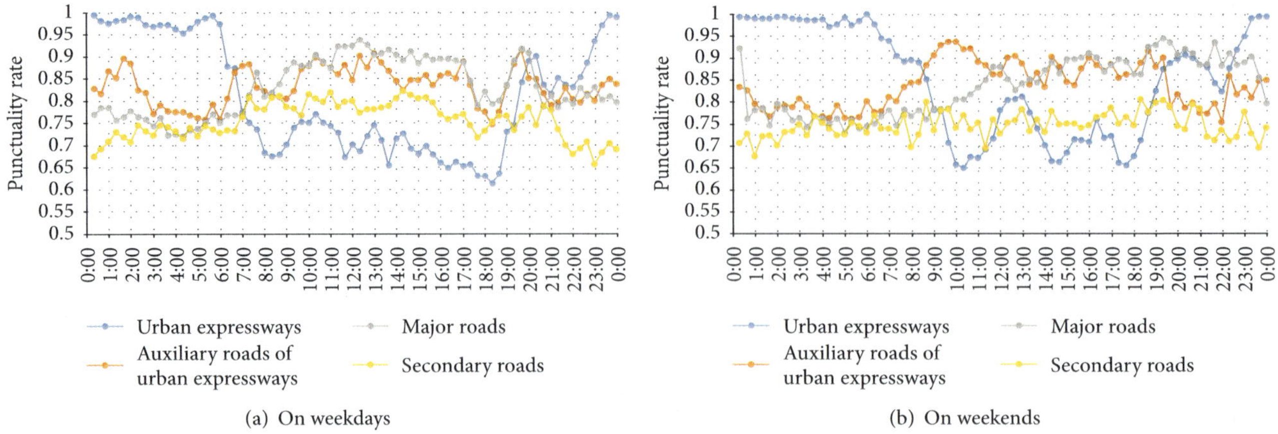

FIGURE 5: Punctuality rates through the day.

in the late evening, for example, around 20:00–0:00. Hence, we may need to consider the changes of multiple indicators comprehensively to accurately measure the road performance.

5.4. Punctuality Rate. In this study, the punctuality rate (PR) was defined as the probability that the travel times were less than or equal to 1.1 times as much as the average travel time of road type X within time window w during N days. To a certain extent, PR can reflect the operational performance of different road types. The higher the PR is, the more desirable the road performance is. The reason we prefer average travel time to median travel time when calculating PR is that the skew statistics of most links during most time periods are larger than 1 indicating that the median travel time is smaller than average travel time. As a result, median-based PR may underestimate the operational performance. The PR can be expressed as

$$\mathrm{PR}_X = \sum_{n=1}^{N} P\left\{\mathrm{TT}_{X,n,w} \le 1.1 \times \overline{\mathrm{TT}_{X,w}}\right\}, \qquad (4)$$

where PR_X is the punctuality rate for road type X within time window w during N days.

The PRs of travel times for each road type are shown in Figure 5. It is noteworthy that from late evening till early evening, for example, 22:00–7:00, on both weekdays and weekends, the PRs of urban expressways were over 0.9 and remained the highest among four types of roads. During the same periods, the PRs of auxiliary roads of urban expressways and major roads fluctuated between 0.7 and 0.9, also indicating desirable operational performance with less traffic demand. By comparison, secondary roads had relatively lower PRs, that is, less than 0.75.

It is worth noting that during the daytime after 7:00 the PRs of urban expressways began to decline rapidly and remained the lowest among four road types. On the other hand, the PRs of auxiliary roads of urban expressways and major roads began to rise slightly and remained the top two among four road types. For secondary roads, the PRs remained relatively stable through the day, and even higher than urban expressway during the daytime period of 7:00–19:00 on weekdays and 9:00–19:00 on weekends. It implies that, with the increasing commuting traffic demand, most of travelers might prefer to choose urban expressways first and they increased the possibility of traffic congestion on urban expressways. Meanwhile, the PRs of auxiliary roads of urban expressways and major roads became higher. Furthermore, it is interesting to notice that auxiliary roads of urban expressways and major roads have similar changing tendency of PRs as well as unit distance travel time and BTI and thus had similar travel time variation patterns.

6. Summary and Conclusions

In order to investigate the volatile characteristics of travel time in an urban network, this study explored the urban TTD and the variability patterns of different road types by using the probe vehicle data collected inside the Third Ring Road in Beijing. Different from the previous studies, in total, 200 links covering four road types, that is, urban expressways, auxiliary roads of urban expressways, major roads, and secondary roads, were exploited to provide more detailed analysis of travel time characteristics for each specific road type. The major achievements are summarized below, which may be beneficial to travelers for making reliable route choices and to traffic engineers for deploying sophisticated management and control.

(1) Link TTDs are characterized using probe vehicle data. Four common probability distributions, including normal, lognormal, gamma, and Weibull, are subjected to standard statistical tests such as K-S, A-D, and chi-squared test. The periods of interest were divided into four groups, that is, peak hours on weekdays, off-peak hours on weekdays, peak hours on weekends, and off-peak hours on weekends.

(2) Lognormal distribution is superior compared with other three types of distribution. The average acceptance rates on weekdays were always less than on weekends for each road type, and those during peak

hours were always less than during off-peak hours within each time period except for the weekends of urban expressway. The travel times on auxiliary roads of urban expressways and major roads fitted standard distributions better than the other types of road. It can be implied that travel time characteristics vary from different traffic states and road types. Moreover, fitting results of link TTDs can provide guidance for determining path TTDs and network modeling.

(3) Four reliability measures, that is, unit distance travel time, CV, BTI, and PR, were used to quantify the day-of-week TTV patterns of different road types. In general, various indicators can reflect the change trend of traffic states collaboratively. However, some reliability measures may be inconsistent in their depictions of TTV, such as the case of BTI that may remain constant for different values of CV. As a result, it is more reasonable and accurate to consider the changes of multiple indicators comprehensively when measuring the road performance.

(4) On weekdays, the morning and afternoon peaks could be distinguished easily for urban expressways, auxiliary roads of urban expressways, and major roads, all of which appeared around 8:00–9:00 and 18:00–19:00, respectively. However, on weekends, all types of roads did not have obvious morning and afternoon peaks, and the peak through the day appeared around 11:00–12:00 or 16:00–17:00, indicating distinctly different traveling behavior and variability patterns compared with weekdays. Moreover, travelers can attempt to avoid traveling during peak hours to save time, and traffic management departments may target peak hours for improving traffic conditions.

(5) Auxiliary roads of urban expressways and major roads had similar changing tendency of PR as well as unit distance travel time and BTI, thus sharing similar TTV patterns. The travel times of auxiliary roads of urban expressways and major roads appeared more stable and reliable than other road types in daylight, and urban expressways had the most reliable travel times at night. It implies that with the increasing commuting traffic demand, most of travelers might prefer to choose urban expressways first and increase the possibility of traffic congestion on urban expressways.

In the future work, to enhance the effectiveness and reliability of the results, it is expected to adopt larger scale probe data in the time-space dimension or to integrate the data from different sources, for example, fusing probe data with other data collected by traffic sensors such as loop detectors and camera [32]. Furthermore, since the probability distribution types in this study only include four standard distributions, examining the goodness-of-fit of other distributions, for example, multimode distribution and truncated distribution, will also be conducted. On the other hand, additional reliability measures are supposed to be incorporated for more comprehensive evaluation. Last but not least, travel time forecast problem can also be delved into [33, 34].

Conflicts of Interest

The authors declare that they have no conflicts of interest.

Acknowledgments

The authors acknowledge the National Nature Science Foundation of China (U1564212 and 51508014) and the Fundamental Research Funds for the Central Universities for kind support of this research.

References

[1] SHRP2, *Second Strategic Highway Research Program*, 2010, http://www.trb.org/StrategicHighwayResearchProgram2SHRP2/Blank2.aspx.

[2] P. Chen, G. Yu, X. Wu, Y. Ren, and Y. Li, "Estimation of red-light running frequency using high-resolution traffic and signal data," *Accident Analysis & Prevention*, vol. 102, pp. 235–247, 2017.

[3] M. Martchouk, F. Mannering, and D. Bullock, "Analysis of Freeway Travel Time Variability Using Bluetooth Detection," *Journal of Transportation Engineering*, vol. 137, no. 10, pp. 697–704, 2011.

[4] W. Zeng, T. Miwa, Y. Wakita, and T. Morikawa, "Application of Lagrangian relaxation approach to α-reliable path finding in stochastic networks with correlated link travel times," *Transportation Research Part C: Emerging Technologies*, vol. 56, pp. 309–334, 2015.

[5] X. Wang, H. Liu, R. Yu, B. Deng, X. Chen, and B. Wu, "Exploring operating speeds on urban arterials using floating car data: Case study in Shanghai," *Journal of Transportation Engineering*, vol. 140, no. 9, Article ID 04014044, 2014.

[6] P. Chen, K. Yin, and J. Sun, "Application of finite mixture of regression model with varying mixing probabilities to estimation of urban arterial travel times," *Transportation Research Record*, vol. 2442, pp. 96–105, 2014.

[7] K. Lyman and R. L. Bertini, "Using travel time reliability measures to improve regional transportation planning and operations," *Transportation Research Record*, no. 2046, pp. 1–10, 2008.

[8] P. Chen, C. Ding, G. Lu, and Y. Wang, "Short-term traffic states forecasting considering spatial-temporal impact on an urban expressway," *Transportation Research Record*, vol. 2594, pp. 61–72, 2016.

[9] F. Lei, Y. Wang, G. Lu, and J. Sun, "A travel time reliability model of urban expressways with varying levels of service," *Transportation Research Part C: Emerging Technologies*, vol. 48, pp. 453–467, 2014.

[10] J. W. C. Van Lint and H. J. Van Zuylen, "Monitoring and predicting travel time reliability: using width and skew of day-to-day travel time distribution," *Transportation Research Record*, vol. 1917, pp. 54–62, 2005.

[11] M. Yazici, C. Kamga, and K. Mouskos, "Analysis of travel time reliability in New York City based on day-of-week and time-of-day periods," *Transportation Research Record*, no. 2308, pp. 83–95, 2012.

[12] M. A. Yazici, C. Kamga, and K. Ozbay, "Highway versus urban roads: Analysis of travel time and variability patterns based on facility type," *Transportation Research Record*, vol. 2442, pp. 53–61, 2014.

[13] M. A. P. Taylor, "Travel time variability - the case of two public modes," *Transportation Science*, vol. 16, no. 4, pp. 507–521, 1982.

[14] E. Mazloumi, G. Currie, and G. Rose, "Using GPS data to gain insight into public transport travel time variability," *Journal of Transportation Engineering*, vol. 136, no. 7, Article ID 006007QTE, pp. 623–631, 2010.

[15] E. B. Emam and H. Al-Deek, "Using real-life dual-loop detector data to develop new methodology for estimating freeway travel time reliability," *Transportation Research Record*, no. 1959, pp. 140–150, 2006.

[16] L.-M. Kieu, A. Bhaskar, and E. Chung, "Public transport travel-time variability definitions and monitoring," *Journal of Transportation Engineering*, vol. 141, no. 1, Article ID 04014068, 2015.

[17] Y. Ji and H. Zhang, "Travel Time Distributions on Urban Streets: Estimation with Hierarchical Bayesian Mixture Model and Application to Traffic Analysis with High-Resolution Bus Probe Data," in *Transportation Research Board 92nd Annual Meeting*, 2013.

[18] F. Guo, Q. Li, and H. Rakha, "Multistate travel time reliability models with skewed component distributions," *Transportation Research Record*, vol. 2315, pp. 47–53, 2012.

[19] F. Guo, H. Rakha, and S. Park, "Multistate model for travel time reliability," *Transportation Research Record*, no. 2188, pp. 46–54, 2010.

[20] P. Cao, T. Miwa, and T. Morikawa, "Modeling Distribution of Travel Time in Signalized Road Section Using Truncated Distribution," *Procedia - Social and Behavioral Sciences*, vol. 138, pp. 137–147, 2014.

[21] Y. Wang, W. Dong, L. Zhang et al., "Speed modeling and travel time estimation based on truncated normal and lognormal distributions," *Transportation Research Record*, no. 2315, pp. 66–72, 2012.

[22] Y. Nie, X. Wu, J. F. Dillenburg, and P. C. Nelson, "Reliable route guidance: A case study from Chicago," *Transportation Research Part A: Policy and Practice*, vol. 46, no. 2, pp. 403–419, 2012.

[23] M. Yildirimoglu, Y. Limniati, and N. Geroliminis, "Investigating empirical implications of hysteresis in day-to-day travel time variability," *Transportation Research Part C: Emerging Technologies*, vol. 55, pp. 340–350, 2015.

[24] R. B. Noland and J. W. Polak, "Travel time variability: a review of theoretical and empirical issues," *Transport Reviews*, vol. 22, no. 1, pp. 39–54, 2002.

[25] E. Moylan, "Performance of Reliability Metrics on Empirical Travel Time Distributions," in *Transportation Research Board 93rd Annual Meeting*, 2014.

[26] H. Rakha, I. El-Shawarby, M. Arafeh, and F. Dion, "Estimating path travel-time reliability," in *Proceedings of the ITSC 2006: 2006 IEEE Intelligent Transportation Systems Conference*, pp. 236–241, September 2006.

[27] W. Pu, "Analytic relationships between travel time reliability measures," *Transportation Research Record*, vol. 2254, pp. 122–130, 2011.

[28] R. Chase, B. Williams, and N. Rouphail, "Detailed Analysis of Travel Time Reliability Performance Measures from Empirical Data," in *Transportation Research Board 92nd Annual Meeting*, 2013.

[29] P. Alvarez and M. Hadi, "Time-variant travel time distributions and reliability metrics and their utility in reliability assessments," *Transportation Research Record*, no. 2315, pp. 81–88, 2012.

[30] Florida Department of Transportation, *The Florida Reliability Method. In Florida's Mobility Performance Measures Program*, 2000.

[31] Federal Highway Administration (FHWA), *Travel Time Reliability: Making It There On Time, All The Time*, 2006.

[32] Y. Wiseman, "Real-time monitoring of traffic congestions," in *Proceedings of the 2017 IEEE International Conference on Electro Information Technology, EIT 2017*, pp. 501–505, USA, May 2017.

[33] Z. Zhang, Y. Wang, P. Chen, Z. He, and G. Yu, "Probe data-driven travel time forecasting for urban expressways by matching similar spatiotemporal traffic patterns," *Transportation Research Part C: Emerging Technologies*, vol. 85, pp. 476–493, 2017.

[34] J. J. Tang, F. Liu, Y. J. Zou, W. B. Zhang, and Y. H. Wang, "An improved fuzzy neural network for traffic speed prediction considering periodic characteristic," *IEEE Transactions on Intelligent Transportation Systems*, vol. 18, pp. 2340–2350, 2017.

14

Time-Dependent Transportation Network Design considering Construction Impact

Yi He,[1] Ziqi Song,[1] and Lihui Zhang[2]

[1]*Department of Civil and Environmental Engineering, Utah State University, Logan, UT 84322, USA*
[2]*Institute of Transportation Engineering, Zhejiang University, Hangzhou, Zhejiang 310058, China*

Correspondence should be addressed to Lihui Zhang; lihuizhang@zju.edu.cn

Academic Editor: Martin Trépanier

A traditional discrete network design problem (DNDP) always assumes transportation infrastructure projects to be one-time events and ignores travelers' delays caused by construction work. In fact, infrastructure construction usually lasts for a long time, and the impact on traffic can be substantial. In this paper, we introduce time dimension into the traditional DNDP to explicitly consider the impact of road construction and adopt an overtime policy to add flexibility to construction duration. We address the problem of selecting road-widening projects in an urban network, determining the optimal link capacity, and designing the schedule of the selected projects simultaneously. A time-dependent DNDP (T-DNDP) model is developed with the objective of minimizing total weighted net user cost during the entire planning horizon. An active-set algorithm is applied to solve the model. A simple example network is first utilized to demonstrate the necessity of considering the construction process in T-DNDP and to illustrate the trade-off between the construction impact and the benefit realized through capacity extension. We also solve the T-DNDP model with data from the Sioux Falls network, which contains 24 nodes, 76 links, and 528 origin-destination (O-D) pairs. Computational results for the problem are also presented.

1. Introduction

Transportation plays an essential role in economic development and in the overall improvement of people's living standard. Unfortunately, rapid population growth and the accompanying accelerated economic development in urban areas are usually accompanied by sharp increases in traffic demand, which may induce all sorts of traffic-related issues, such as congestion, air pollution, and noise pollution. Among these issues, traffic congestion is particularly a prominent problem, especially in large cities. According to the 2015 Urban Mobility Scorecard, the cumulative travel delay caused by congestion was about 7 billion hours and the wasted fuel totaled almost 3 billion gallons in 2014 alone in the United States [1]. To maintain a reasonable level of service and alleviate current and future traffic congestion, transportation agencies are often faced with the need to expand the existing road network. However, generally they must work with a limited amount of resources. Therefore, given all of the potential candidate road-widening projects, selecting appropriate projects becomes a crucial decision in maximizing the benefit of expansion projects. This kind of problem is usually referred to in the literature as the network design problem (NDP).

Over the past few decades, NDP has been widely studied. Most of the literature related to NDP has focused on either modeling or new algorithms for network design models. However, these early studies regarded road construction work as a one-time event and did not consider the gradual improvement of the network until researchers introduced the time dimension to the traditional NDP (see [2–4]). Lo and Szeto [4] claimed that the road network is improved year by year before the completion of the improvement project, which makes the NDP model more realistic. Nevertheless, even though they considered network improvement to be gradual in their model, they still assumed the construction process to be a one-off procedure. Actually, capacity expansion work usually involves work zones and lane closures, which may reduce the current link capacity during construction and result in congestion and delays for road users. Furthermore, road infrastructure construction generally lasts for months or even years, and the impact of construction may greatly affect planners' decisions. For example, when multiple projects are simultaneously underway, planners may choose to adjust the schedule of some projects to avoid excessive delays in a region. Therefore, the impact of construction work should not simply be ignored.

This study explicitly considers construction impact in conjunction with the benefits brought about by capacity expansion as the two primary factors that govern the network design problem. Furthermore, in light of the fact that the construction process may have a tremendous impact on the road network, shortening the construction period represents a possible method for mitigating the impact. Thus, the proposed model also allows the construction period to be flexible, which means the planners can choose to speed up construction to shorten its duration by paying overtime to construction personnel.

Compared with existing NDP models, the proposed model has the following advantages:

(1) The construction impact is clearly evaluated so that the selection and schedule of road infrastructure projects will be optimized.

(2) This model adopts an overtime policy in the candidate projects, which allows planners to choose whether or not to accelerate a project by paying overtime. Thus, the construction durations of the candidate projects are flexible.

(3) This model is able to address the problem of selecting road-widening projects from several candidate projects, simultaneously determining the optimal amount of increased capacity and designing the optimal schedule for the chosen projects.

2. Background

2.1. Network Design Problem. The transportation NDP aims to achieve certain objectives, for example, reducing traffic congestion, energy consumption, and environmental pollution, by choosing improvements or additions to an existing network [5]. A common methodology used to formulate the NDP is bilevel programming. The upper level is the system level, which optimizes the system benefits subject to limited resources, while the lower level is the users' level, which models users' route choice behavior in the network. The upper level can be formulated with different decision variables and objective functions. The decision variables can be merely continuous or discrete or can contain both continuous and discrete elements. Based on the types of decision variables, network design problems are generally divided into three categories. The network design problem with only continuous variables is called the continuous network design problem (CNDP) (see [6–11]). In road network design problems, continuous variables are usually introduced in order to simplify computation. For example, the capacity expansion of a roadway can be continuous (see [12, 13]). However, continuous variables do not necessarily indicate the changes that are practical, because road capacity is normally measured by the number of lanes. Hence, despite the fact that it may be more computationally expensive, the discrete network design problem (DNDP) with solely discrete variables (see [14–22]) and the mixed network design problem (MNDP) with both continuous and discrete variables (see [23–26]) are still worth investigating.

Previous studies have made substantial contributions to the understanding and applications of DNDP. Some have studied various applications associated with DNDP. For instance, Drezner and Wesolowsky [18] formulated a DNDP for the purpose of selecting the best distribution of one-way and two-way routes in a road network. Wu et al. [27] solved the DNDP of choosing the location of pedestrian-only streets in a multimodel network. Song et al. [28] developed a DNDP model that settled the problems of selecting locations for high-occupancy vehicle (HOV)/high-occupancy vehicle (HOT) lanes and determining toll rates on HOT lanes. Miandoabchi and Farahani [29] determined street orientations and expansions, as well as lane allocations, based on the reserve capacity concept in a DNDP model. Others have developed different kinds of approaches to solve DNDP. It is well known that solving a bilevel network design problem is very difficult because the problem is NP-hard and nonconvex. After Leblanc [15] proposed a branch-and-bound algorithm to solve this bilevel problem, many researchers began to seek better approaches to assess the trade-off between computation of speed and solution accuracy. For example, Dantzig et al. [6] transformed the nonconvex programming problem to a convex problem using system equilibrium flow to replace user equilibrium flow. Poorzahedy and Turnquist [30] utilized approximation to transform the bilevel problem into a single-level problem. Solanki et al. [31] decomposed the highway network design problem in a sequence of small subproblems and limited the search using heuristics to reduce computation time. Poorzahedy and Abulghasemi [20] adapted metaheuristic algorithms to solve NDP for the Sioux Falls network. Poorzahedy and Rouhani [32] improved the metaheuristic algorithm and designed the hybrid metaheuristic algorithm. A genetic algorithm is also widely used (see [19, 33, 34]). Gao et al. [21] transformed the upper-level programming of the traditional DNDP to a nonlinear problem based on the support function concept. Zhang et al. [35] developed the active-set algorithm, which eliminates complementarity constraints in the DNDP by assigning initial values and solving binary knapsack problems. Farvaresh and Sepehri [36] revised the branch-and-bound algorithm proposed by Leblanc [15] for bilevel DNDP.

2.2. Time-Dependent Network Design Problem. In recent years, the time varying evolution of road networks began to gain interest in transportation network design problems. Different time scales were studied in the literature, ranging from the smallest day-to-day dynamics [2, 3, 37] to network upgrades spanning years [38–40]. Lo and Szeto [4] introduced the time dimension to CNDP and built a comprehensive and practical model that considered not only user equilibrium (UE), but also travel demand and land-use patterns as time-dependent. In conjunction with other researchers, they further studied a series of time-dependent NDP problems, including budget sensitivity analysis among users, private toll road operators, and the government [41]; the trade-off between the social and financial aspects of three possible network improvement strategies under demand and the value of time uncertainty [42]; the trade-off between social benefit and intergeneration equity [38]; cost recovery issues over time [40, 43]; land-use transport interaction over time [44]; sustainability with land-use transport interaction over time [45]; health impacts attributable to network construction [46]; and a multiobjective time-dependent model to determine the sequence of link expansion projects and link construction projects [47]. Time dimension was also introduced in other studies. For instance, Kim et al. [48] formulated a time-dependent DNDP framework to address the project scheduling problem; Ukkusuri and Patil [49] developed a multiperiod flexible network design model with demand uncertainty and demand elasticity; Hosseininasab and Shetab-Boushehri [50] integrated project selection and scheduling into a single time-dependent DNDP model.

However, in the literature referenced above, the road network is optimized for a certain future time without considering the construction impact. In practice, modifications to a network are gradual processes rather than one-off events. Hence, the construction process, which results in a negative impact to traffic, should also be considered. The construction process is explicitly modeled in this paper.

3. Basic Considerations

This paper considers the problem of simultaneously determining the selection and scheduling of road expansion projects for a transportation network. The evaluation of a design is based on system performance throughout a given planning horizon, which includes the construction process. Below, we summarize our basic considerations and assumptions for the modeling and analysis of the construction process of road expansion projects.

(1) Within the planning horizon, a road segment has at most one expansion project. This consideration is not overly restrictive, as we can always divide a road segment into several parallel links and assign each link with a project.

(2) The construction procedure of an expansion project spans a continuous period of time.

(3) Throughout the planning horizon, the route choice behaviors of drivers in the network follow the UE principle. Considering the construction process, the traffic network will change, as will the UE pattern.

(4) The potential demand growth over time is known.

(5) The interest and inflation rates are constant within the planning horizon.

For the convenience of readers, below we list some notations frequently used in the paper.

Sets

N: set of nodes

L: set of links

L_1: set of links with a potential expansion project

L_2: set of links without a potential expansion project

W: set of O-D pairs

Parameters

a: link $a = (i, j) \in L$

w: O-D pair $w \in W$

M: the total number of unit time intervals for the planning horizon

M^C: the total number of unit time intervals for the construction time window

m: time interval $m \in \{1, 2, \ldots, M\}$

d_m^w: travel demand between O-D pair $w \in W$ in time interval $m \in \{1, 2, \ldots, M\}$

D_a^0: fixed time cost for the expansion project on link $a \in L_1$

D_a^1: variable time cost per additional lane for the expansion project on link $a \in L_1$

c_a: average cost per time interval during construction for the expansion project on link $a \in L_1$ without overtime work

Variables

$x_{a,m}^w$: traffic flow on link a for O-D pair $w \in W$ in time interval $m \in \{1, 2, \ldots, M\}$

$v_{a,m}$: aggregate traffic flow on link $a \in L$ in time interval $m \in \{1, 2, \ldots, M\}$

$t_{a,m}$: travel time of link $a \in L$ in time interval $m \in \{1, 2, \ldots, M\}$

$C_{a,m}$: capacity of link $a \in L$ in time interval $m \in \{1, 2, \ldots, M\}$

$y_{a,m}$: a binary variable, representing whether link $a \in L_1$ is under construction in time interval $m \in \{1, 2, \ldots, M\}$. If yes, $y_{a,m} = 1$; otherwise, $y_{a,m} = 0$

$z_{a,m}$: a binary variable, representing whether construction has been finished on link $a \in L_1$ in time interval $m \in \{1, 2, \ldots, M\}$. If yes, $z_{a,m} = 1$; otherwise, $z_{a,m} = 0$

$S_{a,m}$: a binary variable, representing whether time interval $m \in \{1, 2, \ldots, M\}$ is the start date of construction on link $a \in L_1$. If yes, $S_{a,m} = 1$; otherwise, $S_{a,m} = 0$

$E_{a,m}$: a binary variable, representing whether time interval $m \in \{1, 2, \ldots, M\}$ is the end date of construction on link $a \in L_1$. If yes, $E_{a,m} = 1$; otherwise, $E_{a,m} = 0$

l_a: number of newly added lanes on link

D_a^e: the estimated construction duration for the expansion project on link $a \in L_1$ without overtime work

D_a^r: reduced construction duration for the expansion project on link $a \in L_1$ through overtime work.

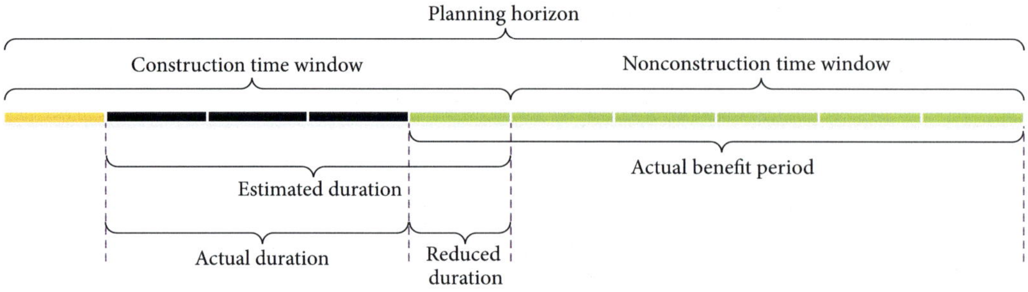

FIGURE 1: An example of the timeline of a road expansion project.

4. Problem Formulation

Consider a general transportation network $G(N, L)$, where N and L are the set of nodes and the set of directed links, respectively. The latter are represented as a node pair (i, j), where $i, j \in N$ and $i \neq j$, or a single letter a. There are two types of links in the network, the links with a potential road-widening project, and the links without a potential project, denoted as L_1 and L_2, respectively. In this study, the planning horizon $[0, T]$ is equally divided into M unit intervals. The unit interval could be a month, a season, or another reasonable time interval. Note that the unit interval is the unit of measurement of the time cost of the construction process. The planning horizon includes a construction time window and a nonconstruction time window. All construction projects are supposed to be completed within the construction time window; the nonconstruction time window is designed to evaluate the continuing benefits realized through the finished road expansion projects. Planners determine the lengths of these two time windows. Approximately, the duration of the nonconstruction time window represents the service life of the improved roads before requiring extensive renovation. For an individual project, the benefit period begins immediately after the completion of the project. Therefore, the benefit period should be at least as long as the nonconstruction time window. Let M^C denote the number of intervals in the construction time window, $M^C < M$.

Figure 1 shows an example of the timeline of one road expansion project. In this example, the planning horizon is divided into 10 intervals, among which the former 5 intervals belong to the construction time window, and the latter 5 intervals belong to the nonconstruction time window. This project is scheduled to start at the beginning of the second time interval, and the estimated construction duration is four unit intervals. The planner decides to shorten the construction duration by one interval through overtime work. Therefore, the actual construction duration is reduced to 3 unit intervals, and the benefit lasts for 6 unit intervals (the detailed description of the flexible construction duration will be presented in the following model).

4.1. Time-Dependent Traffic Assignment Constraints

Feasible Region. To describe the feasible flow distributions of a network, let A be the node-arc incidence matrix associated with the network, and let E^w be an "input-output" vector indicating the origin and destination of O-D pair w. E^w has exactly two nonzero components: one has the value 1 corresponding to the origin node of the O-D pair w, and the other's value is −1, corresponding to the destination node. For all other nodes in this O-D pair, E^w equals 0. The flow distributions are said to be feasible if and only if the following constraints hold for x_m^w:

$$Ax_m^w = E^w d_m^w \quad \forall w \in W,\ m \in \{1, 2, \ldots, M\} \tag{1}$$

$$x_m^w \geq 0 \quad \forall w \in W,\ m \in \{1, 2, \ldots, M\} \tag{2}$$

$$v_m = \sum_w x_m^w \quad \forall m \in \{1, 2, \ldots, M\}, \tag{3}$$

where $x_m^w \in R^{|L|}$ is a vector whose components, $x_{a,m}^w$, represent a link flow on link a for O-D pair w in interval m, and v_m is a vector whose components, $v_{a,m}$, represent an aggregate link flow on link a in interval m. d_m^w represents the travel demand between O-D pair w in interval m. For simplicity, the travel demand of each O-D pair is assumed to be increasing at a constant rate. For an O-D pair $w \in W$, given the travel demand in the first interval, that is, d_1^w, the demand in interval $m \in \{1, 2, \ldots, M\}$ is calculated as

$$d_m^w = d_1^w \cdot (1 + \varepsilon^w)^{m-1} \quad \forall w \in W,\ m \in \{1, 2, \ldots, M\}, \tag{4}$$

where ε^w is the growth factor of demand between O-D pair w.

To make the subsequent expressions more easily discernable, we introduce a set V_m^F for each period m to cover all of the feasible flow distributions:

$$V_m^F = \left\{ v_m : v_m = \sum_w x_m^w,\ Ax_m^w = E^w d_m^w,\ x_m^w \geq 0,\ \forall w \in W \right\} \quad \forall m \in \{1, 2, \ldots, M\}. \tag{5}$$

Time-Dependent Link Capacity. Within the planning horizon, if a link is selected for expansion, its capacity will be time-dependent. During construction, the capacity of a link may be reduced due to the impact of construction; after construction, the capacity of a link will be improved due to added lanes. Two binary variables, $y_{a,m}$ and $z_{a,m}$, are introduced to indicate the status of a link $a \in L_1$ with a potential widening project in time interval $m \in \{1, 2, \ldots, M\}$. $y_{a,m}$ represents whether link $a \in L_1$ is under construction in time interval $m \in \{1, 2, \ldots, M\}$. If yes, $y_{a,m} = 1$; otherwise, $y_{a,m} = 0$. $z_{a,m}$ represents whether construction has been finished on link $a \in L_1$ in time interval $m \in \{1, 2, \ldots, M\}$. If yes, $z_{a,m} = 1$; otherwise, $z_{a,m} = 0$. Note that if link $a \in L_1$ is not selected for expansion, there will be no construction process on link a, and $y_{a,m} = 0$, $z_{a,m} = 0$, $\forall m \in \{1, 2, \ldots, M\}$. The time-dependent capacity function can be formulated in (6)–(8):

$$C_{a,m} = C_a^0 - y_{a,m} \cdot C_a^r + z_{a,m} \cdot l_a \cdot C_a^1 \quad \forall a \in L_1,\ m \in \{1, 2, \ldots, M\} \tag{6}$$

$$C_{a,m} \le C_a^{\max} \quad \forall a \in L_1,\ m \in \{1, 2, \ldots, M\} \tag{7}$$

$$C_{a,m} = C_a^0 \quad \forall a \in L_2,\ m \in \{1, 2, \ldots, M\}, \tag{8}$$

where C_a^0, C_a^r, C_a^1, and $C_a^{\max}$ are the initial capacity, the reduced capacity during construction, the capacity of a single lane, and the maximum allowable capacity of link a, respectively. l_a denotes the number of lanes added after construction, which is a decision variable to be optimized in our model. l_a is an integer variable. Equation (7) restricts the capacity of a link to be less than its maximum allowable capacity.

Travel Time. In this paper, the Bureau of Public Roads (BPR) function is used to define the link travel time. The travel time of an existing link in each period, $t_{a,m}$, is determined by the link travel flow, $v_{a,m}$, and the link capacity, $C_{a,m}$.

$$t_{a,m} = t_a^0 \left[1 + 0.15 \left(\frac{v_{a,m}}{C_{a,m}} \right)^4 \right] \quad \forall a \in L,\ m \in \{1, 2, \ldots, M\}, \tag{9}$$

where t_a^0 is the free flow travel time on link a.

User Equilibrium Assignment. For each time interval $m \in \{1, 2, \ldots, M\}$, the user's route choice behavior is assumed to follow Wardrop's first principle [51], which is ensured by

$$\begin{aligned} \min_{(v_m)} \quad & \sum_{a \in L} \int_0^{v_{a,m}} t_{a,m}(\omega)\, d\omega, \\ \text{s.t.} \quad & v_m \in V_m^F \\ & \text{Definitional constraints } (6), (8), (9). \end{aligned} \tag{10}$$

The KKT conditions of this user equilibrium model are shown as follows:

$$\begin{aligned} & t_{i,j,m}\left(v_{i,j,m}, y_{i,j,m}, z_{i,j,m}, l_{i,j}\right) - \left(\rho_{i,m}^w - \rho_{j,m}^w\right) \ge 0 \\ & \qquad \forall w \in W,\ (i,j) \in L,\ m \in \{1, 2, \ldots, M\} \\ & x_{i,j,m}\left[t_{i,j,m}\left(v_{i,j,m}, y_{i,j,m}, z_{i,j,m}, l_{i,j}\right) - \left(\rho_{i,m}^w - \rho_{j,m}^w\right)\right] = 0 \\ & \qquad \forall w \in W,\ (i,j) \in L, \end{aligned} \tag{11}$$

where the multipliers $\rho_{i,m}^w$ and $\rho_{j,m}^w$ are associated with (1) and are called "node potentials" [52].

4.2. Time-Dependent Construction Constraints

Design Constraints with Flexible Construction Duration. In practice, for each expansion project, the workload can be estimated based on the planner's experience. We assume that the normal working hours per day are fixed, for example, 8 hours, and the work efficiency of a crew team is stable. The construction duration for a project can then be roughly estimated according to the workload of that project. The estimated construction duration, denoted as D_a^e, can be expressed as a function of the number of newly added lanes l_a, given by

$$D_a^e = f^a(l_a). \tag{12}$$

In this model, we assume that D_a^e is linearly related to l_a for simplicity. Other functional forms can be adopted in our model framework without difficulty:

$$D_a^e = D_a^0 + D_a^1 \cdot l_a \quad \forall a \in L_1 \tag{13}$$

$$l_a \in \mathbb{Z} \quad \forall a \in L_1, \tag{14}$$

where D_a^0 represents the fixed time cost of the project on link a regardless how many lanes are added, for example, the required time for construction preparation and quality control, and D_a^1 denotes the extra time cost for each additional lane.

In practice, planners may choose to pay extra money for overtime work to accelerate a project if necessary. In this paper, we introduce an integer variable, D_a^r, to denote the reduced component of the construction duration. The actual duration for the project on link a should then be $D_a^e - D_a^r$. Even though overtime work can speed up the process, project duration cannot be infinitely shortened. Let $D_a^{\max}$ denote the maximum allowable shortened duration for a project on link a.

$$D_a^r \le D_a^{\max} \quad \forall a \in L_1 \tag{15}$$

$$D_a^r \in \mathbb{Z}\text{âŊĹ}\forall a \in L_1. \tag{16}$$

Within the planning horizon, the construction process on link $a \in L_1$ should be a continuous period of unit intervals. To properly model the timeline of the construction process, we introduce two additional binary variables, $S_{a,m}$ and $E_{a,m}$. $S_{a,m} = 1$ implies that the construction process on link a starts at the beginning of interval m, and $S_{a,m} = 0$ otherwise. $E_{a,m} = 1$ implies that the construction process on link a ends by the end of interval m, and $E_{a,m} = 0$ otherwise. Note that if a link $a \in L_1$ is not selected for expansion, there will be no construction process on link a, and $S_{a,m} = 0$, $E_{a,m} = 0$, $\forall m \in \{1, 2, \ldots, M\}$. Moreover, there should be only one starting time and one ending time for each chosen project. As shown in Figure 2, we use the same road expansion project used in Figure 1 to illustrate the values of $S_{a,m}$, $E_{a,m}$, $y_{a,m}$, and $z_{a,m}$ throughout the entire planning horizon.

Variables $S_{a,m}$, $E_{a,m}$, $y_{a,m}$, and $z_{a,m}$ are not mutually independent. Based on their definitions and the fact that they are all binary variables, the relationships among them

S	0	1	0	0	0	0	0	0	0	0
E	0	0	0	1	0	0	0	0	0	0
y	0	1	1	1	0	0	0	0	0	0
z	0	0	0	0	1	1	1	1	1	1

FIGURE 2: An example to illustrate the values of $S_{a,m}$, $E_{a,m}$, $y_{a,m}$, and $z_{a,m}$ throughout the entire planning horizon.

can be specified by a series of conditional constraints. Let the construction time window be $[1, M^C]$. Subsequently, the nonconstruction time window is $[M^C+1, M]$. This yields the following constraints:

$$S_{a,m}, E_{a,m}, y_{a,m}, z_{a,m} \in \{0,1\}, \quad \forall a \in L_1,\ m \in \{1,2,\ldots,M\} \tag{17}$$

$$S_{a,m}, E_{a,m}, y_{a,m} = 0 \quad \forall a \in L_1,\ m \in \left\{M^C+1,\ldots,M\right\} \tag{18}$$

$$z_{a,m} = z_{a,M^C+1} \quad \forall a \in L_1,\ m \in \left\{M^C+2,\ldots,M\right\}. \tag{19}$$

Equation (17) requires variables $S_{a,m}$, $E_{a,m}$, $y_{a,m}$, and $z_{a,m}$ to be binary. Equation (18) ensures that no project can start, end, or be under construction in the nonconstruction time window. Equation (19) ensures that the completion status of the potential expansion project on link $a \in L_1$ will not change in the nonconstruction time window.

The logical relationship between $S_{a,m}$ and $y_{a,m}$ can then be given by the following conditional constraints:

$$S_{a,m} \le y_{a,m} \quad \forall a \in L_1,\ m \in \left\{1,2,\ldots,M^C\right\} \tag{20}$$

$$S_{a,m} \le 1 - y_{a,m-1} \quad \forall a \in L_1,\ m \in \left\{2,3,\ldots,M^C\right\} \tag{21}$$

$$S_{a,m} \ge y_{a,m} - y_{a,m-1} \quad \forall a \in L_1,\ m \in \left\{2,3,\ldots,M^C\right\}. \tag{22}$$

Equation (20) ensures that if the project on link a starts at time interval m, that is, $S_{a,m} = 1$, this project must be under construction at interval m: that is, $y_{a,m} = 1$. Equation (21) guarantees that if link a is under construction at time interval $m-1$, that is, $y_{a,m-1} = 1$, it cannot start at time interval m: that is, $S_{a,m} = 0$. Equation (22) ensures that if the project on link a is not under construction at interval $m-1$ and is under construction at time interval m, that is, $y_{a,m} = 1$, $y_{a,m-1} = 0$, then interval m must be the starting time of the project: that is, $S_{a,m} = 1$. These three equations cover all possible relationships between $S_{a,m}$ and $y_{a,m}$.

Similarly, the relationships between $E_{a,m}$ and $y_{a,m}$ are specified by the following constraints:

$$E_{a,m} \le y_{a,m} \quad \forall a \in L_1,\ m \in \left\{1,2,\ldots,M^C\right\} \tag{23}$$

$$E_{a,m} \le 1 - y_{a,m+1} \quad \forall a \in L_1,\ m \in \left\{1,2,\ldots,M^C-1\right\} \tag{24}$$

$$E_{a,m} \ge y_{a,m} - y_{a,m+1} \quad \forall a \in L_1,\ m \in \left\{1,2,\ldots,M^C-1\right\}. \tag{25}$$

Equation (23) means that if the project on link a is to end at time interval m, that is, $E_{a,m} = 1$, this project must be under construction at interval m: that is, $y_{a,m} = 1$. Equation (24) ensures that if the project on link a is to end at time interval m, that is, $E_{a,m} = 1$, this project cannot be under construction at time interval $m+1$: that is, $y_{a,m+1} = 0$. Additionally, (25) guarantees that if the project on link a is under construction at interval m and is no longer under construction at time interval $m+1$, that is, $y_{a,m} = 1$, $y_{a,m+1} = 0$, then interval m must be the ending time of the project; that is, $E_{a,m} = 1$.

Likewise, the logical relationships between $E_{a,m}$ and $z_{a,m}$ are given as follows:

$$E_{a,m} \le z_{a,m+1} \quad \forall a \in L_1,\ m \in \left\{1,2,\ldots,M^C\right\} \tag{26}$$

$$E_{a,m} \le 1 - z_{a,m} \quad \forall a \in L_1,\ m \in \left\{1,2,\ldots,M^C\right\} \tag{27}$$

$$E_{a,m} \ge z_{a,m+1} - z_{a,m} \quad \forall a \in L_1,\ m \in \left\{1,2,\ldots,M^C\right\}. \tag{28}$$

Equation (26) ensures that if time interval m is the ending time of the project on link a, that is, $E_{a,m} = 1$, then in the next interval $m+1$, the project must have been finished: that is, $z_{a,m+1} = 1$. Equation (27) indicates that if in time interval m the project on link a has already been finished, that is, $z_{a,m} = 1$, then time interval m cannot be the ending time: that is, $E_{a,m} = 0$. Equation (28) ensures that if in time interval $m+1$ the project on link $a \in L_1$ has already been finished, that is, $z_{a,m+1} = 1$, but in interval m the project has not been finished, that is, $z_{a,m} = 0$, then interval m must be the ending time of the project: that is, $E_{a,m} = 1$.

Moreover, $S_{a,m}$ and $E_{a,m}$ should satisfy the following constraints:

$$\begin{gathered}\sum_{m=1}^{M^C} S_{a,m} \le 1 \quad \forall a \in L_1 \\ \sum_{m=1}^{M^C} E_{a,m} \le 1 \quad \forall a \in L_1 \\ \sum_{m=1}^{M^C} S_{a,m} = \sum_{m=1}^{M^C} E_{a,m} \quad \forall a \in L_1.\end{gathered} \tag{29}$$

Equations (29) ensure that the potential expansion project on link $a \in L_1$ either has no starting time and no ending time, that is, the project is not selected, or has exactly one starting time and one ending time.

Finally, the following two constraints must also hold:

$$\sum_{m=1}^{M^C} y_{a,m} = (D_a^e - D_a^r) \cdot \sum_{m=1}^{M^C} S_{a,m} \quad \forall a \in L_1 \tag{30}$$

$$\sum_{m=1}^{M^C} y_{a,m} + \sum_{m=1}^{M} z_{a,m} = M \cdot \sum_{m=1}^{M^C} S_{a,m} - \sum_{m=1}^{M^C} (S_{a,m} \cdot (m-1)) \quad \forall a \in L_1. \tag{31}$$

The value of $\sum_{m=1}^{M^C} S_{a,m}$ can represent whether the potential expansion project on link $a \in L_1$ is selected. If the project is selected, $\sum_{m=1}^{M^C} S_{a,m} = 1$, and $\sum_{m=1}^{M^C} S_{a,m} = 0$ otherwise. Therefore, (30) guarantees that if the potential expansion project on link $a \in L_1$ is selected, that is, $\sum_{m=1}^{M^C} S_{a,m} = 1$, the total length of the time intervals under construction equals the actual construction duration. Equation (31) ensures that if the potential expansion project on link $a \in L_1$ is selected, that is, $\sum_{m=1}^{M^C} S_{a,m} = 1$, the total duration of the construction period and benefit period is equal to the planning horizon, subtracting the duration before construction. Note that $\sum_{m=1}^{M^C}(S_{a,m} \cdot (m-1))$ can represent the duration before construction if the potential expansion project on link $a \in L_1$ is selected.

These above constraints, that is, (13)–(31), ensure that if the potential expansion project on link $a \in L_1$ is selected, different phases of the project (i.e., before construction, under construction, and after construction) occur in correct sequence.

In order to reduce the complexity of our model and improve computational speed, the nonlinear constraint (30), together with constraints (7) and (15), can be equivalently replaced by the following linear constraints:

$$\sum_{m=1}^{M^C} y_{a,m} = D_a^0 \cdot \sum_{m=1}^{M^C} S_{a,m} + D_a^1 \cdot l_a - D_a^r \quad \forall a \in L_1 \tag{32}$$

$$l_a \cdot C_a^1 \le \left(C_a^{\max} - C_a^0\right) \cdot \sum_{m=1}^{M^C} S_{a,m} \quad \forall a \in L_1 \tag{33}$$

$$D_a^r \le D_a^{\max} \cdot \sum_{m=1}^{M^C} S_{a,m} \quad \forall a \in L_1. \tag{34}$$

We briefly prove the equivalence by examining both the selected and unselected projects. If the potential expansion project on link $a \in L_1$ is not selected, $\sum_{m=1}^{M^C} S_{a,m} = 0$. Equations (14) and (33) imply that $l_a = 0$. Equations (16) and (34) imply that $D_a^r = 0$. Consequently, (32) implies that $\sum_{m=1}^{M^C} y_{a,m} = 0 = (D_a^e - D_a^r) \cdot \sum_{m=1}^{M^C} S_{a,m}$, which is identical to (30). If the potential expansion project on link $a \in L_1$ is selected, $\sum_{m=1}^{M^C} S_{a,m} = 1$. Equation (33) is reduced to (7), and (34) is reduced to (15). Equations (13) and (32) imply that $\sum_{m=1}^{M^C} y_{a,m} = D_a^e - D_a^r = (D_a^e - D_a^r) \cdot \sum_{m=1}^{M^C} S_{a,m}$, which is identical to (30).

Budget and Resource Constraints. We assume that the government allocates a certain amount of construction budget B_τ at the beginning of time interval τ. This time interval τ does not have to be the same as the predefined time interval m. We introduce a conversion factor β to express the ratio between τ and m. For example, if the unit of m is month, and the unit of τ is year, then β should be 12. In this study, we assume that the remaining budget in period τ is available for use in period $\tau + 1$. Similar assumptions have been employed in [4]. Apart from the budget limitation, we should also consider other resource limitations, for example, the limitation of construction personnel and the limited number of specialized construction equipment. For the sake of simplicity, we only consider the construction personnel limitation in this paper. We assume that all construction teams have the same construction capability, and each ongoing project requires one construction team. The total number of available construction teams is limited and denoted by $R^{\max}$. The budget and construction personnel constraints are then given as follows:

$$\mathrm{TC}_1 + \mathrm{RB}_1 = B_1 \tag{35}$$

$$\mathrm{TC}_\tau + \mathrm{RB}_\tau = \mathrm{RB}_{\tau-1} + B_\tau \quad \forall \tau > 1,\ \tau \in \left\{1, 2, \ldots, \frac{M^C}{\beta}\right\} \tag{36}$$

$$\sum_{a \in L_1} y_{a,m} \le R^{\max} \quad \forall m \in \left\{1, 2, \ldots, M^C\right\}, \tag{37}$$

where TC_τ is the total construction cost generated in period τ and RB_τ is the cumulative remaining budget in period τ. Equations (35) and (36) represent the budget constraints. M^C/β converts the construction time to the same time unit as τ. If the government allocates the entire budget at the beginning of the planning horizon, the budget constraints will be reduced to (35). Equation (37) specifies that, for each time interval $m \in \{1, 2, \ldots, M^C\}$, the total construction teams at work should not exceed the number of available teams.

The total construction cost of a project consists of two components: basic costs to complete the project (e.g., equipment cost, material cost, and labor cost) and extra costs for overtime work. Based on the previous assumption that a construction team works a fixed number of hours per day under normal condition, we assume that, without overtime work, each construction time interval for the expansion project on link $a \in L_1$ includes an identical and fixed basic cost, denoted as c_a. c_a includes all of the wages for workers and other costs (e.g., material costs, equipment costs) needed in a normal construction time interval. If the planner wants to shorten the duration of a project, workers may choose any period to work overtime as long as they meet the time

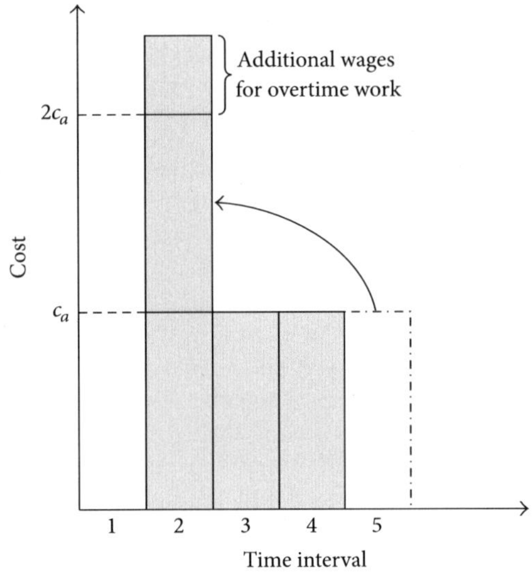

FIGURE 3: Illustration of construction costs for a project.

limit requirement. To simplify our model, we assume the overtime cost to be placed in the starting time interval of any project (Figure 3). The example in Figure 3 corresponds to the example in Figures 1 and 2. The expansion project originally lasts for four months, and it is reduced to three months through overtime. Therefore, both the basic cost in the fourth construction interval and the additional wages attributable to overtime work are allocated to the first construction interval when applying overtime work.

The total construction cost in period τ can be formulated as

$$\mathrm{TC}_\tau = \left[\sum_{m=\beta(\tau-1)+1}^{\beta\tau} \sum_{a\in L_1} y_{a,m} \cdot c_a + \sum_{a\in L_1} \mathrm{OC}_{a,\tau} \right] \cdot (1+\theta_1)^{\tau-1} \quad \forall \tau \in \left\{1,2,\ldots,\frac{M^C}{\beta}\right\}, \tag{38}$$

where $\mathrm{OC}_{a,\tau}$ is the overtime cost for the project on link a in period τ. θ_1 represents the inflation rate.

$$\sum_{\tau=1}^{M^C/\beta} \mathrm{OC}_{a,\tau} = \lambda \cdot c_a \cdot (1+\mu) \cdot D_a^r + (1-\lambda) \cdot c_a \cdot D_a^r \quad \forall a \in L_1 \tag{39}$$

$$\mathrm{OC}_{a,\tau} \geq 0 \quad \forall a \in L_1,\ \forall \tau \in \left\{1,2,\ldots,\frac{M^C}{\beta}\right\} \tag{40}$$

$$\mathrm{OC}_{a,\tau} \leq \sum_{m=\beta(\tau-1)+1}^{\beta\tau} S_{a,m} \cdot Q \quad \forall a \in L_1,\ \forall \tau \in \left\{1,2,\ldots,\frac{M^C}{\beta}\right\}, \tag{41}$$

where λ denotes the percentage of the workers' salary in the total construction cost, μ is the increased rate of overtime salary, and Q is a large constant value. The first term on the right-hand side of (39) represents the salary portion of the overtime cost, and the second term represents the remaining portion. Equations (39)–(41) ensure that the overtime cost for the expansion project on link $a \in L_1$ is placed in the starting period.

4.3. Objective Function. As aforementioned, during the construction period, the system performance may deteriorate due to work zones or closure of lanes. Different decision makers may have different preferences when selecting road expansion projects. Some may focus more on future benefits of the projects, while others may focus more on reducing adverse impacts of the projects during construction. To provide a flexible model, the objective function is to minimize the weighted sum of the total travel time during construction period and the total travel time during benefit period. This can be stated as

$$\mathrm{NPV} = \alpha_1 \sum_{\tau=1}^{M^C/\beta} \frac{\mathrm{TT}_\tau}{(1+\theta_2)^{\tau-1}} + \alpha_2 \sum_{\tau=M^C/\beta+1}^{M/\beta} \frac{\mathrm{TT}_\tau}{(1+\theta_2)^{\tau-1}}, \tag{42}$$

where α_1 and α_2 are the weighting factors for the construction period and benefit period, respectively, TT_τ is the total travel time occurring in period τ, θ_2 is the discount rate, and $(1+\theta_2)^{\tau-1}$ represents the discount factor for period τ. TT_τ is given in

$$\mathrm{TT}_\tau = \sum_{m=\beta(\tau-1)+1}^{\beta\tau} \sum_{a\in L} t_{a,m} v_{a,m} \quad \forall \tau \in \left\{1,2,\ldots,\frac{M}{\beta}\right\}. \tag{43}$$

4.4. The Base Model. Based on the above notations, the time-dependent discrete network design problem considering construction impact and flexible duration can be formulated as the following mathematical program (P1).

$$\min \quad \mathrm{NPV} = \alpha_1 \sum_{\tau=1}^{M^C/\beta} \frac{\sum_{m=\beta(\tau-1)+1}^{\beta\tau} \sum_{a\in L} t_{a,m} v_{a,m}}{(1+\theta_2)^{\tau-1}} + \alpha_2 \sum_{\tau=M^C/\beta+1}^{M/\beta} \frac{\sum_{m=\beta(\tau-1)+1}^{\beta\tau} \sum_{a\in L} t_{a,m} v_{a,m}}{(1+\theta_2)^{\tau-1}}$$

$$\text{s.t.} \quad v_m \in V_m^F \quad \forall m \in \{1,2,\ldots,M\}$$

Time-dependent definitional constraints, equations (6), (8), (9);

Traffic assignment constraints, equations (11);

Design constraints, equations (13), (14), (16) – (29), (31) – (34);

Budget and resource constraints, equations (35) – (41).

(P1)

This formulation involves two integer variables (i.e., l_a and D_a^r). As is generally known, it is much more difficult to solve optimization problems with integer variables, especially for large-scale networks. Hence, we introduce two sets of binary variables, p_a^{b1} and q_a^{b2}, to replace l_a and D_a^r as follows:

$$l_a = \sum_{b1=1}^{B1} 2^{(b1-1)} p_a^{b1} \quad \forall a \in L_1 \tag{44}$$

$$D_a^r = \sum_{b2=1}^{B2} 2^{(b2-1)} q_a^{b2} \quad \forall a \in L_1. \tag{45}$$

According to (44), the number of newly built lanes l_a can take the value 0 to $(2^{B1} - 1)$. For example, if we use three binary variables to represent l_a, that is, $B1 = 3$, then $l_a = p_a^1 + 2p_a^2 + 4p_a^3$, ranging from 0 to 7. Similarly, the reduced value of construction interval D_a^r can range from 0 to $(2^{B2} - 1)$. Note that the binary variables can be written in the form of complementarity constraints so that the binary variable can be treated as continuous variables as follows:

$$\begin{aligned} 0 \le p_a^{b1} \le 1 \quad &\forall a \in L_1 \\ p_a^{b1}\left(1 - p_a^{b1}\right) = 0 \quad &\forall a \in L_1 \\ 0 \le q_a^{b1} \le 1 \quad &\forall a \in L_1 \\ q_a^{b1}\left(1 - q_a^{b1}\right) = 0 \quad &\forall a \in L_1. \end{aligned} \tag{46}$$

Then, l_a and D_a^r in previous equations can be replaced by $\sum_{b1=1}^{B1} 2^{(b1-1)} p_a^{b1}$ and $\sum_{b2=1}^{B2} 2^{(b2-1)} q_a^{b2}$, respectively. Equations (6), (13), (32), (33), (34), and (39) are replaced by (47), (48), (49), (50), (51), and (52), respectively.

$$C_{a,m} = C_a^0 - y_{a,m} \cdot C_a^r + z_{a,m} \cdot \sum_{b1=1}^{B1} 2^{(b1-1)} p_a^{b1} \cdot C_a^1 \quad \forall a \in L,\ m \in \{1, 2, \dots, M\} \tag{47}$$

$$D_a^e = D_a^0 + D_a^1 \cdot \sum_{b1=1}^{B1} 2^{(b1-1)} p_a^{b1} \quad \forall a \in L_1 \tag{48}$$

$$\sum_{m=1}^{M^C} y_{a,m} = D_a^0 \cdot \sum_{m=1}^{M^C} S_{a,m} + D_a^1 \cdot \sum_{b1=1}^{B1} 2^{(b1-1)} p_a^{b1} - \sum_{b2=1}^{B2} 2^{(b2-1)} q_a^{b2} \quad \forall a \in L_1 \tag{49}$$

$$\sum_{b1=1}^{B1} 2^{(b1-1)} p_a^{b1} \cdot C_a^1 \le \left(C_a^{\max} - C_a^0\right) \cdot \sum_{m=1}^{M^C} S_{a,m} \quad \forall a \in L_1 \tag{50}$$

$$\sum_{b2=1}^{B2} 2^{(b2-1)} q_a^{b2} \le D_a^{\max} \cdot \sum_{m=1}^{M^C} S_{a,m} \quad \forall a \in L_1 \tag{51}$$

$$\sum_{\tau=1}^{M^C/\beta} \mathrm{OC}_{a,\tau} = \lambda \cdot c_a \cdot (1 + \mu) \cdot \sum_{b2=1}^{B2} 2^{(b2-1)} q_a^{b2} + (1 - \lambda) \cdot c_a \cdot \sum_{b2=1}^{B2} 2^{(b2-1)} q_a^{b2} \quad \forall a \in L_1. \tag{52}$$

We also use the following complementarity constraints to replace (17) so that the binary variables $S_{a,m}$, $E_{a,m}$, $y_{a,m}$, and $z_{a,m}$ can be treated as continuous variables:

$$\begin{aligned} &0 \le S_{a,m} \le 1 \quad \forall a \in L_1,\ m \in \left\{1, 2, \dots, M^C\right\} \\ &S_{a,m}\left(1 - S_{a,m}\right) = 0 \quad \forall a \in L_1,\ m \in \left\{1, 2, \dots, M^C\right\} \\ &0 \le E_{a,m} \le 1 \quad \forall a \in L_1,\ m \in \left\{1, 2, \dots, M^C\right\} \\ &E_{a,m}\left(1 - E_{a,m}\right) = 0 \quad \forall a \in L_1,\ m \in \left\{1, 2, \dots, M^C\right\} \\ &0 \le y_{a,m} \le 1 \quad \forall a \in L_1,\ m \in \left\{1, 2, \dots, M^C\right\} \\ &y_{a,m}\left(1 - y_{a,m}\right) = 0 \quad \forall a \in L_1,\ m \in \left\{1, 2, \dots, M^C\right\} \\ &0 \le z_{a,m} \le 1 \end{aligned}$$

$$\forall a \in L_1,\ m \in \{1,2,\ldots,M^C+1\}$$

$$z_{a,m}(1 - z_{a,m}) = 0$$

$$\forall a \in L_1,\ m \in \{1,2,\ldots,M^C+1\}. \tag{53}$$

Based on the above discussions, the base model can be reformulated as follows:

$$\begin{aligned}
\min \quad & \text{NPV} = \alpha_1 \sum_{\tau=1}^{M^C/\beta} \frac{\sum_{m=\beta(\tau-1)+1}^{\beta\tau} \sum_{a\in L} t_{a,m} v_{a,m}}{(1+\theta_2)^{\tau-1}} + \alpha_2 \sum_{\tau=M^C/\beta+1}^{M/\beta} \frac{\sum_{m=\beta(\tau-1)+1}^{\beta\tau} \sum_{a\in L} t_{a,m} v_{a,m}}{(1+\theta_2)^{\tau-1}} \\
\text{s.t.} \quad & v_m \in V_m^F \quad \forall m \in \{1,2,\ldots,M\} \\
& (46) - (53); \\
& \text{Time-dependent definitional constraint, equations } (8),\ (9); \\
& \text{Traffic assignment constraints, equations } (11); \\
& \text{Design constraints, equations } (18) - (29),\ (31); \\
& \text{Budget and resource constraints, equations } (35) - (38),\ (40) - (41).
\end{aligned} \tag{P2}$$

5. Solution Algorithm

(P2) is a mathematical program with complementarity constraints (MPCC). It is well known that MPCC problems are difficult to solve because the feasible region of an MPCC problem is not convex, and the Mangasarian-Fromovitz constraint qualification (MFCQ) fails to hold [53]. Several previous efforts have been undertaken to make the problem easier to solve (see [54, 55]). In this paper, we extend the active-set algorithm (ASA) proposed by Zhang et al. [35] to solve the MPCC problem. Figure 4 shows the fundamental concepts of our model and the conceptual solution procedure of the ASA.

Instead of solving an MPCC directly, the ASA solves two simpler problems sequentially. The first problem is a restricted version of (P2), in which the variables p_a^{b1}, q_a^{b1}, $S_{a,m}$, $E_{a,m}$, $y_{a,m}$, and $z_{a,m}$ have a set of predefined values. The initial set of values is only a feasible solution, not necessarily the best solution. Therefore, we need to make adjustments to find a better solution, which is realized through the second problem, a (SUB) problem. For each variable, that is, p_a^{b1}, q_a^{b1}, $S_{a,m}$, $E_{a,m}$, $y_{a,m}$, or $z_{a,m}$, there are only two possible values: 0 and 1. We divide all components of each variable into two active sets, where one set stores the components with a value of 0 and the other set stores the components with a value of 1. Then, (46) and (53) can be replaced by the following equations:

$$p_a^{b1} = 0 \quad \forall (a, b1) \in \Omega_{p,0},\ b1 \in B1 \tag{54}$$

$$p_a^{b1} = 1 \quad \forall (a, b1) \in \Omega_{p,1},\ b1 \in B1 \tag{55}$$

$$q_a^{b2} = 0 \quad \forall (a, b2) \in \Omega_{q,0},\ b2 \in B2 \tag{56}$$

$$q_a^{b2} = 1 \quad \forall (a, b2) \in \Omega_{q,1},\ b2 \in B2 \tag{57}$$

$$S_{a,m} = 0 \quad \forall (a, m) \in \Omega_{S,0} \tag{58}$$

$$S_{a,m} = 1 \quad \forall (a, m) \in \Omega_{S,1} \tag{59}$$

$$E_{a,m} = 0 \quad \forall (a, m) \in \Omega_{E,0} \tag{60}$$

$$E_{a,m} = 1 \quad \forall (a, m) \in \Omega_{E,1} \tag{61}$$

$$y_{a,m} = 0 \quad \forall (a, m) \in \Omega_{y,0} \tag{62}$$

$$y_{a,m} = 1 \quad \forall (a, m) \in \Omega_{y,1} \tag{63}$$

$$z_{a,m} = 0 \quad \forall (a, m) \in \Omega_{z,0} \tag{64}$$

$$z_{a,m} = 1 \quad \forall (a, m) \in \Omega_{z,1}. \tag{65}$$

The restricted version of (P2) can be formulated as

$$\begin{aligned}
\min_{(v,p,q,S,E,y,z,\rho)} \quad & \text{NPV} = \alpha_1 \sum_{\tau=1}^{M^C/\beta} \frac{\text{TT}_\tau}{(1+\theta_2)^{\tau-1}} + \alpha_2 \sum_{\tau=M^C/\beta+1}^{M/\beta} \frac{\text{TT}_\tau}{(1+\theta_2)^{\tau-1}} \\
\text{s.t.} \quad & v_m \in V_m^F \quad \forall m \in \{1,2,\ldots,M\}
\end{aligned}$$

(47) – (52)

(54) – (65)

Time-dependent definitional constraint, equations (8), (9);

Traffic assignment constraints, equations (11);

Design constraints, equations (18) – (29), (31);

Budget and resource constraints, equations (35) – (38), (40) - (41).

(P3)

Let $\overline{v}_m$ denote the solution of the UE problem for time period m ($m \in \{1, 2, \ldots, M\}$) for a given feasible design $(\overline{p}, \overline{q}, \overline{S}, \overline{E}, \overline{y}, \overline{z})$, and combine $\overline{v}_m$ into one vector denoted as $\overline{v}$ (i.e., $\overline{v} = \{\overline{v}_1, \overline{v}_2, \ldots, \overline{v}_{m=M}\}$). According to Proposition 1.1 in a study conducted by Facchinei and Pang [56], there must exist a multiplier vector $\overline{\rho}$ associated with (1) so that $(\overline{p}, \overline{q}, \overline{S}, \overline{E}, \overline{y}, \overline{z}, \overline{v}, \overline{\rho})$ is also optimal to (P3). Therefore, instead of solving (P3) directly, we can obtain the optimal solution for (P3) by solving a set of corresponding UE problems. Let δ_a^{b1} and γ_a^{b1} denote the multipliers associated with (54) and (55), respectively, and let $\sigma_{a,m}$, $\xi_{a,m}$, $\chi_{a,m}$, and $\varrho_{a,m}$ denote the multipliers associated with (62), (63), (64), and (65), respectively. The feasible design and the active sets can then be improved based on the information obtained from these multipliers. For example, if $\delta_a^{b1} < 0$ for some specific $(a, b1) \in \Omega_{p,0}$, shifting the $(a, b1)$ from $\Omega_{p,0}$ to $\Omega_{p,1}$ may reduce the objective function value. And if $\gamma_a^{b1} > 0$ for some specific $(a, b1) \in \Omega_{p,1}$, it may be beneficial to shift the $(a, b1)$ from $\Omega_{p,1}$ to $\Omega_{p,0}$. Similarly, the multipliers $\sigma_{a,m}$, $\xi_{a,m}$, $\chi_{a,m}$, and $\varrho_{a,m}$ provide information on updating $\Omega_{y,0}$, $\Omega_{y,1}$, $\Omega_{z,0}$, and $\Omega_{z,1}$, respectively. The switching process, however, may make the budget and crew constraints unsatisfactory. For this reason, the following (SUB) problem is used to complete the switching process as well as preventing problem (P3) from becoming infeasible.

$$\min_{(g,h,\eta)} \sum_{(a,b1)\in\Omega_{p,0}} K^{b1}\delta_a^{b1} g_a^{b1} + \sum_{(a,m)\in\Omega_{y,0}} \sigma_{a,m}h_{a,m} + \sum_{(a,m)\in\Omega_{z,0}} \chi_{a,m}\eta_{a,m} - \sum_{(a,b1)\in\Omega_{p,1}} K^{b1}\gamma_a^{b1} g_a^{b1} - \sum_{(a,m)\in\Omega_{y,1}} \xi_{a,m}h_{a,m} - \sum_{(a,m)\in\Omega_{z,1}} \varrho_{a,m}\eta_{a,m} \quad \text{(SUB)}$$

s.t. Design constraints, equations (17) – (29), (31), (47) – (52);

Budget and resource constraints, equations (35) – (38), (40) - (41), (52);

$$p_a^{b1} = g_a^{b1} \quad \forall (a, b1) \in \Omega_{p,0} \quad (66)$$

$$p_a^{b1} = 1 - g_a^{b1} \quad \forall (a, b1) \in \Omega_{p,1} \quad (67)$$

$$y_{a,m} = h_{a,m} \quad \forall (a, m) \in \Omega_{y,0} \quad (68)$$

$$y_{a,m} = 1 - h_{a,m} \quad \forall (a, m) \in \Omega_{y,1} \quad (69)$$

$$z_{a,m} = \eta_{a,m} \quad \forall (a, m) \in \Omega_{z,0} \quad (70)$$

$$z_{a,m} = 1 - \eta_{a,m} \quad \forall (a, m) \in \Omega_{z,1} \quad (71)$$

$$\sum_{b1=1}^{B1} g_a^{b1} \le 1 \quad \forall a \in L_1 \quad (72)$$

$$g_a^{b1} \in \{0, 1\} \quad \forall a \in L_1,\ b1 \in B1 \quad (73)$$

$$h_{a,m}, \eta_{a,m} \in \{0, 1\} \quad \forall a \in L_1,\ m \in \left\{1, 2, \ldots, M^{\mathrm{CP}}\right\} \quad (74)$$

$$q_a^{b2} \in \{0, 1\} \quad \forall a \in L_1,\ b2 \in B2 \quad (75)$$

$$\sum_{(a,b1)\in\Omega_{p,0}} K^{b1}\delta_a^{b1} g_a^{b1} + \sum_{(a,m)\in\Omega_{y,0}} \sigma_{a,m}h_{a,m} + \sum_{(a,m)\in\Omega_{z,0}} \chi_{a,m}\eta_{a,m} - \sum_{(a,b1)\in\Omega_{p,1}} K^{b1}\gamma_a^{b1} g_a^{b1} - \sum_{(a,m)\in\Omega_{y,1}} \xi_{a,m}h_{a,m} - \sum_{(a,m)\in\Omega_{z,1}} \varrho_{a,m}\eta_{a,m} > \varphi, \quad (76)$$

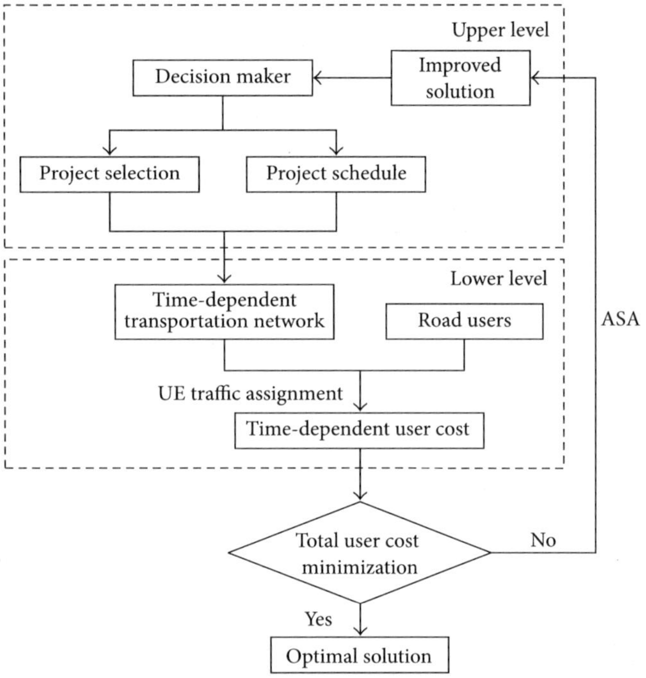

FIGURE 4: Framework of the T-DNDP model and ASA.

TABLE 1: Link characteristics of the Nguyen-Dupuis network.

Link	Free flow travel time t_a^0 (min)	Initial capacity C_a^0
1-5	11.17	177
1-12	14.36	104
4-5	14.36	163
4-9	18.88	235
5-6	4.79	245
5-9	14.36	121
6-7	7.98	295
6-10	70.75	213
7-8	7.98	183
7-11	14.36	291
8-2	14.36	275
9-10	12.32	221
9-13	11.25	278
10-11	12.76	241
11-2	14.36	283
11-3	12.77	169
12-6	11.17	164
12-8	5.05	179
13-3	11.25	278

where binary variables g_a^{b1}, $h_{a,m}$, and $\eta_{a,m}$ are "switch" variables, indicating whether to move the corresponding design variable to the complementary set. Equation (72) ensures that only one digit of variable p_a^{b1} can be changed at a time to prevent too much fluctuation in iterations. Equation (76) gives a predetermined lower bound to the objective function value of the (SUB) problem. A vector of constant K^{b1} is introduced to ensure that changes are always made to the smallest digit possible, because the multipliers generated by the CONOPT solver [57] are linear in magnitude with respect to its digit $b1$. Note that although we only introduce "switch" variables for variables p_a^{b1}, $y_{a,m}$, and $z_{a,m}$, due to the dependency relationships among p_a^{b1}, q_a^{b1}, $S_{a,m}$, $E_{a,m}$, $y_{a,m}$, and $z_{a,m}$, variables q_a^{b1}, $S_{a,m}$, and $E_{a,m}$ will also be determined.

The procedure to solve the T-DNDP is as follows.

Step 0. Choose an initial feasible design (p_a^{b1}, q_a^{b1}, $S_{a,m}$, $E_{a,m}$, $y_{a,m}$, and $z_{a,m}$) and solve the UE problem. Initialize sets $\Omega_{p,0}$, $\Omega_{p,1}$, $\Omega_{q,0}$, $\Omega_{q,1}$, $\Omega_{S,0}$, $\Omega_{S,1}$, $\Omega_{E,0}$, $\Omega_{E,1}$, $\Omega_{y,0}$, $\Omega_{y,1}$, $\Omega_{z,0}$, and $\Omega_{z,1}$.

Step 1. Solve (P3) and denote the optimal objective function value as TT. Obtain multipliers δ_a^{b1}, γ_a^{b1}, $\sigma_{a,m}$, $\xi_{a,m}$, $\chi_{a,m}$, and $\varrho_{a,m}$.

Step 2. Set $\varphi = -\infty$ and let ($\overline{p}_a^{b1}$, $\overline{q}_a^{b1}$, $\overline{S}_{a,m}$, $\overline{E}_{a,m}$, $\overline{y}_{a,m}$, and $\overline{z}_{a,m}$) solve the (SUB) problem. Denote the optimal objective function value as $\overline{\varphi}$. If $\overline{\varphi} = 0$, stop, as ($\overline{p}_a^{b1}$, $\overline{q}_a^{b1}$, $\overline{S}_{a,m}$, $\overline{E}_{a,m}$, $\overline{y}_{a,m}$, and $\overline{z}_{a,m}$) is the best solution found. Otherwise, go to Step 3.

Step 3. Solve the UE problem with ($\overline{p}_a^{b1}$, $\overline{q}_a^{b1}$, $\overline{S}_{a,m}$, $\overline{E}_{a,m}$, $\overline{y}_{a,m}$, and $\overline{z}_{a,m}$). If the total travel time associated with the UE distribution is greater than TT, set $\varphi = \overline{\varphi} + \varepsilon$, where $\varepsilon > 0$ is sufficiently small, and return to Step 2. Otherwise, use ($\overline{p}_a^{b1}$, $\overline{q}_a^{b1}$, $\overline{S}_{a,m}$, $\overline{E}_{a,m}$, $\overline{y}_{a,m}$, and $\overline{z}_{a,m}$) to update the current design (p_a^{b1}, q_a^{b1}, $S_{a,m}$, $E_{a,m}$, $y_{a,m}$, and $z_{a,m}$) and sets $\Omega_{p,0}$, $\Omega_{p,1}$, $\Omega_{q,0}$, $\Omega_{q,1}$, $\Omega_{S,0}$, $\Omega_{S,1}$, $\Omega_{E,0}$, $\Omega_{E,1}$, $\Omega_{y,0}$, $\Omega_{y,1}$, $\Omega_{z,0}$, and $\Omega_{z,1}$. Return to Step 1.

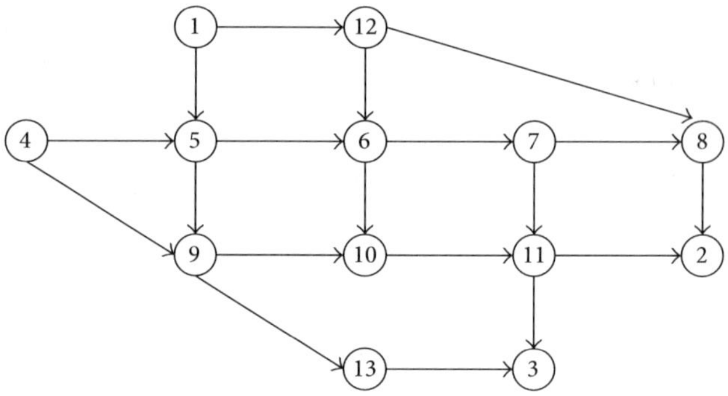

FIGURE 5: Nguyen-Dupuis network.

6. Numerical Studies

In this section, two numerical examples are presented to demonstrate the proposed model and solution algorithm.

6.1. Example 1: Nguyen-Dupuis Network. To illustrate the usefulness and advantages of our model, we first solve it for the Nguyen-Dupuis network [58] with two different scenarios. As shown in Figure 5, the Nguyen-Dupuis network consists of 13 nodes, 19 links, and four O-D pairs. Table 1 reports the link characteristics of the network. The travel

TABLE 2: Parameters for candidate projects in the Nguyen-Dupuis network.

Candidate link	Initial capacity C_a^0	Lane capacity C_a^1	Maximum allowable capacity $C_a^{\max}$	Number of closed lanes k_a	Fixed construction duration (month)	Extra duration for adding one lane D_a^1
5-9	121	121	242	0	3	3
7-11	291	291	582	1	4	2
9-10	241	241	482	0	6	6
9-13	278	278	556	1	7	8
10-11	241	241	482	1	5	7
12-6	164	164	328	0	3	5
13-3	278	278	556	0	6	9

TABLE 3: System performance comparison for Scenario 1.

	Net user cost in construction period	Net user cost in benefit period	Total weighted net user cost NPV
Separate optimization	105952239	132304481	119128360
Joint optimization	24516051	143304748	83910400
		Improvement	29.6%

demand is given by [58]: $q_{1\to2} = 400$ veh/h; $q_{1\to3} = 800$ veh/h; $q_{4\to2} = 600$ veh/h; $q_{4\to3} = 200$ veh/h.

Scenario 1: Considering Construction Impacts during Project Selection. In this scenario, we show that our model can consider the construction impacts during the project selection process and thus provide better solutions than conventional methods (i.e., separately optimizing the selection and schedule of road expansion projects).

In order to clearly illustrate the results without being distracted by other factors, this scenario does not consider overtime policy. Assume that there are six candidate road expansion projects on links (5, 9), (7, 11), (9, 10), (10, 11), (9, 13), and (13, 3). The parameters for the candidate projects are given in Table 2. Other parameters are given as follows:

(1) Planning horizon: 20 years (the planning horizon is equally divided into 240 design periods, namely, 240 months); construction period: two years; benefit period: 18 years

(2) Weighting for construction period $\alpha_1 = 0.5$; weighting for benefit period $\alpha_2 = 0.5$

(3) Budget: $B_1 = 20$, $B_2 = 20$

(4) Number of available crew teams: $R^{\max} = 2$

(5) Inflation rate $\theta_1 = 0.01$; discount rate $\theta_2 = 0.05$

(6) Conversion factor: $\beta = 12$

The ASA solution procedure is implemented using GAMS [59] and CONOPT solver [57] on a Dell computer with a 3.4 GHz processor and 16.0 GB RAM. It takes 23 minutes and 48 seconds to solve the model. The project selection and scheduling results are shown in Figure 6. To show the benefits of our model, we separately optimize the selection and schedule of road expansion projects. Figure 7 presents the results with separate optimization, and Table 3 compares the system performance under the two different approaches. It can be observed that the joint optimization approach improves the overall system performance by 29.6%. Compared with the separate optimization approach, the joint optimization results have much better performance in the construction period and a little bit worse performance in the benefit period.

Through further comparing the selected projects in the two approaches, we have the following observations: first, when other conditions keep the same, the project on link (9, 13) will have the same benefit as the project on link (13, 3); second, when other conditions keep the same, the project on link (10, 11) will have slightly higher benefit than the project on link (9,10); third, the projects on links (13, 3) and (9, 10) will have no adverse construction impact because they do not require lane closures, while the projects on links (9, 13) and (10, 11) will have great adverse construction impacts because they both require lane closures. Based on these observations, the results of the joint and separate optimization approaches can be further analyzed. Due to the fact that the separate optimization approach only considers the benefits but neglects the construction impacts when selecting road expansion projects, the projects on links (9, 13) and (10, 11) are selected. Nevertheless, because the joint optimization approach can explicitly consider the potential construction impact during project selection and scheduling, the projects on links (13, 3) and (9, 10) that have better overall performance are selected. Therefore, the proposed joint optimization approach has the potential of providing better solutions for planners.

Compared with the conventional planning approach that separately selects and schedules road expansion projects, the proposed time-dependent joint optimization approach can help planners choose the projects that not only have significant benefits after completion but also yield relatively less adverse impacts during construction. As shown in the above numerical experiment, this joint optimization approach is beneficial, especially when there are projects with similar potential benefits but quite different construction impacts.

TABLE 4: System performance comparison for Scenario 2.

	Net user cost in construction period	Net user cost in benefit period	Total weighted net user cost NPV
Separate optimization	105952239	132304481	127034033
Joint optimization	29964650	132304481	111836515
		Improvement	12.0%

TABLE 5: Parameters for candidate projects in the Sioux Falls network.

Link	Lane capacity C_a^1	Maximum allowable capacity $C_a^{\max}$	Number of closed lanes k_a	Maximum allowable shortened duration $D_a^{\max}$	Fixed duration D_a^0	Extra duration for adding one lane D_a^1
(1, 2)	13.0	40	1	4	8	8
(9, 8)	3.0	12	1	2	3	3
(11, 10)	5.0	15	1	1	3	3
(12, 13)	13.0	50	1	2	4	4
(14, 15)	3.0	9	1	1	2	2
(15, 19)	8.0	32	1	1	2	2
(16, 18)	10.0	40	1	1	2	3
(18, 20)	12.0	36	1	2	4	4
(23, 22)	2.4	8	1	0	1	1
(24, 21)	2.4	8	1	0	1	1

Scenario 2: Focusing More on Future Benefits. In Scenario 1, we consider a 20-year planning horizon with a two-year construction period and an 18-year benefit period and assume that the weightings for construction period and benefit period are the same (i.e., $\alpha_1 = \alpha_2 = 0.5$). This assumption is preferred for planners who focus on the near-term overall performance of a transportation network. For planners who focus more on future benefits of road expansion projects, they can choose relatively higher weighting for the benefit period.

In this scenario, the weighting factors are given by $\alpha_1 = 0.2$ and $\alpha_2 = 0.8$ and other parameters are the same as Scenario 1. Note that, with these weighting factors, it is approximately equivalent to considering a 72-year benefit period. The new results from our joint optimization model are shown in Figure 8. The selection and scheduling results of the separate optimization approach will not change. Table 4 compares the system performance under the two different approaches. It can be observed that the joint optimization approach improves the overall system performance by 12.0%. Compared with the separate optimization approach, the joint optimization results have the same performance in the benefit period but have better performance in the construction period. Compared with Scenario 1, this scenario selects the project on link (10, 11) instead of the project on link (9, 10) because the project on link (10, 11) will lead to better overall system performance. We should note that, because the system performance in the benefit period for the two results is the same, the separate optimization approach may obtain the same optimal solution as the joint optimization approach under the best-case situation. However, because the separate optimization approach cannot consider the construction impact during project selection, it has a high chance to obtain the less optimal solutions.

This scenario first shows the flexibility of our model in considering planners with different preferences. It also further demonstrates that the proposed time-dependent joint optimization approach can provide better solutions than a separate optimization approach because it considers construction impacts during project selection.

6.2. Example 2: Sioux Falls Network. To further demonstrate the real-world applicability of our model, we solve it for the transportation network of the city of Sioux Falls. Figure 9 shows the network of Sioux Falls. The yellow lines represent the links with candidate projects. The network data are derived from a study conducted by Leblanc et al. [15], and the attributes of all 10 candidate projects are given in Table 5. Other parameters are given as follows:

(1) Planning horizon: 20 years (the planning horizon is equally divided into 240 design periods, namely, 240 months); construction period: two years; benefit period: 18 years

(2) Weighting for construction period $\alpha_1 = 0.5$; weighting for benefit period $\alpha_2 = 0.5$

(3) Budget: $B_1 = 15$, $B_2 = 20$

(4) Number of available crew teams: $R^{\max} = 2$

(5) Inflation rate $\theta_1 = 0.01$; discount rate $\theta_2 = 0.05$

(6) Conversion factor: $\beta = 12$

(7) Percentage of the workers' salary in the total construction cost: $\lambda = 0.1$

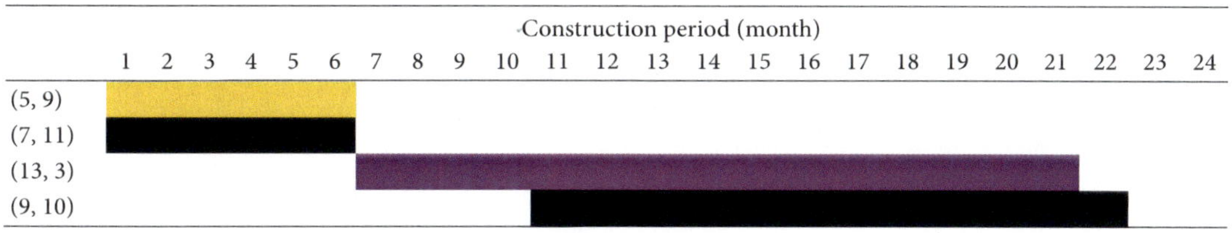

FIGURE 6: Selection and scheduling results with joint optimization for Scenario 1.

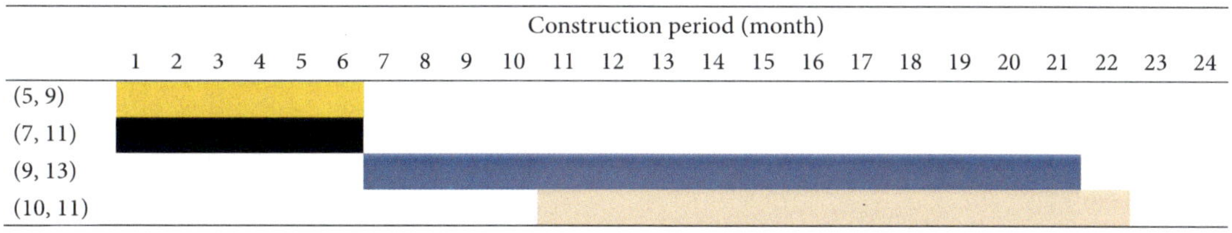

FIGURE 7: Selection and scheduling results with separate optimization for Scenario 1.

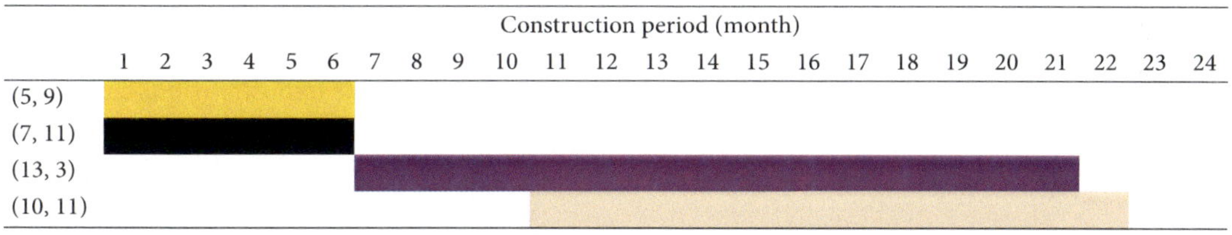

FIGURE 8: Selection and scheduling results with joint optimization for Scenario 2.

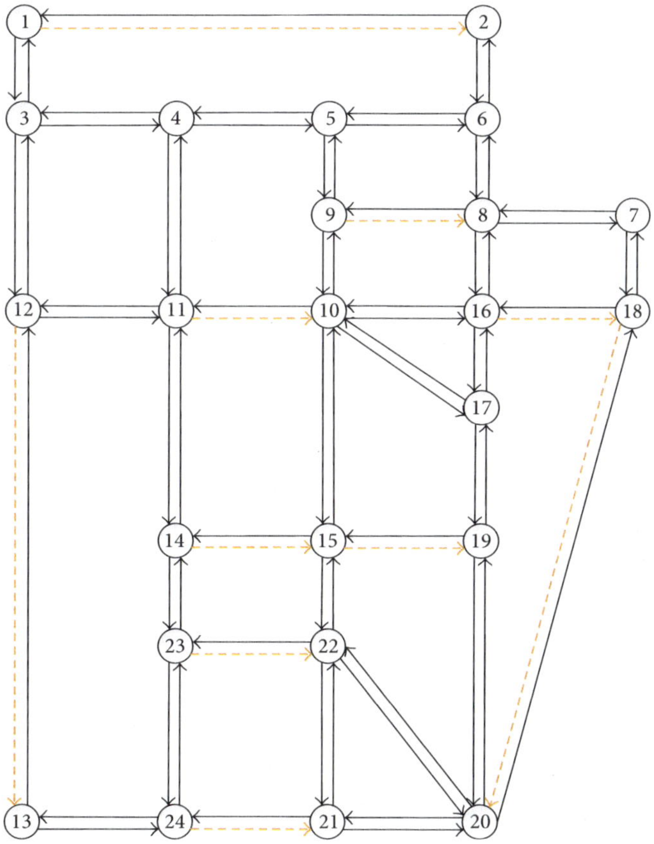

FIGURE 9: Network of Sioux Falls.

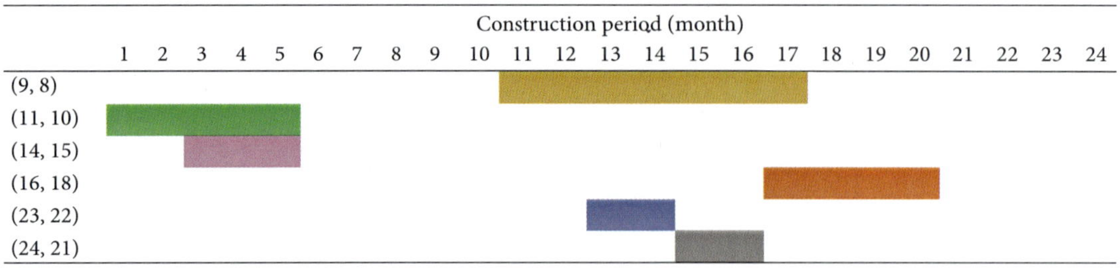

Figure 10: Illustration of the scheduling results for Example 2.

Table 6: Selection and scheduling results for Example 2.

	Stating time	Ending time	Newly added lanes	Reduced construction duration
(9, 8)	11	17	2	2
(11, 10)	1	5	1	1
(14, 15)	3	5	1	1
(16, 18)	17	20	1	1
(23, 22)	13	14	1	0
(24, 21)	15	16	1	0

(8) Overtime salary parameter: $\mu = 0.5$

(9) Normal costs per period without overtime work: $c_a = 1$

The ASA solution procedure is implemented using GAMS [59] and CONOPT solver [57] on a Dell computer with a 3.4 GHz processor and 16.0 GB RAM. It takes 4 hours, 13 minutes, and 30 seconds to solve the model. The project selection and scheduling results are provided in Table 6. To make the scheduling results more readable, Figure 10 provides a graphical representation. We can observe that six projects are chosen, among which four are chosen to be shortened by overtime work. The construction duration of the project on link (9, 8) is shortened by two months, and for the other three projects on links (11, 10), (14, 15), and (16, 18), the construction duration is shortened by one month. Total construction costs generated in the first and second years are 14.526 and 14.132, respectively, which are within the budget. According to the scheduling results, no more than two projects are under construction simultaneously. Hence, the resource constraint is also met. Without any road expansion projects, the total weighted net user cost will be 5.662×10^{10}. The selected road expansion projects will reduce the total weighted net use cost to 4.274×10^{10}. The overall system performance within the planning horizon is improved by 24.5%.

7. Concluding Remarks

The primary significance of the model developed in this paper is that it introduces time dimension into the traditional NDP to consider the impact of road construction work and applies overtime policy to further improve the design. The proposed model can solve the capacity expansion project selection and project scheduling problems simultaneously. The proposed T-DNDP model also allows for the addition of time-dependent resource constraints. We employ the active-set algorithm to solve this problem and test two numerical examples to demonstrate the effectiveness of the proposed model. The results show that the proposed T-DNDP model has the potential of providing better solutions than the conventional approach that separately optimizes the selection and scheduling of road expansion projects.

A number of research extensions can be considered in the future. For instance, the objective function of the proposed T-DNDP formulation only takes into account total system travel time. In future studies, we plan to integrate multiple objectives that are often considered by decision makers.

Disclosure

The views expressed are those of the authors and do not reflect the official policy or position of the project's sponsors. Part of this study has been presented as a poster at the Transportation Research Board (TRB) 97th Annual Meeting.

Conflicts of Interest

The authors declare that there are no conflicts of interest regarding the publication of this paper.

Acknowledgments

The study was partially sponsored by Mountain-Plains Consortium, a Regional University Transportation Center sponsored by the U.S. Department of Transportation, the National Natural Science Foundation of China under Projects 71401025 and 51338008, and Lloyd's Register Foundation that helps to protect life and property by supporting engineering-related education, public engagement, and the application of research.

References

[1] D. Schrank, B. Eisele, T. Lomax, and J. Bak, Urban mobility scorecard, 2015.

[2] T. L. Friesz, D. Bernstein, N. J. Mehta, R. L. Tobin, and S. Ganjalizadeh, "Day-to-day dynamic network disequilibria and idealized traveler information systems," *Operations Research*, vol. 42, no. 6, pp. 1120–1136, 1994.

[3] T. L. Friesz, D. Bernstein, and R. Stough, "Dynamic systems, variational inequalities and control theoretic models for

predicting time-varying urban network flows," *Transportation Science*, vol. 30, no. 1, pp. 14–31, 1996.

[4] H. K. Lo and W. Y. Szeto, "Planning transport network improvements over time," *Urban and Regional Transportation Modeling: Essays in Honor of David Boyce*, pp. 157–176, 2004.

[5] M. Abdulaal and L. J. LeBlanc, "Continuous equilibrium network design models," *Transportation Research Part B: Methodological*, vol. 13, no. 1, pp. 19–32, 1979.

[6] G. B. Dantzig, R. P. Harvey, Z. F. Lansdowne, D. W. Robinson, and S. F. Maier, "Formulating and solving the network design problem by decomposition," *Transportation Research Part B: Methodological*, vol. 13, no. 1, pp. 5–17, 1979.

[7] H. Z. Aashtiani and T. L. Magnanti, "Equilibria on a congested transportation network," *Society for Industrial and Applied Mathematics. Journal on Algebraic and Discrete Methods*, vol. 2, no. 3, pp. 213–226, 1981.

[8] C. Suwansirikul, T. L. Friesz, and R. L. Tobin, "Equilibrium decomposed optimization: a heuristic for the continuous equilibrium network design problem," *Transportation Science*, vol. 21, no. 4, pp. 254–263, 1987.

[9] T. L. Friesz, H.-J. Cho, N. J. Mehta, R. L. Tobin, and G. Anandalingam, "Simulated annealing approach to the network design problem with variational inequality constraints," *Transportation Science*, vol. 26, no. 1, pp. 18–26, 1992.

[10] Q. Meng, H. Yang, and M. G. H. Bell, "An equivalent continuously differentiable model and a locally convergent algorithm for the continuous network design problem," *Transportation Research Part B: Methodological*, vol. 35, no. 1, pp. 83–105, 2001.

[11] Q. Meng and H. Yang, "Benefit distribution and equity in road network design," *Transportation Research Part B: Methodological*, vol. 36, no. 1, pp. 19–35, 2002.

[12] D. H. Lee, "Planning Transport Network Improvements over Time," in *Urban and Regional Transportation Modeling: Essays in Honor of David Boyce*, Edward Elgar Publishing, Inc., USA, 2004.

[13] Y. Yin and S. Lawphongpanich, "A robust approach to continuous network designs with demand uncertainty," *Transportation and Traffic Theory*, pp. 110–126, 2007.

[14] P. A. Steenbrink, "Transport network optimization in the Dutch integral transportation study," *Transportation Research*, vol. 8, no. 1, pp. 11–27, 1974.

[15] L. J. Leblanc, "An algorithm for the discrete network design problem," *Transportation Science*, vol. 9, no. 3, pp. 183–199, 1975.

[16] M. Chen and A. S. Alfa, "A network design algorithm using a stochastic incremental traffic assignment approach," *Transportation Science*, vol. 25, no. 3, pp. 215–224, 1991.

[17] C. Lee and K. Yang, "Network design of one-way streets with simulated annealing," *Papers in Regional Science*, vol. 73, no. 2, pp. 119–134, 1994.

[18] Z. Drezner and G. O. Wesolowsky, "Selecting an optimum configuration of one-way and two-way routes," *Transportation Science*, vol. 31, no. 4, pp. 386–394, 1997.

[19] Z. Drezner and G. O. Wesolowsky, "Network design: selection and design of links and facility location," *Transportation Research Part A: Policy and Practice*, vol. 37, no. 3, pp. 241–256, 2003.

[20] H. Poorzahedy and F. Abulghasemi, "Application of Ant System to network design problem," *Transportation*, vol. 32, no. 3, pp. 251–273, 2005.

[21] Z. Gao, J. Wu, and H. Sun, "Solution algorithm for the bi-level discrete network design problem," *Transportation Research Part B: Methodological*, vol. 39, no. 6, pp. 479–495, 2005.

[22] Q. Meng and H. L. Khoo, "Optimizing contraflow scheduling problem: Model and algorithm," *Journal of Intelligent Transportation Systems: Technology, Planning, and Operations*, vol. 12, no. 3, pp. 126–138, 2008.

[23] G. E. Cantarella, G. Pavone, and A. Vitetta, "Heuristics for urban road network design: lane layout and signal settings," *European Journal of Operational Research*, vol. 175, no. 3, pp. 1682–1695, 2006.

[24] G. E. Cantarella and A. Vitetta, "The multi-criteria road network design problem in an urban area," *Transportation*, vol. 33, no. 6, pp. 567–588, 2006.

[25] M. Gallo, L. D'Acierno, and B. Montella, "A meta-heuristic approach for solving the urban network design problem," *European Journal of Operational Research*, vol. 201, no. 1, pp. 144–157, 2010.

[26] P. Luathep, A. Sumalee, W. H. K. Lam, Z. Li, and H. K. Lo, "Global optimization method for mixed transportation network design problem: a mixed-integer linear programming approach," *Transportation Research Part B: Methodological*, vol. 45, no. 5, pp. 808–827, 2011.

[27] Z. X. Wu, W. H. K. Lam, and K. S. Chan, "Multi-model network design: Selection of pedestrianisation location," *Journal of Eastern Asia Society for Transportation Studies*, vol. 6, pp. 2275–2290, 2005.

[28] Z. Song, Y. Yin, and S. Lawphongpanich, "Optimal Deployment of Managed Lanes in General Networks," *International Journal of Sustainable Transportation*, vol. 9, no. 6, pp. 431–441, 2015.

[29] E. Miandoabchi and R. Z. Farahani, "Optimizing reserve capacity of urban road networks in a discrete Network Design Problem," *Advances in Engineering Software*, vol. 42, no. 12, pp. 1041–1050, 2011.

[30] H. Poorzahedy and M. A. Turnquist, "Approximate algorithms for the discrete network design problem," *Transportation Research Part B: Methodological*, vol. 16, no. 1, pp. 45–55, 1982.

[31] R. S. Solanki, J. K. Gorti, and F. Southworth, "Using decomposition in large-scale highway network design with a quasi-optimization heuristic," *Transportation Research Part B: Methodological*, vol. 33, no. 2, pp. 127–140, 1998.

[32] H. Poorzahedy and O. M. Rouhani, "Hybrid meta-heuristic algorithms for solving network design problem," *European Journal of Operational Research*, vol. 182, no. 2, pp. 578–596, 2007.

[33] Y. Yin, "Genetic-algorithms-based approach for bilevel programming models," *Journal of Transportation Engineering*, vol. 126, no. 2, pp. 115–119, 2000.

[34] K. Jeon, J. S. Lee, S. Ukkusuri, and S. T. Waller, "Selectorecombinative genetic algorithm to relax computational complexity of discrete network design problem," *Transportation Research Record*, no. 1964, pp. 91–103, 2006.

[35] L. Zhang, S. Lawphongpanich, and Y. Yin, "An active-set algorithm for discrete network design problems," *Transportation and Traffic Theory*, pp. 283–300, 2009.

[36] H. Farvaresh and M. M. Sepehri, "A branch and bound algorithm for bi-level discrete network design problem," *Networks and Spatial Economics*, vol. 13, no. 1, pp. 67–106, 2013.

[37] T. L. Friesz and S. Shah, "An overview of nontraditional formulations of static and dynamic equilibrium network design," *Transportation Research Part B: Methodological*, vol. 35, no. 1, pp. 5–21, 2001.

[38] W. Y. Szeto and H. K. Lo, "Transportation network improvement and tolling strategies: the issue of intergeneration equity," *Transportation Research Part A: Policy and Practice*, vol. 40, no. 3, pp. 227–243, 2006.

[39] W. Y. Szeto and H. K. Lo, "Time-dependent transport network improvement and tolling strategies," *Transportation Research Part A: Policy and Practice*, vol. 42, no. 2, pp. 376–391, 2008.

[40] O. Liam and Y. W. Szeto, "The discrete network design problem over time," *HKIE Transactions*, vol. 14, no. 4, pp. 47–55, 2007.

[41] H. Lo and W. Y. Szeto, "Time-dependent transport network design: a study on budget sensitivity," *Journal of the Eastern Asia Society for Transportation Studies*, vol. 5, 2003.

[42] W. Y. Szeto and H. K. Lo, "Strategies for road network design over time: Robustness under uncertainty," *Transportmetrica*, vol. 1, no. 1, pp. 47–63, 2005.

[43] H. K. Lo and W. Y. Szeto, "Time-dependent transport network design under cost-recovery," *Transportation Research Part B: Methodological*, vol. 43, no. 1, pp. 142–158, 2009.

[44] W. Y. Szeto, X. Jaber, and M. O'Mahony, "Time-dependent discrete network design frameworks considering land use," *Computer-Aided Civil and Infrastructure Engineering*, vol. 25, no. 6, pp. 411–426, 2010.

[45] W. Y. Szeto, Y. Jiang, D. Z. Wang, and A. Sumalee, "A sustainable road network design problem with land use transportation interaction over time," *Networks and Spatial Economics*, vol. 15, no. 3, pp. 791–822, 2015.

[46] Y. Jiang and W. Y. Szeto, "Time-dependent transportation network design that considers health cost," *Transportmetrica A: Transport Science*, vol. 11, no. 1, pp. 74–101, 2015.

[47] E. Miandoabchi, F. Daneshzand, R. Zanjirani Farahani, and W. Y. Szeto, "Time-dependent discrete road network design with both tactical and strategic decisions," *Journal of the Operational Research Society*, vol. 66, no. 6, pp. 894–913, 2015.

[48] B. J. Kim, W. Kim, and B. H. Song, "Sequencing and scheduling highway network expansion using a discrete network design model," *Annals of Regional Science*, vol. 42, no. 3, pp. 621–642, 2008.

[49] S. V. Ukkusuri and G. Patil, "Multi-period transportation network design under demand uncertainty," *Transportation Research Part B: Methodological*, vol. 43, no. 6, pp. 625–642, 2009.

[50] S.-M. Hosseininasab and S.-N. Shetab-Boushehri, "Integration of selecting and scheduling urban road construction projects as a time-dependent discrete network design problem," *European Journal of Operational Research*, vol. 246, no. 3, pp. 762–771, 2015.

[51] J. Wardrop, "Some theoretical aspects of road traffic research," *ICE Proceedings: Engineering Divisions*, vol. 1, no. 3, pp. 325–362, 1952.

[52] R. K. Ahuja, T. L. Magnanti, and J. Orlin, *Network Flows: Theory, Algorithms, and Applications*, Prentice-Hall, 1993.

[53] H. Scheel and S. Scholtes, "Mathematical programs with complementarity constraints: stationarity, optimality, and sensitivity," *Mathematics of Operations Research*, vol. 25, no. 1, pp. 1–22, 2000.

[54] G. Bouza and G. Still, "Mathematical programs with complementarity constraints: convergence properties of a smoothing method," *Mathematics of Operations Research*, vol. 32, no. 2, pp. 467–483, 2007.

[55] A. U. Raghunathan and L. T. Biegler, "An interior point method for mathematical programs with complementarity constraints (MPCCs)," *SIAM Journal on Optimization*, vol. 15, no. 3, pp. 720–750, 2005.

[56] F. Facchinei and J.-S. Pang, *Finite-dimensional variational inequalities and complementarity problems. Vol. I*, Springer Series in Operations Research, Springer-Verlag, NY, USA, 2003.

[57] A. S. Drud, "CONOPT—A Large-Scale GRG Code," *ORSA Journal on Computing*, vol. 6, no. 2, pp. 207–216, 1994.

[58] S. Nguyen and C. Dupuis, "An efficient method for computing traffic equilibria in networks with asymmetric transportation costs," *Transportation Science*, vol. 18, no. 2, pp. 185–202, 1984.

[59] R. E. Rosenthal, *GAMS-A User's Guide*, GAMS Development Corporation, Washington, DC, USA, 2012.

15

Design Method of ADAS for Urban Electric Vehicle based on Virtual Prototyping

Katarzyna Jezierska-Krupa and Wojciech Skarka

Institute of Fundamentals of Machinery Design, Faculty of Mechanical Engineering, Silesian University of Technology, Gliwice, Poland

Correspondence should be addressed to Wojciech Skarka; Wojciech.Skarka@polsl.pl

Academic Editor: Rongxin Cui

Since 2012, the Smart Power Team has been actively participating in the Shell Eco-marathon, which is a worldwide competition. From the very beginning, the team has been working to increase driver's safety on the road by developing Advanced Driver Assistance Systems. This paper presents unique method for designing ADAS systems in order to minimize the costs of the design phase and system implementation and, at the same time, to maximize the positive effect the system has on driver and vehicle safety. The described method is based on using virtual prototyping tool to simulate the system performance in real-life situations. This approach enabled an iterative design process, which resulted in reduction of errors with almost no prototyping and testing costs.

1. Introduction

Conducting the design process of Advanced Driver Assistance Systems (ADAS) in an optimum way requires a specific approach to defining and solving problem [1]. Profound research and analysis have to be carried in order to prepare well for this task. Sometimes it is impossible though to identify and avoid certain mistakes in the process, without using a prototype [2–4]. Currently, there are more and more tools available for engineers that enable creating virtual prototypes and therefore optimizing the design process in terms of its effectiveness and fault minimization [5]. Nevertheless, in order to use these tools efficiently, there needs to be a proper approach to design process adopted. The aim of this article is to present the approach to ADAS design process, which includes the use of simulation methods and virtual prototyping. Preceding the actual method description, there are some considerations included, whether there are reasonable grounds for using virtual prototyping tool and what other tools should be considered before making the final decision. The subsequent sections of this article include the method description and the use case, that is, design process of Blind Spot Information System for urban vehicle. This article is an extended version of the article [6] presented at the International Conference on Transdisciplinary Engineering 2016 in Curitiba, Brazil.

2. Consideration of the Rationale for Using Virtual Prototyping Tool

2.1. Defining the Needs. The first step that needs to be taken into system of design process needs definition [4, 7, 8]. What must be identified are the problems that a driver is facing and possibilities of finding the right solutions for them. The most effective ways of finding them are interviews with actual drivers and analysis of traffic situations of great risk for driver, vehicle, or third parts safety. The most common conditions in which the driving task is performed [1, 5] need to be analysed in this step. For vehicles dedicated for specific goals, such as competitions, these conditions can be described very precisely, which enables finding most current problems and most effective solutions. The choice of prototyping tool, which would be used in further steps, needs to be based on tool possibilities in terms of reflecting real environmental and situational conditions.

2.2. Defining the External Constraints. Before deciding on using virtual prototyping tool for ADAS design, it needs to be considered whether there is such a need at all. To answer this question, the external constraints must be considered and it must be verified how the design process will be affected by using simulation methods.

There are multiple constraints that must be considered, when thinking about designing ADAS. One of the most basic ones are costs, time, and quality. According to a well-known business rule of thumb, it is accepted that only two of these three constraints can be fulfilled at the same time.

2.3. Discussion of Possible ADAS Design Methods. There are various methods of ADAS design available and it is crucial to choose the one that fits the needs of the project best. ADAS can be designed in the following ways:

(1) No prototype: which does not enable us to verify its performance in real-life situations until the system is implemented

(2) Virtual prototype: which, accordingly to a chosen tool, enables us to simulate the natural conditions in really accurate but not perfectly accurate way, generates low costs, and enables the iterative design process [3, 5]

(3) Real prototype: which enables absolute verification but has also many drawbacks such as costs, production, and construction time or difficult interaction [3]

When considering the virtual prototyping tool, the above-mentioned aspects must be included.

3. Design Method

3.1. Project Analysis. In order to minimize the need for changes in further steps of the ADAS design process, it is recommended to define all internal constraints before working on an actual model. This approach enables us to precisely set the goal of the designed ADA system and define the required constraints early enough to be included in the model. Using a virtual prototyping tool for such complicated systems as ADAS requires splitting considerations about the whole system into subsystems.

3.2. Designing and Testing. Testing part of designed system is a step that must not be neglected. Thorough tests are the last step before real system implementation that allow error detection without generating additional costs in manufacturing process. Correctly conducted testing enables also very accurate error identification and makes its correction easier.

In order to properly define test cases for ADAS virtual prototype testing, three separate system aspects must be considered, in which an error could appear:

(1) Data processing: software

- (1.1) What are the system inputs?
- (1.2) What are the system outputs?
- (1.3) What is the path of converting inputs to outputs?
- (1.4) What values are independent from the system design and may not be tested?

(2) Functional requirements: UI and usability

- (2.1) What is the system supposed to do?

FIGURE 1: Bytel vehicle during Shell Eco-marathon race in Rotterdam (2015).

- (2.2) What information is it supposed to communicate to the driver?
- (2.3) What must the system not do?

(3) Models design: the idea

- (3.1) How should the system work?
- (3.2) What inputs and outputs should it accept/generate?
- (3.3) What functions are the system described by?

Such analysis enables us to create test cases sheet that should be used in test experiments outcome evaluation. Only when every test case will end successfully, that is, the values will be corresponding with template values and the visual verification will be fulfilled, the system can be implemented.

For correct testing process of ADAS virtual prototype, it is also needed to cover all possible situations in simulations, in which an error could appear. The test experiments should include both typical traffic situations of a higher risk and nonrisky situations in order to test the system against false alarms.

4. Use Case: BLIS for Urban Electric Vehicle

The above-described design method was used in design process of Blind Spot Information System (BLIS) [2, 6, 9, 10] for urban electric vehicle, Bytel [6, 11, 12] (Figure 1). Bytel is a vehicle created to participate in Shell Eco-marathon (SEM) [13] competitions, the event which aims to encourage design of highly efficient vehicles by students and scientific organizations. Bytel has participated in SEM in 2014 and 2015 in two power source categories: battery electric and hydrogen (hydrogen fuel cell stack).

The steps described above were taken to successfully finish the process at minimal time and maximum quality, keeping the costs reasonable at the same time.

4.1. Historical Determinants of BLIS System Development. The concept of this BLIS system is based on the team's experience in this field, taking into account specific conditions and upon existing solutions. The basic assumption of the system is to inform the driver about the appearance of another vehicle in

FIGURE 2: Vehicle MuSHELLka at Shell Eco-marathon 2013 in Rotterdam. Placement of BLIS system sensors in MuSHELLka.

the blind spot. The final form of the system was influenced by experience with previous ADAS systems, including the BLIS system, which was fitted with another vehicle [11] (MuSHELLka) developed by the team (Figure 2).

Previous technical solution of BLIS system [6, 11] is composed of a controller which is based on Microcontroller ATmega8 and 3 sensors. The main task of the controller is to handle sensors and to transmit information about the detected threat to ACS controller using RS232. Two ultrasonic sensors HCSR-04 and diffusive photoelectric sensor Datalogic S300 are used for scanning area behind the vehicle. The HC-SR04 ultrasonic sensor uses sonar to determine the distance from an object in the range of 2 cm–400 cm, with 3 mm resolution. Sound wave that is used for measurement has a frequency of 40 kHz [11]. Datalogic S300 is an advanced photoelectric sensor with a detection range of 0–500 cm for white objects and 0–350 cm for black objects. Infrared radiation and triangulation are used for measurement. The sensor response is binary, and the detection range can be adjusted in the full range of measurement. Due to the use of the sensor, a critical parameter is the response time. Maximum response time for this sensor is shorter than 2 ms [11]. All system components are placed in the rear fairing, which was possible thanks to the compact design and minimum weight.

This solution proved to be a much smaller vehicle because of its simple construction, low cost, and high operational reliability. It is also suitable for small size vehicles and small mass. It is important to use the BLIS virtual prototyping method in the development of this system. As part of the development of the previous system, a virtual race track system was used, a vehicle model in the TASS PreScan [14] environment, and a numerical simulation model of the BLIS system in this environment using the MATLAB/Simulink environment. As part of the development of the previous system, a full simulation test was conducted in the simulation environment and subsequent verification was performed on the actual site. These tests confirmed the compatibility of the system parameters with the assumptions and the experience of the design process, virtual testing, and racing operation are the basis for the development of the new BLIS system.

As the next designed vehicle [15, 16] (Bytel) is only equipped with side mirrors and does not have a rear window or a central rear-view mirror, it is not only the monitoring of the area on the side of the vehicle, but also the area immediately behind the vehicle. A number of conceptual solutions have been analysed, of which three are presented below (Figure 3).

The first concept (Figure 3(a)) was based on the use of photoelectric sensors. The experience of using this type of sensor in previous solutions indicated the potential for such a system, but ultimately, due to the need to use a large number of such sensors, the vehicle aerodynamic effects were adversely affected, and its relatively low efficiency and range, especially in variable weather conditions, and insufficient information obtained from such a system has been rejected.

The second concept (Figure 3(b)) was based on the use of a vision system based on the Kinect. The design assumes the use of the depth camera included in the kit to determine the distance of the approaching vehicles. If larger areas than a single sensor have to be scanned, more Kinect devices can be used. This, however, complicates the connection and requires data analysis from two or more different sensors. Creating a system, based on two or a few of this type of sensors, which would recognize an object with satisfying accuracy and calculates position, is a difficult task and thus it requires great processing power. This affects not only the recognition of the image, but also the angle at which the vehicles will be detected. Taking into account the parameters of a commonly available Kinect device, a sensor placement scheme was used when using four devices. The advantages of this solution are the innovation and the use of ready-made modules and the large potential of the system. The disadvantages include the large dimensions of the device, relatively small scanning angle

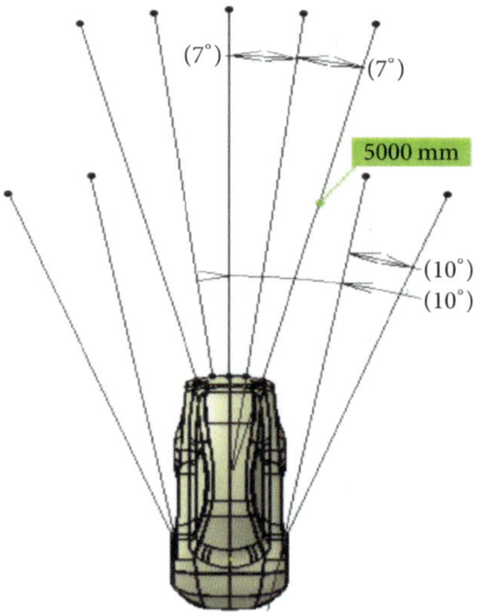

(a) BLIS concept based on photoelectric sensors

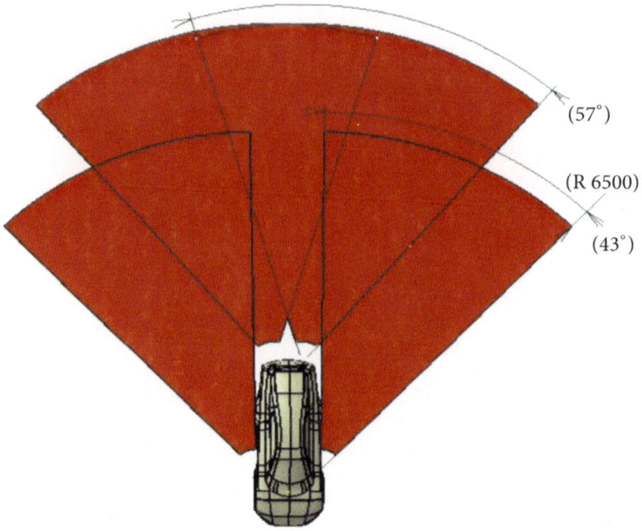

(b) BLIS concept based on visual system

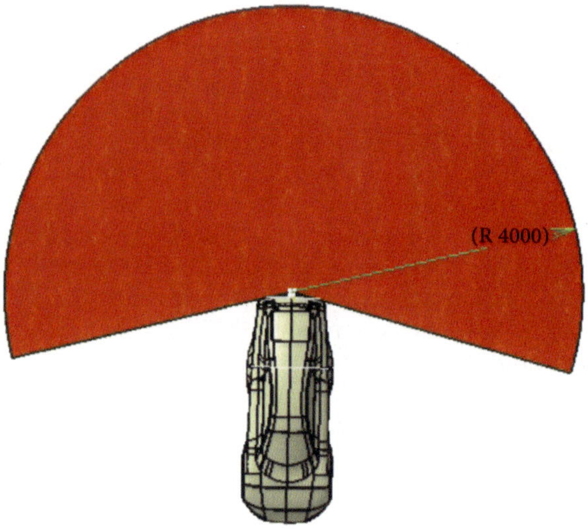

(c) BLIS concept based on Lidar sensor

FIGURE 3: Alternative concepts of BLIS system.

in case of a single sensor, the possibility of disturbance of measurements in real conditions (sun rays), and the need to use a PC.

The third concept (Figure 3(c)) assumes the use of a Hokuyo laser sensor. This sensor is characterized by high resolution (1080 samples per 270°, corresponding to 4 samples per 1°), high sample rate (40 Hz), and wide scan area (270°) [17]. The laser used in this sensor fulfils the 1st safety class, which allows it to be used during SEM competition and is characterized by low sensitivity to atmospheric agents (sun, rain, snow, and fog). The advantages of this solution are simple assembly, single device, and high reliability, whereas the main disadvantage is the high cost of the sensor and the possibility of problems arising from the analysis of so much data. This, however, provides a field for the use of complex algorithms to analyse the data to eliminate errors and increase operational certainty, which was also used and presented later in this paper. Ultimately, the laser sensor solution has been used in further development work.

The reflections and experience gained from previous development work led to the new BLIS solution to modify both the system itself and the design methodology. The new design method was shortened by abandoning the cost-intensive and time-consuming verification of the real object. Each of the newly constructed vehicles is treated as a prototype vehicle and a shortened test of the actual object is justified. These tests are designed to be carried out on the track and during races. This does not mean that the system will not be verified. The emphasis was placed on virtual prototyping, a formal verification phase was established during these studies, and new methods were introduced to verify the results of the simulation tests by introducing new independent computational methods to duplicate the results of standard simulations.

As far as the concept was concerned, it was decided to change the concept of the system completely by introducing a new interface in the driver's field of vision, providing much more information but not distracting the driver. The use of a single Lidar sensor with a range of measurable resistance to ambient conditions and at the same time to meet the rigorous requirements of the vehicle aerodynamic performance is critical to ensure high reliability of the proposed system of the development method using virtual prototyping methods and it provides the possibility of using this sensor and integration of the BLIS system into other ADAS systems [18, 19].

4.2. Consideration of the Rationale for Using Virtual Prototyping Tool. The decision to develop BLIS for urban vehicle was made based on low viewing range, which is characteristic for common conditions for these kinds of vehicles [18], that is, vehicles designed specifically for Shell Eco-marathon competitions. BLIS is the system that informs the driver about vehicles coming from the back of the vehicle, which are invisible to the driver due to its low viewing range. The need of creating such a system has been recognized based on analysis of many accidents and dangerous situations happening during the race and also after discussing the problem with drivers participating in SEM.

While deciding to use virtual prototyping tool, the crucial reason for using it was the necessity for parallel production process. As the vehicle was still in development phase during the design of BLIS, there was no possibility of using real prototype in order to conduct tests and work on data processing improvement. This is equivalent to accepting time as the main external constraint that applies to this project.

The second constraint was quality of the system, as the vehicle purpose was to take part in SEM competition; that is, in the race, it was considered extremely important to provide the driver with ADAS of the best possible quality. The point was to help the driver keep safe and avoid active or passive participation in accidents.

Last but not least, the adequate tool needed to be chosen for designing and conducting BLIS simulations. There were a few virtual prototyping tools especially dedicated to ADAS design which were considered. The final choice was TASS PreScan [14] due to its wide range of predefined sensors, very friendly user interface, and great visualization capabilities.

4.3. Project Analysis

4.3.1. The Scope of Development Works Performed Using Virtual Prototyping Methods. As part of the development work on the new BLIS system for the Bytel vehicle, the following scope of work is planned:

(i) Analytical analysis of previous versions of BLIS and competing solutions

(ii) Development of detailed functional requirements and requirements for integration with vehicle and other vehicle systems, as well as internal and external constraints

(iii) Development of a detailed BLIS concept

(iv) Construction of a simulation environment involving simulation models of an example race track, designed vehicle, and other vehicles

(v) Construction of the mathematical simulation model of the BLIS system

(vi) Identification and development of simulation scenarios

(vii) Development of alternative algorithms for methods of verification of virtual prototyping simulations

(viii) Carrying out a series of simulations with verification tests and improving the parameters and structure of the developed BLIS solution in iterative mode until the results are achieved

(ix) Performance analysis and production and integration plan

According to the designers of the vehicle, a set of detailed planning simulation tests and independent allowing algorithms to verify the results of simulation studies are particularly important for the new method. Correct planning and carrying out of this part of the research guarantee the possibility of doing without the physical prototype. Therefore, further attention was paid to the detailed description of these fragments of the research.

4.3.2. Defining the Internal Constraints. Following the described method, the first step after recognizing the need of creating BLIS and defining the external constraints that lead to decision of using virtual prototyping tool was defining the internal constraints of the designed system.

(1) The system cannot affect actively the breaking and/or steering system.

(2) The system functions must be restricted to information/warning functions.

The self-imposed constraints of the project were also directly connected to Shell Eco-marathon assumptions, as the purpose of Bytel vehicle was to achieve the best result possible in this competition (SEM is about developing highly energetic vehicles). In order to satisfy this goal, there were the following constraint sets:

(1) The weight of the system must be minimized.

(2) The system should not consume relevant amount of power.

These constraints enabled us to define the general approach to designed system.

At this point, the decision regarding system parts has been made, which was vital to create the system model in TASS PreScan. There were three possible kinds of sensors considered: photoelectric sensors, Lidar, and depth camera. These kinds of sensors are available in predefined form in TASS PreScan, with a possibility of their configuration in very wide range. The decision was made to use Hokuyo Lidar (Figure 3(c)).

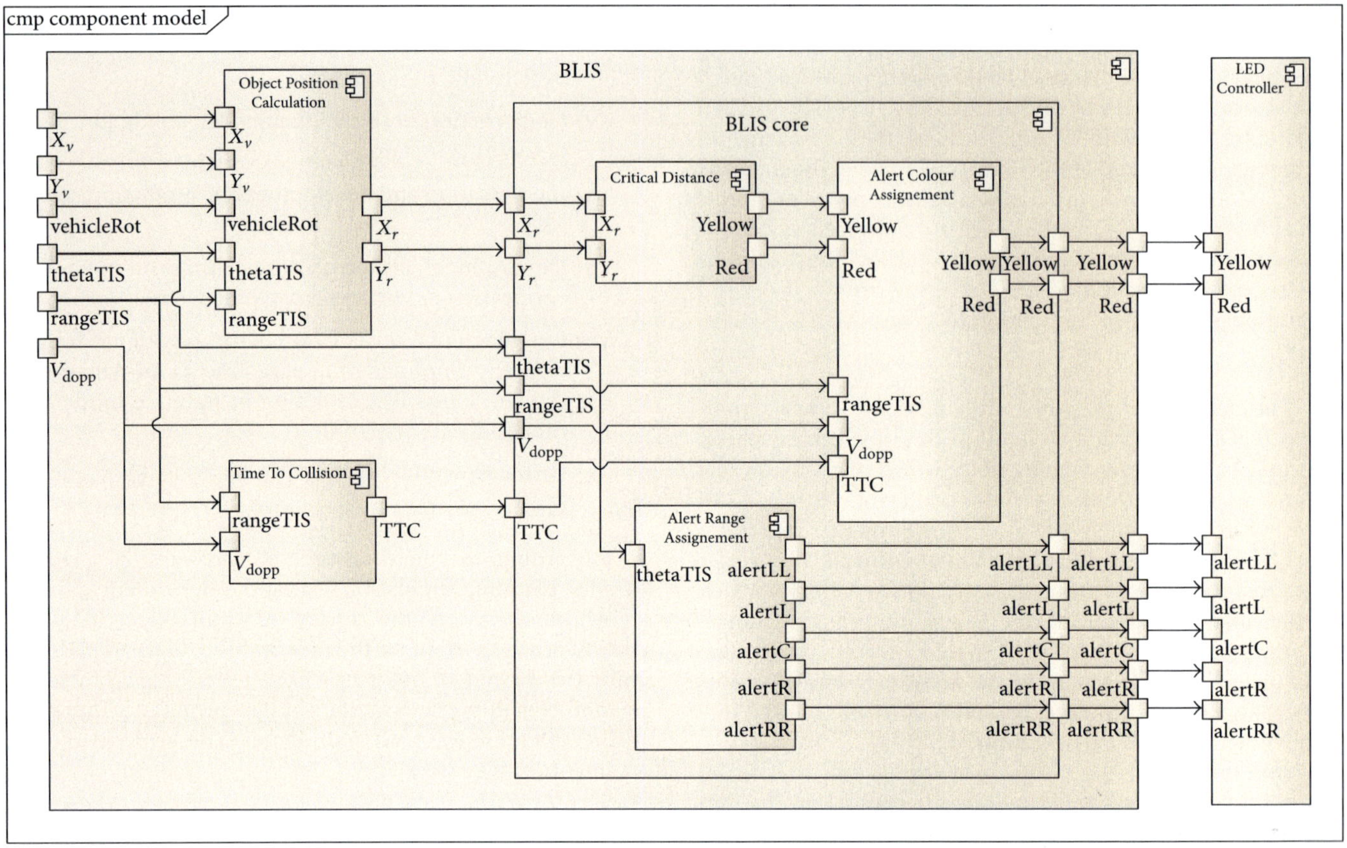

Figure 4: Flowchart of BLIS for Bytel vehicle.

What was crucial when deciding about rejecting the photoelectric sensors option was their low resistance to environmental conditions; a high risk of inaccurate results was identified. Also, in case of photoelectric sensors, there would be a need for using at least 9 sensors in order to cover the area invisible to the driver sufficiently, whereas there was only one Lidar needed to enable scanning of this range.

However, the difficulties in configuring and designing a system based on depth camera and the size of the system that interfere with the aerodynamic requirements have resulted in rejection of this solution.

4.3.3. Splitting System into Subsystems. Defining the subsystems and splitting the system into them when developing a virtual prototype not only made it easier to follow the design process step by step but also created good ground for tests analysis.

The main determinants of dividing line which would be used for splitting the system were assumed real-life subsystems. This means that it was needed to consider the separate parts of BLIS system, as they were meant to be implemented in real life. This is why the decision was made to consider BLIS in two subsystems: data processing subsystem and control subsystem.

The data processing system included also a verification subsystem, which uses the data generated in PreScan in order to compare the values between ideal ones and the ones that were received via data processing based on assumed input values. An example of such comparison can be seen in Figure 9.

4.4. Operational Model of the System. Operational model of the system can be seen in Figure 4. The base of the system is the use of one laser sensor with a large angular range and coverage.

Input variables of the system are as follows:

(i) X_v: X coordinate of the Bytel vehicle in the absolute coordinate system

(ii) Y_v: Y coordinate of the Bytel vehicle in the absolute coordinate system

(iii) vehicleRot: rotation angle of the vehicle in the z-axis

(iv) thetaTIS: the angle at which the vehicle was detected

(v) rangeTIS: the distance of the detected vehicle to the sensor

(vi) V_{dopp}: vehicles relative velocity

Outputs of the system are as follows:

(i) yellowAlert: low (yellow) risk information

(ii) redAlert: high (red) risk information

(iii) alertLL: information about the appearance of an object in the extreme left area

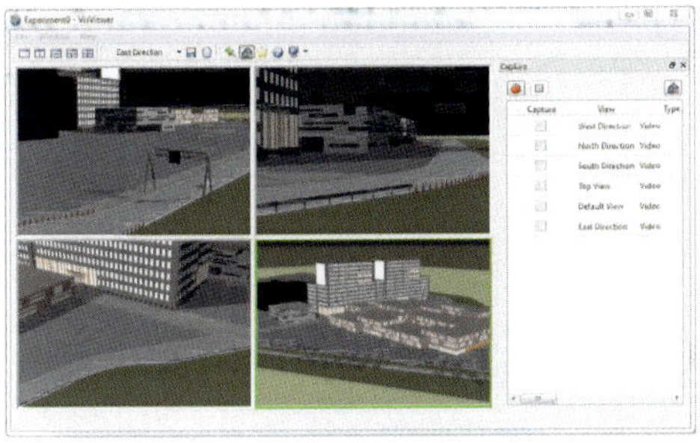

FIGURE 5: Model of the environment made in TASS PreScan.

(iv) alertL: information about the appearance of an object in the left area

(v) alertC: information about the appearance of an object in the center area

(vi) alertR: information about the appearance of an object in the right area

(vii) alertRR: information about the appearance of an object in the extreme right

4.5. Building Scenario: Building Environment

4.5.1. Simulation Environment Modelling. In order to carry out the relevant simulations to verify the correctness of operation of the designed system, it was necessary to design a simulation environment, vehicles with the dynamics, and trajectory, as well as modeling the sensor which is the basis of the system. Arrangement of roads and buildings is based on the imported Street Map model of the Ahoy Arena area in Rotterdam, where the Shell Eco-marathon 2014 competition took place. The designed model of the environment (Figure 5) was supposed to reflect the actual conditions as faithfully as possible.

4.5.2. Vehicle Movement Parameters Modelling. After building the environment, we have to upload the vehicle model for which Advanced Driver Assistance System is designed. It is necessary to assign a trajectory and to define the parameters of vehicle movement.

TASS PreScan software allows implementation of any velocity profile and trajectory. First trajectories have to be defined and then velocity profile for the two vehicles has to be determined. As a result, speeds and acceleration and paths of the vehicles during the simulation are described in detail. Figure 6 shows the speed profile for a vehicle overtaking a leading Bytel vehicle.

4.5.3. Sensor Parameters Modelling. An important point of the simulation design process is accurate mapping of parameters of used sensor. The PreScan allows the use of the Lidar model; however, due to the large angular range and high frequency sampling, a better solution was to use a sensor of TIS type (Technology Independent Sensor) (Figure 7). TIS, thanks to the wide possibilities of modification of parameters, has enabled a very accurate representation of the Hokuyo sensor to be used in the BLIS assistance system.

PreScan software enables visual verification of the range of the proposed sensor (Figures 13, 14, and 15), which helps in controlling the distance and angular range. With this visualization, the right area of operation for the designed sensor can be easily chosen. By controlling the moment of appearance of another vehicle in sensor range, it is possible to verify operation during simulation of the system as early as at design stage (Figures 13, 14, and 15).

4.6. System Model: Data Processing System. After developing all the details of the simulation, it is allowed to start modelling of the system. TASS PreScan software cooperates with the MATLAB/Simulink software. When installing TASS PreScan the reference to MATLAB/Simulink is established. After starting PreScan Sim module user has ready-made models of sensors, vehicles, and other control elements visible as Simulink models. These models are generated by the TASS PreScan individually for each of the simulations taking into account the modifications described earlier.

4.6.1. System Main Window. The main window of system visible after opening the PreScan Sim module contains two models of vehicles and components fulfilling control warning LEDs functions in the simulation.

The main block model represented in the main window of PreScan Sim module contains modifications resulting in BLIS system design. The output from the model of the first Bytel vehicle that is a vehicle equipped with a sensor provides brief information on occurrence of alarms yellow and red and the area of detection of the object. At the input of each diode is given the information about activity in its assigned area and information about yellow or red alarm occurrence.

4.6.2. Vehicle Model. GUI window of vehicle model includes four groups of blocks:

(i) After entering the vehicle model (Figure 8), in the left side of the window, you can see models generated by the PreScan software describing it. In the case of a Bytel vehicle, it is a trajectory model, parameters motion model, and model of sensors assigned to the vehicle, TIS. Based on previously developed block diagram showing the planned method of the system operation, you can immediately pull out of these variables which are necessary when developing a BLIS model: x and y coordinates of the absolute position of the Bytel vehicle, rotation of the vehicle in the absolute coordinate system (vehicleRot), speed of the vehicle (V_v), the value of distance of detected obstacle (rangeTIS), and the relative speed of vehicles were determined using the TIS sensor (V_{dopp}) and the angle in which the obstacle appears (thetaTIS).

(ii) Blocks containing programs designed to perform the calculations of parameters used in the BLIS system: Object Position Calculation, Time To Collision, and Object Velocity Calculation. The blocks are linked to necessary input variables and output variables are derived.

(iii) Block of BLIS system together with its inputs and outputs

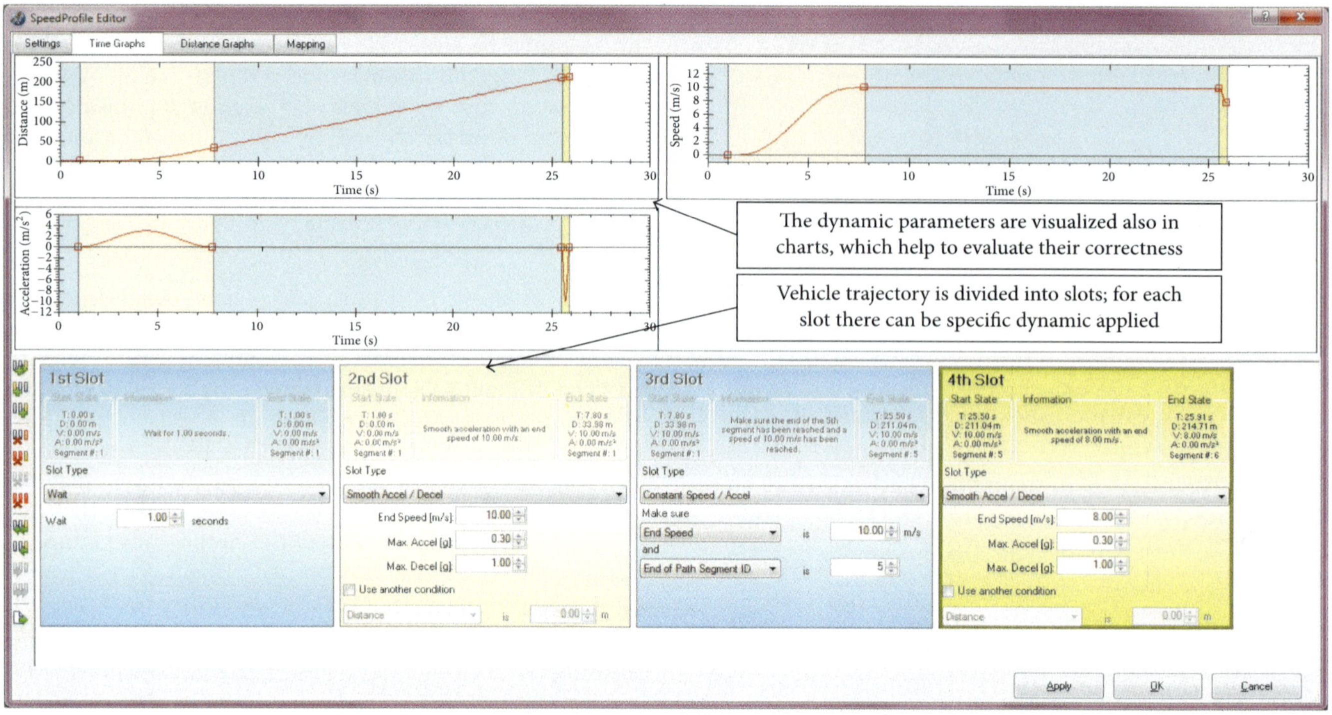

FIGURE 6: Speed profile for overtaking vehicle.

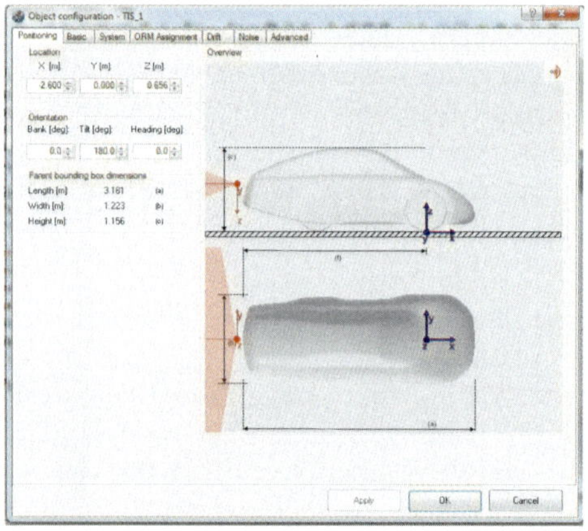

FIGURE 7: Modelling Hokuyo Lidar in BLIS.

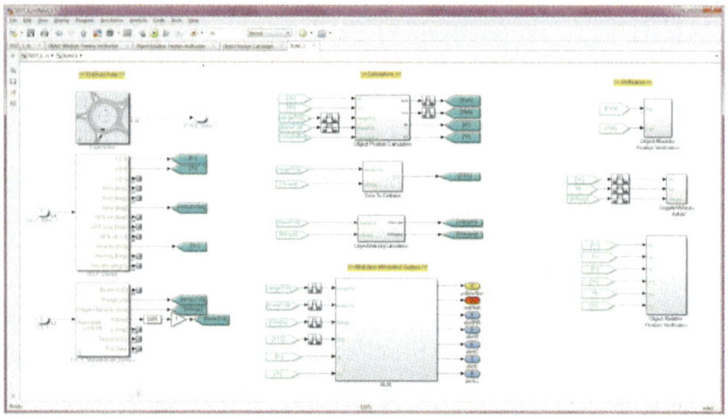

FIGURE 8: General models and blocks of the BLIS system.

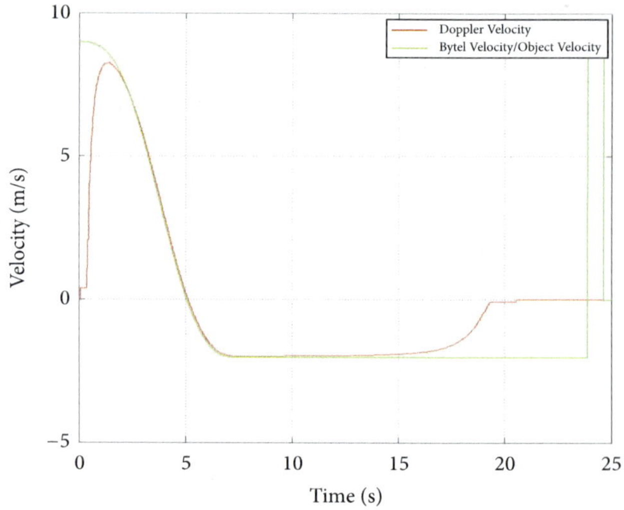

FIGURE 9: Comparison of two values: ideal one (Doppler Velocity) and the one achieved in data processing (Bytel Velocity, Object Velocity).

(iv) Blocks including programs used to verification calculations

4.6.3. Part I: Calculations. Based on the block diagram of the BLIS system, it had to be determined which calculations should be performed to obtain the necessary to define the logic of the system variables. Directly from the PreScan software, the following variables were extracted:

(i) X and Y coordinates of Bytel vehicle in the absolute coordinate system
(ii) Rotation angle of the vehicle in absolute coordinate system (vehicleRot)
(iii) Vehicle velocity (V_v),
(iv) The distance of the detected obstacle (rangeTIS)
(v) Relative velocity determined by TIS sensor (V_{dopp})
(vi) The angle at which obstacle was detected (thetaTIS)

These variables do not provide direct way to determine the logic of the system, which was based on three basic pieces of information:

(i) Comparison of the relative position of the object with the value of YCD (Yellow Critical Distance) or RCD (Red Critical Distance), which are functions defining the threshold value of position of the object in the x axis, for which there is raised up risk of collision
(ii) Value of relative velocity of vehicles V_{dopp}
(iii) TTC value (Time To Collision)

On the basis of these variables, the following result variables were obtained in the way of calculations:

(i) X_r: x coordinate of the object relative position relative to Bytel vehicle determined based on the value rangeTIS and thetaTIS
(ii) Y_r: Y coordinate of the object relative position relative to the Bytel vehicle determined based on the value rangeTIS and thetaTIS
(iii) Yellow Critical Distance: function determined based on the value Y_r
(iv) Red Critical Distance: function determined based on the value Y_r
(v) Time To Collision: time to collision, determined based on the value rangeTIS and V_{dopp}

4.6.4. Object Position Calculation. The activities carried out in Object Position Calculation block are divided into two subsystems: the calculation of the value of X and Y coordinates relative position of the detected object and calculation of the coordinate values of X and Y absolute position of the object.

Program used to calculate the values of the X and Y coordinates relative position of the detected object accepts as input values thetaTIS and rangeTIS. Values X_r and Y_r determining the relative position of the vehicle in the x and y axis are obtained by transforming the coordinate system of polar to Cartesian coordinate system.

The second part of the Object Position Calculation block represents calculations regarding the position of the object in the absolute coordinate system. The program uses the input values: thetaTIS, vehicleRot, rangeTIS, coordinates X_v and Y_v, point rotation matrix in the two-dimensional space, and value of coordinates and the value o displacement vector of the center of the local coordinate system of the Bytel vehicle from the edge detected in the x and y axis.

An important assumption was rangeTIS 6 = 0, because the value 0 is assigned to variable rangeTIS at a time when there are any object detected in sensor range. In this case, variables X_{ob} and Y_{ob} are assigned the value 0.

4.6.5. Time To Collision. Time To Collision block task is to calculate the value of time, from the temporary situation of the temporary traffic parameters to the potential collision. This is an important element of calculations, because the value of TTC is directly influenced by the decision to activate the yellow or red alarm.

Also, here switch block was used making passed signal dependent on the value of the variable V_{dopp}. The variable takes the value 0 when there are no objects detected in the sensor range sensor. In this case, the variable TTC is assigned the value 20 as the maximum time for which the threat of a collision is practically nonexistent.

4.6.6. Object Velocity Calculation. Then, the third block of the calculation is an Object Velocity Calculation. It contains code to calculate the coordinates relative speed values V_{dopp}. Therefore, a distribution of the vector velocity V_{dopp} on the components of X and Y depends on the angle thetaTIS. Output values are V_{doppx} and V_{doppy}.

4.6.7. Part II: Verification. The second part of the prepared program was verification part. Before modelling BLIS system, it had to be checked whether calculations delivering the required input values to the system are correct. For this purpose, the values were compared with values generated by the simulation environment (PreScan).

Verification of the relative speed values V_{dopp}: relative speed V_{dopp} is called the Doppler Velocity, which results in the fact that its value is not always unequivocal from the expected one. In order to check the difference between the variable V_{dopp} and the actual vehicle speeds difference verification program was elaborated. The example results can be seen in Figure 9.

Verification of X and Y coordinates relative position of the object is as follows.

To verify the calculations of the value of X and Y coordinates relative position of the object (relative to the Bytel vehicle), the computed variables were compared with variables X_r and Y_r with the difference of the values of the coordinates of the vehicle Bytel (X_v, Y_v) and the object (X_o, Y_o) obtained artificially by PreScan software, taking into account the offset point to be the center of the coordinate system of the vehicle, to whom they are given the coordinates of absolute position.

Verification of coordinates X and Y absolute position of the object is as follows. In order to verify the calculation of the X and Y coordinates of the relative position of the object (relative to the Bytel vehicle), the computed variables X_{ob} and Y_{ob} were compared with the object position coordinates (X_o, Y_o) obtained artificially from the PreScan software.

4.6.8. Part III: BLIS System. The last step was to build the logics of the BLIS system. This part of the system consists of three parts: Data, Alert Range Assignment, and Alert Colour Assignment.

Data. This section of the BLIS window contains ordered model input values, blocks of recording of signals to the workspace.

Alert Range Assignment. The part called Alert Range Assignment is responsible for assigning to the appropriate areas of the alarm (LL, L, C, R, RR) angular ranges in which the object appears.

(i) Range LL is assigned for thetaTIS ⟨48; 87.5⟩ deg.
(ii) Range L is assigned for thetaTIS ⟨7; 49⟩ deg.
(iii) Range C is assigned for thetaTIS ⟨−8; 8⟩ deg.
(iv) Range R is assigned for thetaTIS ⟨−49; −7⟩ deg.
(v) Range RR is assigned for thetaTIS ⟨−87.5; −48⟩ deg.

Alert Assignment Colour. Code fragment entitled Alert Colour Assignment is responsible for assigning the alarm corresponding colour in accordance with the degree of threat, yellow or red.

Decisions on the colour of the alarm are taken on the basis of three pieces of information:

(i) Comparison of the position of the object relative to the value of YCD (Yellow Critical Distance) or RCD (Red Critical Distance), which are functions defining the threshold value of position of the object in the x axis, for which there is an increased risk of collision
(ii) Values V_{dopp}
(iii) The value of TTC (Time To Collision)

The YCD function (Yellow Critical Distance) is given by

$$\mathrm{YCD} = -4 * \left|Y_r\right| + 12. \quad (1)$$

The RCD (Red Critical Distance) is given by

$$\mathrm{RCD} = -\frac{4}{3} * \left|Y_r\right| + 4. \quad (2)$$

Yellow alarm occurs when the following conditions are met:

$$\left(X_r \leq \mathrm{YCD}\ \ (\mathrm{yellowFun}) \cap X_r \neq 0 \cap V_{\mathrm{dopp}} \leq 0\right) \cup 4 < \mathrm{TTC} \leq 10. \quad (3)$$

Red alarm occurs when the following conditions are met:

$$\left(((X_r \leq \mathrm{RCD}\ \ (\mathrm{redFun}) \cap X_r \neq 0) \cup 0 < X_r \leq 1) \cap V_{\mathrm{dopp}} \leq 0\right) \cup 0{,}1 < \mathrm{TTC} \leq 4. \quad (4)$$

Figure 10 visualizes the conditions for activating the alarms.

X_r: X coordinate of relative position of the object
YCD: Yellow Critical Distance
RCD: Red Critical Distance
V_{dopp}: relative velocity of the object
TTC: Time To Collision

4.7. System Model: Control System. Having data processing system, we can begin to design the control system of warning LEDs. In the previous section, signals have been mentioned that we receive at the output of the processing unit. The system of transmission between the BLIS system and the LED is shown in Figure 11 and is the same for each of the diodes.

On the output from the Bytel vehicle block, we have seven binary signals:

(i) yellowAlert: the state of yellow alarm
(ii) redAlert: the state of red alarm
(iii) alertRR: the state of alarm for the RR range
(iv) alertR: the state of alarm for the R range
(v) alertC: the state of alarm for the C range
(vi) alertL: the state of alarm for the L range
(vii) alertLL: the state of alarm for the LL range

At the input of each of the LEDs blocks, the following signals are given:

(i) yellowAlert: state of yellow alarm
(ii) redAlert: state of red alarm
(iii) alertX: state of alarm for the corresponding diode range (RR, R, C, L, and LL)

For each of the five LEDs used to display warning information, the same control was developed.

4.7.1. Part I: System Activation. The logics of the system is based on a switch block which, depending on the condition of the simultaneous occurrence of any of the alarms and

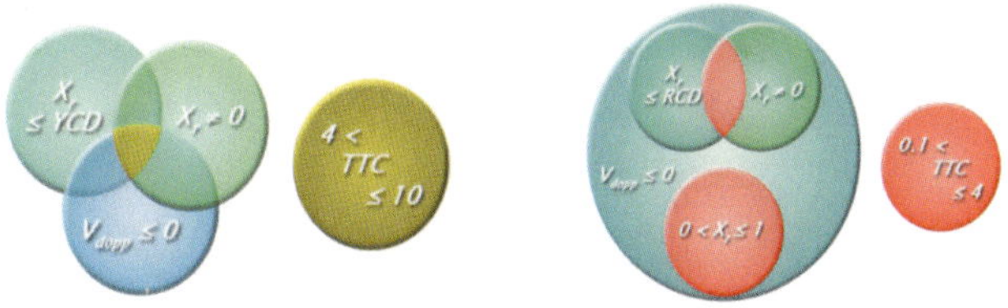

FIGURE 10: Logic of the red and yellow alarm.

the activation of the range corresponding to LED (in this case the LL range), transmits a signal specifying the colour of the alarm (equivalent to the activation of the alarm) or transmits the signal from the default, LEDs gray colour. The same condition checked in switch block acts as a trigger of alertColour block.

4.7.2. Part II: Determination of Alarm Colour. The condition mentioned above checked in switch block acts simultaneously as a trigger of alertColour block.

In the case where you get any of the alarms (yellowAlarm or redAlarm) and at the same time the object was detected in the corresponding range, the block alertColour is executed. This block is to assign a diode suitable for the degree of danger colour: the colour red for a higher degree of threat and the yellow colour for a lower degree of threat. In the case where there is information about the appearance of red alert and activating the appropriate range, signal carrying information about the red alarm is passed, if not, yellow. This logic makes the red alarm the priority.

5. Testing

The test phase of the design process is often treated in a disrespectful way, but properly planned and carried out tests allow you to determine the imperfections and errors of the project very accurately and at the same time to facilitate their improvement greatly. During test planning, the first step is to verify the correctness of the calculations made in the model. In order to determine what we should investigate properly, it is necessary to answer the following questions:

(i) What are the inputs to the model?

(ii) What are the outputs from the model?

(iii) How outputs were from the model obtained?

(iv) Which of these activities relate to the independent variables of the operation of the system, which are the physical representation of unknown values?

Answering these questions will allow the identification of transformation, which should be explored in terms of their compliance with the physical values that do not belong to a set of values of the input model. PreScan software provides many of these values, which makes the comparison of the model with the actual data possible without an actual field testing.

The next step in the test planning is to return to the design assumptions and answer the questions:

(i) What is the designed system supposed to do? What are the requirements?

(ii) What information does the system provide the user with?

(iii) What should not the system do?

The above questions relate to the direct effects of the system, its functionality, and purpose but they do not specify requirements regarding the project completely. The tests are considered for remaining elements, which are invisible to the user and dependent largely on the designer of the system. These elements are the basis of the system. In order to determine them, it is necessary to answer the following questions:

(i) How should the designed system operate?

(ii) What inputs and outputs should it generate/set?

(iii) What functions are its main characteristics?

The answers to these questions should be clarified on the basis of the design phase, the software built, and the expected behaviour of a system. Having already collected information about the system sorted out according to the above scheme, it is possible to begin to plan test case studies. What is most important in the operation of the safety system is its correct operation and efficiency. If the system is to comply with the quite simple functions (e.g., warn the driver about a negative temperature), it is not necessary to develop complex test scenarios.

The case is complicated, however, when the system has a more extensive set of functionalities, when it has to react differently in different situations, when the output value depends on the value of larger quantities of inputs, and where there is the possibility of false alarms. In this case, it is good to previously develop scenarios to test and test cases for each scenario. After tests, a summary of the verification of the correctness of the system should be made. In the case where the system does not meet one (or several) of the designated test case studies, it is necessary to carry out repair procedures. This phase resembles the design phase, but instead of building the system from the beginning, repairs are carried out and changes to the existing system on the basis of data obtained in the test phase are introduced. Well-conducted tests allow us to identify errors in the system quickly and often arise the way for solutions. In this chapter, the test phase for BLIS system design for the Bytel vehicle will be described.

5.1. Tests Plan

5.1.1. Calculations Verification. Based on the previously elaborated flowchart, the first step of the procedure of Blind Spot Information System test planning for Bytel vehicle is the answer to the question about the correctness of the calculations implemented in the model.

(i) What are inputs to the model?

In the model, the following inputs are used:

(a) X_v [m]: X coordinate of Bytel vehicle in absolute coordinate system

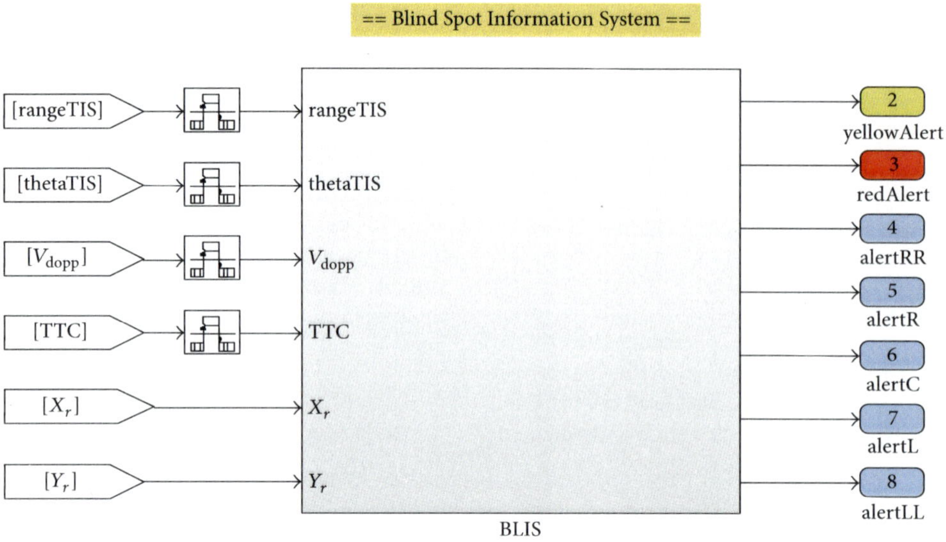

FIGURE 11: BLIS block.

(b) Y_v [m]: Y coordinate of Bytel vehicle in absolute coordinate system

(c) vehicleRot [°]: rotation angle related to z axis of Bytel vehicle

(d) V_v [m/s]: Bytel vehicle velocity

(e) rangeTIS [m]: distance from the sensor to detected object

(f) V_{dopp} [m/s]: relative vehicles velocity

(g) thetaTIS [°]: angle at which obstacle was detected

(ii) What are the outputs of model?

In the model, the following outputs are generated:

(a) yellowAlert (0/1): yellow alarm status

(b) redAlert (0/1): red alarm status

(c) alertLL (0/1): LL alarm status

(d) alertL (0/1): L alarm status

(e) alertC (0/1): C alarm status

(f) alertR (0/1): R alarm status

(g) alertRR (0/1): RR alarm status

(iii) How are the above outputs generated?

Decisions on the classification of the alarm due to the level of risk (yellow or red) system are taken on the basis of three pieces of information:

(a) Comparison of relative position of object with value of YCD (yellow Critical Distance) or RCD (Red Critical Distance), which are functions determining threshold values of object along x axis for which increased risk of collision exists

(b) V_{dopp} value

(c) TTC value (Time To Collision)

Decision on selection of the area, which is assigned to an alarm, depends only on the value of the angle thetaTIS.

(iv) Which of these activities relate to the independent variables from the operation of the system and which are the physical representation of unknown values?

Variable determined by calculation are as follows:

(a) X_r: X coordinate of object position related to Bytel vehicle determined on the basis of values of rangeTIS and thetaTIS

(b) Y_r: Y coordinate of object position related to Bytel vehicle determined on the basis of values of rangeTIS and thetaTIS

(c) Yellow Critical Distance: function determined on the basis of Y_r value

(d) Red Critical Distance: function determined on the basis of Y_r value

(e) Time To Collision: time to collision determined on the basis of rangeTIS and V_{dopp} values

Based on this information, it was decided to examine the above-mentioned dependence by comparing them with the values of the corresponding variables artificially generated by the PreScan software. An example of such a comparison can be seen in Figure 12.

5.1.2. Verification from the User Perspective: Visual Verification. The next step is to answer the questions about system importance from the perspective of the user (Figures 13 and 14).

(i) What should the designed system do?

The purpose of the system is to inform the user of the location of another vehicle behind the driven one, which is not visible to the driver, or its notification

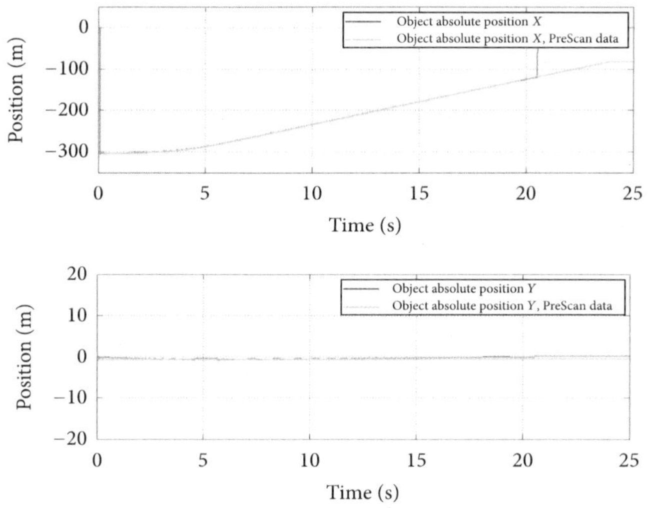

FIGURE 12: Comparison of corresponding variables determined by virtual BLIS system and directly by PreScan.

requires too much effort of the driver. The information should be communicated in a visual way, using LEDs.

(ii) What information should the system provide the user with?

System is to inform the user of the observation of the vehicle in one (or more) of the five ranges designated for the driven vehicle and the degree of risk associated with the event. For low-level threats, the alarm should accept yellow colour and for high threat red.

(iii) What should not the system do?

The system should not react to vehicles in the area of the sensor, but in a place visible to the driver of the vehicle. The system also should not generate false positives: that is, the alarm is not caused by the appearance of the object in the field of sight of the driver.

5.1.3. Numerical Verification. The next step is to analyse important questions from the perspective of the creators of the system (Figure 16):

(i) How the designed system should operate?

The system is intended to operate on the basis of information from the vehicle (vehicle location, vehicle speed), and the information from the sensor (the distance of the detected object from the sensor, the angle at which the object appears, Doppler speed). The system should set, on the basis of the data, relative location information of vehicles to each other and the speed difference between vehicles. The system should assign one of the five predesignated areas to a given position. Based on the information about the speed difference between vehicles and their relative position system, it should set TTC (Time To Collision). Depending on the value of TTC and the distance between the vehicles, the system should assign areas of the degree of risk: no risk, the LED is not lit; low threat, LED yellow; high risks, the LED lights red.

(ii) What input and output values should it take?

The system should set the following inputs and outputs:

(a) Distance of detected object from driven vehicle: 0–40 [m]
(b) Velocity of driven vehicle: 0–15 [m/s]
(c) Relative velocity of vehicles: 0–15 [m/s]
(d) Position of the vehicle on x, y axis [m]
(e) Angle of appearance of object: $-87{,}5^\circ$–$87{,}5^\circ$

The system should generate the following output values:

(a) Value 0/1 on output corresponding to LL, L, C, R, and RR LEDs
(b) Value 0/1 on output corresponding to yellow alarm
(c) Value 0/1 on output corresponding to red alarm

(iii) What functions characterize it?

(a) TTC (Time To Collision) for every moment when object is within the sensor range
(b) RangeTIS for every moment when object is in the sensor range
(c) Relative object and Bytel vehicle position: X_r, Y_r
(d) ThetaTIS angle, in which the object detected remains for each moment of time in which the object is located within the sensor range
(e) Relative object and Bytel vehicle velocity

5.1.4. The Development of a Set of Test Cases. Based on the information collected, Table 1 was prepared containing test cases, which were used to verify the operation of the system BLIS.

5.2. Test Case Scenario: Vehicle Overtaken from the Left Side

5.2.1. Description of the Scenario. The object comes from the left side of the vehicle Bytel. The ratio of speed of an object to Bytel vehicle speed is in the range of 0 to 1.5 for each moment of virtual experiment. The speed difference between the object and the Bytel vehicle is not greater than 10 m/s. The speed of the object and the vehicle does not exceed 12 m/s (Figures 13, 14, and 15).

5.2.2. Results Verification. In order to investigate the abovementioned cases, we have to use different methods. For testing cases (1) (Table 1) and (3) (Table 1), it was necessary to analyse the relevant variables for cases (2) (Table 1) and visual inspection simulation is sufficient.

Calculation Verification. In order to check the correctness of the algorithms implemented in the BLIS system model comparative charts were used comparing variables with variables generated by the PreScan software.

FIGURE 13: Example of visual verification, scenario 1, overtaking phase 1.

FIGURE 14: Example of visual verification, scenario 1, overtaking phase 2.

(i) Comparison of V_{dopp} speed with the difference in velocity of the Bytel vehicle and the object overtaking Bytel; both speeds are available in the PreScan software.

Based on a comparison graph of the Doppler Velocity and difference of vehicles velocities, it can be concluded that the mapping of Doppler Velocity does not fully coincide with the actual vehicles speeds difference. The biggest differences can be seen for the time intervals: 0–2 s and 16–19 s. This is connected with the moment of occurrence of the object in a short distance from the sensor. Despite these cases, it is concluded that the proposed function of the velocity can be used in the design of system.

(ii) Comparison of the relative position in the x and y axis of overtaking object with respect to the Bytel vehicle, calculated using information about the distances of overtaking object and the angle in which it is located, with a difference of location coordinates of the vehicle and the object with a leading Bytel; coordinates are available in the PreScan software.

Based on the comparison charts X and Y calculated using data on the distance and angle of the object overtaking with the actual difference in the position of the Bytel vehicle and the overtaking object, it was found that the adopted model calculation of the relative position of the object is correct. In addition, the graphs present compliance of variables only until reaching 20th second. It is associated with the period when the object is within the range of the sensor, which enabled the calculation of its relative position.

(iii) Comparison of the absolute position of the x and y axis overtaking object calculated using information about the distances of overtaking object, the angle at which it is located and the coordinates of the Bytel vehicle, the absolute position of the object, which is made available in the PreScan software.

Based on the comparison charts X and Y calculated using data on the distance, angle and coordinates of the object overtaking the Bytel vehicle with the actual position of the absolute object, it was found that the adopted computational model of the absolute position of the object is correct. The graphs show compliance only until reaching 20th second. It is associated with the period when the object is within the range of the sensor, which enabled the calculation of its relative position.

FIGURE 15: Example of visual verification, scenario 1, overtaking phase 3.

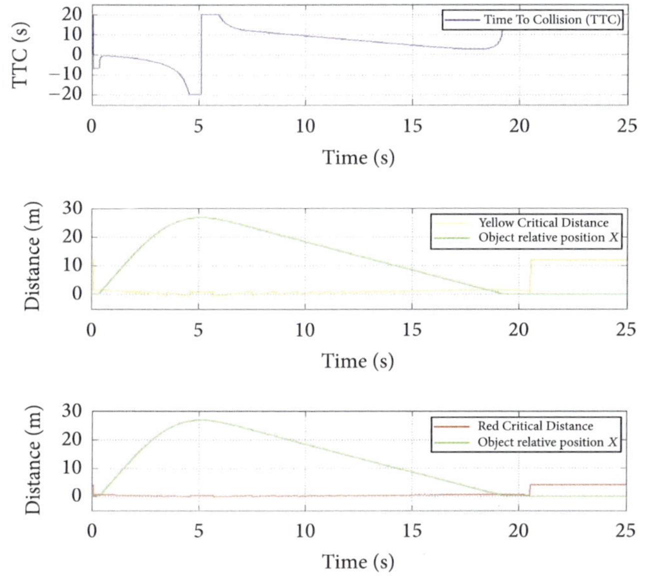

FIGURE 16: TTC (Time To Collision), comparison of YCD (Yellow Critical Distance), and RCD (Red Critical Distance) with object relative position X.

Visual Verification (Figures 13, 14, and 15)

(i) The object is coming from the direction established in the description of the scenario: object is coming from the left.

Based on the simulation results, the system has been found to meet this condition.

(ii) The object is located within the sensor range: it was possible to carry out tests of the BLIS system and the object must be in range of the sensor.

Based on the simulation results, the system has been found meeting the above condition.

(iii) Operation of warning LL LEDs, since the object is overtaking on the left side of the Bytel vehicle after fulfilment of additional conditions LED should light up depending on the object distances in red and/or yellow.

Based on the simulation results, the system has been found meeting the above condition.

(iv) Operation of warning L LEDs, since the object is overtaking on the left side of the Bytel vehicle after fulfilment of additional conditions LED should light up depending on the object distances in red and/or yellow.

Based on the simulation results, the system has been found to meet the above condition.

(v) Operation of warning C LED, since the overtaking object at any moment in time is located centrally behind the Bytel vehicle the LED should not light up.

Based on the simulation results, the system has been found to meet the above condition.

(vi) Operation of warning R LED, since the overtaking object at any moment in time is located from the right side of Bytel vehicle the LED should not light up.

Based on the simulation results, the system has been found to meet the above condition.

(vii) Operation of warning RR LED, since the overtaking object at any moment in time is located from the right side of Bytel vehicle, the LED should not light up.

Based on the simulation results, the system has been found to meet the above condition.

Numerical Verification

(i) Inspection of Bytel vehicle velocity:

According to the assumptions of this scenario, Bytel vehicle velocity at any moment in time should not be greater than 12 m/s. Based on the velocity chart of the Bytel vehicle the system has been found to meet this assumption.

(ii) Inspection of overtaking object velocity:

TABLE 1: Test cases used to verify the operation of the BLIS system.

Number	Test description	Initial conditions	Expected value	Outcome
(1)	Calculation verification			
(1.1)	X_r position	rangeTIS $\neq 0$	$X_r \cong X_v - X_o$	
(1.2)	Y_r position	rangeTIS $\neq 0$	$Y_r \cong Y_v - Y_o$	
(1.3)	Yellow Critical Distance	rangeTIS $\neq 0$	$\text{YCD} = -4 * \text{abs}(Y_r) + 12$	
(1.4)	Red Critical Distance	rangeTIS $\neq 0$	$\text{RCD} = -4/3 * \text{abs}(Y_r) + 4$	
(1.5)	Time To Collision	$V_{\text{dopp}} \neq 0$	$\text{TTC} = -\text{rangeTIS}/V_{\text{dopp}}$	
(2)	Visual tests			
(2.1)	Object approaches from direction assumed in the scenario description	-	Left/right side, behind the object	
(2.2)	Object is in the sensor range	-	The object is in the sensor range	
(2.3)	Correct functioning of the LL warning LED	Object in the LL area, danger	LED lights up Y/R	
(2.4)	Correct functioning of the L warning LED	Object in the L area, danger	LED lights up Y/R	
(2.5)	Correct functioning of the C warning LED	Object in the C area, danger	LED lights up Y/R	
(2.6)	Correct functioning of the R warning LED	Object in the R area, danger	LED lights up Y/R	
(2.7)	Correct functioning of the RR warning LED	Object in the RR area, danger	LED lights up Y/R	
(3)	Computational tests			
(3.1)	Object velocity/Bytel vehicle velocity	$V_v \neq 0$	≥ 1	
(3.2)	Object velocity/Bytel vehicle velocity	$V_v \neq 1$	≤ 10 m/s	
(3.3)	Object velocity	-	≤ 12 m/s	
(3.4)	Bytel vehicle velocity	-	≤ 10 m/s	
(3.5)	Turning on yellow alert	$4 \leq \text{TTC} \leq 10 \cup (X_r \leq \text{YCD} \cap V_{\text{dopp}} < 0 \cap X_r \neq 0)$	yellowAlert = 1	
(3.6)	Turning on red alert	$0.1 \leq \text{TTC} \leq 4 \cup (X_r \leq \text{RCD} \cap V_{\text{dopp}} < 0 \cap X_r \neq 0)$	redAlert = 1	
(3.7)	LL area control	$48 \leq \text{thetaTIS} \leq 87.5$	alertLL = 1	
(3.8)	L area control	$7 \leq \text{thetaTIS} \leq 49$	alertL = 1	
(3.9)	C area control	$8 \leq \text{thetaTIS} \leq -8$	alertC = 1	
(3.10)	R area control	$-49 \leq \text{thetaTIS} \leq -7$	alertR = 1	
(3.11)	RR area control	$-87.5 \leq \text{thetaTIS} \leq -48.5$	alertRR = 1	

According to the assumptions of this scenario overtaking vehicle velocity at any moment in time should not be greater than 12 m/s. Based on the velocity chart of the overtaking vehicle the system has been found to meet this assumption.

(iii) Inspection of object and Bytel vehicle velocities ratio:

According to the assumptions of this scenario, ratio of velocities of object to Bytel vehicle velocity at any moment in time should not be less than 1. Based on the velocities ratio chart, the system has been found to meet this assumption.

(iv) Inspection of object and Bytel vehicle velocities difference:

According to the assumptions of this scenario, the difference of overtaking vehicle velocity and Bytel vehicle velocity at any moment in time should not be greater than 10 m/s. Based on the vehicles velocities difference chart system has been found to meet this assumption.

(v) Inspection of alarm activation:

In order to inspect correctness of activation of yellow and red alarm, the following charts were used: TTC (Time To Collision), YCD (Yellow Critical Distance), and RCD (Red Critical Distance) compared to X_r (relative position of object in x axis). Charts are shown in Figure 16.

(vi) Inspection of yellow alarm activation:

According to the assumptions of the design, the yellow alarm should be activated under the following conditions:

(a) TTC (Time To Collision) ranges from 4 to 10 s or

(b) the value of the relative position of the object in the x-axis representing the longitudinal axis of the Bytel vehicle is less than the value of the function YCD (Yellow Critical Distance).

Based on the charts of relative position of the object in the x-axis and functions of Yellow Critical Distance, charts of TTC and yellow alarm activity system have been found to meet this assumption.

(vii) Inspection of red alarm activation:

According to the assumptions of the design, the red alarm should be activated under the following conditions:

(a) TTC (Time To Collision) ranges from 0,1 to 4 s or

(b) the value of the relative position of the object in the x-axis representing the longitudinal axis of the Bytel vehicle is less than the value of the function RCD (Red Critical Distance)

Based on the charts of relative position of the object in the x-axis and functions of Red Critical Distance, the charts of TTC and red alarm activity system have been found to meet this assumption.

(viii) Inspection of operation of LL alarm corresponding to extreme left scanning area:

As designed LL alarm should be activated when the vehicle is detected in an angular range from 48° to 87.5°. Based on the charts, TIS sensor range, the angle of appearance of the object, and graph LL alarm activity system have been found to meet this assumption.

(ix) Inspection of operation of L alarm corresponding to left scanning area:

As designed L alarm should be activated when the vehicle is detected in an angular range from 7° to 49°. Based on the charts, TIS sensor range, the angle of appearance of the object, and graph L alarm activity system have been found to meet this assumption.

(x) Inspection of operation of C alarm corresponding to central scanning area:

As designed, C alarm should be activated when the vehicle is detected in an angular range from −8° to 8°. Based on the charts, TIS sensor range, the angle of appearance of the object, and graph C alarm activity system have been found to meet this assumption.

(xi) Inspection of operation of R alarm corresponding to right scanning area:

As designed R alarm should be activated when the vehicle is detected in an angular range from −49° to −7°. Based on the charts, TIS sensor range, the angle of appearance of the object, and graph R alarm activity system have been found to meet this assumption.

(xii) Inspection of operation of RR alarm corresponding to extreme right scanning area:

As designed RR alarm should be activated when the vehicle is detected in an angular range from −87,5° to −48°. Based on the charts, TIS sensor range, the angle of appearance of the object, and graph RR alarm activity system have been found to meet this assumption.

5.2.3. Final Results of Virtual Verification. Based on the above results, Table 1 of test cases was completed, and it was found that tests for scenario I were completed successfully as can be seen in Table 2.

6. Consideration of Possible Improvements

Although the final result of applying the described method was satisfying, there were still areas for the improvement. First of all there should be stronger emphasis put on the iterative approach to design, combining design phase with testing phase.

In this context, it is important to choose the scenarios of simulation tests carefully and, on the basis of subsequent experience, expand the list of these scenarios to the new situations observed in the race conditions.

What also would be of a great value is real-life implementation system and comparison of real-life values with outputs received from simulation. This could help to identify the areas where the described method does not prove correct or needs improvement.

Similarly, as TASS, PreScan enables also creating experiments for various weather conditions and it would be valuable to test the designed system for different atmospheric conditions. It is especially true for systems that assume using sensors which output values are weather-dependent (e.g., sonars, Lidars).

There is considerable potential for the transfer of work results related to the BLIS system as well as its methodology for commercial development. The particular advantage of the developed system is its modularity and the ability to adapt to other types of vehicle and other operating conditions.

7. Conclusions

The described method of Advanced Driver Assistance System design, with the use of virtual prototyping tool, has been applied in real-life project. The outcome of taking this approach can be evaluated as successful, since the reasons for using this kind of method have been proven and the final results have been considered satisfying.

Using simulation method in ADAS design shortens work time, makes the iterative approach to design easier and

TABLE 2: Approved results of test cases.

Number	Test description	Initial conditions	Expected value	Outcome
(1)	Calculation verification			Approved
(1.1)	X_r position	rangeTIS $\neq 0$	$X_r \cong X_v - X_o$	Approved
(1.2)	Y_r position	rangeTIS $\neq 0$	$Y_r \cong Y_v - Y_o$	Approved
(1.3)	Yellow Critical Distance	rangeTIS $\neq 0$	YCD $= -4 * \text{abs}(Y_r) + 12$	Approved
(1.4)	Red Critical Distance	rangeTIS $\neq 0$	RCD $= -4/3 * \text{abs}(Y_r) + 4$	Approved
(1.5)	Time To Collision	$V_{\text{dopp}} \neq 0$	TTC $= -\text{rangeTIS}/V_{\text{dopp}}$	Approved
(2)	Visual tests			Approved
(2.1)	Object approaches from direction assumed in the scenario description	-	Left/right side, behind the object	Approved
(2.2)	Object is in the sensor range	-	The object is in the sensor range	Approved
(2.3)	Correct functioning of the LL warning LED	Object in the LL area, danger	LED lights up Y/R	Approved
(2.4)	Correct functioning of the L warning LED	Object in the L area, danger	LED lights up Y/R	Approved
(2.5)	Correct functioning of the C warning LED	Object in the C area, danger	LED lights up Y/R	Approved
(2.6)	Correct functioning of the R warning LED	Object in the R area, danger	LED lights up Y/R	Approved
(2.7)	Correct functioning of the RR warning LED	Object in the RR area, danger	LED lights up Y/R	Approved
(3)	Computational tests			Approved
(3.1)	Object velocity/Bytel vehicle velocity	$V_v \neq 0$	≥1	Approved
(3.2)	Object velocity/Bytel vehicle velocity	$V_v \neq 1$	≤10 m/s	Approved
(3.3)	Object velocity	-	≤12 m/s	Approved
(3.4)	Bytel vehicle velocity	-	≤10 m/s	Approved
(3.5)	Turning on yellow alert	$4 \leq \text{TTC} \leq 10 \cup (X_r \leq \text{YCD} \cap V_{\text{dopp}} < 0 \cap X_r \neq 0)$	yellowAlert = 1	Approved
(3.6)	Turning on red alert	$0.1 \leq \text{TTC} \leq 4 \cup (X_r \leq \text{RCD} \cap V_{\text{dopp}} < 0 \cap X_r \neq 0)$	redAlert = 1	Approved
(3.7)	LL area control	48 ≤ thetaTIS ≤ 87.5	alertLL = 1	Approved
(3.8)	L area control	7 ≤ thetaTIS ≤ 49	alertL = 1	Approved
(3.9)	C area control	8 ≤ thetaTIS ≤ −8	alertC = 1	Approved
(3.10)	R area control	−49 ≤ thetaTIS≤ −7	alertR = 1	Approved
(3.11)	RR area control	−87.5 ≤ thetaTIS ≤ −48.5	alertRR = 1	Approved

faster, enables early error detection and identification, and encourages parallel work. Therefore, it can be stated that using this method results also in costs reduction and system quality improvement.

The elimination of the need to verify the system in practice is due to the use of algorithms that verify the simulation results without allowing uncritical reliance on simulation results solely.

The high reliability of the BLIS system itself was achieved by taking into account not only the position of the adjacent vehicles, but also their direction of travel as well as the relative speed of the vehicles. This made it possible to calculate the potential collisions and predict the time it could take place (TTC). This approach eliminates false alarms of most nonroad traffic situations, despite the relatively close proximity of the vehicle to other objects on the track or in the immediate vicinity.

Conflicts of Interest

The authors declare that they have no conflicts of interest.

References

[1] D. Ammon and W. Schiehlen, "Advanced road vehicles: Control technologies, driver assistance," in *Dynamical Analysis of Vehicle Systems. Theoretical Foundations and Advanced Applications*, W. Schiehlen, Ed., pp. 283–304, Springer, New York, NY, USA, 2009.

[2] F. Coşkun, Ö. Tuncer, E. Karsligil, and L. Güvenç, "Vision system based lane keeping assistance evaluated in a hardware-in-the-loop simulator," in *Proceedings of the ASME 2010 10th Biennial Conference on Engineering Systems Design and Analysis, ESDA2010*, pp. 103–114, Istanbul, Turkey, July 2010.

[3] O. J. Gietelink, J. Ploeg, B. De Schutter, and M. Verhaegen, "Development of a driver information and warning system with vehicle hardware-in-the-loop simulations," *Mechatronics*, vol. 19, no. 7, pp. 1091–1104, 2009.

[4] A. D. Hall, *A methodology for Systems Engineering*, Van Nostrand, 1962.

[5] A. Belbachir, J. Smal, J. Blosseville, and D. Gruyer, "Simulation-driven validation of advanced driving-assistance systems," *Transport Research Arena – Europe*, vol. 48, pp. 1205–1214, 2012.

[6] K. Jezierska-Krupa and W. Skarka, "Using Simulation Method for Designing ADAS Systems for Electric Vehicle. Proceedings of the 23rd ISPE Inc. International Conference on Transdisciplinary Engineering Location: Fed Univ Technol, Curitiba, Brazil, October, 2016," in *Transdisciplinary Engineering: Crossing Boundaries*, M. Borsato, N. Wognum, and M. Peruzzini, Eds., vol. 4 of *Advances in Transdisciplinary Engineering*, pp. 595–604, 2016.

[7] E. Hull, K. Jackson, and J. Dick, *Requirements Engineering*, Springer Science & Business Media, 2005.

[8] G. Pahl and W. Beitz, *Engineering Design: A Systematic Approach*, Design Council/Springer, 1988.

[9] W. Schiehlen and D. Ammon, *Advanced Road Vehicles: Control Technologies, Driver Assistance*, Institute of Engineering and Computational Mechanics, University of Stuttgart, Stuttgart, Germany.

[10] B.-F. Wu, H.-Y. Huang, C.-J. Chen, Y.-H. Chen, C.-W. Chang, and Y.-L. Chen, "A vision-based blind spot warning system for daytime and nighttime driver assistance," *Computers and Electrical Engineering*, vol. 39, no. 3, pp. 846–862, 2013.

[11] A. Cholewa, W. Skarka, K. Sternal, and M. Targosz, "Electric vehicle for the students' shell eco-marathon competition. design of the car and telemetry system," in *Telematics in the Transport Environment, Proceedings of the 12th International Conference on Transport Systems Telematics, TST 2012, Katowice-Ustroń, Poland, October, 2012*, J. Mikulski, Ed., Communications in Computer and Information Science, pp. 26–33, Springer, Berlin, Germany.

[12] W. Skarka, "Reducing the Energy Consumption of Electric Vehicles," in *Transdisciplinary Lifecycle Analysis of Systems, Proceedings of the 22nd ISPE-Inc International Conference on Concurrent Engineering Location: Delft Univ Technol, Delft, Netherlands, July, 2015*, R. Curran, N. Wognum, M. Borsato et al., Eds., vol. 2 of *Advances in Transdisciplinary Engineering*, pp. 500–509, IOS Press, Amsterdam, Netherlands, 2015.

[13] Shell, Shell Eco-marathon, referenced at: https://www.shell.com/energy-and-innovation/shell-ecomarathon.html.

[14] TASS International, *PreScan: Simulation of ADAS and Active Safety*, 2015, https://www.tassinternational.com/prescan.

[15] A. Jałowiecki and W. Skarka, "Modeling in Ultra-Efficient Vehicle Design. Proceedings of the 23rd ISPE Inc. International Conference on Transdisciplinary Engineering Location: Fed Univ Technol, Curitiba, Brazil October, 2016," in *Transdisciplinary Engineering: Crossing Boundaries*, M. Borsato, N. Wognum, M. Peruzzini et al., Eds., vol. 4 of *Advances in Transdisciplinary Engineering*, pp. 999–1008, 2016.

[16] W. Skarka and M. Wąsik, "Aerodynamic Features Optimization of Front Wheels Surroundings for Energy Efficient Car. Proceedings of the 23rd ISPE Inc. International Conference on Transdisciplinary Engineering, Fed Univ Technol, Curitiba, Brazil, October, 2016," in *Transdisciplinary Engineering: Crossing Boundaries*, M. Borsato, N. Wognum, M. Peruzzini et al., Eds., vol. 4 of *Advances in Transdisciplinary Engineering*, pp. 483–492, 2016.

[17] LTD Hokuyo Automatic Co. Scanning laser range finder utm-30lx-ew specification. 2012.

[18] K. Cichoński and W. Skarka, "Innovative Control System for High Efficiency Electric Urban Vehicle. Proceedings of the 15th International Conference on Transport Systems Telematics (TST): Wrocław University of Science and Technology, Wroclaw, Poland, April, 2015," in *Mikulski, J*, OF. TOOLS, Ed., vol. 531 of *Communications in Computer and Information Science*, pp. 121–130, 2015.

[19] S. A. Rodríguez Flórez, V. Frémont, P. Bonnifait, and V. Cherfaoui, "Multi-modal object detection and localization for high integrity driving assistance," *Machine Vision and Applications*, vol. 25, no. 3, pp. 583–598, 2014.

16

Analysis of Nonsevere Crashes on Two- and Four-Lane Urban and Rural Highways: Effects of Wet Pavement Surface Condition

Aschalew Kassu [1] and Michael Anderson [2]

[1] *Department of Mechanical, Civil Engineering and Construction Management, Alabama A&M University, Normal, AL 35762, USA*
[2] *Department of Civil & Environmental Engineering, University of Alabama in Huntsville, Huntsville, AL 35899, USA*

Correspondence should be addressed to Aschalew Kassu; aschalew.kassu@aamu.edu

Academic Editor: David F. Llorca

This study examines the effects of wet pavement surface conditions on the likelihood of occurrences of nonsevere crashes in two- and four-lane urban and rural highways in Alabama. Initially, sixteen major highways traversing across the geographic locations of the state were identified. Among these highways, the homogenous routes with equal mean values, variances, and similar distributions of the crash data were identified and combined to form crash datasets occurring on dry and wet pavements separately. The analysis began with thirteen explanatory variables covering engineering, environmental, and traffic conditions. The principal terms were statistically identified and used in a mathematical crash frequency models developed using Poisson and negative binomial regression models. The results show that the key factors influencing nonsevere crashes on wet pavement surfaces are mainly segment length, traffic volume, and posted speed limits.

1. Introduction

According to FHWA report [1], 12.6 percent of the total accidents for the ten-year period between 2000 and 2009 in the United States occurred on wet roadway pavement condition with insufficient pavement friction. Based on this report, until 2009, eight states have implemented successful and formal wet weather crash program to reduce the crash associated with wet pavement skidding. These states were California, Florida, Kentucky, Maryland, Michigan, New Jersey, New York, and Virginia. The procedures followed by each state to identify and mitigate wet crash locations were different. For instance, State of California uses the statistical relationship to evaluate the number of crashes required for significance average wet crash rate and identify wet weather crash locations based on Poisson's distribution one-tail test for 99.5 percent confidence interval. The statistical analysis considers the average number of crashes, ADT, the length of the highway segment under consideration (typically 0.2 mile), analysis period, the percentage of the wet time, and weather condition. In the State of Florida, wet weather crashes locations were identified using five years of crash data. For a site to be considered as wet weather crash location, out of a minimum of four wet weather crashes, at least 25% are due to wet weather or 50 percent of the crashes in that specific segment for the last five years have to be wet weather-related. In the State of Virginia, a location is classified as a potential wet accident hotspot (PWAH), if a minimum of three wet weather crashes were reported with at least 20 percent of which were more than the ratio for all roads in the area. Kentucky adopts a minimum of eight crashes, and at least 35 percent of the crashes in that location were wet pavement related crashes. In New York State, a location is designated as wet weather crash location if a minimum of six (in rural locations) and ten (urban areas) wet weather crashes with a minimum of 35 percent of the crash are in the specific segment for at least two years.

2. Literature Review

The impact of wet pavement on highway crash involves the drivers' behavior and condition, the vehicle, pavement surface, environmental and geometric characteristics of the road, travel speed, visibility, the drivers' behavior, and other conditions. The relationship between these factors and integration of these variable makes wet pavement related crash a complex problem to solve [2, 3]. El Halim et al. [2] reported the results of four statistical models: exponential, logarithmic, power-based, and linear based models. The variables considered included regions, collision severity classes (fatal, injury, and property damage), pavement surface (dry, ice, snow, and wet), environment (clear, drifting snow, fog, freezing rain, rain, snow, and wind), visibility (clear, fog, rain, and snow), and roadway location (intersection, mainline, off the road, and shoulder). Based on the findings of the analysis, one model which was used to best fit the data for one of the road surface conditions may result in low correlation for the other variables. For instance, the logarithmic model was found to best fit wet surface condition with an R^2 value of 0.822 whereas exponential model best fits dry and snow road surface conditions with R^2 values of 0.855 and 0.900, respectively [2]. Recently, Leard and Roth [4] studied the impacts of snowfall, rainfall, and temperature on the three types of accidents: property damage only (PDO), injury, and fatality were analyzed. A Poisson regression model was used to predict the frequency of PDO weather-related accidents, injuries, and fatalities; and a linear probability model was used to model the relationship between weather variation and fatality rates. To account for deferred travel arrangements, the model included lag variables for weather and temperature. Based on the regression results, there was about a 97% increase in PDO accidents during the coldest day as compared with 50°F temperature. Similarly, it was found that a 3 cm rainfall and snowfall of 3 cm per day increases the PDO accident by 18.8% and 43.3%, respectively, as compared with days without rainfall event.

Using a five-year data from 2004 to 2009 obtained from Wisconsin DOT (Department of Transportation) crash database, Jung et al. [5] used negative binomial regression and sequential logistic regression for statistical modeling of crash frequency and crash severity, respectively. The study found that high rainfall and less sunlight in winter, the existence of off-ramp, and change in left lane width increased multivehicle crash. On the other hand, it was shown that wet pavement crash decreased due to the drivers paying more attention as a result of low visibility during strong wind and rainfall. It was also reported that rainfall crash severity is low on Mondays and Fridays due to commuting peak hour traffic which makes the driver more cautious as well. However, Levine et al. [6] reported that, in Honolulu, low accident rates were reported on Fridays.

Dense-graded pavements and wider roadway sections were reported to be more favorable to produce hydroplaning-crashes [7]. To reduce wet pavement crashes, Florida Department of Transportation limits the maximum number of lanes in one direction with a cross slope in one direction to three lanes. If the design requires wider than three lanes in one direction, the cross slope of the inner lane will be toward the median. The hydroplaning related crash analysis was based on the data obtained from the Florida Department of Transportation Crash Analysis Reporting System (CARS) database. The roadway sections considered were sections with "standing water." During peak hours, the driving speed was usually low which reduces the chances of a hydroplaning crash. Based on this study, during light rainfall (≤0.025 cm/hr) with visibility of 1500 ft, for a free-flow speed of 70 mph, the speed reduction reported was in the range of 1.4 mph to 6.4 mph. During heavy rainfall, for FFS and visibility, the speed reduction reported ranges from 3.7 mph to 20.3 mph. As a result, the analysis excludes the off-peak period crash data. Wet weather crash data at intersections, on curves and superelevated sections, were not considered due to other contributing factors to crash. The analysis was made to determine the correlation of wet weather crashes to pavement conditions (rutting, cracks, International Roughness Index), pavement surfaces (open graded friction course, dense-graded asphalt, Portland cement concrete, and dense-graded asphalt concrete), and interlane hydroplaning potential of multilane facilities. The study concluded the following: (i) the impact of pavement rutting and cracks in weather-related crash is insignificant; (ii) the International Roughness Index (IRI) of the pavement surface was found to have a significant impact on wet weather crash [7]; (iii) on multilane highways, in general, the higher the travel speed in any lane, the higher the hydroplaning related crash; and (iv) the hydroplaning related crash on dense-graded asphalt surface for two- and three-lane highway was higher than the crash rate on Portland cement concrete.

Another work by Villiers et al. [8] in cooperation with the State of Florida DOT, used a driving simulator for evaluating drivers behavior to hydroplaning and recommended the safe speed limits during rainfall events for two categories of rainfall: light intensity rainfall (0.025-0.635 cm/hr) and heavy rainfall (rainfall intensity ≥0.635 cm/hr). The study was conducted on six Florida highway sections with 55 mph or higher posted speed limit and segments close to airport locations where rainfall data from NOAA's rain gauge stations could be utilized. They found that heavy rain which reduces visibility was the prime cause of operating speed reduction. Based on the study, the average speed reductions due to light and heavy rainfall were 2 mph and 5 mph, respectively. During nighttime rainfall events, 8 mph drop in speed was reported.

Chan et al. [9] used the 2006 Tennessee accident database collected from 55 mph posted speed limit divided interstates to calibrate regression models to predict the relationship between crash frequency and pavement conditions (mainly pavement distress) in the State of Tennessee for daytime, nighttime, different weather conditions (normal and rainy), peak hour, and nonpeak hour accidents. The variables used in the negative binomial regression model were rut depth (RD), present serviceability index (PSI), and IRI. The results found that only 20% of the accidents occur during rainy weather condition and about 77% of the accidents happened during normal weather condition, and IRI and PSI were found to be key predictive factors.

Another possible factor which influences traffic accident on wet pavements is pavement skid resistance. To determine the quantitative relationship between pavement skid resistance and crash risk, Wu et al. [10] used a method called Crash

Rate Ratio (CRR) and developed CRR-SN model (curve), which could be used at the network level. They claimed that the CRR-SN has several benefits over the general regression models. Some of the advantages mentioned include ease of model calibration and a better quantitative relationship between skid resistance and crash risk for the data. Several earlier studies reported that there was a clear relationship between rainfall intensity and traffic volume. As the rainfall intensity increases, the traffic volume tends to decrease [11]. The report indicated a reduction of 1.86 and 2.16 percent in traffic volume on rainy days in winter and spring, respectively. At nighttime the reduction in traffic volume in winter and spring was 0.87 and 2.91 percent. Levine et al. [6] suggested that rainfall traffic increases accidents which were not only due to the weather but also as a result of the fluctuations on the daily traffic accident count resulting from the interaction between the weather and traffic volume, and other variables such as weekday versus holiday were influencing factors. One interesting finding by the team was that the interaction effect between the rainfall and the time of the day, specifically the afternoons, elevates the crash count. Another finding by the same team was that, in Honolulu, Hawaii, more accidents occurred on Fridays and Saturdays than the other days of the week. However, the observation on Fridays does not agree with another study by Jung et al. [5], which claimed that Mondays and Fridays had the lowest crash rate in Wisconsin.

One of the other factors elevating traffic accidents during rainfall is visibility. As the rainfall intensity increases, the drivers' visibility decreases, which in turn increases the crash risk [12–15]. Based on these studies, the influence of rainfall intensity on drivers' visibility was associated with the water film on the windshield [16], the condition of the wipers, and splash from one vehicle to another. The other major factor reducing the visibility of the drivers was the interaction between rainfall intensity and vehicle speed [13]. Traffic accidents were found to be significantly higher on the first day of rainfall occurring after a prolonged period of dry weather [11, 17]. One of the reasons suggested was the existence of road grime accumulated on the pavement surface, which created a slippery surface. Sudden onset of rainfall especially on the first day of the rainfall event significantly increased the traffic accident count as compared with the second and third day of the rainy days [6].

Negative binomial regression model has been extensively used in traffic safety studies. However, recent studies suggest that Generalized Warning (GW) model, which is based on Generalized Warning distribution [18] outperforms the traditional negative binomial model in terms of providing both a better fit and information on crash variances. On the other hand, Ma et al. [19] compared the adequacy of Tobit model with hurdle models developed using Poisson, Gamma, Weibull, and lognormal distributions. The study found that the lognormal hurdle model with flexible scale parameter is superior to the other models. More recently, Ye et al. [20] proposed seminonparametric (SNP) Poisson model which accounts for unobserved heterogeneity in crash data. Using ten years of crash data collected from California rural roads, the authors suggested that the overall performance of SNP-Poisson model is better than the traditional negative binomial model. The remainder of the paper is organized as follows. The next section covers a review of the literature on traffic safety modeling and crash contributing factors, followed by the methodologies adopted for extracting and preparation of dataset used for crash analysis and the investigation of the effects of wet pavement surface conditions on traffic accidents. Next, the Results and Discussion presents the results of the statistical analysis with the discussions. Finally, the key findings of the research are summarized and concluding remarks are provided.

3. Methodology

The study area covers a wide range of locations in the State of Alabama. For model formulation and comparison of crash rates recorded during normal weather condition (dry pavement) with crash rate occurring on wet pavement surface condition, the crash record of the State of Alabama route system, ranging from 2010 to 2014, was extracted from Critical Analysis Reporting Environment (CARE) managed by the Center for Advanced Public Safety (CAPS) at the University of Alabama and was used. The selected segments represent different geographic locations and counties in the State, excluding intersection areas, bridge locations, and ramps. The crash data from CARE have a detailed information on the location, day of the week, time of the day, speed limit, surface condition of the pavement (wet, dry), visibility, the number of vehicles involved, if the accident was at intersection or not, the distance from intersection, and more. However, it does not include information on the traffic volume, roadway data like horizontal and vertical curvatures, superelevation, and pavement roughness conditions. Other variables, including AADT (annual average daily traffic), the percentage of trucks, IRI, pavement rutting, grade, and cross slope were linked to the CARE data. After combining the two datasets, explanatory factors essentially considered for further analysis were filtered. The scope of this study was to analyze the nonsevere crashes on two- and four-lane urban and rural highways in Alabama. With this in mind, the data were prepared as follows. As the majority of the crashes occurred on two- and four-lane rural and urban highway segments, the data for these sections were filtered, excluding three-, five-, and six-lane crash data. The crash data recorded on freeway segments and at intersections were also excluded from the analysis.

The number of crash counts within the segments were converted to crash rates using AADT and vehicles-miles-traveled (VMT) for each roadway segment using the Federal Highway Administration guideline [21]: $R = (C \times 10^8)/(AADT \times 365 \times N \times L)$; and $VMT = AADT \times 365 \times L$. Here, R is roadway crashes rate for the segment expressed as crashes per 100 million vehicle- miles of travel; C is crash count for the segment; N is number of years of data; and L is length of roadway segment.

Analysis of individual highways and tests of homogeneity of the routes for combining the crash data of individual highways have rarely been reported in the literature. To create a homogenous highway section, Cafiso and Graziano [22] and Thomas [23] used the comparison of the mean values of individual segments. The approach used t-test statistics to determine the statistical significance of the differences between the mean values of the adjacent highway segments. The highway segments with statistically insignificant differences in the mean values were aggregated and used for further analysis. In this study, the approach adopted was to combine the crash data of the homogenous segments and examine the statistical significance of each of the categorical variables independently across the dummy coded groups. These included the number of lanes (two versus four), the significance of urban versus rural sections of the segments under considerations, the lighting (dark or light), and weather (rainfall or normal) condition. Initially, sixteen rural and urban state highways (AL 001 to AL 017, except AL 011) representing the major geographic locations of the State of Alabama were selected. Since the crash data for three-, five-, and six-lane urban and rural highways were insignificant as compared with the two- and four-lane highways, these crash data were excluded from the analysis. Here it is well understood that highways with different roadway geometric features and traffic characteristics may yield similar crash rates. However, the variables such as the traffic volume, the geometric design elements of the highways, and the environmental factors considered in the modeling and analysis of the individual segments considered, where the crash rates were observed, were assigned corresponding to each data point within the individual segment length. The relationship between the explanatory variables and the crash rates for each data point is estimated using their respective explanatory factors as independent variables in the negative binomial regression model.

In order to identify the homogenous highways with similar characteristics in terms of homogeneity of variances, equality of means, and similarity in distribution of the data, the crash data were tested with respect to the following three null (H_0) and alternate (H_1) hypotheses. (1) H_0: the mean crash rates for each highway were the same; H1: at least one of the highways was different; (2) H_0: the variances of the highways were equal; H1: at least one of the highways had different variance; (3) H_0: the distributions of crash rates were the same across the highways; H1: at least one highway had different distribution of the crash rate. Equality of means of the crash rates across the sixteen highways was tested using Welch's test followed by Games-Howell simultaneous test for differences of means (*post hoc* test). As the data were not normally distributed, a test of homogeneity of variances was performed using nonparametric Levene's test. Independent samples Kruskal-Wallis test was used to test the similarity of the distributions of the crash data across the different highways.

Initially, fourteen candidate engineering, environmental, and traffic conditions influencing the crash rates were identified. The variables selected include segment length, AADT, TADT (truck average daily traffic), grade, IRI, rut depth, macrotexture, cross slope, number of lanes, posted speed limit, urban/rural designation of the segment, and weather (rainfall/normal), lighting (dark/light), and pavement surface condition (wet/dry). To determine the impact of the categorical variables and ascertain the significance of the differences in the means, variances, and distributions of the crash rates across the categories of the individual variable, statistical tests were performed with respect to the null and alternate hypothesis described above. For the crash rates whereby the categories of the variable were statistically significant, the categorical variables were retained by assigning dummy codes 0 and 1. For the crash rates which were not statistically significantly different across the assigned categories, it is suggested that there was no need to include the variable as a predictive variable. The categorical variables used to test the crash rates across the categories include urban versus rural sections, two lanes versus four lanes, dark versus light, rainfall versus normal weather, dry versus wet pavement conditions.

Traffic crash data are characterized as count data mostly following overdispersion distribution [24, 25]. To address the overdispersion of the count data, mixed-Poisson regression model such as a negative binomial or Poisson-gamma model whereby the mean of the Poisson follows a gamma distribution is widely utilized. Due to this, the mixed-Poisson regression model such as negative binomial regression model (a mixture of Poisson-gamma distribution) has been widely used for analyzing highway crashes [25–28]. However, there were cases where crash data follow an underdispersed distribution. For such distribution and when the sample mean of the distribution is low, the negative binomial model performs poorly as well [26]. For the negative binomial distribution described by the parameter, μ, the mean of the distribution is given by $\ln(\mu_i) = \beta_0 + \sum_j x_{ij}\beta_j + \varepsilon$ [25, 26]. The models developed were evaluated by using goodness-of-fit tests including, Deviance, Log Likelihood, Akaike's Information Criterion (AIC), and Bayesian Information Criterion (BIC).

4. Results and Discussions

Using the tests outlined in the methodology section, for dry and wet pavement conditions, out of the sixteen highways initially considered, eight and ten of the highways, respectively, were found to be homogenous and are not found to be statistically significantly different in their mean values, variances, and distributions of the crash rates (Tables 1 and 2). Hence, the homogenous highways shown in these tables were combined to form two different datasets one representing the dry pavement crashes and another one representing wet pavement accidents. To combine the dry and wet pavement crashes, the common six routes from each group were selected (Table 3) and dummy coded to identify the dry and wet pavement crashes forming a single crash dataset for the aggregate data. Based on both Shapiro-Wilk (S-W) and Kolmogorov-Smirnov (K-S) tests of normality, all the crash rate at the individual and combined routes were not normally distributed at a significance level of $\alpha = .001$. Similar descriptive analyses and test of normality of the crash data across the categorical variables were performed for dry and wet weather and aggregate datasets and presented in Tables 4, 5, and 6, respectively.

To determine whether crash rates across the categorical variables were significantly different or not, statistical tests including tests of homogeneity of variances, equality of

TABLE 1: Descriptive statistics of the nonsevere crash rates occurring on dry pavement surfaces and tests of normality for the selected eight routes.

Route ID	Length (mi)	⁺N	Mean	St. Dv.	Normality Test		
					Skewness	Kurtosis	*Sig.
AL 001	353	420	481	694	2.751	8.940	0.001
AL 005	198	161	364	551	3.698	21.575	0.001
AL 006	240	172	375	509	2.813	10.802	0.001
AL 008	218	141	371	456	2.381	7.048	0.001
AL 010	231	122	345	579	4.327	22.885	0.001
AL 012	232	170	368	545	3.133	11.862	0.001
AL 016	77	84	480	695	2.730	9.350	0.001
AL 017	347	178	357	522	4.424	28.303	0.001
¥Combined	1896	1448	405	592	3.223	13.629	0.001

*Significance level for both S-W and K-S tests, with the level of significance <0.01. ¥All routes combined. ⁺Number of segments.

TABLE 2: Descriptive statistics of the nonsevere wet-weather crash rates and tests of normality for the ten selected routes.

Route ID	Length (mi)	N	Mean	St. Dv.	Normality Test		
					Skewness	Kurtosis	Sig.
AL 001	353	408	64	81	1.529	1.784	0.001
AL 003	374	347	70	85	1.217	0.577	0.001
AL 004	192	101	75	88	1.000	0.008	0.001
AL 005	198	152	67	91	1.468	1.294	0.001
AL 006	240	161	50	69	1.830	3.924	0.001
AL 010	231	131	59	82	1.667	2.213	0.001
AL 012	232	170	55	79	1.624	2.036	0.001
AL 014	218	135	69	77	1.189	0.944	0.001
AL 015	227	98	69	84	1.215	0.971	0.001
AL 016	77	69	69	94	1.424	1.101	0.001
Combined	2342	1772	68	82	1.417	1.341	0.001

TABLE 3: Descriptive statistics of the nonsevere aggregate (dry and wet) crash rates and tests of normality for the ten selected routes.

Route ID	Length (mi)	N	Mean	St. Dv.	Normality Test		
					Skewness	Kurtosis	Sig.
AL 001	353	774	164	224	2.051	4.290	0.001
AL 005	198	296	157	238	2.261	5.202	0.001
AL 006	240	317	154	226	2.233	5.161	0.001
AL 010	231	244	141	197	2.131	4.745	0.001
AL 012	232	325	145	206	2.347	6.419	0.001
AL 016	77	142	181	262	1.799	2.529	0.001
Combined	1331	2098	157	224	2.149	4.750	0.001

TABLE 4: Descriptive statistics of the nonsevere crash rates occurring on dry pavement surfaces across the categorical variables and tests of normality for the selected eight routes.

Variable		N	Mean	St. Dv.	Skewness	Kurtosis	Sig.
Rural/Urban	Rural	712	278	439	4.922	31.890	0.001
	Urban	736	527	688	2.496	8.285	0.001
No. of Lanes	Two	732	422	610	4.092	25.215	0.001
	Four	716	386	573	2.672	8.533	0.001
Lighting	Dark	434	332	440	2.866	10.970	0.001
	Light	1014	435	644	3.113	12.255	0.001
Weather	Cloudy	293	412	648	3.973	20.168	0.001
	Clear	1155	403	577	2.944	10.889	0.001

TABLE 5: Descriptive statistics of the nonsevere wet-weather crash rates across the categorical variables and tests of normality for the selected ten routes.

Variable		N	Mean	St. Dv.	Skewness	Kurtosis	Sig.
Rural/Urban	Rural	810	56	73	1.606	2.341	0.001
	Urban	962	72	87	1.253	0.678	0.001
No. of Lanes	Two	936	60	80	1.458	1.545	0.001
	Four	836	69	85	1.371	1.126	0.001
Lighting	Dark	642	57	77	1.544	1.876	0.001
	Light	1130	68	85	1.347	1.076	0.001
Weather	Rain	1412	64	82	1.433	1.415	0.001
	No Rain	360	66	84	1.364	1.100	0.001

TABLE 6: Descriptive statistics of the nonsevere aggregate crash rates across the categorical variables and tests of normality for the ten selected routes.

Variable		N	Mean	St. Dv.	Skewness	Kurtosis	Sig.
Surface Cond.	Dry	1032	262	273	1.293	1.126	0.001
	Wet	1066	54	71	1.455	1.462	0.001
Rural/Urban	Rural	1022	133	180	2.326	7.048	0.001
	Urban	1072	179	256	1.890	3.024	0.001
No. of Lanes	Two	1084	167	235	2.088	4.193	0.001
	Four	1010	145	210	2.210	5.466	0.001
Lighting	Dark	683	137	196	2.281	5.816	0.001
	Light	1411	166	235	2.071	4.238	0.001
Weather	Rain	864	55	74	1.613	2.459	0.001
	No Rain	1230	228	262	1.509	1.829	0.001

means, and similarity in distributions across the categories were performed. For the crash rates which were not statistically significantly different, the result suggested that the particular categorical variable did not have any influence on the likelihood of the occurrences of crashes and hence needs to be excluded from further analysis. For those significantly different, the variable was retained, dummy coded, and used as categorical variables in the mathematical models.

In terms of evaluating the crash rate with respect to equality of mean, homogeneity of variances, and similarity in distribution across each of the categorical variables, we can see that for the accidents occurring on dry pavements (Table 7) the crashes that occurred on two-lane highways were found to have similar distribution (p-value=0.007) and significantly different means (p-value=0.927) and inhomogeneous variances (p-value=0.246). Similarly, the mean, variances, and distributions of crashes occurring during clear and cloudy weather conditions were not significantly different with p-values of 0.981, 0.829, and 0.814, respectively. However, the mean, variances, and the distributions of the crashes occurring in rural and urban sections of the highway segments and across the lighting condition (light versus dark) were significantly different, which warrants dummy coding and the inclusion of these categorical variables in statistical modeling. This indicates that the mean values, the variances, and the distributions of the crashes occurring during dark time driving was significantly different from the crashes occurring during nighttime driving with no lighting condition.

For crashes that occurred on wet pavement surfaces (Table 8), the mean, variances, and distributions of crashes occurring during rainfall condition and normal weather conditions were not significantly different with p-values of 0.573, 0.686, and 0.788, respectively. However, the mean, variances, and the distributions of the crashes that occurred in rural and urban sections of the highway segments and across the lighting condition (light versus dark) were significantly different, which suggests dummy coding and the inclusion of these categorical variables in statistical modeling. Test of homogeneity of variances across the number of lanes showed that the variances of crashes across the two- and four-lane highways were not significantly different (p-value = 0.305). However, the mean and distribution of the data across the categories were significantly different with a p-value of 0.029 and 0.004, respectively. This result also suggests the inclusion of a number of lanes as an explanatory variable in the model. The crash rate across the category of lighting, on the other hand, was found to be significantly different with respect to the variances (p-value = 0.021), the mean (p-value = 0.001), and the distribution (p-value = 0.008). The variances, the means, and the distribution of the crash rates across the categories of weather (rainfall versus normal weather condition) were not found to be significantly different, suggesting that the inclusion of this categorical variables was not necessary.

For the aggregate dataset consisting of dry pavement and wet weather crashes the mean, variances, and distributions of the crash data across the pavement surface conditions (wet

TABLE 7: Test for equality of mean, homogeneity of variances, and similarity in distribution of the dry pavement crashes across the categorical variables.

Variables		Levene's Test	Welch's Test	Kruskal-Wallis Test	Games-Howell Test
Rural/Urban	Test Stat.	70.430	67.450	49.500	8.210
	P-Value	0.001	0.001	0.001	0.001
No. of Lanes	Test Stat.	0.010	1.350	7.150	1.160
	P-Value	0.927	0.246	0.007	0.246
Lighting	Test Stat.	10.430	12.400	3.100	3.520
	P-Value	0.001	0.001	0.078	0.001
Weather	Test Stat.	0.000	0.050	0.060	0.220
	P-Value	0.981	0.829	0.814	0.820

TABLE 8: Test for equality of mean, homogeneity of variances, and similarity in distribution of the wet pavement crashes across the categorical variables.

Variables		Levene's Test	Welch's Test	Kruskal-Wallis Test	Games-Howell Test
Rural/Urban	Test Stat.	21.160	17.440	8.610	4.180
	P-Value	0.001	0.001	0.003	0.001
No. of Lanes	Test Stat.	1.050	4.800	8.510	2.190
	P-Value	0.305	0.029	0.004	0.028
Lighting	Test Stat.	5.350	17.440	6.730	2.720
	P-Value	0.021	0.001	0.008	0.007
Weather	Test Stat.	0.320	0.160	0.070	0.400
	P-Value	0.573	0.686	0.788	0.686

and dry) were all found to be significantly different at 0.001 level (Table 9). This warrants the inclusion of the pavement surface condition as a categorical variable in the model. Across the rural-urban designation of the segments and the lighting conditions, the means and variances appeared to be significantly different, but the distributions of the crash rates were similar. However, Games-Howell post hoc tests suggested that the pairs of the crash data across these categories belong to different groups. This again suggested the inclusion of these categorical variables in the model. Similarly, the crash rates across the number of lanes and weather were found to be significantly different.

The results of the final negative binomial models developed using the key variable identified were shown in Table 10. For wet pavement condition, the principal variables influencing the likelihood of occurrences of crashes were segment length (p-value = 0.001), AADT (p-value = 0.002), TADT (p-value = 0.001), and posted speed limits (p-value = 0.01). Other highway design parameters such as number of lanes, pavement cross slope, grade, IRI, rutting depth, and macrotexture of the pavement surfaces of the highway segments studied were not found to have significant impact on the occurrences of crashes on wet pavement surfaces, which agreed well with previous work by Li et al. [29]. The urban-rural designation of the segments, weather, and lighting conditions were not also the key factors associated with the crashes recorded on wet pavement surfaces.

For the aggregate dataset consisting of crashes recorded on both dry and wet pavement surfaces, it appeared that pavement surface condition (B=-1.624 and IRR = 0,197) was one of the significant factors (p-value =0.001) for the crash occurrence. The results of the analysis suggested that the likelihood of occurrences of the crash on wet pavement surface was 80.3 percent less than on dry pavement surfaces. Some of the possible explanations for this are as follows: (1) short exposure time on wet pavement as compared with dry pavement condition within a year; (2) drivers change in travel plan due to bad weather condition, which alters the traffic volume and reduces the likelihood of occurrences of accidents; (3) reduction in driving speed due to rainfall, visibility, and other factors associated with inclement weather conditions [7]; and (4) the case when the weather condition is not conducive and drivers tend to take appropriate precautions to avoid crashes [5, 30].

For the aggregate dataset, the key factors influencing traffic accidents were segment length, AADT, TADT, posted speed limits, and the pavement surface conditions all at a significance level of 0.001. As can be seen, the increase in segment length and AADT values increased the likelihood of occurrences of crashes. But, the more the number of heavy trucks on the segments is, the higher the posted speed limits decrease the likelihood of occurrences of crashes on the segments. For every unit increase in TADT (B = -0.035, IRR = 0.966) there was a likelihood of 3.4 percent decrease in the crash rate, which agrees with previous reports by Ma et al. [19] and Usman et al. [31]. Similarly, for every unit increase in the posted speed limit (B=-0.014, IRR=0.987) along the study segments, the crash rate reduces by 1.3 percent. This

TABLE 9: Test for equality of mean, homogeneity of variances, and similarity in distribution of the aggregate crashes across the categorical variables.

Variables		Levene's Test	Welch's Test	Kruskal-Wallis Test	Games-Howell Test
Pavement Surf. Cond.	Test Stat.	555.250	558.790	430.090	23.640
	P-Value	0.001	0.001	0.001	0.001
Rural/Urban	Test Stat.	33.140	22.330	1.580	4.730
	P-Value	0.001	0.001	0.209	0.001
No. of Lanes	Test Stat.	2.210	4.800	7.900	2.190
	P-Value	0.137	0.029	0.005	0.028
Lighting	Test Stat.	8.820	9.050	3.440	3.010
	P-Value	0.003	0.003	0.064	0.003
Weather	Test Stat.	381.900	478.790	293.590	21.880
	P-Value	0.001	0.001	0.001	0.001

TABLE 10: Estimated parameters for nonsevere crashes on two- and four-lane rural and urban highways using the significant variables.

Parameter	Wet-Pavement Crash			Aggregate Crash		
	Estimate	Sig.	IRR	Estimate	Sig.	IRR
Constant	4.74	0.001	114.3	6.112	0.001	451
Seg. Length	0.347	0.001	1.415	0.279	0.001	1.322
AADT	0.000021	0.002	1.000	0.00002	0.001	1.000
TADT	-0.044	0.001	0.957	-0.035	0.001	0.966
Speed Limit	-0.016	0.01	0.984	-0.014	0.004	0.987
Surface Cond.						
Wet=1	–	–	–	-1.624	0.001	0.197
Dry=0						
Disp. Coeff.	5.23			4.00		
Goodness-of-Fit Statistics						
AIC	15084			21350		
BIC	15117			21390		
LL	-7536			-10668		
Dev./DF	1.02			1.11		

DF=degree of freedom.

complies with the result suggested by Choi et al. [32]. Prior studies showed that, during heavy rainfall, the drivers tend to reduce their operating speed in the ranges from 3.7 mph to 20.3 mph [7]. During light and heavy rainfall, for a free-flow speed of 70 mph, the speed reduction reported was in the range of 1.4 mph to 6.4 mph and the range from 3.7 mph to 20.3 mph, respectively. This reduction in driving speed below the posted speed limits along the study segments could be one possible factor for the reduction of the chances of wet weather crash along the segments with higher posted speed limits. Rainfall intensity also has a relationship with traffic volume which is one of the influencing factors for traffic accidents. As the rainfall intensity increases, the traffic volume tends to decrease by 1.86 and 2.16% in winter and spring, respectively [11]. This is another potential factor for the lower crash rates on wet pavement surfaces as compared with the dry surface condition.

Based on the results shown in Table 10, a negative binomial model for nonsevere crashes that occurred on wet pavement (CR_w) as well as for the aggregate data consisting of both the crashes that occurred on dry and wet pavements (CR_a) per 100 million VMT can be represented by the following regression equations:

$$CR_w = \exp(4.74 + 0.347\ SL + 2.1x10^{-5}\ AADT - 0.044\ TADT - 0.016\ PSL).$$

$$CR_a = \exp(6.112 + 0.279\ SL + 2x10^{-5}\ AADT - 0.035\ TADT - 0.014\ PSL + 0.0\ DRY - 1.624\ WET).$$

Here, SL is segment length, PSL is posted speed limit, DRY is dry pavement surface, and WET is wet pavement surface.

The models developed were evaluated by using goodness-of-fit tests including Deviance, Log Likelihood, Akaike's Information Criterion (AIC), and the Bayesian Information Criterion (BIC). To assess the relative improvements of the models, the AIC and the deviance values of both the negative binomial model (NBM) and Poisson regression model (PRG) models (not shown here) for crashes occurring only on wet pavements (degree of freedom, DM=1767) and the aggregate crash data (DF=2088) were compared. The model with a smaller value of AIC indicates better model to approximate the statistical fit to the crash rates [25]. The AIC and deviance values show sizeable drop indicating that, in both crashes, the negative binomial models show significant improvements over the Poisson models fitting the crash involvement rates on wet pavement surfaces as well as for the aggregate dataset including the crashes on dry and wet pavement surface conditions.

5. Conclusions

In this paper, the effect of wet pavement surface conditions on the likelihood of occurrences of nonsevere crashes in two- and four-lane urban and rural highways in Alabama was analyzed. The significance of fourteen explanatory variables including segment length, AADT, TADT, grade, IRI, rut depth, macrotexture, cross slope, number of lanes, posted speed limit, urban/rural designation of the segment, weather, lighting, and pavement surface condition was tested. The principal terms were statistically identified and used in a mathematical crash frequency models developed using negative binomial regression models. The performance of the models developed was evaluated using the goodness-of-fit test statistics such as AIC, BIC, Deviance, and Log Likelihood ratio tests. As can be seen from the goodness-of-statistics values, the negative binomial regression model developed for wet pavement condition has superior prediction power than the model developed using the aggregate dataset. The results show that the key factors influencing nonsevere crashes on wet pavement surfaces were mainly segment length, traffic volume, and posted speed limits. For the aggregate dataset consisting of dry and wet pavement crashes, the principal variables were segment length, AADT, TADT, posted speed limits, and the pavement surface conditions. In both models, geometric design elements such as grade, cross slope, and number of lanes of the highways and the urban-rural classification of the highways were not the critical factors contributing to the likelihood of occurrences of crashes. The results of the study will benefit stakeholders engaged in transportation safety in the State of Alabama.

Disclosure

This research did not receive specific funding.

Conflicts of Interest

The authors declare that there are no conflicts of interest regarding the publication of this article.

Acknowledgments

The authors would like to acknowledge the Center for Advanced Public Safety (CAPS) at the University of Alabama for providing access to the CARE data and the Alabama Department of Transportation (ALDOT) for providing the data on traffic volume as well as geometric and pavement conditions of the highways.

References

[1] FHWA, "State Practices to Reduce Wet Weather Skidding Crashes," Tech. Rep. FHWA-SA-11-21, U.S Department of Transportation, Federal Highway Administration, May 2011.

[2] A. A. El Halim, S. Tighe, and T. Klement, "Development of a simplified approach for assessing the level of safety of a highway network associated with pavement friction," in *Proceedings of the TRB 88th Annual Meeting*, Transportation Research Record, January 2009.

[3] "Influence of Roadway Surface Discontinuities on Safety," State of the Art Report, Transportation Research Circular E-C134, May 2009.

[4] B. Leard and K. Roth, "Weather, traffic accidents, and climate change. Resource for the future," Discussion Paper RFF DP 15-19, May 2015.

[5] S. Jung, X. Qin, and D. A. Noyce, "Modeling highway safety and simulation in rainy weather," *Transportation Research Record*, no. 2237, pp. 134–143, 2011.

[6] N. Levine, K. E. Kim, and L. H. Nitz, "Daily fluctuations in Honolulu motor vehicle accidents," *Accident Analysis & Prevention*, vol. 27, no. 6, pp. 785–796, 1995.

[7] M. Gunaratne, Q. Lu, J. Yang et al., "Hydroplaning on multi lane facilities," Final Report, Florida Department of Transportation, 2012.

[8] C. Villiers, D. Guo, B. Augustin, R. Tarr, and L. Hernandez, *Evaluation of Driver Behavior to Hydroplaning in the State of Florida Using Driving Simulation*, Florida Department of Transportation, Tallahassee, Fla, USA, 2012.

[9] C. Y. Chan, B. Huang, X. Yan, and S. Richards, "Investigating effects of asphalt pavement conditions on traffic accidents in Tennessee based on the pavement management system (PMS)," *Journal of Advanced Transportation*, vol. 44, no. 3, pp. 150–161, 2010.

[10] H. Wu, Z. Zhang, and M. Murphy, "Quantitative relationship between crash risks and pavement skid resistance," Tech. Rep., The University of Texas at Austin, Center for Transportation Research, May 2014.

[11] K. Keay and I. Simmonds, "The association of rainfall and other weather variables with road traffic volume in melbourne, australia," *Accident Analysis Prevention*, vol. 37, no. 1, pp. 109–124, 2005.

[12] D. L. Ivey, L. I. Griffin III, T. M. Newton, R. L. Lytton, and K. C. Hankins, "Predicting wet weather accidents," *Accident Analysis & Prevention*, vol. 13, no. 2, pp. 83–99, 1981.

[13] L. D. Ivey, K. E. Lehtipuu, and W. J. Button, "Rainfall and visibility-the view from behind the wheel," Research Report 135-3, Texas A&M University, February 1975.

[14] V. D. Bhise, J. F. Meldrum, L. M. Forbes, T. H. Rockwell, and E. D. McDowell, "Predicting driver seeing distance in natural rainfall," *Human Factors*, vol. 23, no. 6, pp. 667–682, 1981.

[15] J. Andrey and R. Olley, "Relationship between weather and road safety: Past and future research directions," *Climatological Bulletin*, vol. 24, no. 3, pp. 123–137, 1990.

[16] F. Bernardin, R. Bremond, V. Ledoux et al., "Measuring the effect of the rainfall on the windshield in terms of visual performance," *Accident Analysis & Prevention*, vol. 63, pp. 83–88, 2014.

[17] D. Eisenberg, "The mixed effects of precipitation on traffic crashes," *Accident Analysis & Prevention*, vol. 36, no. 4, pp. 637–647, 2004.

[18] Y. Peng, D. Lord, and Y. Zou, "Applying the Generalized Waring model for investigating sources of variance in motor vehicle crash analysis," *Accident Analysis & Prevention*, vol. 73, pp. 20–26, 2014.

[19] L. Ma, X. Yan, and J. Weng, "Modeling traffic crash rates of road segments through a lognormal hurdle framework with flexible scale parameter," *Journal of Advanced Transportation*, vol. 49, no. 8, pp. 928–940, 2015.

[20] X. Ye, K. Wang, Y. Zou, and D. Lord, "A semi-nonparametric Poisson regression model for analyzing motor vehicle crash data," *PLoS ONE*, vol. 13, no. 5, Article ID e0197338, 2018.

[21] "Highway Performance Monitoring System, Field Manual," Tech. Rep. FHWA Publication 5600.1A, U.S. Department of Transportation, Washington, DC, USA, 1990.

[22] S. Cafiso and A. Di Graziano, "Definition of Homogenous Sections in Road Pavement Measurements," *Procedia - Social and Behavioral Sciences*, vol. 53, pp. 1069–1079, 2012.

[23] F. Thomas, "Generating homogeneous road sections based on surface measurements: available methods," in *Proceedings of the 2nd European Pavement and Asset Management Conference*, Berlin, Germany, March 2004.

[24] M. J. Maher, "A bivariate negative binomial model to explain traffic accident migration," *Accident Analysis & Prevention*, vol. 22, no. 5, pp. 487–498, 1990.

[25] G. R. Wood, "Generalised linear accident models and goodness of fit testing," *Accident Analysis & Prevention*, vol. 34, no. 4, pp. 417–427, 2002.

[26] D. Lord and F. Mannering, "The statistical analysis of crash-frequency data: a review and assessment of methodological alternatives," *Transportation Research Part A: Policy and Practice*, vol. 44, no. 5, pp. 291–305, 2010.

[27] L.-Y. Chang, "Analysis of freeway accident frequencies: Negative binomial regression versus artificial neural network," *Safety Science*, vol. 43, no. 8, pp. 541–557, 2005.

[28] Y. Zhang, Z. Ye, and D. Lord, "Estimating dispersion parameter of negative binomial distribution for analysis of crash data: Bootstrapped maximum likelihood method," *Transportation Research Record*, no. 2019, pp. 15–21, 2007.

[29] Y. Li, C. Liu, and L. Ding, "Impact of pavement conditions on crash severity," *Accident Analysis & Prevention*, vol. 59, pp. 399–406, 2013.

[30] S. A. Nassar, F. F. Saccomanno, and J. H. Shortreed, "Road accident severity analysis: a micro level approach," *Canadian Journal of Civil Engineering*, vol. 21, no. 5, pp. 847–855, 1994.

[31] T. Usman, L. Fu, and L. F. Miranda-Moreno, "Injury severity analysis: Comparison of multilevel logistic regression models and effects of collision data aggregation," *Journal of Modern Transportation*, vol. 24, no. 1, pp. 73–87, 2016.

[32] J. Choi, R. Tay, and S. Kim, "Effects of changing highway design speed," *Journal of Advanced Transportation*, vol. 47, no. 2, pp. 239–246, 2013.

17

Study on Gasoline Vehicle Emission Inventory Considering Regional Differences in China

Dong Guo,[1] Zhan-gu Wang,[2] Liang Sun,[1] Kai Li,[1] Juan Wang,[1] Feng Sun,[1] and Hai Zhang[1]

[1] *School of Transportation and Vehicle Engineering, Shandong University of Technology, Zibo, Shandong 255000, China*
[2] *Jilin University State Key Laboratory of Automotive Simulation and Control, Changchun, Jilin 130000, China*

Correspondence should be addressed to Feng Sun; sunfeng0904@sina.com

Academic Editor: Monica Menendez

Rapid growth of China's urban road vehicles, in particular, the increase in the number of gasoline vehicles, leads to an increase in the traffic congestion and problems pertaining to air pollution. The establishment of the emission inventory of gasoline vehicles is influenced by several factors, like environmental characteristics, vehicle conditions, road conditions, and so on. In order to obtain gasoline vehicle emission inventory in accordance with the actual situation in different regions, this study proposed a method of establishing a list of gasoline vehicles with regional differences. Comprehensive consideration and evaluation of various factors that affect the vehicle emissions were carried out and the corresponding correction factors were obtained. According to the formula of comprehensive emission factor for Zibo city, the emission inventory of gasoline vehicle was established. This method can be effectively utilized to obtain the emission inventory of gasoline vehicles in different cities more accurately and provide theoretical support for control strategies of gasoline vehicle emissions.

1. Introduction

With the development of urban economy in China, pollution due to vehicle's emission has become a significant environmental problem. By the end of 2015, the gasoline vehicle population in China reached 172 million, for which the quantity of emission accounted for 65.5% of the total emission of vehicles [1]. Therefore, a strengthened control on gasoline vehicle's exhaust emission is a significant measure to improve the urban air quality. To determine more targeted energy saving and emission reduction strategies, the approach to determine a reasonable and accurate emission inventory of gasoline vehicles becomes one of the most important and challenging tasks.

A series of related studies about vehicle emission inventory is available at home and abroad. Sun et al. [2] took into account the dynamic changes of emission standards and fuel quality, and considering the differences of driving conditions for each vehicle type in urban road, suburban road, and highway, they established a high-resolution vehicle emission inventory based on COPERT model and GIS technology. Li et al. [3] selected Chang-Zhu-Tan urban agglomeration as the research region and established the vehicle emission inventory in this region. Further, they analyzed the space-time distribution characteristics and contribution rate of regional vehicle emissions. Yao et al. [4] selected 12 typical cities in China to establish the vehicle emission inventory from 1990 to 2009 and analyzed the historical evolution trend of vehicle emission in various cities. They finally identified the gasoline vehicle as the main source of the city's carbon monoxide (CO) and volatile organic compounds (VOCs) emissions and provided the results of the contribution rate of different gasoline models to specific pollutants. A high-resolution inventory of greenhouse gas (GHG) emission distribution of the road transport sector in Argentina was established based on GIS map by Puliafito et al. [5] after obtaining regional spatial distribution by DMSP satellite.

Commonly, the studies of vehicle emission inventory do not take into account the comprehensive consideration of the influence of environmental factors, vehicle situations, and road conditions in different regions. Based on relevant studies of low-carbon urban planning in Beijing, Wang et al. [6] put forward methods for low-carbon urban planning, including preparation of current and proposed carbon emission inventories, analysis of proposed carbon emission scenarios, city low-carbon development madmap, selection of strategies, and formulation of policies for low-carbon development. Guo et al. [7] evaluated the International Vehicle Emissions (IVE) model by utilizing a dataset available from the remote sensing measurements on a large number of vehicles at five different sites in Hangzhou, China, in 2004 and 2005. They contrasted the remote sensing data of three pollutants and proposed the adjustment of the basic emission factors from the local study to improve the model. Li et al. [8] investigated the emission profile of exhaust $PM_{2.5}$ of 12 light gasoline vehicles by using vehicle test bench and particle dilution sampling system and analyzed the $PM_{2.5}$ emission characteristics of light gasoline vehicle exhaust. Ni et al. [9] tested gasoline vehicle emissions and fuel consumption at different altitudes. They further studied the mechanism of gasoline vehicle emissions and fuel consumption at different altitudes through emission characteristics of CO, total hydrocarbon (THC), and nitrogen oxides (NOx) and also through changes in fuel consumption. Zhou et al. [10] explored the real-world emission status of light gasoline vehicles by using roadside remote sensing measurement and calculated the emission factors based on fuel and distance by mass balance method. Guo et al. [11–13] analyzed the relationship between light duty electronic fuel injection (EFI) gasoline vehicles emissions and useful life using the vehicle emissions test data from motor vehicle exhaust monitoring center.

In this study, the gasoline vehicle was selected as the research object and the environmental parameters, the average vehicle speed, the load coefficient, and the deterioration coefficient were adopted for correction. Furthermore, the comprehensive emission factor suitable for the regional characteristics was obtained. Finally, the gasoline vehicle's emission inventory was computed and analyzed based on the established model.

2. Calculation Framework

2.1. Basic Emission Factors. Since 1999, China has begun to adopt the European emission standard (National Phase-I emission standard corresponding to the emission standard of Euro I), and by 2015 it has implemented the five national standards. In light of the investigation results released by the National Bureau of statistics (China Statistical Yearbook 2015), China's vehicles that have been implementing different emission standards from National Standard I to National Standard V accounted for 2, 4, 12, 54, and 28%, respectively. Based on the national average environmental condition, vehicle condition, and road condition, and related industry standards, the basic emission factors of gasoline vehicle were obtained. Table 1 lists the basic emission factor data of various types of gasoline vehicles under different emission standards.

2.2. Model. First, the regional vehicle population and the annual average distance of various types of gasoline vehicles were obtained pursuant to the ***Yearbook of National Bureau of Statistics and Local Bureau of Statistics***. Second, the basic emission factors of gasoline vehicle were established based on the displacement of various gasoline vehicles and related industry standards. Then the emission correction factors model of gasoline vehicle affected by regional differences were established based on the basic emission factors with a comprehensive consideration of such factual factors affecting gasoline vehicle emission as environmental conditions, vehicle type, engine running status, fuel properties, and vehicle load. Finally, emission inventory of regional gasoline vehicles was acquired by combination of emission factors, local motor vehicle population, and driving mileage of various types of vehicles. The model calculation framework is shown in Figure 1.

Vulcanization correction factors were used to reduce or eliminate the effect of sulfur content in gasoline on other polluting gases.

3. Analysis of Influencing Factors

Environmental condition and vehicle situation are the two standing out factors influencing the gasoline vehicle exhaust. On the one hand, significant regional differences are observed in temperature, humidity, and altitude, which directly affect the engine operating conditions [14]. On the other hand, running velocity, deterioration factor, fuel quality, and vehicle load coefficient also have an immediate implication on its exhaust. For these reasons, the emission correction factors of gasoline vehicle considering regional differences are conducive not only to a more accurate and real local vehicles emission condition, but also to a more realistic emission inventory.

3.1. Environmental Parameter Correction Factors. China is a country with a vast territory of around 9.6 million square kilometers. The far-flung land area and diverse geomorphological characteristics shape a huge environmental difference in different regionals of China, making the hot and humid southern area in contrast with the cold and dry northern area. Based on the information released in ***China Climate Bulletin in 2015***, the representative higher and lower values of altitude, temperature, and humidity are listed in Table 2.

Differences in environmental conditions exert direct impacts on engine operating conditions. In the high altitude and frigid region with thin air, engine excess air coefficient φ_a becomes larger beset by less oxygen, lower atmospheric pressure, and larger intake resistance. Moreover, CO and HC exhaust, on the premise of oxygen deficit and incomplete combustion, increases significantly, which is also aggravated due to low environmental temperature, poor gasoline atomization effect, and insufficient burning of combustible gas mixture in hot and humid environment. In contrast, the engine becomes inefficient, troubled by high environmental temperature, slow engine cooling and heat dissipation, and high operating temperature of combustion chamber,

Table 1: Emission factors for different types of gasoline vehicles.

Vehicle type	Emissions factor (g km^{-1})				
	CO	HC	NOx	$PM_{2.5}$	PM_{10}
Minivan					
GI	6.71	0.663	0.409	0.026	0.029
GII	2.52	0.314	0.324	0.011	0.012
GIII	1.18	0.191	0.100	0.007	0.008
GIV	0.68	0.075	0.032	0.003	0.003
GV	0.46	0.056	0.017	0.003	0.003
Middle-sized Coach					
GI	21.43	2.567	1.781	0.060	0.067
GII	15.37	1.443	1.461	0.018	0.020
GIII	4.33	0.373	0.474	0.011	0.012
GIV	1.98	0.107	0.196	0.006	0.007
GV	1.98	0.107	0.147	0.006	0.007
Light Duty Truck					
GI	26.16	3.324	2.006	0.060	0.067
GII	21.54	2.210	1.656	0.018	0.020
GIII	5.61	0.61	0.534	0.011	0.012
GIV	2.37	0.169	0.229	0.006	0.007
GV	2.37	0.169	0.172	0.006	0.007
Other Gasoline Vehicles					
GI	16.12	1.368	1.89	0.066	0.064
GII	8.25	0.869	1.52	0.044	0.049
GIII	5.456	0.613	1.089	0.026	0.016
GIV	3.77	0.418	0.775	0.015	0.009
GV	3.77	0.418	0.582	0.005	0.006

GI, GII, GIII, GIV, and GV, respectively, represent the national emission standards of Phase-I, Phase-II, Phase-III, Phase-IV, and Phase-V.

Table 2: Extreme statistics of environmental parameters in China.

	Higher value	Representative	Lower value	Representative	D-value
Altitude (M)	4025	Shigatse	33	Tianjin	3992
Temperature (°C)	46	Chongqing	−52.3	Mohe	98.3
Humidity (%)	92	Guiyang	32	Yinchuan	60

Data from China Climate Bulletin in 2015.

ultimately leading to aggravation in pollutant emission with increasing fuel consumption. In high relative humidity environment, water molecules present in the air enter into combustion chamber through the engine intake system. This results in incomplete combustion of combustible gas mixture and carbon deposition. This easily leads to surface ignition under high temperature, inducing preignition deflagration and some other tough operating conditions of engine, and thereby dramatically increasing the emission amount of NO_x and $PM_{2.5/10}$.

According to ***Technical Guidelines for Preparation of Air Pollutants Emission Inventory of Road Vehicles*** [15], based on influencing factors such as temperature, humidity, sulfur content, and altitude, the correction factors with similar environmental conditions were integrated based on actual situations, and the environmental parameter correction factors in high state and low state are listed in Table 3.

Definitions of the low state and high state of various environmental factors are summarized in Table 4.

3.2. Average Velocity Correction Factors. Vehicle operating conditions in different provinces and cities in China vary significantly owing to the different landforms, population densities, and road conditions of various regions. According to the data released in the ***Investigation Report for National Urban Automobiles Driving Conditions in 2015*** by National Bureau of Statistics, vehicle operating conditions of six representative cities are listed in Table 5.

Average velocity exerts direct influences on engine operating condition and emissions. When the vehicle runs in idle low load states and low velocity, less amount of mixed gas would get into the engine through its narrower throttle opening, leading to the serious dilution of the mixed gas by residual gas. Besides, slow engine velocity leads to slow inlet

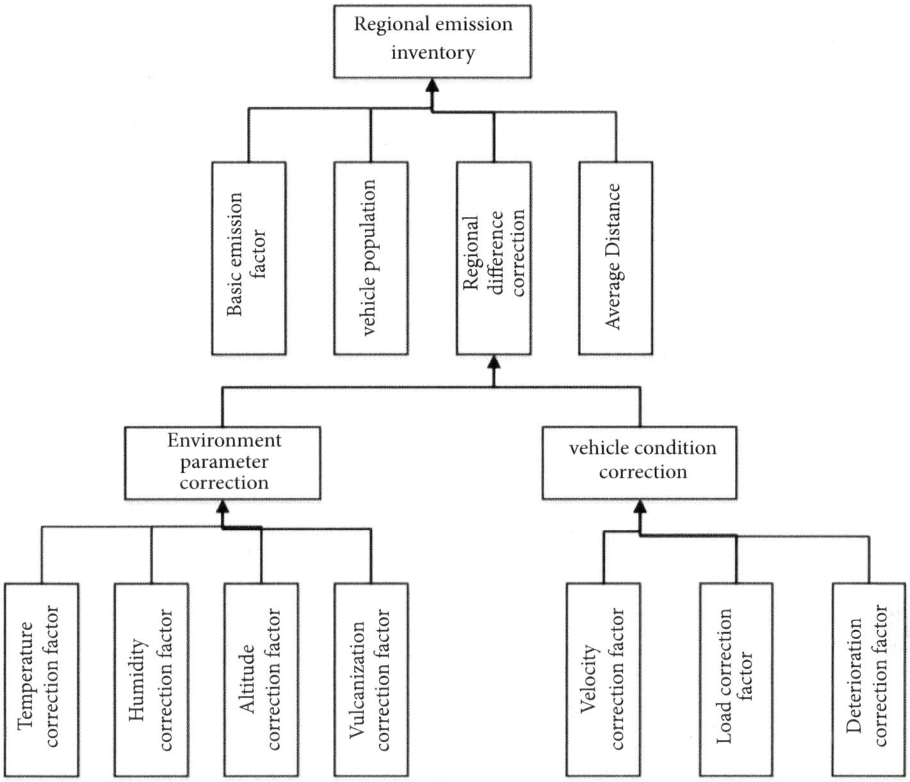

FIGURE 1: Model calculation flowchart.

TABLE 3: Gasoline vehicle environmental parameter correction factor.

Classification of Pollutants	CO		HC		NOx		$PM_{2.5}$ /PM_{10}	
	Low state	High state	Low state	High state	Low state	High state	Low state	High state
Temperature Correction Factor	1.36	1.23	1.47	1.08	1.15	1.31	1.45	0.67
Humidity Correction Factor	0.97	1.04	0.99	1.01	1.13	0.87	1.00	1.00
sulfur Content Correction Factor	0.90	1.80	0.96	1.41	0.95	2.08	0.56	1.21
Altitude Correction Factor								
Light/Small-sized	1.00	1.58	1.00	2.46	1.00	3.15	1.00	1.00
Middle-/Large-sized	1.00	3.95	1.00	2.26	1.00	0.88	1.00	1.00

Data from Technical Guidelines for Preparation of Air Pollutants Emission Inventory of Road Vehicles.

TABLE 4: The boundary between the high and low states.

	Temperature	Humidity	Sulfur content	Altitude
Low state	<10°C	<50%	<10ppm	<1500m
High state	>+25°C	>50%	>500ppm	>1500m

velocity, poor gasoline atomization and evaporation effect, and uneven mixed gas. Thus, in this case, the rich mixture is required. However, an overrich mixture in turn arouses incomplete combustion, thus exacerbating exhaust emission.

According to the guide, the gasoline vehicle velocity involved in the correction method has been divided into the following five intervals: 0–20, 20–30, 30–40, 40–80, and 80 km h^{-1}. Moreover, the average vehicle velocity correction factors of the gasoline vehicle were worked out and listed in Table 6.

3.3. Load Correction Factors. Owing to a sound growth momentum, the transportation industry in China confronts a noticeable problem of spatial development imbalance. For

TABLE 5: Typical urban vehicle operating conditions in China.

Unit	maximum velocity (km h^{-1})	Average velocity (km h^{-1})	Idling Time Proportion (%)
Beijing	70.5	27.7	20.5
Shanghai	52.3	21.6	30.2
Guangzhou	55.4	25.6	17.3
Dalian	76.3	40.5	8.6
Wuhan	80.5	36.4	15.3
Lasa	105.4	60.7	5

instance, China's eastern coastal area, which boasts prosperous commodity economy, perfect transportation infrastructure, and advanced road network structure, possesses high transport efficiency with an average load coefficient around 65%. China's midwestern areas, in contrast, being limited by such factors of policy, economy, and landform, have relatively low transport efficiency with an average load coefficient of around 45%.

Engine operating condition is directly influenced by vehicle load. With the increase in the loading weight, the vehicle suffers more running resistance and ramp resistance in the course of moving. In order to maintain the average velocity and reserve power, the richer mixed gas should be provided to the engine, which will generate a smaller excess air coefficient of chamber. Under the circumstances, the mixed gas, in which gasoline accounts for a large proportion, burns rapidly with less heat loss, contributing to a higher effective power of engine. However, still its combustion remains incomplete due to the insufficient inlet, thus producing large amounts of CO and HC and leading to carbon deposition. In this case, the probability of ignition and detonation of engine surface gets significantly increased, as also the exhaust amount of $PM_{2.5}/_{10}$.

With reference to the index of load correction factors of gasoline vehicle in the guide, the load correction factors were obtained and listed in Table 7.

3.4. Degradation Correction Factors. Exposure of the running vehicle to the external environment such as sunshine, air, wind, sand, rain, and snow is inevitable. Moreover, the interaction of internal parts of vehicle also leads to the heating, wear, and corrosion. Vehicle degradation is usually divided into tangible wear, invisible wear, and general wear. Similar to other mechanical equipment, the vehicle also degrades under the technical conditions and its performance also reduces after a period of use. This is well represented by the reduction of mechanical transmission efficiency and the increase of fuel consumption, as well as poor exhaust ternary catalytic effect. These degradation results lead to direct increase in exhaust pollutant emission.

The selection of service life as a deterioration correction parameter can facilitate the calculation and analysis of emission inventories. Therefore, referring to the data information in the guide, the correction parameters of gasoline vehicle deterioration were obtained as summarized in Table 8.

4. The Vehicle Emission Inventory

4.1. The Combined Emission Factors. The combined emission factors are the most important parameters for establishing the gasoline vehicle emission inventory. The combined correction emission factors of gasoline vehicle were calculated in accordance with the actual situation by correcting the basic emission factors in different regions, like environmental characteristics, vehicle conditions, and so on. These factors were calculated by using formula (1) as follows:

$$EF_{m,i,j} = \sum_{k=3}^{5} \left(BEF_{m,i,j} \times \varphi_j \times \psi_j \times \lambda_j \times \alpha_i \times \zeta_i \times \eta_i \times \chi_j \times \frac{n_{i,j,k}}{n_{i,j}} \right) \quad (1)$$

The meaning of each symbol in the formula is summarized in Table 9.

$EF_{m,i,j}$ indicates the m type emission factor of i vehicle in the j region; $BEF_{m,i,j}$ denotes the m type basic emission factor of i vehicle in the j region; $n_{i,j}$ represents the i type vehicle population in the j region; and $n_{i,j,k}$ indicates the i type vehicle pollution in the j region that conforms to the k emission standard.

The calculation method of gasoline vehicle emission inventory is represented as follows:

$$Q_{m,i,j} = EF_{m,i,j} \times VKT_{i,j} \times N_{i,j} \times 10^{-6} \quad (2)$$

where $Q_{m,i,j}$ indicates the m type emission of i vehicle in the j region; $EF_{m,i,j}$ represents the m type emission factor of i vehicle in the j region; $VKT_{i,j}$ denotes the annual average driving distance of i vehicle in the j region; and $N_{i,j}$ indicates the i type vehicle emission in the j region.

4.2. Application Analysis

4.2.1. Regional Parameters of Zibo. According to ***2015 Statistical Yearbook of Zibo Statistical Bureau*** [16], Zibo's vehicle population was 719,000 in 2015. The huge emission from vehicles resulted in great pressure to the atmospheric environment in Zibo. Pursuant to statistical data of atmospheric environmental quality released by Zibo Environmental Protection Bureau in 2015, there were only 41 days with good air quality. Therefore, establishment of the gasoline vehicle emission inventory, which is in accordance with the actual situation of Zibo city, was expected to lay the foundation for formulation of a targeted emission control strategy.

The basic environmental parameters and average vehicle running state of Zibo city were obtained by referring to the relevant data released by the Zibo Environmental Protection Bureau and the Zibo Traffic Administration Bureau. The annual average temperature in Zibo is 13.2°C, the annual relative humidity is 57%, and the average altitude is 34.7 m. The average running velocity of gasoline vehicle is 43 km h^{-1}, the average service life of gasoline vehicle is 3–5 years, the average load factor is 50%, and the sulfur content of gasoline

TABLE 6: Gasoline vehicle average speed correction factor.

Vehicle emission	Velocity Range (km h^{-1})				
	<20	20–30	30–40	40–80	>80
CO	1.69	1.26	0.79	0.39	0.62
HC	1.68	1.25	0.78	0.32	0.59
NO_x	1.38	1.13	0.90	0.86	0.96
$PM_{2.5}/PM_{10}$	1.68	1.25	0.78	0.32	0.59

Data from Chinese Research Academy of Environmental Sciences—Technical Guidelines for Preparation of Air Pollutants Emission Inventory of Road Vehicles [15].

TABLE 7: Gasoline vehicle load correction factor.

Load Coefficient	CO	HC	NO_x	$PM_{2.5}/PM_{10}$
0	0.87	1.00	0.83	0.90
50%	1.00	1.00	1.00	1.00
60%	1.07	1.00	1.09	1.05
75%	1.16	1.00	1.21	1.13
100%	1.33	1.00	1.43	1.26

Data from Chinese Research Academy of Environmental Sciences—Technical Guidelines for Preparation of Air Pollutants Emission Inventory of Road Vehicles [15].

TABLE 8: Gasoline vehicle deterioration correction factor.

Pollutant	1–4 years	4–7 years	7–10 years	Above 10 years
CO	1.02	1.14	1.27	1.48
HC	1.01	1.09	1.19	1.34
NOx	1.00	1.12	1.34	1.47
$PM_{2.5}/PM_{10}$	1.01	1.15	1.26	1.45

Data from Chinese Research Academy of Environmental Sciences—Technical Guidelines for Preparation of Air Pollutants Emission Inventory of Road Vehicles [15]

is less than 50 ppm. The regional emission correction factors of Zibo, as listed in Table 10, were obtained according to Tables 3–8.

4.2.2. Regional Vehicle Characteristics. The annual average driving distance and local vehicle population are important indicators for measuring the degree of transportation development in a region, as well as the important basis for establishing regional emission inventory. Through consulting ***2015 Statistical Yearbook of Zibo Transportation Administration Bureau***, the annual average driving distance and population of different types of gasoline vehicles in Zibo were obtained. The specific data is presented in Table 11, where the numbers 1 to 4 represent minivan, middle-sized coach, light duty truck, and other gasoline vehicles, respectively.

4.2.3. The Regional Emission Inventory. The regional combined emission factors can be obtained by combining basic emission factors with correction factors in Zibo. Then the gasoline vehicle emission inventory can be obtained by combining regional vehicle characteristics. The results are listed in Table 12.

Figure 2 reveals the following conclusions:

(1) For minivan and light duty truck, the emission amount of CO and HC, respectively, accounts for 94 and 96% of the total emission amount, more than 90% of the total emissions of these two pollutants.

(2) The number of small-sized trucks is only 10% of small passenger vehicles; however, its NOx emissions account for about 40% of total NOx emissions.

(3) Large number of small passenger vehicles account for 70% of the total $PM_{2.5}/PM_{10}$ emissions and are the key models for $PM_{2.5}/PM_{10}$ emission control.

In order to compare emission estimation using average emission factors versus emission correction factors in Zibo, we calculated a new emission inventory by using average emission factors. The results are listed in Table 13.

Table 13 summarizes that the estimated total emission from gasoline vehicles in Zibo city is 140.3% higher than the actual value calculated from the average emission factor in the standard. This huge error is attributed to the different traffic conditions in different regions, which results in different revision coefficients of emission factors. For Zibo city, the average velocity is the main factor leading to the error. Thus, the total emission from gasoline vehicles in Zibo cannot be estimated by means of the average emission factors, and the corresponding correction factors should be obtained by combining the actual local traffic conditions.

4.3. Uncertainty Analysis. The uncertain factors that affect the emission inventory of traditional gasoline vehicles include the establishment of local statistical data and emission factors. In this paper, Monte Carlo uncertainty analysis method is applied to quantify the potential uncertainty of gasoline vehicle emission inventory. Simultaneity, the uncertainty range of the emission inventory (95% confidence interval) is obtained by repeated sampling method, as shown in Table 14.

By the careful consideration of various factors that may affect the establishment of emission factors, the overall uncertainty of pollutant emissions does not exceed 35%. Among them, $PM_{2.5}/PM_{10}$ has the highest uncertainty, and the relative error of total emissions is ±34.13%; furthermore, there is a significant difference in different vehicle types. This may be due to the fact that there are some differences between the national emission factors and the actual situation in Zibo. When using self-expanding simulation, the uncertainty of

TABLE 9: The meaning of each symbol in the formula.

Symbol	Meaning	Symbol	Meaning
BEF	Basic emission factor	λ	Altitude correction
i	Vehicle type	α	Velocity correction
j	Regional number	ζ	Deterioration correction
m	Pollutants type	η	Sulfide correction
k	Emission standard	χ	Load correction
φ	Temperature correction	n	Vehicle population
ψ	Humidity correction		

TABLE 10: Regional emission correction factor of Zibo City.

Correction type	φ	ψ	η	λ	α	χ	ζ
CO	1.00	1.04	0.90	1.00	0.39	1,00	1.14
HC	1.00	1.01	0.96	1.00	0.32	1.00	1.09
NO_X	1.00	0.87	0.95	1.00	0.86	1.00	1.12
$PM_{2.5}/PM_{10}$	1.00	1.00	0.56	1.00	0.32	1.00	1.15

Data from Tables 3–8.

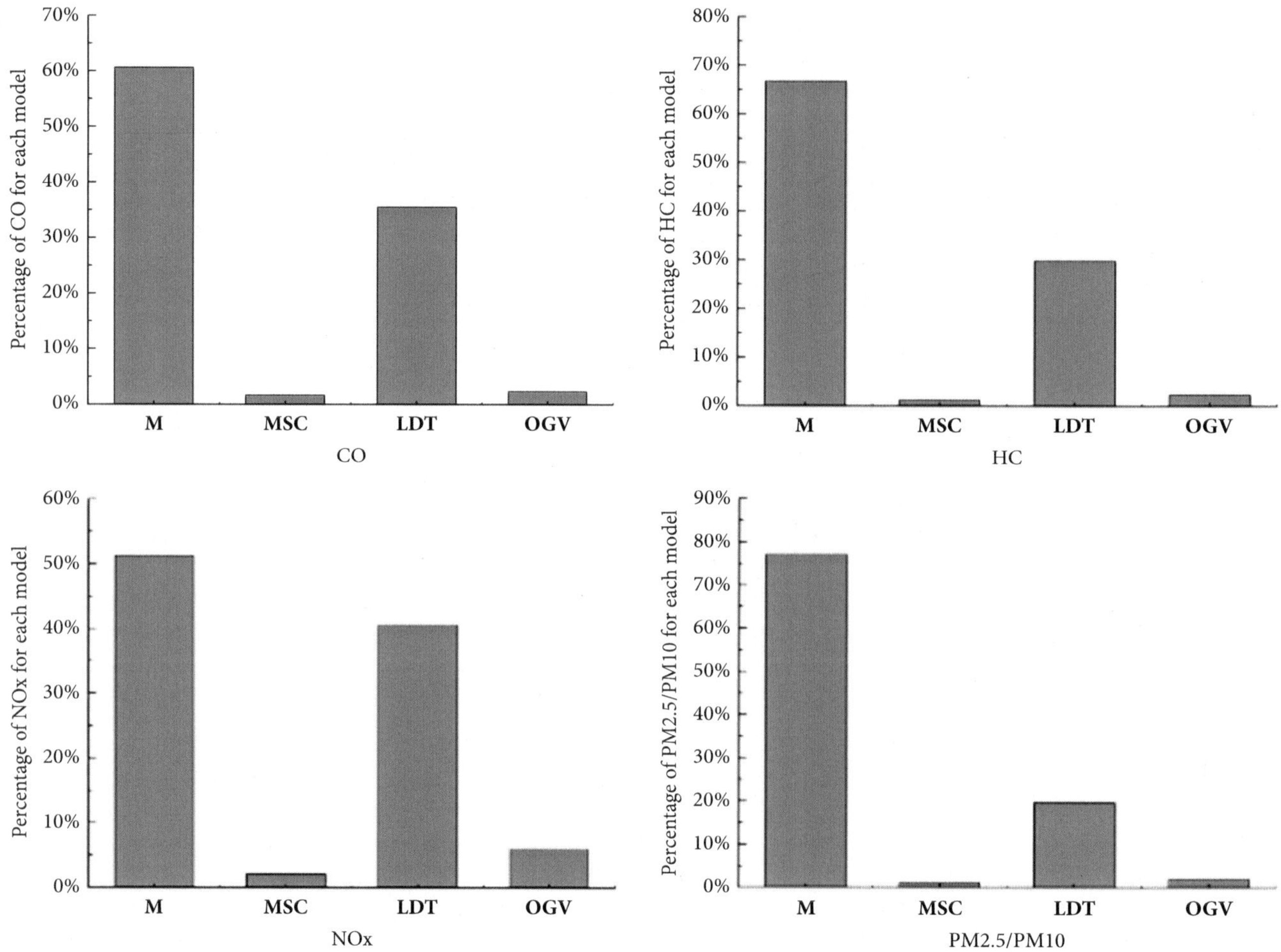

FIGURE 2: Emission sharing rate of each pollutant. Notes: M: minivan; MSC: middle-sized Coach; LDT: light duty truck; MDT: medium duty truck; and OGV: other gasoline vehicles.

TABLE 11: Annual average driving distance and population.

	1	2	3	4
Vehicle population/veh	622782	2716	46930	11446
Driving distance/km	18000	31320	31000	8000

Data from 2015 Statistical Yearbook of Zibo Transportation Administration Bureau [16].

emission factors is increased, and thus the uncertainty range of the inventory is enlarged. Compared with $PM_{2.5}/PM_{10}$, the uncertainty of other pollutants is lower, which may be due to the relatively close emission factors obtained through different ways, which leads to underestimation of its uncertainty of quantitative analysis.

According to the result of uncertainty analysis, the minivan is the most influential for the uncertainty of HC and $PM_{2.5}/PM_{10}$ emissions, and the uncertainty of CO and NOx emissions is mainly determined by light duty trucks.

4.4. Comparison Analysis. The emissions inventory established in this study was compared with the inventory developed by other scholars. We used a similar approach to vehicle types selection and calculation of emissions inventories. The results are shown in Table 15. These data include the results of gasoline vehicle emissions of Chengdu [17] (Chen et al. 2015), Zhengzhou [18] (Gong et al. 2017), Foshan [19] (Zeng et al. 2013), and Tianjin [20] (Zhang et al. 2017).

According to the comparison of the data in Table 15, there are some differences in the pollutant emissions between Zibo and other cities in China. We analyze the reasons for these differences mainly in the following three aspects. Firstly, the based year: China's emission standards have been formulated since 2000 and then updated every 3-4 years; each update brings about a reduction of about 30% of the emission factor. Secondly, the vehicle population: due to differences in urban scale and economic development level among cities, there is a big difference in vehicle population, and there is a positive correlation between vehicle population and emissions. The third is the differences between the emission factors and the correction factors. For example, the basic emission factors used by Zeng et al. are derived from the experimental data of foreign vehicles, and there is a big difference between the emission factors obtained by the Chinese road tests used in this study. Moreover, the correction factors used in the relevant research have some difference, such as Chen et al.'s study, which increased the impact of factors such as slope on vehicle emissions.

4.5. Impact in Air Quality. In this study, there is no total statistics of all atmospheric environmental pollutants; thus, it is impossible to analyze the degree of environmental impact of vehicle emissions. Referring to the conclusions of other similar cities, gasoline vehicles account for total CO, HC, NOx, and PM2.5/PM10 by 20%, 30%, 25%, and 12%, respectively. According to the comparison of the research results, the pollutant emission factors obtained in this study are approximately 41% of the average emission factors.

We are currently conducting larger vehicle road trials to update and obtain more detailed emission factors and develop relevant emissions inventory software to accommodate international research needs, including China.

5. Summary and Outlook

Considering that gasoline vehicle emission conditions differ significantly in different areas, the study determining basic emission factors with reference to the relevant guidelines, correcting factors based on regional differences, combined with driving distance and vehicle population, and other information of local gasoline vehicles finally figures out the method of establishing gasoline vehicle emission inventory, including the following.

(1) The gasoline vehicles are classified into four types based on their usage, and basic emission factors are determined according to the relevant guidelines. Notably, the emission estimation from the revised emission factor and the average emission factor are significantly different. This difference is attributed to the fact that the average velocity of gasoline vehicle in Zibo city is 43 km h^{-1} and the corresponding velocity correction coefficient is 0.39; thus the velocity correction factor is far less than the standard. Therefore, it is important to identify local emission factors in conjunction with local factors when emissions from an area are estimated. Moreover, velocity is an important factor affecting emissions.

(2) After the analysis of the differences in environmental parameters in different regions, average velocity, load coefficients, and deterioration factors, the corrections factors were confirmed. Moreover, their influences on the gasoline vehicle emission were systematically investigated. Comparative analysis of the emission estimations from other provinces or regions in China indicates that the difference in emission factors in different regions is due to different major factors. For example, the major emission factor of Chongqing Province is slope and Zibo city terrain is dominated by plain; therefore, the determination of Zibo city emission factor does not need to consider the slope factor.

(3) The computing method of gasoline vehicle emission inventory for different cities was established, considering Zibo city as an applied research example. Emission share rates of different vehicle types were analyzed, and finally the focus of governance was determined.

This study can extend theoretical support to quantitative evaluation on gasoline vehicle emission conditions in different areas and lay foundation for providing well-targeted regulation measures. In developing countries, medium-sized cities such as Zibo are very common, so this study has practical significance for the international urban emission estimation similar to Zibo urban cluster.

Table 12: Pollutant emission inventory of gasoline vehicle in Zibo City.

Vehicle type	Total emission volume (t/a)			
	CO	HC	NO_X	$PM_{2.5}/PM_{10}$
Minivan	4138.54	497.61	261.71	41.45
Middle-sized Coach	112.80	8.55	10.55	0.62
Light Duty Truck	2422.53	221.98	207.21	10.58
Other Gasoline Vehicles	160.67	17.42	30.89	1.08
Total	6834.55	745.57	510.36	53.73

Table 13: Pollutant emission inventory of gasoline vehicle in Zibo City referring to average emission factors.

Vehicle type	Total emission volume (t/a)			
	CO	HC	NO_X	$PM_{2.5}/PM_{10}$
Minivan	9944.94	1195.76	628.88	99.61
Middle-sized Coach	271.07	20.55	25.34	1.49
Light Duty Truck	5821.36	533.43	497.93	25.43
Other Gasoline Vehicles	386.10	41.86	74.24	2.59
Total	16423.47	1791.60	1226.39	129.11

Table 14: Uncertainly analysis of the emission inventory of gasoline vehicles.

Pollutants	Vehicle type	Estimated value/t	Simulated average/t	95% confidence interval/t	Relative error
CO	Minivan	4138.54	7838.56	[6214.76,9462.36]	±20.72%
	Middle-sized Coach	112.80	185.30	[130.63,239.98]	±29.50%
	Light Duty Truck	2422.53	3198.93	[2128.35,4269.51]	±33.47%
	Other Gasoline Vehicles	160.67	260.80	[178.43,343.17]	±31.58%
	Total	**6834.55**	**11483.59**	**[8652.17,14315.02]**	**±24.66%**
HC	Minivan	497.61	911.55	[581.55,1241.56]	±36.20%
	Middle-sized Coach	8.55	18.26	[13.69,22.83]	±25.03%
	Light Duty Truck	221.98	423.32	[355.85,490.80]	±15.94%
	Other Gasoline Vehicles	17.42	27.45	[19.87,35.02]	±27.61%
	Total	**745.57**	**1380.58**	**[970.96,1790.21]**	**±29.67%**
NO_x	Minivan	261.71	546.78	[431.45,662.11]	±21.09%
	Middle-sized Coach	10.55	20.77	[16.10,25.44]	±22.48%
	Light Duty Truck	207.21	293.43	[190.33,396.52]	±35.14%
	Other Gasoline Vehicles	30.89	67.25	[50.39,84.12]	±25.07%
	Total	**510.36**	**928.23**	**[688.27,1168.19]**	**±25.85%**
$PM_{2.5}/PM_{10}$	Minivan	41.45	80.76	[51.81,109.71]	±35.85%
	Middle-sized Coach	0.62	0.97	[0.77,1.16]	±20.62%
	Light Duty Truck	10.58	16.19	[11.71,20.68]	±27.67%
	Other Gasoline Vehicles	1.08	1.73	[1.35,2.11]	±21.97%
	Total	**53.73**	**99.65**	**[65.64,133.66]**	**±34.13%**

Table 15: Comparison of emission inventory between Zibo and other regions.

Region	Based year	Vehicle population (veh)	Emissions ($\times 10^4$ t)				Sources
			CO	HC	NO_x	$PM_{2.5}/PM_{10}$	
Zibo	2015	68.4×10^4	0.68	0.08	0.05	0.0053	This study
Chengdu	2012	144.3×10^4	2.59	-	0.21	0.0153	Chen et al. 2015
Zhengzhou	2013	123.3×10^4	2.34	0.23	0.19	0.0121	Gong et al.2017
Foshan	2010	70.2×10^4	1.13	-	0.06	0.0069	Zeng et al. 2013
Tianjin	2013	155.6×10^4	1.82	0.20	0.14	0.0143	Zhang et al.2017

Note: "-" indicates no data.

Conflicts of Interest

The authors declare no conflicts of interest regarding the publication of this paper.

Acknowledgments

This study was supported by the National Natural Science Foundation of China (Grant nos. 51508315 and 51608313), the Technological Development Project of Shandong Province (Grant no. 2016GGB01539), and the Natural Science Foundation of Shandong Province of China (Grant no. ZR2015EL046).

References

[1] National Bureau of Statistics of the People's Republic of China, *China Statistical Yearbook 2015*, China Statistics Press, 2016, http://www.stats.gov.cn/.

[2] S.-D. Sun, W. Jiang, and W.-D. Gao, "Vehicle emission inventory and spatial distribution in Qingdao," *Zhongguo Huanjing Kexue/China Environmental Science*, vol. 37, no. 1, pp. 49–59, 2017.

[3] B. R. Li, Z. Liu, X. Y. You, and Y. Huang, "Emission and characteristics of vehicle exhausts in changsha-zhuzhou-xiangtan area of hunan province," *Environmental Science & Technology*, vol. 39, no. 11, pp. 167–173, 2016.

[4] Z.-L. Yao, M.-H. Zhang, X.-T. Wang, Y.-Z. Zhang, H. Huo, and K.-B. He, "Trends in vehicular emissions in typical cities in China," *Zhongguo Huanjing Kexue/China Environmental Science*, vol. 32, no. 9, pp. 1565–1573, 2012.

[5] S. E. Puliafito, D. Allende, S. Pinto, and P. Castesana, "High resolution inventory of GHG emissions of the road transport sector in Argentina," *Atmospheric Environment*, vol. 101, pp. 303–311, 2015.

[6] Y. J. Wang and Y. He, "Methods for low-carbon city planning based on carbon emission inventory," *China Population and Resources and Environment*, vol. 25, no. 6, pp. 72–79, 2015.

[7] H. Guo, Q.-Y. Zhang, Y. Shi, and D.-H. Wang, "Evaluation of the International Vehicle Emission (IVE) model with on-road remote sensing measurements," *Journal of Environmental Sciences*, vol. 19, no. 7, pp. 818–826, 2007.

[8] Y. F. Li, Z. H. Li, J. N. Hu et al., "Emission profile of exhaust $PM_{2.5}$ from light-duty gasoline vehicles," *Research of Environment Sciences*, vol. 29, no. 4, pp. 503–508, 2016.

[9] H. Ni, W. Zhao, L. Liu, M. L. Li, T. Q. Fu, and Y. W. Wang, "A research of effects of altitude on the emissions and fuel consumption of a gasoline vehicle," *Automotive Engineering*, vol. 36, no. 10, pp. 1205–1209, 2014.

[10] Y. Zhou, Y. Wu, S. Zhang, L. Fu, and J. Hao, "Evaluating the emission status of light-duty gasoline vehicles and motorcycles in Macao with real-world remote sensing measurement," *Journal of Environmental Sciences*, vol. 26, no. 11, pp. 2240–2248, 2014.

[11] D. Guo, S. Gao, X. Y. Wang, and Q. Shang, "Analysis of emissions deterioration rule of light-duty EFI gasoline vehicle," *Science Technology and Engineering*, vol. 13, no. 15, pp. 4454–4458, 2013.

[12] D. Guo, F. Sun, and J. B. Zhao, *Method for Estimation of Urban Area Vehicle Emission and Reduction Strategy Analysis*, China Communications Press, Beijing, 2017.

[13] D. Guo, H. Zhang, C. Zheng, S. Gao, and D. Wang, "Analysis of the future development of Chinese auto energy saving and environmental benefits," *Xitong Gongcheng Lilun yu Shijian/System Engineering Theory and Practice*, vol. 36, no. 6, pp. 1593–1599, 2016.

[14] H. Wang, C. Chen, C. Huang, and L. Fu, "On-road vehicle emission inventory and its uncertainty analysis for Shanghai, China," *Science of the Total Environment*, vol. 999, no. 1–3, pp. 60–67, 2009.

[15] Chinese Research Academy of Environmental Sciences, "Technical Guidelines for Preparation of Air Pollutants Emission Inventory of Road Vehicles," 52P, 2015.

[16] Zibo Statistical Bureau, *Statistical Yearbook of Zibo City*, China Statistics Press, 393P, 2015, http://tj.zibo.gov.cn/zbtj/pic/2015tjnj.pdf.

[17] J. Chen, W. Fan, J. Qian, Y. Li, and W. Zhao, "Establishment of the light-duty gasoline vehicle emission inventory in Chengdu by the International Vehicle Emission model," *Huanjing Kexue Xuebao/Acta Scientiae Circumstantiae*, vol. 35, no. 7, pp. 2016–2024, 2015.

[18] M. Gong, S. Yin, X. Gu, Y. Xu, N. Jiang, and R. Zhang, "Refined 2013-based vehicle emission inventory and its spatial and temporal characteristics in Zhengzhou, China," *Science of the Total Environment*, vol. 599-600, pp. 1149–1159, 2017.

[19] X. L. Zeng, H. X. Li, and X. M. Cheng, "Study on traffic exhaust emissions and characteristics of Foshan City," *Environmental Pollution & Control*, vol. 35, no. 11, pp. 51–55, 2013.

[20] Y. Zhang, L. Wu, H. J. Mao, J. Teng, and H. Chen, "Research on vehicle emission inventory and its management strategies in Tianjin," *Acta Scientiarum Naturalium Universitatis Nankaiensis*, vol. 50, no. 1, pp. 90–96, 2017.

Identifying Public Transit Commuters based on both the Smartcard Data and Survey Data: A Case Study in Xiamen, China

Shichao Sun[1] **and Dongyuan Yang**[2]

[1]*College of Transportation Engineering, Dalian Maritime University, Dalian 116026, China*
[2]*Key Laboratory of Road and Traffic Engineering of the Ministry of Education, Tongji University, Shanghai 201804, China*

Correspondence should be addressed to Shichao Sun; sunshichao1988@hotmail.com

Guest Editor: Hamzeh Khazaei

Understanding the travel patterns of public transit commuters was important to the efforts towards improving the service quality, promoting public transit use, and better planning the public transit system. Smartcard data, with its wide coverage and relative abundance, could provide new opportunities to study public transit riders' behaviors and travel patterns with much less cost than conventional data source. However, the major limitation of smartcard data is the absence of social attributes of the cardholders, so that it cannot clearly extract public transit commuters and explain the mechanism of their travel behaviors. This study employed a machine learning approach called Naive Bayesian Classifier (NBC) to identify public transit commuters based on both the smartcard data and survey data, demonstrated in Xiamen, China. Compared with existing methods which were plagued by the validation of the accuracy of the identification results, the adopted approach was a machine learning algorithm with functions of accuracy checking. The classifier was trained and tested by survey data obtained from 532 valid questionnaires. The accuracy rate for identification of public transit commuters was 92% in the test instances. Then, under a low calculation load, it identified the objectives in smartcard data without requiring travel regularity assumptions of public transit commuters. Nearly 290,000 cardholders were classified as public transit commuters. Statistics such as average first boarding time and travel frequency of workdays during peak hours were obtained. Finally, the smartcard data were fused with bus location data to reveal the spatial distributions of the home and work locations of these public transit commuters, which could be utilized to improve public transit planning and operations.

1. Introduction

Public transit systems have long been regarded as an effective way to mitigate the growing urban congestion, exhaust emissions, and energy consumption caused by the excessive use of private automobiles [1, 2]. The effect is very significant especially in China where the level of car ownership has developed rapidly in the past decade [3, 4]. In order to unceasingly improve the performance and promote public transit use, the authorities in metropolises of China have been working for years to obtain a better understanding of passengers' travel characteristics [5–7]. In this context, mining the travel patterns of public transit commuters has received much attention of researchers [1, 4, 8–11]. It results from the fact that the commuters not only represent the frequent public transit users, but also represent the ones who comprise the major component of travel demand during peak hours. The accurate knowledge of their demands and travel patterns may help the agencies adjust route plans, provide appropriate policies to retain the loyal riders and enhance the attractiveness of public transit systems [10]. However, acquiring the precise travel information of public transit commuters used to be challenging. The traditional way to analyze the public transit demand largely depended on travel diary surveys, which was often plagued by its high cost, low data accuracy, low response rates, and also privacy concern. Fortunately, Automatic Fare Collection (AFC) systems have been widely implemented in China as a more efficient way of managing fare over the manual collection method [12]. At the same time, smartcard data can provide more abundant and higher quality travel data with less cost, through which travel patterns could be analyzed based on precise observations of individuals' smartcard usage in a time period [10, 13–18].

Nevertheless, the archived smartcard data collected from AFC systems also have its limitations [19]. For instance, smartcards are usually anonymous in China. Thus, social-demographic attributes of the cardholders were absent, such as gender, age, education, and trip purpose. Therefore, identifying public transit commuters from smartcard data was not a straightforward task [5]. To this end, spatial and temporal regularities of recurring trips on the public transit networks were usually incorporated to determine the public transit commuters among the cardholders [4, 6, 9], while, even if the spatial and temporal characteristics of each cardholder could be mined from the smartcard data, how the identified spatiotemporal travel pattern tied to a real commuter was still not clear [5]. To demonstrate it, a case study in Xiamen, China was given, where the smartcard dataset harvested from 1st to 26th June, 2015 was employed. Specifically, for each cardholder, we calculated the number of workdays where he/she swiped cards during both the morning peak and the evening peak (on a maximum of 20 workdays). Then, we obtained the distribution diagram of cardholders according to the calculated number of workdays, as illustrated in Figure 1. In this context, a large number of related works usually adopted a hypothetical threshold or clustering methods to segment the commuters and noncommuters (e.g., cardholders traveling for at least a number of days are identified as transit commuters) [4, 12, 20]. However, there seemed to be no obvious fluctuation between the number of cardholders corresponding to "9 workdays" and the subsequent counterparts. Thus, it was really hard and confused to determine an exact threshold value for the identification according to the statistics in Figure 1. Moreover, even if a threshold value could be set up, it was still challenging to test the accuracy of results without any further circumstantial evidence. Therefore, in the case study of Xiamen, it was difficult to identify the commuters among cardholders by using the traditional approaches derived from the literature.

To this end, an original machine learning algorithm called Naïve Bayesian Classifier (NBC) was adopted in this paper to identify public transit commuters based on both the smartcard data and survey data. Compared with existing methods which were plagued by the validation of the accuracy of the identification results, the employed approach was a machine learning algorithm with functions of accuracy checking. The rest of the paper was organized as follows. In Section 2, a review of literature on smartcard data application was drawn, and the existing approaches of identifying commuters in smartcard data were discussed. Then, the Naïve Bayesian Classifier was introduced, and the classifier was built, trained, and tested in Section 3. Subsequently, the attribute of "commuter or not" for each cardholder was estimated based on the tested classifier, and travel characteristics and regularities of the identified commuters were analyzed in Sections 4 and 5. In the end, conclusions were drawn, and future directions were under discussion.

2. Literature Review

In recent years, the application of smartcard data mining in the planning and management of public transit systems has received much research attention. The previous works can be grouped into three categories, which can be, respectively, described as tactical-level studies, operational-level studies, and strategic-level studies [10].

In the tactical level, the trip pattern analysis and market segment were frequently addressed, contributing to the operational adjustment in public transit [14, 15]. Zhao et al. [21] developed a methodology to detect individuals' travel pattern changes by using smartcard data from London, UK, over two years. They specified one distribution for each of the three dimensions of travel behavior, and a Bayesian method was developed to identify significant change points in travel patterns. The results show that, compared to the traditional generalized likelihood ratio (GLR) approach, the Bayesian method required less predefined parameters and was more robust. Ma et al. [6] employed a Density-Based Scanning Algorithm with Noise (DBSCAN) to segment the smartcard data by analyzing different travel patterns. Then, spatial and temporal regularity of each cluster was derived through a continuous long-term observation period. The original DBSCAN method was improved by Kieu et al. [8, 9] and then applied to divide the passengers under a much lower calculation complexity. The heterogeneity of public transit riders was analyzed by Langlois et al. [22] through four-week smartcard data, and 11 travel patterns were generated. Legara and Monterola [16] developed a new classification method with promising accuracy by using the concept of eigentravel matrices which captured a commuter's characteristic travel routine, while, in operational-level studies, precise performance indicators such as schedule adherence, the number of transfers, and vehicle-kilometers on a public transit network were discussed [1, 10, 13]. Among them, the study of origin-destination (OD) and interchange inference was a hot topic and received much research attention. Trepanier et al. [23] and Ma et al. [24] combined AFC data and AVL data to infer boarding locations. For the alighting location estimation, Munizaga and Palma [7] set up a disutility function which took account of time and distance and then bus destinations were inferred by minimizing the generalized cost of the function. Sanchez-Martinez [17] extended the disutility function approach to rail networks, and a dynamic programming algorithm was designed to infer destinations. Wang et al. [25] applied the trip-chaining to infer bus passenger OD from AFC and AVL data, while Carrel et al. [26] inferred OD matrices by using smartphone and AVL data.

Regarding the applications on the strategic side, although previous researches have adopted smartcard data for the assessment and planning of public transit systems, however, the data source was somehow fragmentary for the behavior analysis [19]. For example, most of the previous literature on the analysis of cardholders' commuting characteristics oversimplified the definition of a public transit commuter [5]. In this context, Kusakabe and Asakura [19] developed a data fusion method based on the Naïve Bayesian Classifier (NBC) by integrating the smartcard data with the personal travel

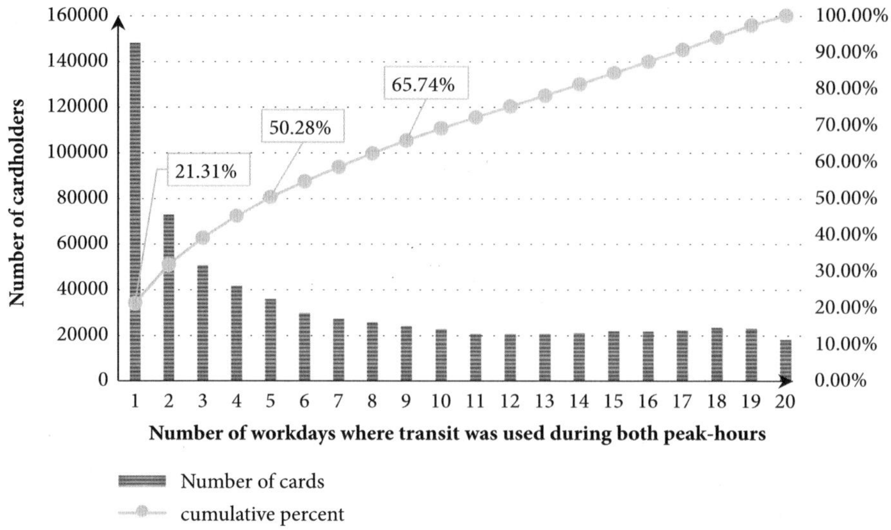

FIGURE 1: Basic statistics of smartcard data in Xiamen, China.

TABLE 1: Examples of smartcard transaction records.

Card id	Date	Transaction Time	Bus Route	Bus Plate Number
326352926	2015/6/22	09:43:47	Bus Route 815	D3789
154380354	2015/6/23	06:40:50	Bus Route 301	D3775
323446469	2015/6/23	20:40:13	Bus Route 713	D9342
051397471	2015/6/24	11:40:23	Bus Route 301	D3764
360122282	2015/6/24	08:42:30	Bus Route 512	D6453

survey data. The approach was intended to estimate absent attributes of trip purpose in smartcard data and enhance the understanding of travelers' behavior. Nevertheless, their study was implemented at only one railway station run by a private company, and the proportion of cardholders only accounted for about 10% of the total ridership. Thus, the applicability, portability, and representativeness of the proposed approach may be affected by the sample size. Moreover, other than estimating the travel purpose for each trip, it was also not clear whether the data fusion method could be applied to infer the attribute of cardholders. Since achieving the identification of public transit commuters in smartcard data was not an easy task in most metropolis of China, this paper extended the application of NBC approach and employed the method to estimate the attribute of cardholders instead of trip purpose. Besides, as an extension to the work of Kusakabe and Asakura [19], the approach was adopted and applied to the whole public transit network in Xiamen, China, with 85% of the transactions completed through smartcards. The spatial distribution of commuters' home and work locations in the case study were also revealed by fusing together smartcard and Automatic Vehicle Location (AVL) data.

3. The Smartcard Data

3.1. A Brief Introduction of Xiamen, China. Xiamen is administered as a subprovincial city of Fujian province with an area of 1,699.39 square kilometers. The population in Xiamen was nearly 3.5 million according to the Census in 2014. Regular bus systems and the Bus Rapid Transit (BRT) are the principal public transportation modes for local residents. There are nearly 320 regular bus routes and 5 BRT lines currently in service in Xiamen.

3.2. Data Preparation. "E-Tong Card" is the only contactless smartcard used for electronic payments in the public transportation system of Xiamen, China, and it started to come into use in 2006. Through the smartcards, passengers can access all the bus routes and BRT lines. Until 2015, the circulation of E-Tong cards has exceeded 6 million, and over 85% of the transactions in the public transit network were completed through the automated fare system. Thus, it provides a good opportunity for researchers to obtain the users' travel activities through smartcard data mining.

The smartcard data analyzed in this paper were harvested from the archived data in the AFC system of "E-Tong Card". The original dataset contained whole transaction records on the public transit network, collected from 22nd to 26th June, 2015 (except elder cards and student cards; 5 consecutive workdays were covered). Since the pricing of the bus system was not distance-based in Xiamen, the passenger only swiped the card once when boarding. Each record included the information of card id, date, transaction time, and bus route as well as plate number, as shown in Table 1.

4. Methods

4.1. The Naïve Bayesian Classifier. The Naïve Bayes classifier was a machine learning technique efficiently utilized for classification/identification applications in data mining environment. The reliability in predicting and decision-making constructed the statistical nature of the NBC. In practice, NBC could estimate the absent attributes of the data by predicting probabilities of class membership. To classify a new instance, the algorithm used the Bayesian rule to calculate the conditional probability of each class value and took the class with the maximum probability as the identified class. The collected survey data were usually split into two parts: training part and test part. The algorithm used the training data to build the classifier and then estimated the required probability values in the test data to test the accuracy of the built classifier.

Let vector $\mathbf{F} = \{f_1, f_2, \ldots, f_N\}$ be a set of behavioral attributes/features, as shown in Figure 2. Each element of vector **F** represented a common attribute or travel characteristic which existed in both survey and smartcard data, such as travel frequency or boarding time. Let C be a variable of the absent classification item in the smartcard data but could be measured by survey data [19]. It was assumed that elements of vector **F** were conditionally independent when C was given. In this study, classification item C and elements of vector **F** were treated as discrete variables. Then, $p(C \mid \mathbf{F})$ could be trained through Bayes' theorem by using the survey data.

$$\begin{aligned} p(C \mid \mathbf{F}) &= \frac{1}{p(\mathbf{F})} p(C)\, p(\mathbf{F} \mid C) \\ &= \frac{1}{p(\mathbf{F})} p(C) \prod_{N=1}^{N'} p(f_N \mid C) \end{aligned} \tag{1}$$

In (1), $p(C)$, $p(\mathbf{F})$, and $p(f_N \mid C)$ could be estimated by survey data. Specifically, $p(C)$ and $p(\mathbf{F})$ were the probabilities derived from the proportion of interviewees having classification item C and vector **F**, respectively. The conditional probability distribution $p(f_N \mid C)$ represented the composition rate of interviewees having attribute f_N corresponding to classification item C.

The attribute vector **F** could also be measured for each cardholder in the smartcard dataset, since it was shared in both of the two data sources. Then, the trained classifier could be adopted to estimate the absent attribute C of cardholders. It could be expressed as the following equation:

$$\hat{c}(\mathbf{F}) = \arg\max_{c \in C} p(c \mid \mathbf{F}), \quad C = \{c_1, c_2, \ldots, c_M\} \tag{2}$$

where c was one item of the classification attribute C.

4.2. Survey Design. The flow chart of the employed NBC method was illustrated in Figure 3. The core of machine learning in this study was to estimate absent elements of smartcard data by using the survey data to train and test the classifier. The survey was designed as shown in Figure 3. The classification item C could only be observed in the survey data, and it could be measured through the interviewees' response to "whether the bus service is your first choice when commute". The attribute vector **F** represented behavioral attributes that could be observed by both of the two data sources. In this study, the vector **F** contained three independent attributes, and it was relatively easy to measure them by using any one of the two datasets: (f_1) "average first boarding time on the workdays in last week"; (f_2) "average last boarding time on the workdays in last week"; and (f_3) "the number of workdays where commuting by bus in the last week", which represented the number of workdays where the cardholders swiped cards during both of the two peak-hours in the last week.

5. Experimental Analysis

5.1. The Trained and Tested NBC. The NBC model was built and tested by the use of survey data harvested from Xiamen, China. The research area in this paper was not broad with about 150 square kilometers; it can be assumed that the characteristics of public transit users were homogenous all over the city. The survey was implemented in the major industrial parks, software parks, and CBDs of Xiamen on 29th, June, 2015. Since these areas nearly provided more than 70% of job positions in Xiamen [27], the samples could be used to serve as the representative of typical commuters in Xiamen. The interviewees were selected at their workplaces with random sampling by meeting the conditions: (1) permanent staff with smartcards and (2) usually commute between home and their workplaces. Thus, the definition of public transit commuters as well as its response indicator "whether the bus service is your first choice when commute" in the survey could be better understood by the selected interviewees in a standard way. The survey was designed as above mentioned, through which the three independent attributes and the classification item could be measured. A total of 900 questionnaires were issued. Eventually, 532 valid samples were collected with 62% (330) of which stated themselves as public transit commuters. The reliability of the survey data was acceptable, since the value of Cronbach's Alpha was greater than 0.8.

Regarding the descriptive statistics of the valid samples (Table 2), 54.1% of the interviewees were female corresponding with 45.9% of which were male. It was in line with the census data of Xiamen in 2014. The age of the valid interviewees mostly fell in the range between 18 and 60. Almost half of the passengers were between 21 and 30 years old (49.6%). Thus, it was also consistent with the characteristics of commuters and meet the feature of smartcard dataset excluding student cards and elder cards. Statistics related to the occupation and income of the interviewees were also calculated. It indicated that company employees (46.7%) accounted for almost half of the samples. Correspondingly, more than half of the interviewees earned less than 4500 yuan (RMB) per month, whereas there was still 38.6% of the ones had a monthly income of less than 3,000 yuan (RMB). The average monthly salary of Xiamen in 2015 was 3,508 yuan (RMB), according to the salary report of employees in 2016. Thus, it implied that there was no potential bias from the selected interviewees regarding the income.

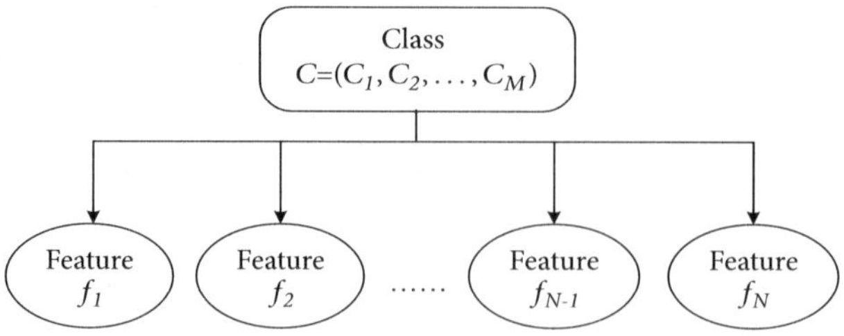

FIGURE 2: Basic structure of the NBC.

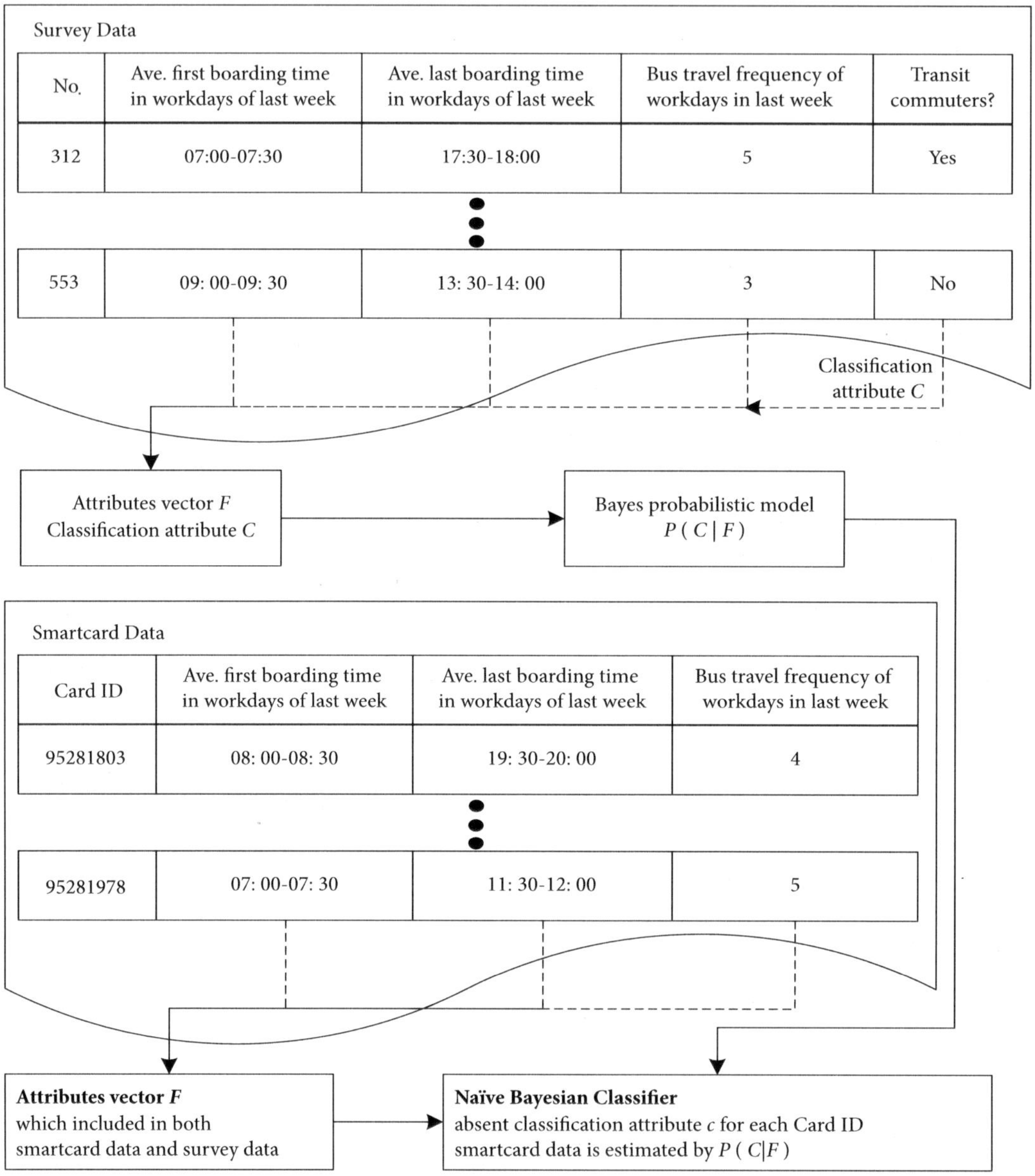

FIGURE 3: Flow chart of NBC for identifying commuters in smartcard data.

Then, the survey data were divided into two datasets: a training dataset and a testing dataset. Each dataset contained 266 valid samples, respectively, 165 of which were stated public transit commuters. By using the training dataset, the conditional probability distribution $p(f_N \mid C)$ could be derived from the proportion of interviewees having attribute f_N (N = 1, 2, 3) corresponding to each classification C (public transit commuters or not), shown in Figures 4(a), 4(b), and 4(c). Since $P(\mathbf{F})$ did not depend on C, it could be regarded as a constant value when $\mathbf{F}$ was given. The distribution $P(C)$

TABLE 2: Descriptive statistics of valid samples.

Attributes	Statistics
Gender	Male (45.9%), female (54.1%)
Age (years)	<=20 (1.2%), 21-30 (49.7%), 31-40 (24.0%), 41-50 (11.7%), 51-60 (11.9%), ≥60 (1.5%)
Occupation	Company employees (46.7%), civil servant (6.8%), educator and researcher (8.8%), service worker (20.7%), self-employed worker (14.6%), other (2.4%)
Monthly income (Unit: RMB)	<=3000 (38.6%), 3001-4500 (20.1%), 4501-6000 (25.6%), >6000 (15.7%)

TABLE 3: Test Results of the trained model.

	Transit Commuters (165 samples)	Non-Transit Commuters (101 samples)
Correctly estimated	151	87
Failure number	14	14
Accuracy Rate	92%	86%

could also be easily derived from the training data, which was calculated here as follows: $P(C_{commuter}) = 62\%$.

Based on (1) and (2), when a new attribute vector **F** was given, the conditional probability of each classification item could be calculated. The item with the maximum probability could be considered as the identified class. For example, the probability of an interviewee being a public transit commuter could be calculated as $p(C_{commuter} \mid \mathbf{F}) = 97.4\%$, if his/her attribute vector could be denoted as

$$\mathbf{F} = \{f_1 = 07:00:00 - 07:30:00, f_2 = 16:30:00 - 17:00:00, f_3 = 5\} \quad (3)$$

Then, the trained NBC model was tested by using the testing dataset. Firstly, only the attribute vector **F** of each interviewee was utilized in the NBC model to predict their class values. Then the accuracy would be checked by comparing the estimated class membership with the actual classification of the interviewee [19]. Test results were reported in Table 3.

For a total of 266 samples in the testing dataset, the average accuracy rate could be calculated as [1-(14+14)/266]∗100%=90%, including 14 failures in each of classification items. Specifically, 92.0% of public transit commuters (151 samples) and 86.0% of nonpublic transit commuters (87 samples) were correctly identified. Since the Root Mean Square Error for the identification of public transit commuters was less than 0.3 (0.291), the accuracy rate of the trained NBC could be considered acceptable. Regarding a smartcard dataset which contained M cardholders, when P of them were identified as commuters by using the NBC ($P<M$), then the actual number of public transit commuters among cardholders could be calculated according to (4), taking the accuracy rate into account.

$$\text{Actual Number} = [P * 0.92 + (M - P) * 0.14] \quad (4)$$

5.2. Results and Discussions. The NBC was employed to identify the commuters in the smartcard dataset, which was trained and tested by using survey data. The identification results and some other statistics were listed in Table 4. Eventually, nearly 290,000 cardholders (41.94% of the total) were identified as public transit commuters. According to the error analysis based on test instances in the last section, the accuracy rate for this result of commuter-identification should be around 92%, corresponding with the Root Mean Square Error of 0.291. Regarding the identified commuters, the statistics reflected that they averagely began their trips at 7:30-8:00 and swiped cards for their last-boarding at 17:30-18:00, 87% of which swiped cards twice in a typical workday. This was likely to represent a typical commuting trip chain, where public transit riders took a bus from home to their workplaces in the morning and then returned home in the evening [6].

Then, in order to better improve and test the accuracy of results, we cross-checked the bus route information of each identified commuter who swiped cards twice per workday. Due to the bus routes overlap, the demand of cardholders' first travel may be fulfilled by different bus routes in a period time. Nevertheless, during the 5-workday period, the statistical results indicated that the bus routes corresponding to commuters' first boarding were stable, more than 93% of which depended on the same bus route. In addition, we also examined whether the bus route corresponding to each commuter's last trip per workday was contained in his/her bus route dataset of first boarding. Since a typical commuting chain reflected a home-work-home journey, the bus route/metro line used in commuter's last trip on the workday should have a strong correlation with the ones of his/her first boarding. The results indicated that 96% of the assumed home-work-home journeys (2 trips per workday) satisfied the above conditions. Thus, the primary OD demand of public transit commuters can be obtained by connecting the first-boarding locations with the last-boarding locations of the identified commuters who swiped cards twice per workday. Since 87% of the identified commuters swiped cards twice per workday, it implied that the planning and the network structure of bus lines in Xiamen were really well to fulfill the demand of commuters. However, the average bus line overlap factor in Xiamen was 5.15 [27]. Thus, in

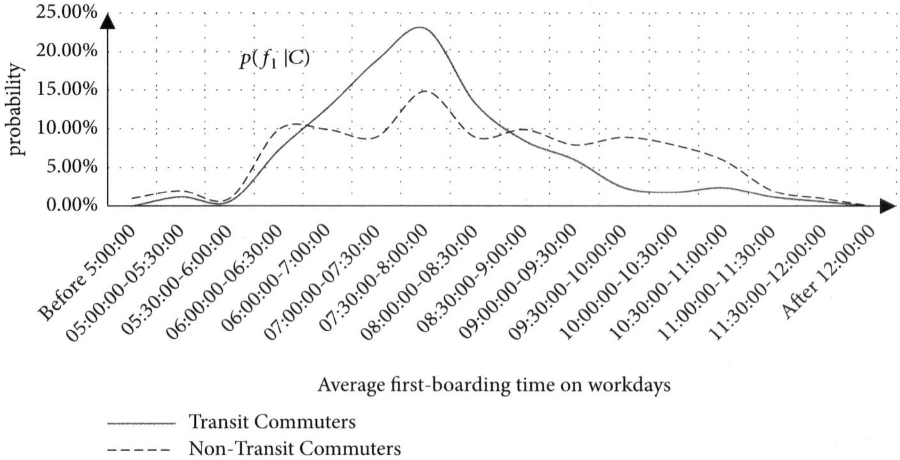

(a) Probabilistic distribution of average first-boarding time on workdays

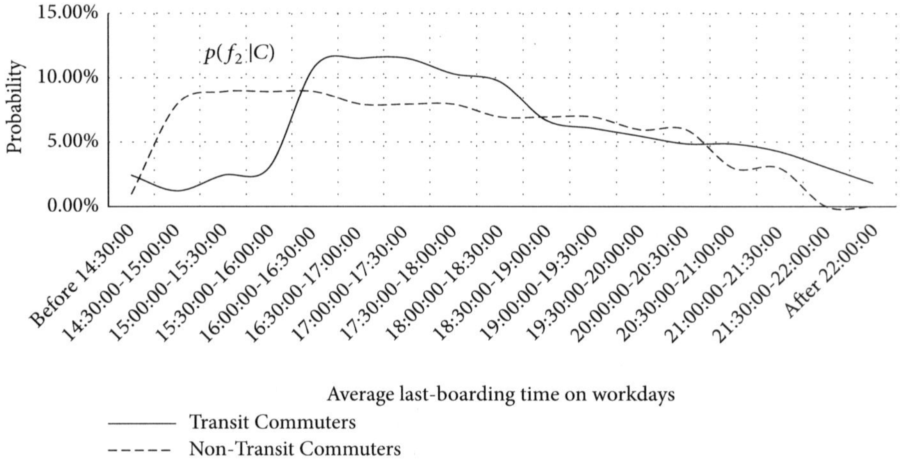

(b) Probabilistic distribution of average last-boarding time in workdays

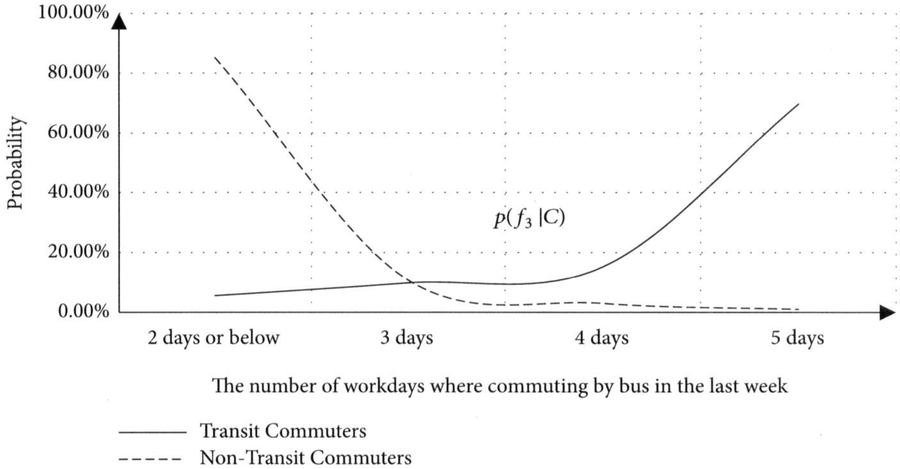

(c) Probabilistic distribution of the number of workdays where commuting by bus

Figure 4

Table 4: Identification results of the NBC model.

Transit Commuters	Number of cards	Ave. first-boarding time	Ave. last-boarding time	Travel frequency of workdays in a week
	288153	7:30-8:00	17:30-18:00	4.2 days

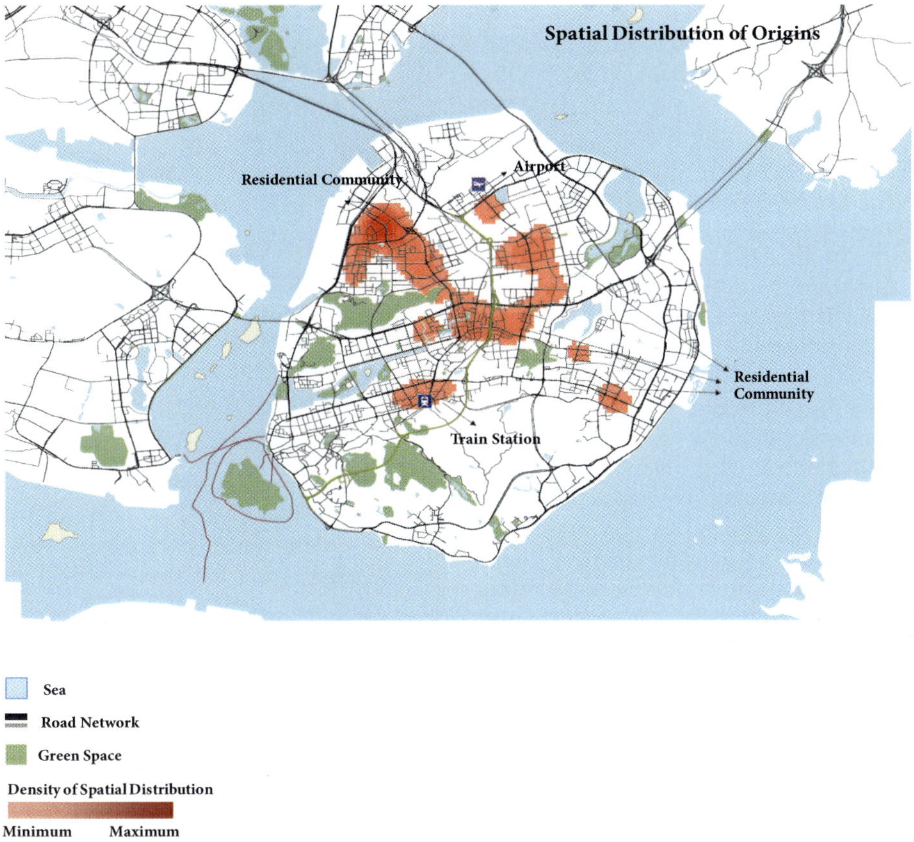

(a) Spatial distribution of origins of commute activities

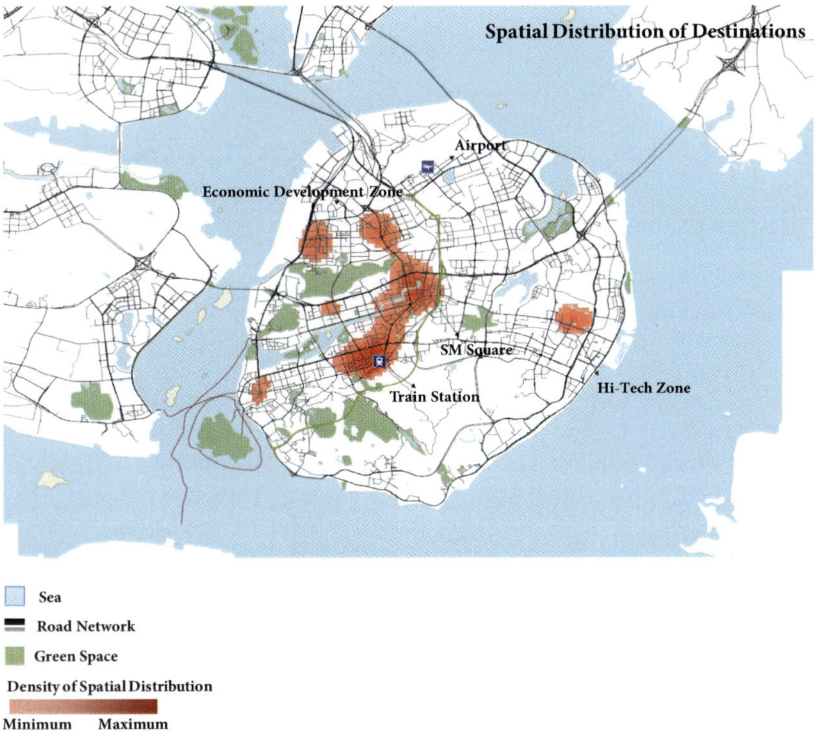

(b) Spatial distribution of destinations of commute activities

FIGURE 5

contrast, the above results also reflected that the existing bus overlap lines have not played an effective role in supporting the majority of commuting activities, which may lead to an extreme crowding of bus routes on main public transit corridors.

As for the analysis of primary OD demand of public transit commuters, however, the AFC system was not initially integrated with AVL system in Xiamen, China. Thus, the boarding-location should be estimated by integrating the two separate databases. In this paper, we employed the estimation method of boarding location proposed in the literature of Ma et al. [6]. It matched the transaction time of smartcard data with the time of GPS data generated from the bus with the same plate number and then inferred the location where he/she got on board. Finally, it obtained the station information by minimizing the distance between the boarding-location and bus stations on the same direction of the route. About 94% of the transaction records were finally matched with the GPS data by minimizing the time difference and then obtained the boarding-location information. Then, based on the boarding-location information, the distribution of commuters' origins and destinations of commuting trips could be obtained, shown in Figure 5. It was likely to depict a whole picture of the distributions of their residence and workplaces. Specifically, the results indicated that the spatial distribution of public transit commuters' origins was disperse when compared with the distribution of their destinations. Most of the public transit commuters lived in several core residential communities. However, workplaces of public transit commuters were more concentrated in downtown and several development zones. For instance, a large proportion of public transit commuters worked at the downtown area near the train station as well as the SM Square which was the CBD of Xiamen. Another two concentration areas of employees were located in the economic development zone in the northwest and the Hi-Tech parks in the east of Xiamen.

6. Conclusion and Future Researches

This paper employed the Naïve Bayesian Classifier to identify public transit commuters based on both the smartcard data and the survey data. A case study was given in Xiamen, China, and the classifier was trained and tested by related survey data collected from 532 valid samples. Then, it was applied to the smartcard cardholders for the identification of public transit commuters. The results indicated the following:

(i) The success rate of the identification of public transit commuters was 92% in the test case. Nearly 290,000 cardholders were classified as public transit commuters. However, considering the error of the NBC for identification of commuters, the actual number of public transit commuters with smartcards in Xiamen should be around 323,000.

(ii) Regarding the identified commuters, the statistics reflected that they averagely began their trips at 7:30-8:00 and swiped cards for their last-boarding at 17:30-18:00, 87% of which swiped cards twice in a typical workday.

(iii) The travel pattern reflected by the commuters who swiped cards twice in a typical workday was likely to represent a typical commuting trip chain, where public transit riders took a bus from home to their workplaces in the morning and then returned home in the evening.

(iv) Through few transfers, bus lines in Xiamen were really well to fulfill the demand of commuters. Conversely, it also reflected that the existing bus overlap lines have not played an effective role in supporting the majority of commuting activities, which may lead to an extreme crowding of bus routes on main public transit corridors.

(v) The primary OD demand of public transit commuters was obtained. It was found that home locations of public transit commuters were more disperse than their work locations in Xiamen. Most of the public transit commuters lived in several core residential communities. However, workplaces of public transit commuters were more concentrated in downtown and several development zones.

Nevertheless, based on the identified public transit commuters, travel pattern analysis of commuting activities could be conducted in further studies. More and more smartcard data related challenges may need to be addressed when researching on deep analysis of public transit rider's travel behavior or fusion method with other data sources.

Conflicts of Interest

The authors declare that there are no conflicts of interest in this paper.

Acknowledgments

This research is supported by National Natural Science Foundation (Project: 51478350) and the Fundamental Research Funds for the Central Universities (3132018305; 3132016301).

References

[1] J. Y. Park, D.-J. Kim, and Y. Lim, "Use of smart card data to define public transit use in Seoul, South Korea," *Transportation Research Record*, no. 2063, pp. 3–9, 2008.

[2] Y. Shiftan, Y. Barlach, and D. Shefer, "Measuring passenger loyalty to public transport modes," *Journal of Public Transportation*, vol. 18, no. 1, pp. 1–16, 2015.

[3] S. Sun, "Public Transit Loyalty Modeling Considering the Effect of Passengers' Emotional Value: A Case Study in Xiamen,

China," *Journal of Advanced Transportation*, vol. 2018, pp. 1–12, 2018.

[4] J. Zhou, E. Murphy, and Y. Long, "Commuting efficiency in the Beijing metropolitan area: An exploration combining smartcard and travel survey data," *Journal of Transport Geography*, vol. 41, pp. 175–183, 2014.

[5] X. Ma, C. Liu, H. Wen, Y. Wang, and Y. Wu, "Understanding commuting patterns using transit smart card data," *Journal of Transport Geography*, vol. 58, pp. 135–145, 2017.

[6] X. Ma, Y. J. Wu, Y. Wang, F. Chen, and J. Liu, "Mining smart card data for transit riders' travel patterns," *Transportation Research Part C: Emerging Technologies*, vol. 36, pp. 1–12, 2013.

[7] M. A. Munizaga and C. Palma, "Estimation of a disaggregate multimodal public transport Origin-Destination matrix from passive smartcard data from Santiago, Chile," *Transportation Research Part C: Emerging Technologies*, vol. 24, pp. 9–18, 2012.

[8] L.-M. Kieu, A. Bhaskar, and E. Chung, "A modified density-based scanning algorithm with noise for spatial travel pattern analysis from smart card AFC data," *Transportation Research Part C: Emerging Technologies*, vol. 58, pp. 193–207, 2015.

[9] L. M. Kieu, A. Bhaskar, and E. Chung, "Passenger segmentation using smart card data," *IEEE Transactions on Intelligent Transportation Systems*, vol. 16, no. 3, pp. 1537–1548, 2015.

[10] M.-P. Pelletier, M. Trépanier, and C. Morency, "Smart card data use in public transit: a literature review," *Transportation Research Part C: Emerging Technologies*, vol. 19, no. 4, pp. 557–568, 2011.

[11] S. Sun, Z. Duan, Q. Xu, and P. Chen, "School bus routing problem in the stochastic and time-dependent transportation network," *PLoS ONE*, vol. 13, no. 8, p. e0202618, 2018.

[12] J. Chen, Y. K. Lv, and M. L. Cui, "Estimating alighting stops of smart card public transportation passengers based on travel patterns," *Xi'an Univ. of Arch. & Tech 1*, vol. 1, pp. 23–29, 2018.

[13] M. Bagchi and P. R. White, "The potential of public transport smart card data," *Transport Policy*, vol. 12, no. 5, pp. 464–474, 2005.

[14] A.-S. Briand, E. Côme, M. Trépanier, and L. Oukhellou, "Analyzing year-to-year changes in public transport passenger behaviour using smart card data," *Transportation Research Part C: Emerging Technologies*, vol. 79, pp. 274–289, 2017.

[15] M. K. El Mahrsi, E. Côme, L. Oukhellou, and M. Verleysen, "Clustering Smart Card Data for Urban Mobility Analysis," *IEEE Transactions on Intelligent Transportation Systems*, vol. 18, no. 3, 2017.

[16] E. F. T. Legara and C. P. Monterola, "Inferring passenger types from commuter eigentravel matrices," *Transportmetrica B*, vol. 6, no. 3, pp. 230–250, 2018.

[17] G. E. Sánchez-Martínez, "Inference of public transportation trip destinations by using fare transaction and vehicle location data: Dynamic programming approach," *Transportation Research Record*, vol. 2652, pp. 1–7, 2017.

[18] D. Tribone, D. Block-Schachter, F. Salvucci, J. Attanucci, and N. Wilson, "Automated, data-driven performance regime for operations management, planning, and control," *Transportation Research Record*, vol. 2415, pp. 72–79, 2014.

[19] T. Kusakabe and Y. Asakura, "Behavioural data mining of transit smart card data: a data fusion approach," *Transportation Research Part C: Emerging Technologies*, vol. 46, pp. 179–191, 2014.

[20] J. Chen and D.-Y. Yang, "Identifying boarding stops of bus passengers with smart cards based on intelligent dispatching data," *Jiaotong Yunshu Xitong Gongcheng Yu Xinxi/Journal of Transportation Systems Engineering and Information Technology*, vol. 13, no. 1, pp. 76–80, 2013.

[21] Z. Zhao, H. N. Koutsopoulos, and J. Zhao, "Detecting pattern changes in individual travel behavior: A Bayesian approach," *Transportation Research Part B: Methodological*, vol. 112, pp. 73–88, 2018.

[22] G. Goulet Langlois, H. N. Koutsopoulos, and J. Zhao, "Inferring patterns in the multi-week activity sequences of public transport users," *Transportation Research Part C: Emerging Technologies*, vol. 64, pp. 1–16, 2016.

[23] M. Trépanier, N. Tranchant, and R. Chapleau, "Individual trip destination estimation in a transit smart card automated fare collection system," *Journal of Intelligent Transportation Systems: Technology, Planning, and Operations*, vol. 11, no. 1, pp. 1–14, 2007.

[24] X.-L. Ma, Y.-H. Wang, F. Chen, and J.-F. Liu, "Transit smart card data mining for passenger origin information extraction," *Journal of Zhejiang University SCIENCE C*, vol. 13, no. 10, pp. 750–760, 2012.

[25] W. Wang, J. P. Attanucci, and N. H. M. Wilson, "Bus passenger origin-destination estimation and related analyses using automated data collection systems," *Journal of Public Transportation*, vol. 14, no. 4, pp. 131–150, 2011.

[26] A. Carrel, P. S. C. Lau, R. G. Mishalani, R. Sengupta, and J. L. Walker, "Quantifying transit travel experiences from the users' perspective with high-resolution smartphone and vehicle location data: Methodologies, validation, and example analyses," *Transportation Research Part C: Emerging Technologies*, vol. 58, pp. 224–239, 2015.

[27] S. Zhang, M. Ding, and Z. F. Shi, "Transport Commission of Xiamen Municipality," Xiamen Transport Annual Report in 2014, Xiamen, China, 2015.

19

Understanding the Effect of an E-Hailing App Subsidy War on Taxicab Operation Zones

Zhixiang Fang, [1,2] **Rongxiang Su**, [1] **and Lian Huang**[3]

[1]*State Key Laboratory of Information Engineering in Surveying, Mapping and Remote Sensing, Wuhan University, Wuhan 430079, China*
[2]*Collaborative Innovation Center of Geospatial Technology, Wuhan 430079, China*
[3]*Shenzhen Transportation Operation Command Center, Shenzhen 518040, China*

Correspondence should be addressed to Zhixiang Fang; zxfang@whu.edu.cn

Academic Editor: Yair Wiseman

Understanding taxicab operation behaviors under various management or market policies (i.e., subsidies) is critical to making informed operating decisions for e-hailing companies and for government surveillance. This paper investigates the change of taxicab operation zones in context of an e-hailing app subsidy war in China, which is an important perspective that reflects changes in taxicab behavior, such as how the operation zones of taxicabs under the e-hailing app subsidy war change and how this change affects their trip distance and cruising time. To investigate this issue, this paper utilizes three indexes to elucidate the change of taxicab operation zones, namely, the repetition ratio of operation zone pairs, the area, and the degree of dispersion in the spatial distribution. A case study using taxicab trajectories during all of the important periods of the e-hailing app subsidy war in Shenzhen, China, was conducted and produced several valuable findings; for example, with respect to taxicabs as a whole, the proportion of habitual operation zone pairs among operation zone pairs in neighboring periods is relatively stable under any subsidy policy, and changes in the operation zones have little effect on changes in the average daily trip distance and average daily cruising time. Four groups of taxicabs divided according to initial change patterns in the operation zones present different change patterns during the subsidy war. By comparing these changes before and after the subsidy war, this paper finds that the subsidy war influences the taxicabs in groups I and II, while it has little influence on the taxicabs in groups III and IV, although all groups were affected during the subsidy war. For the taxicab groups in the period with the highest subsidy, the average daily trip distance and average daily cruising time decreased, whereas, in other periods, they presented different patterns.

1. Introduction

Taxicabs perform an indispensable function in urban transportation systems. Cabdrivers' mobility patterns and operating strategies have been widely studied since large-scale taxi GPS trajectories became obtainable due to the rapid development of information and communication technologies (ICT). For example, previous studies analyzed the distance and direction distributions of intraurban trips [1], urban cabdrivers' activity distribution [2], spatial variations in urban taxi ridership [3], cabdrivers' operation strategies with respect to revenue [4, 5], and alternative taxi services strategies [6].

E-hailing services (e.g., Didi in China and Uber in the US) can connect passengers and cabdrivers directly using communication technology since cabdrivers know the passengers' origins and destinations beforehand. However, several disturbing issues are commonly reported in the taxi industry, including unfair competition, safety concerns, longer working hours, surge pricing, request rejection, and increased difficulties among senior citizens with catching taxis. With the e-hailing app subsidy war that occurred from January to August in 2014 in China, which had a dramatic effect on the taxi industry, it has become feasible to analyze how cabdrivers react to the external policy stimulus and what consequences these different reactions cause. From the perspective of taxicabs, the most direct reflection of the spatial effect of subsidy policy is the change of operation zones. Do the operation zones of taxicabs change under the stimulation of a subsidy war? If so, how does this change affect cabdrivers' trip distance and cruising time? This paper primarily focuses on how the operation zones of taxicabs under the e-hailing app subsidy war changed and how this change affected their trip distance and cruising time. This paper investigates this question in three dimensions, including the repetition ratio of operation zone pairs, the changes in the area, and the changes in the degree of dispersion in the spatial distribution of operation zones. These three dimensions can well describe the change of taxicab operation zones from the perspective of space. Understanding these questions can provide insights into the effects of subsidy policies on taxicabs' operation tactics, helping taxi management authorities to develop precise and targeted management tactics for cabdrivers.

To address the research question, three indexes are calculated to indicate the changes in the operation zones, namely, the repetition ratio, the changes in the area, and the changes in the degree of dispersion in the spatial distribution of operation zones. Moreover, this paper calculates the average daily trip distance and the average daily cruising time for each taxicab to investigate the correlation between the spatial change and these two variables. Selected taxis are used to conduct the investigation of the primary research question.

The remainder of this paper is organized as follows. Section 2 reviews previous studies on spatial patterns of taxicabs and e-hailing services. Section 3 introduces the subsidy war and taxi GPS trajectories data and gives details on three indexes which are used to measure the change of taxicab operation zones and presents the method to access the effect of the change of taxicab operation zones. Section 4 presents the analysis of the experimental results. Section 5 concludes the paper and suggests directions for future study.

2. Related Work

2.1. Spatial Pattern of Taxicabs. Researchers have great interest in studying location data with spatiotemporal information due to the rapid development and increasing availability of various location-acquisition technologies [7, 8]. The large-scale collection of taxicab trajectory data provides a promising tool to investigate operation behavior patterns among cabdrivers.

Researchers have analyzed the spatial patterns of taxicabs using three approaches. The first approach is to uncover pattern variations; for example, Chen et al. [9] discovered frequent trajectories between a given origin and destination. Liu et al. [4] utilized k-means clustering to uncover the spatiotemporal preferences of high-earning drivers and average-earning drivers in Shenzhen. A comparison between these two groups shows that the operation scale of high-earning drivers is much greater than that of ordinary drivers and that ordinary drivers prefer to stay in traditional city centers with the greatest travel demand but ignore potential congestion, while high-earning drivers choose to operate in areas other than the central business district when traffic is congested. Liu et al. [10] investigated the temporal variation in taxi pick-up and drop-off patterns and identified six traffic 'source-sink' areas, which were associated with land use types.

The second approach is to discover the aggregate patterns of taxicabs. For example, Lee et al. [11] performed a temporal analysis to create a time-dependent pick-up pattern within each area and recommended that empty taxis go to the nearest cluster locations derived by the k-means method to pick up customers. Wang et al. [12] investigated the travel patterns to and from hotspots. Kang and Qin [13] studied the habitual operation behaviors of taxicabs in Wuhan City and extracted six dominant types of spatial patterns by utilizing a nonnegative matrix factorization method. Fang et al. [14] discussed the aggregate convergence and divergence patterns of crowd dynamics, which could discover aggregate patterns of various objects, including taxicabs, private cars, and human beings.

The third approach is to discover the spatial operation knowledge of taxicabs. For example, Hu et al. [2] found that the location in which taxi drivers' cruise and search for the next passenger' pick-up location will be near the previous drop-off location by analyzing the statistical results of the distance between each taxi driver's pick-up locations' mean center and drop-off locations' mean center. Dong et al.[5] classified drivers into three groups based on revenue and compared different driver groups in terms of spatiotemporal and strategy aspects. Their conclusion was that top drivers tend to maximize their usage of a limited working time to serve more trips and cover more mileage, which is the main reason that they earn more revenue.

In short, these studies seldom introduce effective indexes to indicate changes in the operation zones, such as the repetition ratio of operation zone pairs and the spatial changes (i.e., changes in the area and the changes in the degree of dispersion in the spatial distribution) of the operation zones.

2.2. E-Hailing Services of Taxicabs. Passengers usually hail a taxi on the street in the traditional manner. However, with advanced information and communication technologies, hailing a taxi through smartphones has become very popular; this approach offers passengers a high level of comfort and efficiency, especially during rush hours and rainy days. With the advent of smartphones, an increasing number of smartphone-based e-hailing applications have emerged in recent years, such as Uber and Lyft in the US and Didi and Kuaidi in China. These e-hailing applications provide an information platform that makes communication between drivers and passengers more efficient and convenient. The e-hailing service of taxicabs, as a new form of communication between drivers and passengers, has aroused scholarly interest.

One group of prior studies investigated the market equilibrium of taxi or related policies. Wang et al. [15] incorporated theories of two-sided markets into studies of the taxi industry, introduced matching functions to model the cross-group externalities between customers and taxi drivers on an e-hailing platform, and proposed a system of nonlinear equations to describe the taxi market equilibrium under hybrid modes of e-hailing and roadside hailing to reveal the impacts of the pricing strategy of an e-hailing platform on taxi service. Zha et al. [16] proposed equilibrium models under different behavioral assumptions about the labor supply in a ride-sourcing market. Qian and Ukkusuri [17] modelled the taxi market as a multiple-leader-follower game at the network

Table 1: Subsidy policies in different periods during the subsidy war (Unit: RMB Yuan).

Period	Duration	Didi		Kuaidi	
		Driver	Passenger	Driver	Passenger
1	Before Jan.9th	0	0	0	0
2	Jan. 10 th -Feb.16 th	10	10	10	10
3	Feb.17 th -Mar. 4 th	10 (50 for new user)	10-20	5-11	10-13
4	Mar. 22nd-May 16 th	10	3-5	5-11	3-5
5	May 17 th -July 8 st	10	0	5-11	0
6	After Aug.10 th	0	0	0	0

level and investigated the equilibrium of the taxi market with competition between traditional taxi services and app-based third-party taxi services. Their conclusion was that the fleet size and pricing policy are closely associated with the level of competition in the market and can have significant impacts on the total passenger cost, average waiting time, and fleet utilization. He and Shen [18] proposed a spatial equilibrium model for balancing the supply and demand of taxi services and depicted the possible adoption of e-hailing applications for drivers and customers in a well-regulated taxi market. Garza et al. [19] analyzed two different government support programs within a dynamic equilibrium model of imperfectly competitive industries in order to investigate their effectiveness.

The second group of studies focuses on pricing strategy. Salnikov et al. [20] compared the cost between Yellow Cab and Uber X, which is the cheapest version of the Uber taxi, in New York City. The authors observed that Yellow Cab appears on average 1.4 US dollars cheaper than Uber X, while Uber X begins to become cheaper only above the threshold of 35 dollars. However, further study by Noulas et al. [21] showed that the Uber X service appeared to be considerably cheaper than Black Cab in London. Interestingly, it was observed that Uber X is much cheaper in London than in New York. Another key issue in e-hailing apps is the surge pricing phenomenon; namely, a trip price fluctuates with time and varies from one area to another in a city. For example, prices can reach 7.5 times the base rate on New Year's Eve [22]. Previous studies have shown that the majority of surge pricing lasts less than 10 minutes and that surge prices have a strong negative impact on passenger demand and a weak positive impact on car supply [23]. Chen [24] also determined that surge pricing increases the supply of rides on Uber. Noulas et al. [25] reported that the driving factor of a higher fee for Uber X in New York is surge pricing.

In addition, a few studies have investigated the effects of subsidy wars on e-hailing services. For example, Wen et al. [26] showed that if subsidies were given to passengers, the difficulty in hailing taxis would not be alleviated, whereas this problem would be alleviated if subsidies were given to the taxi drivers. Leng et al. [27] compared taxi service data collected during a period in a subsidy war with that from another period without subsidies and concluded that, during the period with subsidies, the average number of trips increases, the idle time decreases, and the number of short-distance trips increases, whereas the distribution of long-distance trips is unchanged and the distribution of pick-up and drop-off locations does not change significantly. Su et al. [28] discovered that the changes in pick-up and drop-off location varied greatly in quantity and spatial distribution during the subsidy war.

However, few studies have focused on the question of how the operation zones of taxicabs under an e-hailing app subsidy war change and how this change affects their trip distance and cruising time. Therefore, the present paper investigates these questions.

3. Materials and Methods

This section introduces the subsidy war and experimental data, the measurements of the change of taxicab operation zones, and the average daily trip distance and cruising time derived from taxicab trajectories.

3.1. Subsidy War and Experimental Data. A fierce subsidy war between two e-hailing apps, Didi and Kuaidi, was triggered in 2014 in China. Based on the different subsidy policies available to drivers and passengers, this paper divides the time period of the subsidy war into six subperiods, which are listed in Table 1. No subsidy policy was implemented before Jan. 9th, which is defined as the first period. In the second period, which is from Jan. 10th to Feb. 16th, the subsidies for both cabdrivers and passengers were 10 yuan, since the subsidy war had just begun. The third period extends from Feb. 17th to Mar. 4th, which was a peak period in the subsidy war. During this period, both companies provided the highest subsidies for passengers, which ranged from 10 to 20 yuan. In addition, Didi gave 50 yuan to new drivers who used their app. The fourth period extends from Mar. 22nd to May 16th, during which cabdrivers still received higher subsidies; however, passengers received lower amounts. During the fifth period, which extends from May 17th to July 8th, both e-hailing apps cancelled the subsidies for passengers; however, cabdrivers still received subsidies. Finally, in the sixth period, neither of these apps provided any subsidies to anyone after Aug. 10th.

Human travel patterns on holidays vary greatly from those on nonholidays. For example, an enormous number of people leave Shenzhen City to go back their hometowns to celebrate Spring Festival, which has a considerable effect on the level of taxi service in Shenzhen City. In addition, bad weather, such as rainstorms, can significantly influence the demand for taxi services. To reduce the influence of such factors, holidays and bad-weather days are excluded in our

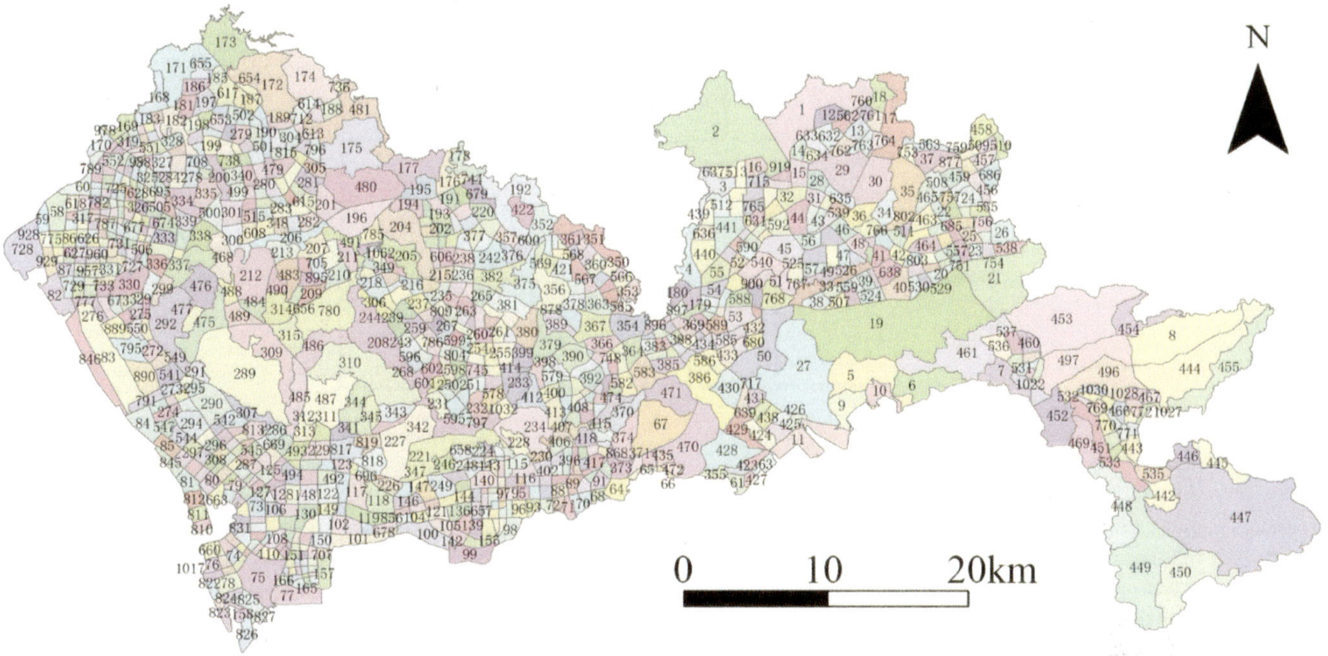

Figure 1: Traffic analysis zones in the city of Shenzhen, China.

datasets. Due to the short time span of the third period, which has only 16 days after excluding holidays and bad-weather days, we choose the same number of days for other periods equally. Finally, Dec. 23-31 in 2013 and Jan. 2, 3-6, 8, and 9 in 2014 were chosen from the first period; Jan. 20, 21, and 23-29 and Feb. 10-16 were selected from the second period; Feb. 17-28 and Mar. 1-4 were chosen from the third period; Mar. 23-28 and Apr. 4, 9, 10, 16, 17, 19-22, and 27 were selected from the fourth period; May 24, 26, 27, and 29 and June 3-5, 11-14, and 26-30 were chosen for the fifth period; and Aug. 11, 15-18, 21, 23-26, and 29 and Sep. 2, 3, 9, 10, and 18 were selected for the sixth period. Let Periods 1, 2, 3, 4, 5, and 6 denote these six periods, respectively.

The data were collected from 9,648 taxicabs equipped with GPS devices that operated in the city of Shenzhen, China, during the selected days in these six periods. All of these taxicab trajectory data cover all six periods. Each record includes the plate number, spatial location (i.e., longitude and latitude), timestamp, operation status (i.e., vacant or occupied), driving direction and velocity of a taxicab. The data do not include the actual identifying information about which taxicab driver uses Didi or Kuaidi app. Therefore, this analysis will reflect this change in the taxicab operation zones under the context of different subsidy policies, which cannot reflect taxicab's behavior changes directly affected by the Didi or Kuaidi app.

In this paper, we extract two datasets from the raw data, which are a passenger carrying dataset called DATASET #1 and a passenger hunting dataset called DATASET #2. For DATASET #1, the plate number, start time and end time of a trip, travel time, pick-up location and drop-off location of a trip, and the travel distance were collected. For DATASET #2, each record has the plate number, end time of the current trip, start time of the next trip, cruising time, drop-off location of the current trip, pick-up location of the next trip, and cruising distance.

Abnormal data are excluded from the dataset, such as, for DATASET #1, (i) records with a travel time less than 0.5 min, based on the reasonable hypothesis that a taxi trip lasts longer than 0.5 min; (ii) records with a travel distance shorter than 500 m based on the reasonable hypothesis that a taxi trip should be longer than 500 m; and (iii) records with the pick-up and drop-off latitude and longitude value outside of the territory of Shenzhen City. For DATASET #2, the following data are excluded: (i) records with a cruising time of more than 60 min based on the reasonable hypothesis that a taxi can find a passenger within 60 min and (ii) records with the pick-up and drop-off latitude and longitude value outside of the territory of Shenzhen City. After these steps, we prepare two preliminary datasets, both of which contain the same 9,605 taxis.

The traffic analysis zone (TAZ) is the analysis unit in this study. TAZ is usually constructed by census block information, where trips begin and end in transportation planning. This study uses TAZs to represent the operation zones of taxicabs from the perspective of taxicab service and the requirement of policy making for transportation planning. As shown in Figure 1, there are 1,067 TAZs in the city of Shenzhen, and each TAZ is given a unique ID. The pick-up and drop-off locations in DATASET #1 and DATASET #2 are aggregated into corresponding TAZs.

3.2. The Measurements of the Change of Taxicab Operation Zones. First, we introduce the definitions of operation zone and operation zone pairs. The operation zone represents a zone where a taxi has ever performed a pick-up or

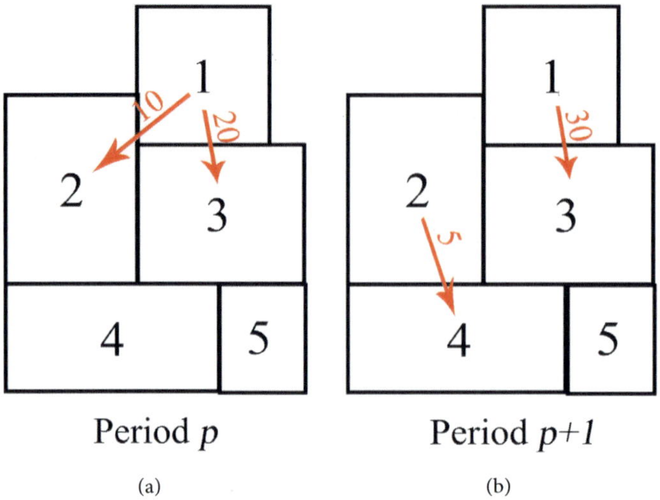

FIGURE 2: Examples of the change of a taxicab's operation zones. The rectangles represent TAZs and each TAZ is given a unique ID ranging from 1 to 5. A red arrow represents this taxicab has completed trips from the start TAZ of the arrow to the end TAZ of the arrow. In addition, the trip number is given beside the arrow in red number. The change of taxicab operation zones between neighboring periods can be derived by comparing the taxicab operation zones in current period (a) and taxicab operation zones in next period (b).

drop-off passengers, which can be described as a sequence $Z = \{x, y, \ldots, n\}$. A taxi's operation zone pair is a pair of origin and destination of a completed trip, which can be described as a sequence $P = \{(o_x, d_y), \ldots, (o_s, d_r)\}$. Figure 2 gives an example of the change of a taxicab's operation zones between neighboring periods. In Period p, the sequence of operation zones of this taxi is $\{1, 2, 3\}$ and the sequence of operation zone pairs is $\{(1, 2), (1, 3)\}$. In Period $p+1$, the sequence of operation zones of this taxi is $\{1, 2, 3, 4\}$ and the sequence of operation zone pairs is $\{(1, 3), (2, 4)\}$

In this section three indexes are introduced to uncover the change of taxicab operation zones in context of the e-hailing app subsidy war, namely, the repetition ratio of operation zone pairs, the area and the degree of dispersion in the spatial distribution of operation zones.

3.2.1. Repetition Ratio of Taxis' Operation Zone Pairs. A repetition ratio indicator is introduced to depict the proportion of repeatedly appearing operation zone pairs of each taxi between two neighboring periods. Each relation among origins and destinations of a taxicab can be described by the adjacency matrix A. The numbers of rows and columns of this matrix are both 1,067 which is the same as the total number of the analysis unit in this paper, or TAZ. If the taxi t_l has trips that start from TAZ s and end in TAZ r, we set a_{sr}, which denotes an element of matrix A, equals to 1; otherwise, $a_{sr} = 0$. To obtain the repeated operation zone pairs between two neighboring periods, the following matrix calculation is implemented. First, we define a matrix-based AND operation, which is denoted by $\oplus$. Consider two matrices, A and A', which denote the two adjacency matrices of two neighboring periods. If both elements in the compared position are 1, then the result comes to 1, namely, $1 \oplus 1 = 1$; otherwise, the result is 0 ($1 \oplus 0 = 0, 0 \oplus 0 = 0$). Thus, the repeated operation zone pairs can be derived from implementing the matrix-based AND operation among two adjacency matrices, as follows:

$$A'' = A' \oplus A \quad (1)$$

where A'' denotes the adjacency matrix of repeated operation zone pairs.

The repetition ratio of the taxis' operation zone pairs is calculated in two different ways, i.e., unweighted and weighted. The unweighted repetition ratio of the taxis' operation zone pairs can be derived by calculating the ratio of the number of repeated operation zone pairs to the total number of distinct pairs:

$$R_w = \frac{\sum_{i=1}^{n}\sum_{j=1}^{m} a''_{ij}}{\sum_{i=1}^{n}\sum_{j=1}^{m} a_{ij} + \sum_{i=1}^{n}\sum_{j=1}^{m} a'_{ij} - \sum_{i=1}^{n}\sum_{j=1}^{m} a''_{ij}} \quad (2)$$

where R_w denotes the unweighted repetition ratio of the taxis' operation zone pairs, a denotes the element of the matrix A, a'_{ij} represents the element of the matrix A', a''_{ij} denotes the element of the matrix A'', and n and m are the numbers of rows and columns of matrices A and A' which are both equal to 1,067 in this case. A taxi's operation zone pairs are similar between two neighboring periods when R_w is large, which means that the taxi prefers to stay in the previous operation zone pairs.

Considering that operation zone pairs with higher visiting frequency should make crucial contributions to the repetition ratio of operation zone pairs, we measure the visiting frequency of each pair as a weight to calculate the weighted repetition ratio of the operation zones:

$$R'_w = \frac{\sum_{i=1}^{n}\sum_{j=1}^{m} a''_{ij}u_{ij} + \sum_{i=1}^{n}\sum_{j=1}^{m} a''_{ij}u'_{ij}}{\sum_{i=1}^{n}\sum_{j=1}^{m} a_{ij}u_{ij} + \sum_{i=1}^{n}\sum_{j=1}^{m} a'_{ij}u'_{ij}} \quad (3)$$

where R'_w denotes the weighted repetition ratio of the taxis' operation zone pairs, u_{ij} denotes the number of trips that start from TAZ i and end in TAZ j in the current period, and u'_{ij} denotes the number of trips that start from TAZ i and end in TAZ j in the next period.

For the convenience of understanding, we calculate this index for the example in Figure 2. First, the adjacency matrices of origin and destination, A and A', are as follows:

$$A = \begin{bmatrix} 0 & 1 & 1 & 0 & 0 \\ 0 & 0 & 0 & 0 & 0 \\ 0 & 0 & 0 & 0 & 0 \\ 0 & 0 & 0 & 0 & 0 \\ 0 & 0 & 0 & 0 & 0 \end{bmatrix}, \quad A' = \begin{bmatrix} 0 & 0 & 1 & 0 & 0 \\ 0 & 0 & 0 & 1 & 0 \\ 0 & 0 & 0 & 0 & 0 \\ 0 & 0 & 0 & 0 & 0 \\ 0 & 0 & 0 & 0 & 0 \end{bmatrix} \tag{4}$$

Next, A'' can be derived by the above defined method:

$$A'' = A' \oplus A = \begin{bmatrix} 0 & 0 & 1 & 0 & 0 \\ 0 & 0 & 0 & 1 & 0 \\ 0 & 0 & 0 & 0 & 0 \\ 0 & 0 & 0 & 0 & 0 \\ 0 & 0 & 0 & 0 & 0 \end{bmatrix} \oplus \begin{bmatrix} 0 & 1 & 1 & 0 & 0 \\ 0 & 0 & 0 & 0 & 0 \\ 0 & 0 & 0 & 0 & 0 \\ 0 & 0 & 0 & 0 & 0 \\ 0 & 0 & 0 & 0 & 0 \end{bmatrix} = \begin{bmatrix} 0 & 0 & 1 & 0 & 0 \\ 0 & 0 & 0 & 0 & 0 \\ 0 & 0 & 0 & 0 & 0 \\ 0 & 0 & 0 & 0 & 0 \\ 0 & 0 & 0 & 0 & 0 \end{bmatrix} \tag{5}$$

Finally, R_w and R'_w can be derived by the above defined method:

$$R_w = \frac{\sum_{i=1}^{5}\sum_{j=1}^{5} a''_{ij}}{\sum_{i=1}^{5}\sum_{j=1}^{5} a_{ij} + \sum_{i=1}^{5}\sum_{j=1}^{5} a'_{ij} - \sum_{i=1}^{5}\sum_{j=1}^{5} a''_{ij}} = \frac{1}{2+2-1} = \frac{1}{3} \tag{6}$$

$$R'_w = \frac{\sum_{i=1}^{5}\sum_{j=1}^{5} a''_{ij}u_{ij} + \sum_{i=1}^{5}\sum_{j=1}^{5} a''_{ij}u'_{ij}}{\sum_{i=1}^{5}\sum_{j=1}^{5} a_{ij}u_{ij} + \sum_{i=1}^{5}\sum_{j=1}^{5} a'_{ij}u'_{ij}} = \frac{20+30}{(10+20)+(30+5)} = \frac{6}{13} \tag{7}$$

3.2.2. Area of the Taxi Operation Zones. The taxi-origin and taxi-destination relations can be described by two adjacency matrices B and B', respectively. The numbers of column of these two matrices are both equal to 1,067 which is the number of TAZs in this study. If taxi t_i has trips that start from TAZ j, we set $b_{ij} = 1$; otherwise, $b_{ij} = 0$. Analogously, we set $b'_{ik} = 1$ if taxi t_i has trips that end in TAZ k and $b'_{ik} = 0$ otherwise. Since the operation zone represents a zone where a taxi has ever performed a pick-up or drop-off passengers, when we collect all of the operation zones of a taxi, the duplicate zones should only count once. To exclude duplicate zones, the following matrix calculation is implemented. First, we define a matrix-based OR operation among two matrices, which is denoted by $\otimes$. Given two matrices, the result in each position is 0 if both elements are 0 ($0 \otimes 0 = 0$), while, otherwise, the result is 1 ($1 \otimes 1 = 1, 1 \otimes 0 = 1$). After this definition, the operation zones can be obtained by performing the matrix-based OR operation on matrix B and matrix B' as follows:

$$B'' = B' \otimes B \tag{8}$$

where B'' denotes the adjacency matrix that describes the relation between the taxis and operation zones.

The area of the taxis' operation zones is calculated in two different ways, i.e., unweighted and weighted. First, the unweighted area of the operation zones of taxi t_i is derived from summing the area of each operation zone as follows:

$$S_i = \sum_{k=1}^{m'} b''_{ik} S_k \tag{9}$$

where S_i denotes the unweighted area of the operation zones of taxi t_i, S_k denotes the area of an operation zone, b''_{ik} denotes an element in matrix B'', and m' denotes the number of columns which is equal to 1,067 in this study.

Considering that operation zones with higher visiting frequency play a more crucial role than other operation zones, this paper sums the number of pick-ups and drop-offs within each operation zone and uses the sum as the weight to calculate the weighted area of the operation zones:

$$S'_i = \frac{\sum_{k=1}^{m'} b''_{ik} \cdot \sum_{k=1}^{m'} b''_{ik} S_k c_k}{\sum_{k=1}^{m'} c_k} \tag{10}$$

where S'_i denotes the weighted area of the operation zones of taxi t_i and c_k denotes the number of pick-ups and drop-offs within an operation zone.

For the convenience of understanding, we calculate this index for the example in Figure 2. First, for Period p, the adjacency matrices of taxi-origin and taxi-destination relations, B and B' are as follow:

$$B = \begin{bmatrix} 1 & 0 & 0 & 0 & 0 \end{bmatrix}, \quad B' = \begin{bmatrix} 0 & 1 & 1 & 0 & 0 \end{bmatrix} \tag{11}$$

Next, B'' can be derived by the above defined matrix-based OR operation:

$$B'' = B' \otimes B = \begin{bmatrix} 1 & 0 & 0 & 0 & 0 \end{bmatrix} \otimes \begin{bmatrix} 0 & 1 & 1 & 0 & 0 \end{bmatrix} = \begin{bmatrix} 1 & 1 & 1 & 0 & 0 \end{bmatrix} \tag{12}$$

Finally, S_i and S'_i can be derived by the above defined method:

$$S_i = \sum_{k=1}^{5} b''_{ik} S_k = 1 \times S_1 + 1 \times S_2 + 1 \times S_3 + 0 \times S_4 + 0 \times S_5 = S_1 + S_2 + S_3$$

$$S'_i = \frac{\sum_{k=1}^{5} b''_{ik} \cdot \sum_{k=1}^{5} b''_{ik} S_k c_k}{\sum_{k=1}^{5} c_k} = \frac{(1+1+1+0+0) \times (1 \times 30 \times S_1 + 1 \times 10 \times S_2 + 1 \times 20 \times S_3 + 0 \times S_4 + 0 \times S_5)}{30+10+20} = \frac{3}{2} S_1 + \frac{1}{2} S_2 + S_3 \tag{13}$$

where S_1, S_2, S_3, S_4, S_5 denote the area of these five operation zones, respectively.

3.2.3. Degree of Dispersion in the Spatial Distribution of the Taxi Operation Zones. Several indexes are widely used to measure the spatial dispersion in landscape studies; for example, the proximity index [29] measures the degree of patch isolation and the degree of fragmentation of the corresponding patch type within the specified neighborhood of the focal patch. Pearce [30] proposed an indicator to assess the degree of dispersion of a forest patch, which considers only the number of specified patch types and ignores the impact of the area of a patch. Nearest-neighbor standard deviation (NNSD) is also a measure of patch dispersion, where a small value implies a more uniform or regular distribution of patches and a large value indicates a more irregular or uneven distribution of patches. This index should be interpreted together with the mean nearest-neighbor distance; otherwise, it can be misleading to rely only on itself. In addition, Ripley's K-function [31] is a widely used analysis tool for study of population spatial pattern. It is used to determine whether features exhibit statistically significant clustering or dispersion over a range of distances. However, it lacks a convenient comparable quality. The nearest-neighbor index (NNI) proposed by Clark and Evans [32] provides a precise measure of the spatial distribution of a pattern, which is more suitable in our case since it accounts for the area, quantity, and spatial locations of the patches comprehensively and could be compared among different cases. For a random arrangement of patches, NNI = 1; NNI < 1 represents more clustered, contiguous conditions that form patches, whereas NNI > 1 indicates a dispersed distribution of patches. Therefore, the NNI is used to investigate the degree of dispersion in the spatial distribution of the taxis' operation zones, as follows:

$$R_{NNI} = \frac{d_o}{d_e} \tag{14}$$

$$d_o = \frac{\sum_{i=1}^{n_m} d_i}{n_m} \tag{15}$$

$$d_e = \frac{1}{2}\sqrt{\frac{\sum_{i=1}^{n_m} \widehat{S_i}}{n_m}} \tag{16}$$

where R_{NNI} denotes the value of NNI, d_o denotes the mean observed nearest-neighbor distance, d_e represents the mean expected nearest-neighbor distance, and d_i is the edge-to-edge straight line distance between zone i and its nearest-neighbor. Based on the spatial proximity among operation zones, adjacent operation zones are merged first to avoid a result of zero for d_i. $\widehat{S_i}$ represents the area of the merged operation zone, and n_m is the number of the merged operation zones.

Similarly, we consider that operation zones with higher visiting frequency should make greater contributions to the degree of dispersion in the spatial distribution of the taxi operation zones. We measure the visiting frequency of each merged zone by summing the numbers of internal pick-ups and drop-offs, and we use the visiting frequency as the weight to calculate the weighted nearest-neighbor index (WNNI):

$$R_{WNNI} = \frac{d'_o}{d'_e} \tag{17}$$

$$d'_o = \frac{\sum_{i=1}^{n_m} c'_i d_i}{\sum_{i=1}^{n_m} c'_i} \tag{18}$$

$$d'_e = \frac{1}{2}\sqrt{\frac{\sum_{i=1}^{n_m} c'_i \widehat{S_i}}{\sum_{i=1}^{n_m} c'_i}} \tag{19}$$

where R_{WNNI} denotes the value of WNNI, d'_o denotes the weighted mean observed nearest-neighbor distance, d'_e represents the weighted mean expected nearest-neighbor distance, and c'_i denotes the number of pick-ups and drop-offs within a merged operation zone.

3.3. The Average Daily Trip Distance and Cruising Time Derived from Taxicab Trajectories. This paper assesses the effect of changes in the taxicab operation zones on their cost (average daily cruising time) and income (average daily trip distance). The average daily trip distance of each taxi in a period can be derived by averaging the sum of the trip distances in the period of subsidy war. The change in the average daily trip distance of each taxi between two neighboring periods can be obtained by subtracting its average daily trip distance in the current period from the next period. Similarly, this paper derived the average cruising time and the change in the average cruising time.

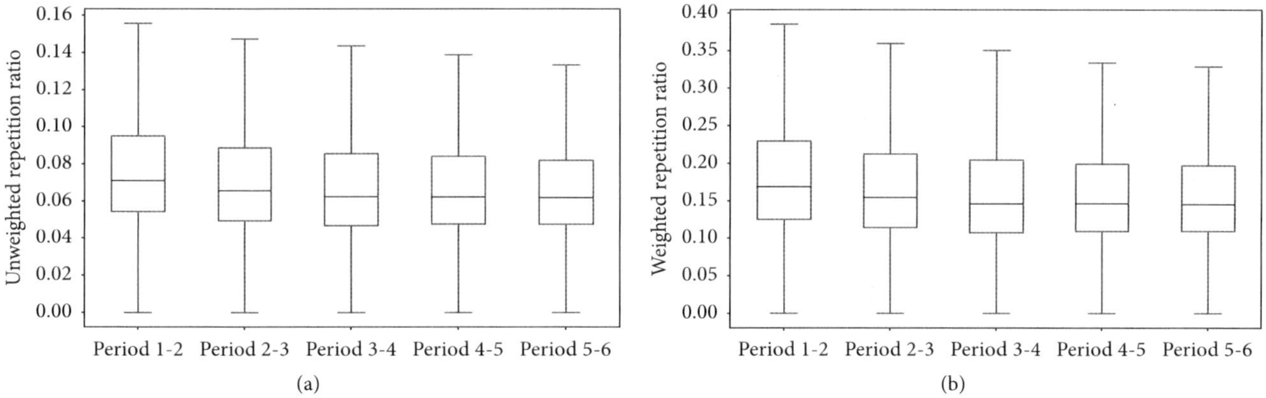

FIGURE 3: Distributions of the unweighted repetition ratio (a) and weighted repetition ratio (b) between different neighboring periods.

TABLE 2: Grouping result of taxicabs.

Group	Area	NNI	Taxicab count	Percentage
I	Increase	Increase	2034	21.18%
II	Increase	Decrease	2608	27.15%
III	Decrease	Increase	2773	28.87%
IV	Decrease	Decrease	2190	22.80%

4. Experimental Results

In this section, the overall changes in the taxicab operation zones in Shenzhen City within the context of the e-hailing app subsidy war are analyzed. Moreover, the change of operation zones in different groups of taxicabs and the influence of this change on their average daily cruising time and average daily trip distance are investigated, and finally, the changes in the patterns of weighted and unweighted areas and the NNI in different periods of this subsidy war are elucidated.

4.1. Overall Repetition Ratio Changes in the Taxicabs' Operation Zones. Figure 3 shows the distributions of the unweighted repetition ratio and weighted repetition ratio of the taxi operation zone pairs between different neighboring periods. It is noteworthy that, with the consideration of the weight, which is represented by the number of trips within an operation zone pair in calculating the repetition ratio, the importance of frequently visited operation zone pairs is highlighted. The unweighted repetition ratios of about half of the taxis are less than 10% for all neighboring periods, and the weighted repetition ratios are less than 20%. These findings indicate that the operation zone pairs of the majority of taxis change greatly. In addition, the results calculated from the weighted method are greater than those by the unweighted method. This finding results from the considerable number of trips that exist among repeated operation zone pairs, although the ratio of repeated operation zone pairs is relatively low. Additionally, the distribution of the weighted repetition ratio is flatter than the distribution of the unweighted repetition ratio, which indicates that although the proportions of repeated operation zone pairs are close among all taxicabs, the numbers of trips within the operation zone pairs vary from cab to cab. In addition, the repetition ratios slightly change in different periods regardless of whether the unweighted or weighted calculation methods are used, which indicates that the proportion of habitual operation zone pairs among all of the operation zone pairs in the neighboring periods is relatively stable under all subsidy policies in this subsidy war. Therefore, the majority of taxicabs have a significant change of their operation zone pairs for all periods, which is not easily affected by different subsidy polices.

4.2. Change Patterns in Different Groups of Taxicabs. This paper analyzes the spatial change in the operation zones from the perspectives of the change in the area and the change in the degree of dispersion in the spatial distribution. To analyze the change of operation zones of taxicabs in the context of the subsidy war, the selected taxicabs are divided into four groups (see Table 2) according to the initial increase or decrease pattern of the area and the degree of dispersion in the spatial distribution of the operation zones of taxicabs during Period 1-2. Here, Period 1 did not have any subsidy, while Period 2 was the first period with a subsidy, and therefore, this division can reflect the direct stimulus of the subsidy policy on the taxicabs' operation zones. Table 2 shows the grouping results, which indicates a relatively close percentage of these four groups.

To investigate the change patterns of these taxicabs in each group during other neighboring periods, this paper lists the taxicab percentages of the four patterns within each group in Table 3. Specifically, pattern A represents taxicabs with increased values for both the area and NNI, and similarly, patterns B, C, and D represent the other combinations of increase or decrease in the area and NNI. There are two findings in the table: (i) the patterns with the largest

TABLE 3: The taxicab percentage of four patterns within each group during neighboring periods.

Group	Pattern	Area	NNI	Period 1-2	Period 2-3	Period 3-4	Period 4-5	Period 5-6
I	A	Increase	Increase	100.00%	8.21%	49.36%	27.29%	8.11%
	B	Increase	Decrease	0	22.86%	30.14%	41.49%	24.63%
	C	Decrease	Increase	0	29.45%	15.63%	20.65%	24.29%
	D	Decrease	Decrease	0	39.48%	4.87%	10.57%	42.97%
II	A	Increase	Increase	0	16.60%	50.61%	25.73%	9.66%
	B	Increase	Decrease	100.00%	13.11%	28.14%	41.72%	23.50%
	C	Decrease	Increase	0	49.46%	17.22%	19.98%	23.73%
	D	Decrease	Decrease	0	20.82%	4.03%	12.58%	43.10%
III	A	Increase	Increase	0	19.58%	3.79%	11.83%	40.53%
	B	Increase	Decrease	0	15.15%	53.62%	25.82%	8.37%
	C	Decrease	Increase	100.00%	40.35%	29.90%	42.84%	20.12%
	D	Decrease	Decrease	0	19.19%	13.09%	19.58%	22.25%
IV	A	Increase	Increase	0	34.98%	50.64%	25.48%	7.90%
	B	Increase	Decrease	0	19.45%	29.73%	42.97%	19.91%
	C	Decrease	Increase	0	32.79%	16.16%	19.68%	23.79%
	D	Decrease	Decrease	100.00%	12.79%	3.47%	11.87%	48.40%

proportions of taxicabs in Period 3-4, Period 4-5, and Period 5-6 are the same for groups I, II, and IV. For example, approximately 50% of the taxicabs in these groups enlarged their operation zones and travelled more dispersed operation zones to service passengers (which is pattern A) due to a decrease in the subsidy for passengers from Period 3 to 4. When the subsidy was cancelled for passengers from Period 4 to 5, approximately 42% of the taxicabs in these groups still enlarged their operation zones, but they travelled relatively less dispersed in operation zones than during the previous neighboring periods (which is pattern B). With the further cancelation of the subsidy for taxi drivers from Period 5 to 6, 42%-48% of the taxicabs decreased their operation zones' area and travelled within these zones with a relatively low dispersed state (which is pattern D). (ii) The taxicabs in group III have a different pattern distribution from the other groups. From Period 2 to 3, 40.35% of the taxicabs reduced their operation zone areas and travelled within concentrated areas (pattern C). From Period 3 to 4, 53.62% of the taxicabs enlarged their operation zones and travelled with less dispersed areas (pattern B) than during the previous periods due to the reduced subsidy on passengers. From Period 4 to 5, 42.84% of the taxicabs readjusted their operation zones within more dispersed areas due to the subsidy cancellation policy for passengers (pattern C). From Period 5 to 6, 40.53% of the taxicabs still needed to travel larger and more dispersed operation zones to sustain their income level (pattern A). (iii) The taxicabs in group III are less sensitive to the subsidy policy than taxicabs in other groups from Period 2 to 3. This could be concluded from the result that 8.21% (group I), 13.11% (group II), and 12.79% (group IV) of taxicabs in these groups keep their initial changes, while 40.35% of taxicabs in group III could keep their initial changes. After this period, very few taxicabs could keep their initial changes. In short, the taxicabs in all groups present a subsidy policy-sensitive changes of their operation zones.

To further investigate the operation changes of these four groups, this paper lists the changes in the operation zones, the average daily trip distance and the cruising time in each group between all neighboring periods in Table 4. The changes in the operation zone areas show that taxicab groups I and II exhibited the same change tendency, as did groups III and IV, and for the changes in the NNI, groups I and III exhibited the same change tendency, as did groups II and IV. The table reveals that (i) there is an obvious stimulus of the subsidy policy in Period 1-2 due to the larger statistical results than in the other periods. (ii) In Period 3, with the highest subsidy, all of the selected taxicabs travelled with decreased areas of operation zones and less passenger hunting time, but they had a decreased average daily trip distance. This finding indicates that the taxicabs could find their passengers efficiently and did not require greater trip distance than in other periods to maintain their incomes because they could receive high subsidies from e-hailing apps. Figure 4(a) illustrates the change trends of each group's average daily cruising distance compared with that in Period 1. The figure shows an obvious decrease in the average daily cruising distance in Period 3, confirming the finding about shorter passenger hunting time. Therefore, this finding provides strong evidence that cabdrivers were more likely to pick up passengers who were nearby during this period. (iii) As the subsidy decreased, all taxicabs had increased areas of operation zones in Periods 3-4 and 4-5, and they had a high dispersion of their operation zones, especially in Period 4; as a result, they increased their average daily trip distance to maintain their incomes. (iv) The average daily cruising time steadily decreased after Period 2. This result indicates that the e-hailing apps improved their passenger hunting efficiencies.

To reflect the changes before and after the subsidy war, this paper compares the change trends of each group's operation zones area (Figure 4(b)) and the NNI (Figure 4(c)) with those in Period 1. Figure 4(b) presents two interesting

TABLE 4: The spatial changes in the operation zones, the changes in the average daily trip distance and the cruising time in each group between all of the neighboring periods.

Group	Period 1-2	Period 2-3	Period 3-4	Period 4-5	Period 5-6
			Unweighted area (Unit: km^2)		
I	30.76	-25.20	41.35	25.80	-34.53
II	40.19	-25.56	40.35	23.85	-33.98
III	-35.61	4.71	46.82	25.59	-41.43
IV	-31.29	4.07	44.92	26.22	-42.17
			NNI		
I	0.27	-0.10	0.16	-0.02	-0.14
II	-0.41	0.15	0.18	-0.04	-0.15
III	0.36	-0.17	0.19	-0.04	-0.18
IV	-0.27	0.18	0.18	-0.03	-0.16
			Average daily trip distance (Unit: km)		
I	20.78	-10.98	9.63	7.69	13.43
II	24.08	-9.36	9.35	6.94	13.71
III	-8.91	-3.11	16.00	3.04	11.22
IV	-9.49	-1.68	12.91	3.65	10.77
			Average daily cruising time (Unit: hour)		
I	0.46	-0.36	-0.21	-0.01	-0.01
II	0.60	-0.38	-0.21	-0.02	-0.07
III	-0.16	-0.23	0.00	-0.06	0.04
IV	-0.08	-0.26	-0.04	-0.05	-0.04

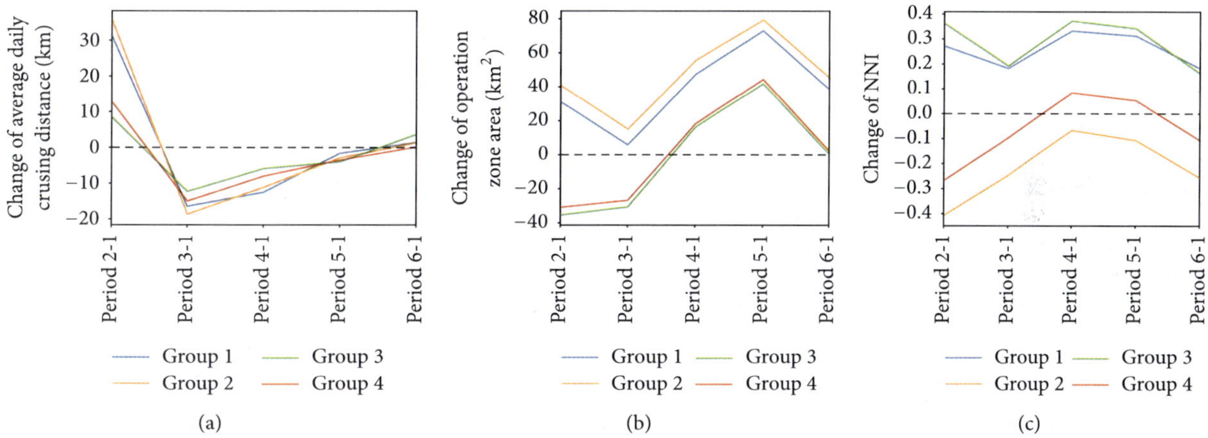

FIGURE 4: The change trends of each group's average daily cruising distance (a), operation zones area (b), and NNI (c) compared with Period 1.

results. The first result is that all taxicabs in groups I and II had more areas of operation zones after the subsidy war than before the war. This result indicates that the subsidy war enlarged their operation zones. The second result is that all taxicabs in groups III and IV reduced their operation zone areas and then increased their operation zone areas to maintain their income, but finally, they returned to same area value of operation zones. These results indicate that when the subsidy war ended, cabdrivers possibly returned to their previous work status before the war. In the aspect of operation zones' dispersion, there are also two results that can be drawn from Figure 4(c). The first result is that the NNI of group I's and group III's operation zones after the beginning of the subsidy war became more dispersed than that before the subsidy war. The second result is that the NNI of group II's and group IV's operation zones became smaller than before the war. This result indicates that the distance between the taxicabs' operation zones became shorter. In addition, when combining these results together, this paper indicates findings about these groups. For example, (i) the taxicabs in group I increased their operation zone areas and increased their NNI of operation zones. This finding suggests that the subsidy war required them to produce a higher workload to change their operation zones. (ii) The taxicabs in group II increased their operation zones' area while decreasing the NNI of their operation zones. This finding suggests that the subsidy war

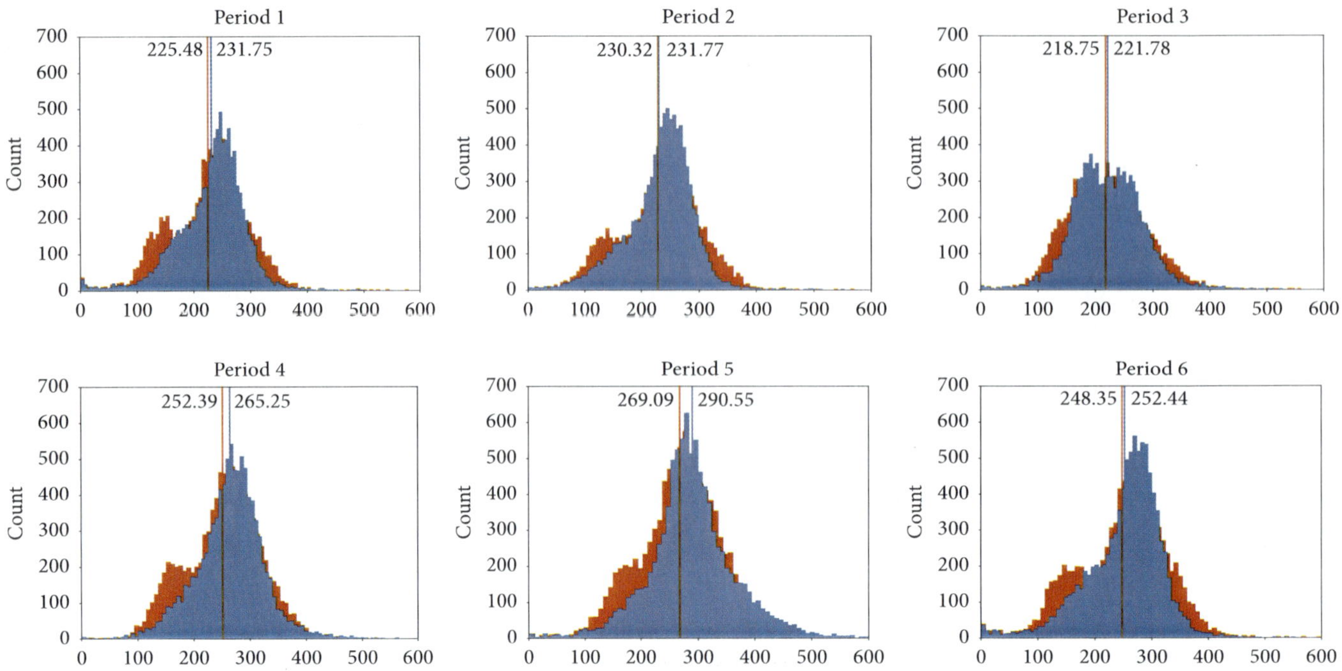

FIGURE 5: Histograms of the unweighted area of the operation zones (in blue) and weighted area of the operation zones (in red) of all of the selected taxis in each period. The blue vertical line represents the mean value of the unweighted area of the operation zones, and the red vertical line represents the mean value of the weighted area of the operation zones.

made them pay a greater attention to certain potential service zones.

4.3. Comparing Weighted and Unweighted Areas and NNIs in Different Periods. To assess the effect of the number of pick-ups and drop-offs on the spatial changes in the operation zones, the weighted area and WNNI are used to compare the difference in the unweighted areas and the NNIs in different periods. This number can help reflect the habitual operating preferences of cabdrivers. Figure 5 shows the histograms of the unweighted areas of operation zones (in blue) and weighted areas of operation zones (in red) of all of the selected taxis in each period. There are three findings in this figure: (i) Compared with the histograms of the unweighted areas of the operation zones, the mean values of the weighted areas of the operation zones are all smaller than the unweighted areas in any period. (ii) The mean values of the weighted areas of the operation zones in Periods 1, 2, 3, and 6 are relatively close to the unweighted values. In other words, from the perspective of aggregate level, there is little significant difference between weighted and unweighted calculation methods in these periods, which implies that the weight distributions of the different operation zones are similar. (iii) The span between the red and blue vertical lines is considerable in Periods 4 and 5, when the subsidies given to the passengers were reduced, which indicates a relatively heterogeneous distribution for the number of pick-ups and drop-offs in the operation zones during these two periods.

The spatial dispersion distribution of the operation zones with a high number of pick-ups and drop-offs is another important characteristic that reflects the change in the operation zones. It is reasonable that the operation zones with high weights could be used to describe the degree of dispersion in the spatial distribution of all of the operation zones. Figure 6 shows the histograms of the maximum number of pick-ups and drop-offs within an operation zone of each taxi in every period. The results show that the majority of pick-ups and drop-offs total less than 150. Thus, this paper establishes thresholds for the number of pick-ups and drop-offs within an operation zone to filter out the dominant operation zones, namely, 0, 25, 50, 75, 100, and 125. Notably, the higher the threshold is, the more dominant the operation zones are. Figure 7 provides statistical results. For example, (i) there are slight changes in the WNNI in different periods when the threshold=0 and the values are all extremely low (Figure 7(a)), which cannot be explained by decreased dispersion of the spatial distribution of the taxicabs' operation zones because it should be influenced by the merging of adjacent operation zones in calculating the WNNI. (ii) Between Period 1 and Period 2, when the subsidy war began and the threshold was ≥25, the dominant operation zones become more dispersed. (iii) The value of the WNNI in Period 1 is equal to that in Period 6, when the threshold is 100 and 125, which means that the degree of dispersion of the dominant operation zones returns to the original levels when the subsidy war ends. These operation zones should be highly visited habitual zones. These results indicate that the distances of the taxicabs' highly visited habitual operation zones were similar before and after the subsidy war.

4.4. The Correlation Analysis between Changes in the Operation Zones and the Change in the Average Daily Trip Distance or Average Daily Cruising Time. Pearson's correlation coefficient R [33] was used to investigate the relationship of

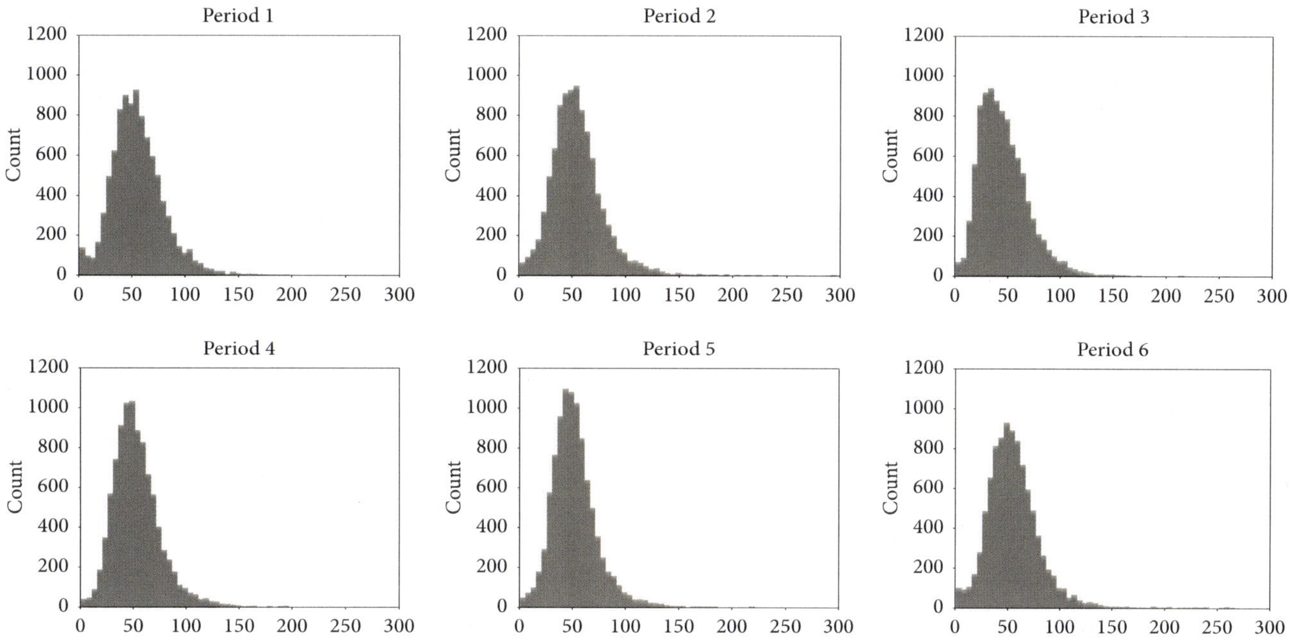

FIGURE 6: Histograms of the maximum number of pick-ups and drop-offs within an operation zone of each taxi in every period.

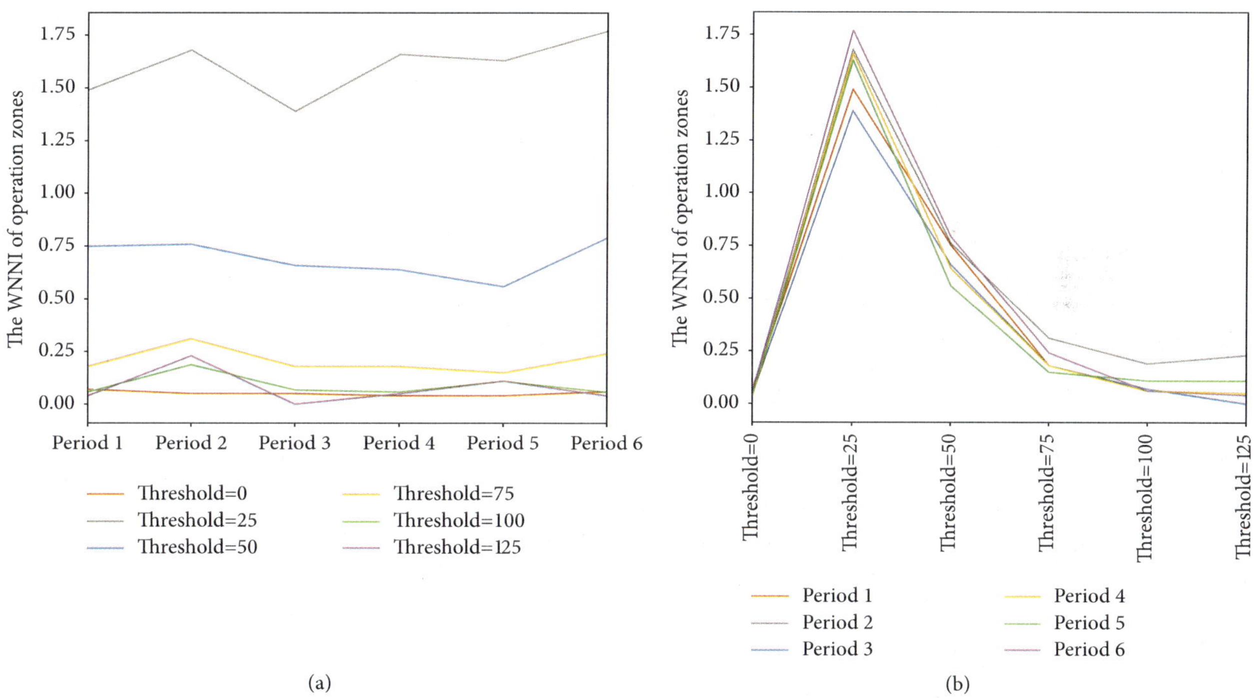

FIGURE 7: The distribution of WNNI with respect to periods (a) and thresholds (b).

the spatial change in the operation zones with the change in the average daily trip distance or the change in the average daily cruising time. The results in Table 5 show that the change in the average daily trip distance or the change in the average daily cruising time and the change in the unweighted area of operation zones are significant. The correlation coefficients are relatively low except that between the change in the average daily trip distance and the change in the unweighted area of the operation zones in Period 1-2, which means that there is a relatively moderate correlation between these factors. In addition, the change in the average daily trip distance or the change in the average daily cruising time and the change in the NNI of the operation zones are significant. However, the correlation coefficients are extremely low, indicating a weak correlation. Therefore, the changes in the operation zones have little effect on the

TABLE 5: The R^2 and p-value for the correlation analysis between the change in the operation zones and the change in the average daily trip distance or average daily cruising time in each period.

	Period 1-2	Period 2-3	Period 3-4	Period 4-5	Period 5-6
		Unweighted area of operation zones and the average daily trip distance			
R^2	0.3493	0.1106	0.1534	0.1884	0.1823
p-value	0^{***}	$1.0096e\text{-}246^{***}$	0^{***}	0^{***}	0^{***}
		Unweighted area of operation zones and the average daily cruising time			
R^2	0.1821	0.0649	0.0466	0.0874	0.0041
p-value	0^{***}	$3.3171e\text{-}142^{***}$	$1.2470e\text{-}101^{***}$	$4.9607e\text{-}193^{***}$	$3.4868e\text{-}10^{***}$
		NNI of operation zones and the average daily trip distance			
R^2	0.0209	0.0015	0.0227	0.0334	0.0352
p-value	$4.3059e\text{-}46^{***}$	$1.5593e\text{-}04^{***}$	$8.0890e\text{-}50^{***}$	$6.1279e\text{-}73^{***}$	$7.7453e\text{-}77^{***}$
		NNI of operation zones and the average daily cruising time			
R^2	0.0190	0.0015	0.0068	0.0077	0.0004
p-value	$6.5916e\text{-}42^{***}$	$1.5817e\text{-}04^{***}$	$5.1891e\text{-}16^{***}$	$6.3006e\text{-}18^{***}$	$4.0837e\text{-}02^{*}$

$*$ denotes the significance at the 0.05 level; $**$ denotes the significance at the 0.01 level; $***$ denotes the significance at the 0.001 level.

changes in the average daily trip distance and average daily cruising time in the aggregate perspective of the entire set of taxicabs.

5. Conclusions

This paper uses the taxicab trajectories in Shenzhen City to elucidate the change of taxicab operation zones within the context of e-hailing apps subsidy war. The results reveal interesting findings about the aggregate level of taxicabs: (i) the subsidy policies for taxicabs and passengers have an influence on the operation zones of taxicabs; (ii) the proportion of habitual operation zone pairs among operation zone pairs in neighboring periods is relatively stable under any subsidy policy during this subsidy war; and (iii) the changes in the operation zones have little effect on the changes in the average daily trip distance and average daily cruising time.

From the perspective of taxicab groups, there are also important findings: (i) the initial increase or decrease in patterns of the area and the degree of dispersion in the spatial distribution of the operation zones of the taxicabs in the beginning of the subsidy war could be used to reflect the difference in further operation zone changes between groups during the subsidy war. This group division rule is helpful for similar investigation tasks. (ii) In Period 3, which had the highest subsidy, all of the selected taxicabs travelled with decreased areas of operation zones and less passenger hunting time, but they had a decreased average daily trip distance at the same time. In this period, taxicabs could find their passengers efficiently and did not require a greater trip distance to maintain their incomes than in other periods because they could receive high subsidies from the e-hailing apps. (iii) When the subsidy to the passengers decreased, all of the groups of taxicabs increased their areas of operation zones in Periods 3-4 and 4-5 and had a high dispersion of their operation zones, especially in Period 4; as a result, they increased their average daily trip distance to maintain their incomes. (iv) When the subsidy war ended, some taxicabs may have returned to their prewar work status.

Based on the findings from all taxicabs and grouped taxicabs, this study could inform the operating decisions of e-hailing companies and government surveillance. For example, when e-hailing companies plan their future specific subsidy policies, the impact of the planned policies on taxis can be estimated to some extent. For the taxi management authority, the derived knowledge of the changes in the taxicabs' operation zones and habitual operating preferences under the stimulus of policies could help to develop precise and targeted management tactics for different groups of taxicabs to improve urban mobility and public transportation services.

This paper focuses only on the influence of the subsidy war on cabdrivers, whereas, in future research, it is necessary to deeply investigate the interactions between cabdrivers and passengers, since the subsidy war not only stimulated cabdrivers' behaviors but also influenced passengers' daily travel patterns. To better understand these patterns, it is necessary to integrate the taxi trajectory data with the detailed order data of the e-hailing apps.

Conflicts of Interest

The authors declare no conflicts of interest.

Acknowledgments

The research was supported in part by the National Natural Science Foundation of China (Grants nos. 41771473 and 41231171), the National Key Research and Development Program of China (2017YFB0503802), and the Innovative Research Funding of Wuhan University (2042015KF0167).

References

[1] Y. Liu, C. Kang, S. Gao, Y. Xiao, and Y. Tian, "Understanding intra-urban trip patterns from taxi trajectory data," *Journal of Geographical Systems*, vol. 14, no. 4, pp. 463–483, 2012.

[2] X. Hu, S. An, and J. Wang, "Exploring urban taxi drivers' activity distribution based on GPS data," *Mathematical Problems in Engineering*, vol. 2014, Article ID 708482, 13 pages, 2014.

[3] X. Qian and S. V. Ukkusuri, "Spatial variation of the urban taxi ridership using GPS data," *Applied Geography*, vol. 59, pp. 31–42, 2015.

[4] L. Liu, C. Andris, and C. Ratti, "Uncovering cabdrivers' behavior patterns from their digital traces," *Computers, Environment and Urban Systems*, vol. 34, no. 6, pp. 541–548, 2010.

[5] Y. Dong, Z. Zhang, R. Fu, and N. Xie, "Revealing New York taxi drivers' operation patterns focusing on the revenue aspect," in *Proceedings of the 12th World Congress on Intelligent Control and Automation (WCICA '16)*, pp. 1052–1057, June 2016.

[6] E. Kourti, C. Christodoulou, L. Dimitriou, S. Christodoulou, and C. Antoniou, "Quantifying demand dynamics for supporting optimal taxi services strategies," *Transportation Research Procedia*, vol. 22, pp. 675–684, 2017.

[7] Y. Lu and Y. Liu, "Pervasive location acquisition technologies: opportunities and challenges for geospatial studies," *Computers, Environment and Urban Systems*, vol. 36, no. 2, pp. 105–108, 2012.

[8] Z. Fang, S.-L. Shaw, W. Tu, Q. Li, and Y. Li, "Spatiotemporal analysis of critical transportation links based on time geographic concepts: a case study of critical bridges in Wuhan, China," *Journal of Transport Geography*, vol. 23, pp. 44–59, 2012.

[9] G. Chen, B. Chen, and Y. Yu, "Mining frequent trajectory patterns from GPS tracks," in *Proceedings of the International Conference on Computational Intelligence and Software Engineering (CiSE '10)*, pp. 1–6, December 2010.

[10] Y. Liu, F. Wang, Y. Xiao, and S. Gao, "Urban land uses and traffic 'source-sink areas': evidence from GPS-enabled taxi data in Shanghai," *Landscape and Urban Planning*, vol. 106, no. 1, pp. 73–87, 2012.

[11] J. Lee, I. Shin, and G. Park, "Analysis of the passenger pick-up pattern for taxi location recommendation," in *Proceedings of the Fourth International Conference on Networked Computing and Advanced Information Management (NCM '08)*, pp. 199–204, Gyeongju, South Korea, September 2008.

[12] H. Wang, H. Zou, Y. Yue, and Q. Li, "Visualizing hot spot analysis result based on mashup," in *Proceedings of the International Workshop on Location Based Social Networks (Lbsn '09)*, pp. 45–48, Seattle, Wash, USA, November 2009.

[13] C. Kang and K. Qin, "Understanding operation behaviors of taxicabs in cities by matrix factorization," *Computers, Environment and Urban Systems*, vol. 60, pp. 79–88, 2016.

[14] Z. Fang, X. Yang, Y. Xu, S.-L. Shaw, and L. Yin, "Spatiotemporal model for assessing the stability of urban human convergence and divergence patterns," *International Journal of Geographical Information Science*, vol. 31, no. 11, pp. 2119–2141, 2017.

[15] X. Wang, F. He, H. Yang, and H. Oliver Gao, "Pricing strategies for a taxi-hailing platform," *Transportation Research Part E: Logistics and Transportation Review*, vol. 93, pp. 212–231, 2016.

[16] L. Zha, Y. Yin, and Y. Du, "Surge pricing and labor supply in the ride-sourcing market," *Transportation Research Procedia*, vol. 23, pp. 2–21, 2017.

[17] X. Qian and S. V. Ukkusuri, "Taxi market equilibrium with third-party hailing service," *Transportation Research Part B: Methodological*, vol. 100, pp. 43–63, 2017.

[18] F. He and Z.-J. M. Shen, "Modeling taxi services with smartphone-based e-hailing applications," *Transportation Research Part C: Emerging Technologies*, vol. 58, pp. 93–106, 2015.

[19] D. Garza, Y. Giat, S. T. Hackman, and D. Peled, "A computational analysis of R&D support programs," *Economics of Innovation and New Technology*, vol. 24, no. 7, pp. 682–709, 2015.

[20] V. Salnikov, R. Lambiotte, A. Noulas, and C. Mascolo, "OpenStreetCab: Exploiting Taxi Mobility Patterns in New York City to Reduce Commuter Costs," 2015.

[21] A. Noulas, V. Salnikov, D. Hristova, C. Mascolo, and R. Lambiotte, "Developing and Deploying a Taxi Price Comparison Mobile App in the Wild: Insights and Challenges," 2017.

[22] S. Harding, M. Kandlikar, and S. Gulati, "Taxi apps, regulation, and the market for taxi journeys," *Transportation Research Part A: Policy and Practice*, vol. 88, pp. 15–25, 2016.

[23] L. Chen, A. Mislove, and C. Wilson, "Peeking beneath the hood of uber," in *Proceedings of the ACM Internet Measurement Conference (IMC '15)*, pp. 495–508, October 2015.

[24] M. K. Chen, "Dynamic pricing in a labor market:surge pricing and flexible work on the uber platform," in *Proceedings of the ACM Conference on Economics and Computation*, pp. 455-455, Maastricht, The Netherlands, July 2016.

[25] A. Noulas, V. Salnikov, R. Lambiotte, and C. Mascolo, "Mining open datasets for transparency in taxi transport in metropolitan environments," *EPJ Data Science*, vol. 4, no. 1, pp. 1–19, 2015.

[26] J. Wen, M. Zou, Y. Ma, and H. Luo, "Evaluating the influence of taxi subsidy programs on mitigating difficulty getting a taxi in basis of taxi empty-loaded rate," *International Journal of Statistics and Probability*, vol. 6, no. 2, p. 9, 2017.

[27] B. Leng, H. Du, J. Wang, L. Li, and Z. Xiong, "Analysis of taxi drivers' behaviors within a battle between two taxi apps," *IEEE Transactions on Intelligent Transportation Systems*, vol. 17, no. 1, pp. 296–300, 2016.

[28] R. Su, Z. Fang, N. Luo, and J. Zhu, "Understanding the dynamics of the pick-up and drop-off locations of taxicabs in the context of a subsidy war among e-hailing apps," *Sustainability*, vol. 10, no. 4, p. 1256, 2018.

[29] E. J. Gustafson and G. R. Parker, "Relationships between landcover proportion and indices of landscape spatial pattern," *Landscape Ecology*, vol. 7, no. 2, pp. 101–110, 1992.

[30] M. C. Pearce, "Pattern analysis of forest cover in southwestern Ontario," *The East Lakes Geographer*, vol. 27, pp. 65–76, 1992.

[31] B. D. Ripley, "Modelling spatial patterns," *Journal of the Royal Statistical Society. Series B*, vol. 39, no. 2, pp. 172–212, 1977.

[32] P. J. Clark and F. C. Evans, "Distance to nearest neighbor as a measure of spatial relationships in populations," *Ecology*, vol. 35, no. 4, pp. 445–453, 1954.

[33] K. Pearson, "Notes on the history of correlation," *Biometrika*, vol. 13, no. 1, pp. 25–45, 1920.

Permissions

All chapters in this book were first published in JAT, by Hindawi Publishing Corporation; hereby published with permission under the Creative Commons Attribution License or equivalent. Every chapter published in this book has been scrutinized by our experts. Their significance has been extensively debated. The topics covered herein carry significant findings which will fuel the growth of the discipline. They may even be implemented as practical applications or may be referred to as a beginning point for another development.

The contributors of this book come from diverse backgrounds, making this book a truly international effort. This book will bring forth new frontiers with its revolutionizing research information and detailed analysis of the nascent developments around the world.

We would like to thank all the contributing authors for lending their expertise to make the book truly unique. They have played a crucial role in the development of this book. Without their invaluable contributions this book wouldn't have been possible. They have made vital efforts to compile up to date information on the varied aspects of this subject to make this book a valuable addition to the collection of many professionals and students.

This book was conceptualized with the vision of imparting up-to-date information and advanced data in this field. To ensure the same, a matchless editorial board was set up. Every individual on the board went through rigorous rounds of assessment to prove their worth. After which they invested a large part of their time researching and compiling the most relevant data for our readers.

The editorial board has been involved in producing this book since its inception. They have spent rigorous hours researching and exploring the diverse topics which have resulted in the successful publishing of this book. They have passed on their knowledge of decades through this book. To expedite this challenging task, the publisher supported the team at every step. A small team of assistant editors was also appointed to further simplify the editing procedure and attain best results for the readers.

Apart from the editorial board, the designing team has also invested a significant amount of their time in understanding the subject and creating the most relevant covers. They scrutinized every image to scout for the most suitable representation of the subject and create an appropriate cover for the book.

The publishing team has been an ardent support to the editorial, designing and production team. Their endless efforts to recruit the best for this project, has resulted in the accomplishment of this book. They are a veteran in the field of academics and their pool of knowledge is as vast as their experience in printing. Their expertise and guidance has proved useful at every step. Their uncompromising quality standards have made this book an exceptional effort. Their encouragement from time to time has been an inspiration for everyone.

The publisher and the editorial board hope that this book will prove to be a valuable piece of knowledge for researchers, students, practitioners and scholars across the globe.

List of Contributors

Sara Ezquerro, Juan P. Romero, Jose L. Moura, Juan Benavente and Ángel Ibeas
G.I.S.T. Group of Research in Transport Systems, University of Cantabria, Cantabria, Spain

Ray Lattarulo, Enrique Martí, José Matute, Mauricio Marcano and Joshue Pérez
Department of Automotive, Industry and Transportation Division, Tecnalia Research and Innovation, 48160 Derio, Vizcaya, Spain

Leonardo González
Department of Cyber-Security, ICT Division, Tecnalia Research and Innovation, 48160 Derio, Vizcaya, Spain

Yipeng Ye
School of Economics and Management, Tongji University, Shanghai 200092, China

Hua Wang
Department of Civil & Environmental Engineering, National University of Singapore, Singapore 117576

Yao Shuang-Shuang
School of Economics and Management, Lanzhou Jiao Tong University, Lanzhou 730070, China

Xiao Qiang
School of Traffic and Transportation, Lanzhou Jiao Tong University, Lanzhou 730070, China

Wenwen Qin and Meiping Yun
Key Laboratory of Road and Traffic Engineering of the Ministry of Education, Tongji University, 4800 Cao'an Highway, Shanghai 201804, China

Z. Y. Liu, C. B. Li and M. Y. Jian
Transportation Institute, Inner Mongolia University, Hohhot 010021, China

Margaret O'Mahony
Trinity Centre for Transport Research, Department of Civil, Structural & Environmental Engineering, Trinity College Dublin, Dublin 2, Ireland

Yingying Yuan, Ke Gong and Maozeng Xu
School of Economics and Management, Chongqing Jiaotong University, Chongqing 400074, China

Yong Wang
School of Management and Economics, University of Electronic Science and Technology, Chengdu 610054, China

Kevin Assogba and Haizhong Wang
School of Civil and Construction Engineering, Oregon State University, Corvallis, OR 97330, USA

Yinhai Wang
Department of Civil and Environmental Engineering, University of Washington, Seattle, WA 98195-2700, USA
Transportation Data Science Research Center, College of Transportation Engineering, Tongji University, Shanghai 201804, China

Jing Guo
Graduate School of Environmental Studies, Nagoya University, D2-1Furo-cho, Chikusa-ku, Nagoya 464-8603, Japan

Fumihiko Nakamura
Institute of Urban Innovation, Yokohama National University, 79-5 Tokiwadai, Hodogaya, Yokohama 240-8501, Japan

Qiang Li
College of Resources Science and Technology, Beijing Normal University, 19 Xinjiekouwai Street, Beijing 100875, China

Yuan Zhou
Department of Geographical Sciences, University of Maryland-College Park, College Park, MD 20742, USA

Lídia Montero, Maria Paz Linares, Josep Casanovas and Esteve Codina
Department of Statistics and Operations Research, Universitat Politècnica de Catalunya (UPC), C5-Carrer Jordi Girona 1-3, 08034 Barcelona, Spain

Gonzalo Recio, Ester Lorente and Juan Salmerón
inLab FIB, Universitat Politècnica de Catalunya (UPC), B6-Carrer Jordi Girona 1-3, 08034 Barcelona, Spain

Josep Casanovas
Barcelona Super Computing Center (BSC-CNS), Universitat Politècnica de Catalunya (UPC), Carrer Jordi Girona 29-31, 08034 Barcelona, Spain

Xuan Zhao, Qiang Yu, Jian Ma, Yan Wu, Man Yu and Yiming Ye
School of Automobile, Chang'an University, Xi'an, China

Ming Cai and Yifan Yao
School of Intelligent Systems Engineering, Sun Yat-sen University, Shenzhen, China
Guangdong Provincial Key Laboratory of Intelligent Transportation System, Guangzhou, China
Guangdong Provincial Engineering Research Center for Traffic Environmental Monitoring and Control, Guangzhou, China

Haibo Wang
School of Civil and Transportation Engineering, Hebei University of Technology, Tianjin, China

Rui Tong
Beijing Key Laboratory for Cooperative Vehicle Infrastructure Systems and Safety Control, School of Transportation Science and Engineering, Beihang University, Xue Yuan Road No. 37, Hai Dian District, Beijing 100191, China

Peng Chen, Guangquan Lu and Yunpeng Wang
Beijing Advanced Innovation Center for Big Data and Brain Computing, Beihang University, Xue Yuan Road No. 37, Hai Dian District, Beijing 100191, China

Yi He and Ziqi Song
Department of Civil and Environmental Engineering, Utah State University, Logan, UT 84322, USA

Lihui Zhang
Institute of Transportation Engineering, Zhejiang University, Hangzhou, Zhejiang 310058, China

Katarzyna Jezierska-Krupa and Wojciech Skarka
Institute of Fundamentals of Machinery Design, Faculty of Mechanical Engineering, Silesian University of Technology, Gliwice, Poland

Aschalew Kassu
Department of Mechanical, Civil Engineering and Construction Management, Alabama A&M University, Normal, AL 35762, USA

Michael Anderson
Department of Civil & Environmental Engineering, University of Alabama in Huntsville, Huntsville, AL 35899, USA

Dong Guo, Liang Sun, Kai Li, Juan Wang, Feng Sun and Hai Zhang
School of Transportation and Vehicle Engineering, Shandong University of Technology, Zibo, Shandong 255000, China

Zhan-gu Wang
Jilin University State Key Laboratory of Automotive Simulation and Control, Changchun, Jilin 130000, China

Shichao Sun
College of Transportation Engineering, Dalian Maritime University, Dalian 116026, China

Dongyuan Yang
Key Laboratory of Road and Traffic Engineering of the Ministry of Education, Tongji University, Shanghai 201804, China

Rongxiang Su
State Key Laboratory of Information Engineering in Surveying, Mapping and Remote Sensing,Wuhan University, Wuhan 430079, China

Zhixiang Fang
Collaborative Innovation Center of Geospatial Technology, Wuhan 430079, China

Lian Huang
Shenzhen Transportation Operation Command Center, Shenzhen 518040, China

Index